Lecture Notes in Networks and Systems 1616

The series “Lecture Notes in Networks and Systems” publishes the latest developments in Networks and Systems—quickly, informally and with high quality. Original research reported in proceedings and post-proceedings represents the core of LNNS.

Volumes published in LNNS embrace all aspects and subfields of, as well as new challenges in, Networks and Systems.

The series contains proceedings and edited volumes in systems and networks, spanning the areas of Cyber-Physical Systems, Autonomous Systems, Sensor Networks, Control Systems, Energy Systems, Automotive Systems, Biological Systems, Vehicular Networking and Connected Vehicles, Aerospace Systems, Automation, Manufacturing, Smart Grids, Nonlinear Systems, Power Systems, Robotics, Social Systems, Economic Systems and other. Of particular value to both the contributors and the readership are the short publication timeframe and the world-wide distribution and exposure which enable both a wide and rapid dissemination of research output.

The series covers the theory, applications, and perspectives on the state of the art and future developments relevant to systems and networks, decision making, control, complex processes and related areas, as embedded in the fields of interdisciplinary and applied sciences, engineering, computer science, physics, economics, social, and life sciences, as well as the paradigms and methodologies behind them.

Indexed by SCOPUS, EI Compendex, INSPEC, WTI Frankfurt eG, zbMATH, SCImago.

All books published in the series are submitted for consideration in Web of Science.

For proposals from Asia please contact Aninda Bose (aninda.bose@springer.com).

Maria Lapina · G S Prakasha ·
Deepanraj Balakrishnan · Natalia Gorbunova ·
Walaa H. Elashmawi · Viktoriya Taran
Editors

DET-2024: International Conference on Distance Education Technologies

Springer

Editors
Maria Lapina
North-Caucasus Federal University
Stavropol, Russia

G S Prakasha
CHRIST (Deemed to be University)
Bengaluru, India

Deepanraj Balakrishnan
Prince Mohammad Bin Fahd University
Al Khobar, Saudi Arabia

Natalia Gorbunova
V.I. Vernadsky Federal University
Simferopol, Russia

Walaa H. Elashmawi
Misr International University
Cairo, Egypt

Viktoriya Taran
V.I. Vernadsky Federal University
Simferopol, Russia

ISSN 2367-3370 ISSN 2367-3389 (electronic)
Lecture Notes in Networks and Systems
ISBN 978-3-032-04364-1 ISBN 978-3-032-04365-8 (eBook)
https://doi.org/10.1007/978-3-032-04365-8

This Springer imprint is published by the registered company Springer Nature Switzerland AG
The registered company address is: Gewerbestrasse 11, 6330 Cham, Switzerland

About Book

This book is based on the best papers accepted for presentation during the DET-2024: International Conference on Distance Education Technologies. The book includes research on education, information technologies, artificial intelligence, digital educational environment, augmented and virtual reality, e-learning technologies, end-to-end technologies in education, and educational transformation. The contributions in this volume highlight the significant advancements and innovative methodologies in digital education. The book provides a consolidated research exploration on contemporary themes in education such as digital teaching technologies, digital pedagogy and resources, gamification, and innovations in education with a special focus on educational technology for present and future educators. The authors of these papers have explored diverse topics, such as artificial intelligence in education, the integration of virtual and augmented reality, the development of online learning platforms, and the impact of digital tools on student engagement and learning outcomes.

This collection aims to provide educators, researchers, and practitioners with insights into the latest trends and challenges in digital education. It also seeks to inspire further research and innovation in applying digital technologies to enhance teaching and learning experiences. As the landscape of education continues to evolve, the integration of digital technologies becomes increasingly critical. This book addresses a wide range of issues and opportunities, offering valuable perspectives on how digital tools can be leveraged to improve educational outcomes and accessibility.

This book serves as a vital resource for comprehending the future of education and the transformative role digital technologies play in its development. By delving into the innovative approaches and cutting-edge research presented within, readers gain a deeper understanding of how digital advancements are redefining educational practices and enhancing learning experiences. The insights provided in this volume help educators develop more effective teaching strategies, incorporate new technologies into their curricula, and ultimately create more engaging and inclusive learning environments. Moreover, the book intends to foster new ideas, discussions, and collaborations that will continue to advance educational technology, thus bringing together a diverse array of research and perspectives to encourage a multidisciplinary approach to solving the complex challenges facing modern education.

The book serves as a valuable platform for scholars, practitioners, and students interested in the latest advancements in information security management and applications. It provides an opportunity to expand knowledge, establish connections with peers, and contribute to the further advancement of the field. The book promises to be an engaging and informative for those passionate about distance learning technologies and its various applications.

Contents

About the Editors

Dr. Maria Lapina is Associate Professor of the Department of Computational Mathematics and Cybernetics at North Caucasus Federal University, Stavropol, Russia. She has more than 20 years of teaching experience. She received PhD in mathematics from Stavropol State University in 2005 and received an associate professor degree in 2009. She obtained her Master's degree in Information security. She defended her master's thesis at the North Caucasus Federal University with honors in 2015.

Her research interests include data science, machine learning, big data, cybersecurity, cryptography, trusted artificial intelligence, edu-tech, teacher education, and higher Ed. She has published in peer-reviewed journals indexed in Scopus and WoS. She has more than 100 research papers. He has written book chapters in international publishing houses like Taylor & Francis, IGI Global, and Springer. She is an active editorial board member for Springer Nature and IGI Global. She is the editor of books titled "Educational Perspectives on Digital Technologies in Modeling and Management", & "Digital Technologies in Modeling and Management: Insights in Education and Industry" published by IGI Global. She is an active reviewer of MDPI publishing journals Educational Sciences, Drones, Sensors, Mathematics, AgriEngineering. She is the organizer, co-chair, and editor of the proceedings of international scientific events: SLET-2019, 2020, 2021, 2022, 2023: International Conference Innovative Approaches to the Application of Digital Technologies in Education, AISMA-2021, 2022, 2023, 2024: International Workshop on Advanced in Information Security Management and Applications, DMAI-2021: International Workshop on Data Mining and Al, YRID-2020: International Workshop on Data Mining and Knowledge Engineering. She has participated in international internships at Sapienza University of Rome (Italy), the University of Ljubljana (Slovenia), the Technical University of Dresden (Germany) and the 2nd ASEF Innovation Laboratory of Higher Education (ASEFInnoLab) at Fudan University in Shanghai, China. She received a grant from the Vladimir Potanin Foundation and an award as the Best Lecturer of Russia. She is a finalist of the All-Russian management competition "Leaders of Russia".

Dr. G S Prakasha is Associate Professor at School of Education, CHRIST (Deemed to be University), Central campus, Bangalore, India. He has 22 years of teaching and research experiences in education. He completed his Master's in Education from RIE, Mysore, and PhD from Bangalore University. He has also completed MSc and MPhil. He cleared National Eligibility Test (NET) and awarded junior research fellowship (JRF). He is a certified IB Educator in MYP and DP from DePaul University, Chicago, USA. His research interest areas are teacher-education, educational-technology, teaching-learning, and higher education. He has published research articles in peer-reviewed journals indexed in Scopus and web of science. He conducts quantitative data analysis workshops and delivers lectures on Research and Publication. He is an active editorial board member for Sage Open, Springer Nature, Taylor & Francis, and IGI Global. He is actively involved in research activities, which have national and international importance. He was the project director of Radio-Namaste—a faculty and student exchange project on Indo-Swiss education in collaboration with Switzerland's Teacher Edu University (hep, Vaud) and Christ Uni. He visited Miami University, Cincinnati, Mount Union, Johns Hopkins, Syracuse, Binghamton, Campbellsville, New York, & the Catholic Uni of America. His areas of interest lie in Contemporary ed-issues, Edu-Tech, Teacher Education, and Higher Ed. He has published in peer-reviewed journals indexed in Scopus and WoS. He has presented papers in national and international conferences at Bangkok, Switzerland, Russia, University of Oxford, UK, Binghamton University, USA. He has visited Germany, France, Switzerland, Canada, and USA. He is an active editorial board member for Sage, Springer Nature, and Elsevier journals. He was involved in research activities, which have national and international emphasis. He carried out a Major Research PROJECT on Competencies of Next generation Teacher Educators and Teachers. He led a Rural Education PROJECT in association with a NGO called GRAMA to set up Informal Education Centre to upgrade rural education in India. He has written books titled "Teaching Internship models around the world" published by Routledge, "Educational Perspectives on Digital Technologies in Modeling and Management", and "Digital Technologies in Modeling and Management: Insights in Education and Industry". He has written book chapters in international publishing houses like Taylor & Francis, IGI Global, Lap Lambert, and Springer.

Dr. Deepanraj Balakrishnan is Assistant Research Professor at the College of Engineering, Prince Mohammad Bin Fahd University, Saudi Arabia, currently on leave from Jyothi Engineering College, India. With over 10 years of experience in teaching and research, he has made significant contributions to his field. Dr. Deepanraj has authored more than 200 research articles in peer-reviewed international journals and conferences and holds four Indian patents. He serves as a guest editor, editorial board member, and reviewer for numerous international peer-reviewed journals, including those published by renowned publishers such as Elsevier, Springer, and Inderscience. Globally recognized for his contributions to science and engineering, Dr. Deepanraj was featured in the prestigious World Top 2% Scientist List by Elsevier and Stanford University in both 2023 and 2024.

Prof. Natalia Gorbunova is a leading scientist in the field of IT technologies in education and holds the position of director of the Humanitarian Pedagogical Academy at Vernadsky Federal University. She actively implements innovative IT solutions in the educational process, contributing to the digitalization of the university. Under her leadership, modern educational programs focused on the use of digital technologies are being developed and implemented. Natalia Gorbunova organizes and conducts international conferences that bring together experts in the field of IT and education to share experiences and best practices. Her research activities include more than 40 publications in publications indexed in Scopus, which highlights the high level of her research. She also directs scientific projects aimed at the research and development of new educational technologies. Under her leadership, more than 10 graduate students successfully defended their dissertations, which testifies to her contribution to the training of scientific personnel. Thanks to her efforts, the Academy has become a platform for large-scale events that promote international cooperation. Her work helps strengthen the university's position as a center of innovation in education. She is a recognized expert in her field, whose ideas and projects resonate with colleagues and students. The organization of international conferences under her leadership allows attracting leading experts from all over the world. Natalia Gorbunova pays special attention to the development of research activities, supporting young scientists and students.

Dr. Walaa H. Elashmawi is Professor at the Faculty of Computer Science, Misr International University, on leave from Suez Canal University, Egypt. She has over 20 years of expertise in computer science, specializing in artificial intelligence, data science and machine learning, meta-heuristic algorithms, and the Internet of things (IoT). With a Ph.D. in Computer Science from Hunan University, China (2013), her groundbreaking research on coalition formation in multi-agent systems has established her as a leading figure in her field. She has earned B.Sc. degrees in Computer Science from Suez Canal University (2001) and M.Sc. in 2008 from Minufiya University in Egypt.

Throughout her career, Dr. Elashmawi has combined academic excellence with technical innovation, applying advanced optimization techniques, predictive analytics, and swarm intelligence to solve complex problems across various domains. Her contributions to research are reflected in her prolific output, with over 50 peer-reviewed journal articles and conference papers WoS and Scopus. She has also contributed to books on cutting-edge AI applications, including federated learning and sustainable development in international publishing houses like Taylor & Francis as well as in Springer. She is an editor of books titled "Sustainable Innovations in Statistics and Data Science" published by IGI Global. She is an active journal reviewer for renowned international publications in the Neural Computing and Applications journal, ensuring the quality of cutting-edge research in her field. She has participated in the International Teaching Week (ITW24) held in Christ University, India, where she shared her expertise and innovative teaching methodologies with global academic communities, serves on the Advisory Board of the Acta Scientific Computer Sciences Journal, and is Fellow Member of Threws. She is also Head of the Control Unit at the Faculty of Computer Sciences, Misr International University, where she spearheads innovative academic and administrative initiatives.

Dr. Victoriya Taran works as Associate Professor in the Department of Mathematics and Informatics Vernadsky Federal University, Russia. In 2012, she defended her PhD thesis. From 2003 to 2013, she worked as Associate Professor in the Department of Information Technologies of the European University; since 2014, she has been working as Associate Professor in the Department of Mathematics and Informatics Vernadsky Federal University. She has more than 150 scientific publications in domestic and international journals/conferences and is the author of two textbooks and three monographs. She has experience in organizing all-Russian and International Conferences; since 2016, she has been an editor of the collection of scientific papers of the conference «Distance Lerning Technologies». Her research interests include distance educational technologies, information security, and modelling of complex systems.

Realization of Mixed Education Model Based on Usage of Network Technologies

Irina M. Agibova, Tatyana A. Kulikova(✉), and Natalya A. Poddubnaya

North-Caucasian Federal University, Pushkin Street, 1, Stavropol 355029, Russian Federation
t_a_kulikova@mail.ru

Abstract. The article discusses the distance learning models presented in the researches of various authors. The realization of a mixed distance learning model based on the usage of network technology is proposed. An algorithm for designing and developing content for a distance course, which allows avoiding wasting time and material resources and get a high-quality result, is considered. A comparative analysis of the didactic capabilities of the most popular platforms for creating distance courses is provided, the choice of LMS Moodle is substantiated. It is shown that the use of interactive elements of the platform makes it possible to track the progress of students' lectures, tests, exercises, and organize group work between students. It was revealed that the variety of elements and the possibility of their flexible customization allows successfully applying LMS Moodle at various levels of education (general, professional, additional), implementing a personal-activity approach in the learning process. The didactic capabilities of LMS Moodle are analyzed, methodological techniques of the distance course elements are presented. The pedagogical technologies, that are successfully used in the distance mode, are presented. It is shown that the choice of pedagogical technology is determined by the purpose of the lesson, individual characteristics, and competencies of students. Examples of pedagogically expedient use of LMS Moodle elements in the implementation of some teaching technologies are given. A variant of the implementation of a mixed education model has been proposed, which has been tested in the development of distance courses in LMS Moodle and has been embedded into the educational process of master's and bachelor's programs based on the North Caucasus Federal University.

Keywords: Education models · educational platforms · distance course' designing · pedagogical technologies · distance education · methodological techniques

1 Introduction

In the modern education system at all levels of training, there is a tendency to practice, individualization and personalization of education [1]. In our opinion, it is possible to resolve problems of education modernization by creating a "hybrid" educational environment, which allows without losing the quality and rhythm of education process, to switch it from one format to another. In this regard, at all levels of educational process,

M. Lapina et al. (Eds.): DET 2024, LNNS 1616, pp. 1–12, 2025.
https://doi.org/10.1007/978-3-032-04365-8_1

the active implementation of mixed education, what means such education system, which includes the traditional formats of the interaction with students and teachers and elements of distance technology. The organization of mixed education process depends a lot from teacher' preparedness to work in the context of digital transformation of education [2].

The usage of multimedia technologies and interactive educational tools in the process of mixed education provides the opportunity to organize the effective interaction of the teacher and students not only during classroom lessons in full-time education, but also online [3]. Since such innovative forms and technologies of education are aimed at improving the educational process, the main goal of which is the ability of the teacher not only to give the student fundamental knowledge, but also ensure all the necessary conditions for him that will contribute to his self-education and self-development.

Blended learning based on multimedia and interactive tools enables the teacher to rationally use the time of lesson, allows the student to prove himself during independent work, make the lesson more meaningful, attracting the attention of students [4].

2 Classifications of Education Models

Analysis of the researchers' works [5, 7–9], shows that the determining factor for the classification of models for the implementation of distance education is the degree of integration of digital technologies and the level of implementation of the distance education in the educational process.

Classification of distance education models in UNESCO: single, double, mixed, consortium, franchising, remote classrooms. In the works I.B.J. Seineen and R.S.J. Tuning [10] The following remote educational models are distinguished: consulting, correspondence, regulated education.

According to Voronina T.P. four models are considered before: correspondence education, open type remote education, tele-education, virtual education (teleconferences and education software) [11]. Andreev A.A. selects the following remote education models: correspondent, case, shift, tele-education, telepresence, network education [12].

Let's consider in details the model of mixed education.

Blended learning in general terms includes the following elements: classic teaching (face to face); online education and integration of the experience of the first and second elements (information and communication technologies and standard educational organizations, as we see them now).

There are three principles on which blended learning is based: (1) distance education - distance support becomes mandatory for mixed education, as it allows the educational process to be continuous; (2) education in the classroom, but the education process is not bounded by classroom, students together with the teacher can out the classroom and use other spaces of the school to obtain knowledge and their practical application, thus a new educational environment is formed, which becomes interactive; (3) the computerization of the usage of information technologies, the benefits of which are hard to overestimate, they allow to automate the educational process [13].

Key concepts of mixed education: personalization, project activity, teamwork, individual approach, dynamics of the educational process.

Possible following models of mixed education (blended learning) [4].

Mixed curriculum. A professional educational organization at its discretion in the right to derive a separate element of the main approximate educational program (hereinafter - the MAEP) in the online format. Disciplines of educational cycle can be included in this format, which have no priority in process of entering a profession, disciplines of the general humanitarian and socio-economic cycle ("Fundamentals of Legal Knowledge"), elective courses of small volume or electives.

Mixed individual curriculum. This model is convenient in term of providing an individual approach to students with special educational needs (often sick students, with disabilities, as well as talented students and adults who have a high level of self -organization; receiving second education, those who combine work and education frequently absent due to participation in various competitions). The restrictions of this model include the lack of the ability to participate in discussions, group projects, business games, means in group forms of work.

Educational menu (self-blend). The implementation of this model of education allows the student to study any discipline in person or remotely. In this case, the higher educational institution should provide the possibility and digital support for the implementation of these forms of education, and the student independently determines his educational trajectory and the format of education.

Face-to-face session. The main part of the educational process is going in online format, and sessions are held in face-to-face mode. The face-to-face session includes intermediate and final control, which can be supplemented by various educational events. Such a model is actively used in the professional education system and in the implementation of additional educational programs.

Rotation of stations (change of working zones). This technology is well known all over the world. The logic of the implementation of this technology is that the teacher divides the audience into three groups that differ by type of activity: the activity of the teacher with students (classical scheme of interaction); the zone of work with the computer; group activity zone.

The activities of the teacher with students include: verification of the degree of assimilation of the material; detailed analysis of typical errors that are found in students of this group; working with increased complexity task for the individual training of students.

During working at the computer students can master new material and carry out self-control by testing; summarize, systematize and repeat the educational material that was in the previous lesson; fix the educational material through simulations; build individual educational trajectories.

Group interaction allows to implement various types of project-based research activities of students. This can be: work on the project, conducting a joint study and analysis of the results; laboratory work; the solution of various Case-s.

The Flex model (flexible education) is that the student independently chooses the types of educational activities at each stage of education (independent work on a computer, consultation with a teacher, work in a group). An important problem in such models is that the organization of the education process is entrusted to the student.

Explanatory class (Face-to-Face Driver). The teacher explains the new material, answer on questions about educational tasks and explains the completion of tasks. After that, students consolidate online new material and independently pass control points.

The considered models allow to connect online (digital) component and the traditional education process. There are factors that can affect the final result of students' education process. The problem of effective integration of mixed education into the traditional full-time form of professional education arises.

3 Realization of Mixed Distance Education Model

When implementing mixed learning, in our opinion, it is advisable to apply the capabilities of network technologies. So, for example, in order to realize convenient and quick access to students to educational content and educational and methodological support, organize joint work using interactive digital tools we offer a model of mixed distance learning. The components of the developed model are presented in Fig. 1.

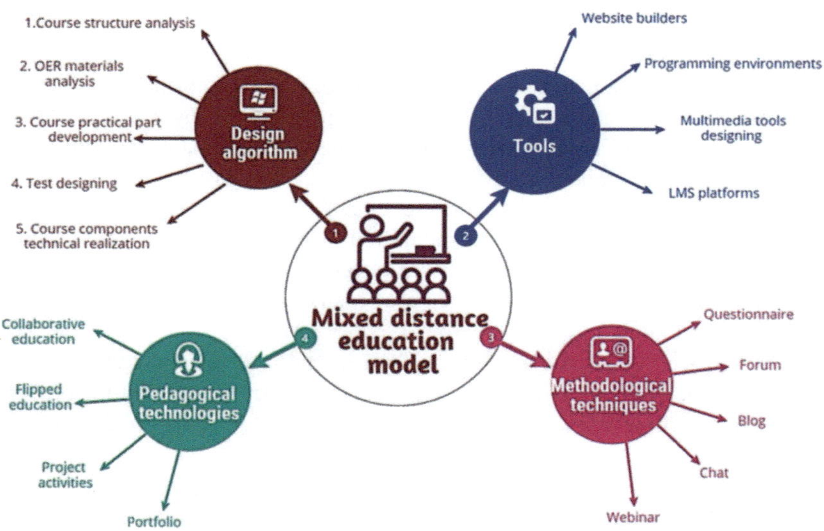

Fig. 1. Realization of mixed distance education model on base of network technology usage.

The model consists of the following four blocks: design and development of the structure and content of the course, digital tools, methodological techniques for using the course elements, pedagogical technologies for organizing the learning process.

Consider the highlighted blocks in more details.

3.1 Distance Course Content Design and Development Algorithm

The algorithm for designing and developing the remote course is presented the next blocks.

The process of developing the structure of the distance course is advisable to start with solving the problem of the distribution of theoretical material, practical tasks, control and measuring materials and justifying the choice of the format of the presentation of digital educational resources (educational video, interactive content, educational multimedia resources, test materials for operational control). When designing, modern digital technologies and tools should be used, recorded didactic principles and ergonomic requirements that educational resources must satisfy, take into account the needs of modern education.

These tasks should be solved in the first block. As a result, the structure of the remote course was developed, the relationship of educational goals and types of educational work has been substantiated, technologies and methods for evaluating learning outcomes are determined.

The second design block involves the analysis of digital educational and methodological resources and ready-made educational content from open sources. As such databases, electronic library systems, digital storages of professional education institutions, electronic archives with an open accesses to the documents of scientific, educational or other purposes can be used. By integrating such resources into a distance course, students will be able to get acquainted with original materials, and the course development will take much less time due to the use of high-quality prepared content.

In the third block, interactive methodological content is developed to organize practical work for students. Methodological recommendations can be presented by various types of digital educational resources, provide tasks of different levels of complexity depending on the goals, objectives and the content of education.

An important aspect of remotely mixed education is the control of students. The organization of control of students and the development of measuring materials is implemented in the fourth block "Test designing". This block provides the following types of students' control organization:

- control of the educational trajectory;
- time control (time spent on the development of elements of educational content and education planning);
- control of the place (the place of education - an educational organization, house, a library, a specially equipped room);
- control of the pace of learning (the speed of performing a specific element of educational content) [14].

This approach to the use of distance technologies in the organization of mixed education allows the following functions: the student's motivation for timely and high-quality development of educational material, adequate assessment of the achieved results, ensure the control of independent work, effective organization and processing of educational analytics data for the tracking of academic performance, progress and differentiation of training [9].

The fifth block determines the technical implementation of the components of the course. During designing, it is necessary to form a modular structure of the course and take into account certain requirements. The main ones are:

- during the transition between the modules, the method of presenting the educational material should remain the same;
- during compiling a module, it is necessary to take into account interdisciplinary relations in specific directions;
- during the modular construction of the course, it is necessary to provide an opportunity to the student to preserve the feeling of the course integrity while studying material;
- the disclosure of module content should be phased. This can be organized through a variety of approaches to the material studying, taking into account its logical sequence;
- the teacher can establish the sequence of assimilation of educational material.

The implementation of the modular structure of the course allows to find the optimal technical solution, to eliminate the redundancy of theoretical material, to determine the possibility of choosing an effective trajectory for the student.

Following the proposed algorithm for designing and developing a remote course for the implementation of a mixed education model, it is possible to optimize material and time costs and realize an economically profitable qualitative result of students' education.

3.2 Remote Course Development Tools

Educational platforms can be used as tools for designing distance courses. A wide range of digital tools allows developing various types of course content.

Figure 2 provides an overview of the capabilities of the most popular platform for planning distance courses.

To develop digital educational resources that make up the content of the designed distance education course, you can use publicly available digital tools. Let's consider some of them. Online boards (virtual boards, VK Board, Yandex Concept, Unidraw, MTS Link Boards, Esboard, Ziteboard, Mural, Miro, Padlet, etc.) are interactive digital tools that can be used to organize interaction between participants in the educational process. A convenient functionality of these services is the ability to collaborate on an educational or research task, a project using various methods of presenting educational information (animation, infographics, mind maps, etc.)

Platforms based on artificial intelligence and machine learning algorithms (YandexGPT, Quillionz, Mathway, Abyssale, Yippity, Brainly, Uchis.ru, Shedevroom, etc.) allow you to:

- generate questions, various types of open and closed tests on the proposed document;
- create texts of any length, format the style of the text, answer problematic questions, search for sources, identify key ideas, etc.

Services for developing interactive [16] and printed tasks (H5P, Wordwall, Wizer.Me, BookWidgets, etc.). Allow teachers to:

- develop interactive educational activities, simulators and game applications that include templates of classic games (quizzes, crosswords, arcades such as Maze Chase and Airplane);
- create multimedia educational materials - add video, audio, images and various types of questions to them;
- share a link to the activity page via email, social networks or other means.

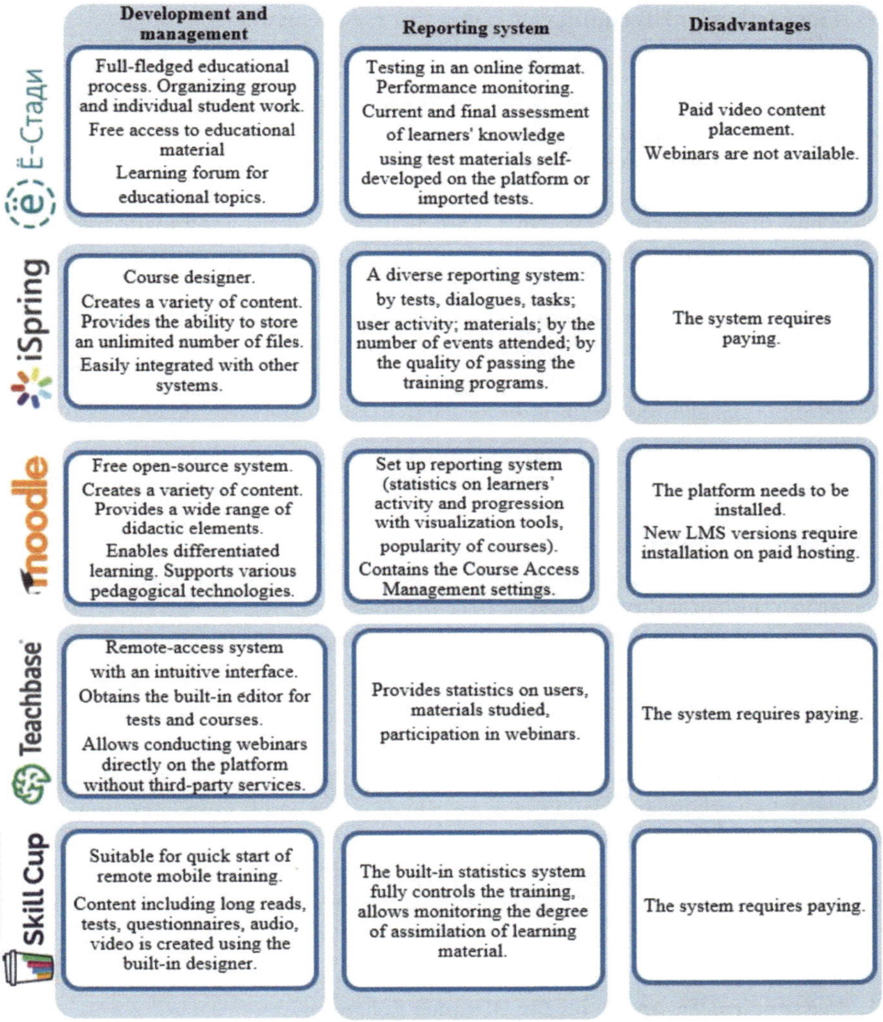

Fig. 2. The overview of the capabilities of the most popular remote course design platforms.

Video content creation programs (Movavi Video Suite, UVScreenCamera, iSpring Suite, Camtasia, Adobe Premiere Pro, Hippo Video, Verse, Vyond, etc.). Used as a tool for developing and editing educational videos. Some services, such as Verse, allows to use the capabilities of artificial intelligence to create interactive videos with polls and action choices, where static content turns into a dynamic digital experience. Other tools, such as Hippo Video, provide access to detailed analytics: analysis of students' work with the viewed video resource.

The advantages of these tools are: simple, intuitive, uncluttered interface, flexibility, portability, private and secure login, technical support [17].

3.3 The Methodical Technique of Using Course Elements

The variety of elements and the possibility of its flexible adjustment make LMS Moodle an effective application on various levels of education (general, professional, additional), enabling to implementation of a personality-centered approach in education.

The didactic capabilities provided by LMS Moodle, on the one hand, are comparable to the functionality of the known distance learning systems, on the other hand, have certain advantages. For one thing, this is an open source-code system with the ability to modify the environment to the requirements of educational content and the introduction of various modules and digital tools. But it is also an intuitive interface and functionality that contributes to the simplicity of mastering, convenience, and implementation of most of the requirements for using this platform in e-learning [18].

When implementing a mixed learning model based on network technologies, LMS Moodle supports all information-training standards available and provides the following features:

- adaptability (improvement of it tools without the system changes and availability of built-in methods for implementation of individual educational routes);
- durability (modifications of the environment without the use of program code);
- accessibility (various options of access to the system and educational content);
- interoperability (ability to use resources of other systems);
- frequency of use (unlimited number of accesses to system components).

The LMS Moodle application makes it possible to envisage all necessary pedagogical requirements related to the organization of e-learning. The environment enables: to switch from informative training methods to active ones, to develop the personal characteristics of the student, to organize collective and project activities, to form the ability of students to work independently and to evaluate own condition and emotions, and the results of their educational activities [19].

In addition, LMS Moodle makes it possible to use different levels of access to educational content: administrator, course developer, teacher, student, or guest. Whereas each of the users differs in functionality and the ability to operate the system.

3.4 Pedagogical Technologies for Distance Learning

The choice of pedagogical technology is determined by the purpose of the training, individual features, and competencies of the students. Modern pedagogical technologies advantageously used in remote mode include "inverted training," project activities, portfolio, training in cooperation, and others.

Here are some examples of how to use LMS Moodle elements in some learning technologies.

In the implementation of distance collaborative learning aimed to form training groups, and discover the wishes and opinions of students on a question, the teacher needs to individualize teaching according to student ability and personal characteristics. To do this, it is advisable to use an LMS Moodle element such as 'Questionnaire'. The 'Questionnaire' element allows you to apply restrictions, for example, to limit the number of topics or discussion participants, the access to the select results, or the process of forming individual and group tasks.

Working in small groups has some advantages; it facilitates to development of skills in information exchange, searching for joint solutions, and carry out coordinated actions [20].

Collective evaluation of group activities contributes to getting a better result, which can be further discussed in such modes as 'Forum', 'Chat', 'Blog' or 'Webinar'.

The teacher can use a variety of the LMS Moodle elements for organizing individual or group work in project activities of distance learning. Thus, the functionality of the 'Forum' element enables discussion of the project research process, analyze the techniques, methods, and means of the project realization, exchanging feedback and opinions on the progress of the design process, etc.

The teacher can discuss the project topic with the project team, make adjustments at various stages of the project activity, initiate student activity, and encourage active work in the project.

In the presentation of the results of the project activity and working on the project probation, it is advisable to use the capabilities of the Wiki element. The statement and presentation of the project can be facilitated using the Webinar element, for conducting the meeting online and determining the degree of students' participation in the project.

To increase the pedagogical efficiency of various teaching technologies in remote education, the teacher is supposed to represent the following components of digital competence: knowledge of the methodological features of organizing interactive studies, ability to design the structure and content of distance courses, be exercised in various methods and techniques for organizing mixed training and have skills to use the didactic capabilities of network technology in distance education.

4 Results

The proposed model of implementing mixed education based on the usage of network technology was tested in the process of developing academic disciplines on the LMS Moodle platform and implementing it in the students' education process among pedagogical directions of different levels of education of the North Caucasus Federal University (see Fig. 3).

The choice of the Moodle system is based on fact that the use of information resources and interactive elements of the platform allows implementing various control and assessment systems, tracking the progress of students in lectures, tests, tasks, organizing communication between the teacher and students using a forum, chat, and a personal messaging system both offline and online. The communication tools of the LMS Moodle allow organizing group work between students in order to implement various learning technologies.

Fig. 3. Examples of e-learning courses of the North Caucasus Federal University.

5 Conclusions

Summarizing the results of the research, we can highlight the positive features of mixed education: a flexible approach to students, development of motivation and self-organization, saving material and time resources, improving soft skills. When implementing a mixed education model based on the use of network technology, it is necessary to follow certain algorithms and take into account both pedagogical and program components.

Thus, blended learning effectively combines traditional and innovative educational technologies, solves practical problems of professional education. Mixed education is based on a taxonomy of educational activities that takes the student from knowledge through understanding to practice, then to analysis, synthesis and evaluation. What allows to improve the universal, general professional and professional competencies of the future teacher.

References

1. Konopko, E.A., Pankratova, O.P., Abdullaev, D.A., Ediev, A.M., Taran, V.N.: Digital education toolkit and an overview of distance learning resources. In: Selected Papers of the V All-Russian Scientific and Practical Conference with International Participation "Distance Learning Technologies" (DLT 2020), Yalta, Crimea, September 22–25, pp. 374–382 (2020)
2. Serole, S.R.: The art of promoting blended learning through blended pedagogies in higher education classrooms. E-J. Hum. Arts Soc. Sci. **6**(3), 274–287 (2025). https://doi.org/10.38159/ehass.2025633
3. Yadav, N.: The impact of digital learning on education. Int. J. Multidisc. Res. Arts Sci. Technol. (IJMRAST). **2**(1), 24–34 (2024). ISSN: 2584–0231). https://doi.org/10.61778/ijmrast.v2i1.34
4. Tabassum, B., Moin, M., Abbas, Q., Kumbhar, M., Khan, M.: The impact of blended learning on student performance. J. Educ. Soc. Stud. **5**, 360–371 (2024). https://doi.org/10.52223/jess.2024.5217
5. Khenner, E.K.: Professional knowledge and professional competencies in higher education. Educ. Sci. J. **2**(20), 9–31 (2018). https://doi.org/10.17853/1994-5639-2018-2-9-31
6. Arinushkina, A.A., Morozov, A.V., Robert, I.V.: Contemporary Challenges in Education: Digitalization, Methodology, and Management, vol. p.392. IGI Global (2023)
7. Men, Q., Gimbert, B., Cristol, D.: The effect of self-regulated learning in online professional training. Int. J. Mobile Blended Learn. **15**, pp.1–17 (2023). https://doi.org/10.4018/IJMBL.318225
8. Mei, X.Y., Aas, E., Medgard, M.: Teachers' use of digital learning tool for teaching in higher education: exploring teaching practice and sharing culture. J. Appl. Res. Higher Educ. **11**(3), 522–537 (2019). https://doi.org/10.1108/JARHE-10-2018-0202
9. Ibrahim, M.M., Nat, M.: Blended learning motivation model for instructors in higher education institutions. Int. J. Educ. Technol. Higher Educ. **16**(1), 1–21, Article 12 (2019). https://doi.org/10.1186/s41239-019-0145-2
10. Tuninga, R.S.J., Seinen, I.B.J.: Demand and supply of distance education in Russia, World Bank. Bureau Cross, 110 (1995)
11. Voronina, T.P., Kashitsin, V.P., Molchanova, O.P.: Education in the Era of NIT. M. AMO (1995)
12. Andreev, A.A.: Distance learning and distance education technologies. Open Educ. **5**(100), 40–46 (2013)
13. Mantulenko, V.V.: Perspectives on the use of the digital footprint in higher education. Teacher XXI Century. **3–1**, 32–42 (2020). https://doi.org/10.31862/2073-9613-2020-3-32-42
14. Shevchenko, G., Kalashnikova, A., Rybakova, A., Brynza, S.: Tasks, functions, and basic activities of an online tutor. CEUR Workshop Proceedings. In: SLET 2020 - Proceedings of the International Scientific Conference on Innovative Approaches to the Application of Digital Technologies in Education, pp. 285–291 (2020)
15. Arancibia Muñoz, M.L., Halal Orfali, C.: The role of learning management systems (LMS) in educational environments: didactic and technical use. In: ICERI2018 Proceedings, pp. 10473–10479 (2018). https://doi.org/10.21125/iceri.2018.0976
16. Nebytova, L.A., Katrenko, M.V., Savin, D.I., ZhuravlevaYu, I.: Augmented reality in the training process of children with hearing impairments. In: Selected proceedings of the IV All-Russian Scientific and Practical Conference with International Participation "Distance Learning Technologies", Yalta, Crimea, September 16–21, pp. 330–339 (2019)
17. Kulikova, T.A., Poddubnaya, N.A.: Training of the future teacher in applying digital tools for the organization of project activities. In: Selected Papers of the V All-Russian Scientific and Practical Conference with International Participation "Distance Learning Technologies" (DLT 2020), Yalta, Crimea, September 22–25, pp. 391–397 (2020)

18. Suhonen, S.: Learning analytics: Combining Moodle, Whatsapp and self-evaluation data for better understanding. In: Popma, W., Francis, S. (eds.) Proceedings of the 6th European Conference on Social Media, pp. 410–413. Academic Publishing, Brighton (2019)
19. Remote educational technologies: design and implementation of training courses. Under the general. Ed. M.B. Lebedeva. St. Petersburg: BHV-Petersburg, p. 336. (2010).
20. Agibova, I.M., Fedina, O.V., Zakinyan, A.R.: Design of science laboratory sessions with magnetic fluids. Int. J. Mech. Eng. Educ. **45**(4), 349–359 (2017)

Research on Effectiveness and Assistants in the Educational Process for Student Learning

Yulia Andrusenko[1], Egor Kosogov[1](✉), Vitaliya Movzalevskaya[1], J. Jenefa[2], and Aleksandr Usanov[3]

[1] North Caucasus Federal University, Stavropol 355000, Russia
egorkosogov@yandex.ru
[2] Christ University, Bangalore, Karnataka 560074, India
[3] Federal State Budgetary Educational Institution of Higher Education Financial University, St. Petersburg, Russia

Abstract. The article discusses the topical issue of the use of AI assistants in the educational process of universities, which is becoming especially important in the context of the rapid development of digital technologies. Artificial intelligence, as one of the key tools of our time, is being actively introduced into the educational sphere, allowing us to optimize routine tasks and increase the effectiveness of the educational process. AI assistants are able not only to automate tasks such as checking assignments, drawing up curricula and analyzing student performance, but also to provide personalized recommendations for students, which significantly improves the quality of learning and creates an immersive environment for students. This article examines the role of AI assistants as a tool to support teachers in the educational process of universities. The main advantages of using them are considered, including improving the effectiveness of teachers, reducing the burden on teaching staff and the possibility of a more flexible approach to student learning. Attention is paid to the difficulties and limitations associated with the introduction of AI technologies, such as the need to adapt existing educational programs and the ethical aspects of using artificial intelligence. The purpose of the study is to analyze the effectiveness of using AI assistants in various educational environments, as well as to identify key factors for the successful integration of these technologies into the educational process. Based on the analysis, recommendations are offered for teachers on the optimal use of AI assistants, taking into account the specifics of educational tasks. The article will be useful both for researchers in the field of educational technologies using IT tools, and for practitioners interested in optimizing and improving the quality of the educational process.

Keywords: Artificial intelligence · AI assistants · digitalization of higher education · learning efficiency · educational technologies (EdTech) · automation of knowledge verification · methods of effectiveness assessment · pedagogical innovations · integration of AI tools

M. Lapina et al. (Eds.): DET 2024, LNNS 1616, pp. 13–31, 2025.
https://doi.org/10.1007/978-3-032-04365-8_2

1 Introduction

With the progress of artificial intelligence technologies, humanity has new opportunities to transform various fields, including education. AI is no longer just a tool for automating routine processes, becoming a powerful mechanism for data analysis, forecasting, and decision-making. In an educational environment, this is a direct way to individualize learning, create individual learning paths, increase student engagement, and optimize teacher performance. One of the key achievements in the field of artificial intelligence has been the emergence of AI assistants who become participants in the educational process.

The theory of cognitive learning focuses on the cognitive activity of the student and the formation of appropriate mental models. Students learn fundamental concepts from their teacher or form learning resources, and then use logical deduction to gain new knowledge [26, 27].

The modern education system faces many challenges that require innovative approaches and integrated solutions. Among them, there is a need to improve the quality of education and integrate new technologies into the educational process.

Artificial intelligence methods and machine learning algorithms are currently used both in manufacturing, to improve automation efficiency, and in medicine, medicine, and financial structures for big data analysis. With the growing popularity of language models and the launch of open source GPT, artificial intelligence methods have entered even more areas of human activity [24, 25, 28].

The study is devoted to AI assistants, which are considered as a tool for supporting teachers in the educational process of universities and are becoming an important element of the modern educational process. The main purpose of the work is to evaluate the effectiveness of the use of AI assistants in educational environments, identify successful cases of their use, as well as identify difficulties in their use. As part of the study, examples of successful integration of AI assistants in universities were analyzed, and a survey was conducted among teachers aimed at identifying the effectiveness of using AI assistants in the educational process for teaching students [22, 23].

2 Literature Review

Artificial intelligence (AI) is the intelligence demonstrated by machines, in particular computer systems [1]. AI refers to a set of technologies and algorithms that enable programs to learn from data and perform tasks previously accessible only to humans.

AI assistants are software platforms or systems based on artificial intelligence technologies that help users solve problems in everyday life.

As noted by David Wilkinson, a researcher in the field of AI in education, who is working on the integration of AI assistants into the educational process, every year more and more teachers turn to AI assistants for help.

Against this background, Carnegie Learning, a company specializing in the development of educational solutions, in particular in the field of mathematics and STEM, conducted a study on the state of artificial intelligence in education for 2024. The results of the Carnegie Learning report are based on the social The survey involved 799 respondents, 72% of whom were teachers, 24% were administrative staff in the field of education, and 4% were other staff [2]. Figure 1 shows the results of a survey on the use of an AI assistant in education. 80% of respondents use AI in the educational process.

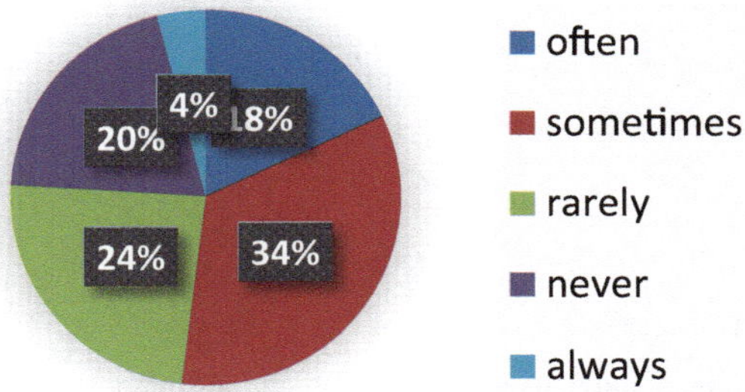

Fig. 1. Survey results: frequency of artificial intelligence use in education.

Artificial intelligence technologies in education offer many advantages that can significantly improve the learning process and enhance the quality of learning. AI tools are capable of performing repetitive and routine work tasks that can be automated. These tasks include data entry and processing, performing simple customer service functions, and creating specific content. Teachers can use AI technologies to automate training courses, form digital portfolios with students' work, and generate assignments [3]. For example, Factspan Analytics Inc. provides a range of services for automating the learning process and helping with the organization of training. For example, tracking class attendance, assessing student knowledge, reporting and auditing, and automatic mailing lists. According to the results of a Carnegie Learning survey (Fig. 2), 42% of respondents noted that the main advantage of using AI is saving time on administrative tasks [2].

Artificial intelligence technologies also allow teachers to create unique personalized learning materials and courses tailored to the needs of each student [3]. For example, Cognii has created the Cognii Virtual Learning Assistant platform. This platform uses machine learning algorithms and neural networks to analyze student responses and provide feedback. The platform also analyzes data, allowing the teacher to track student progress and identify areas that require additional support. The Knewton educational platform uses adaptive learning technologies and aims to improve the learning process by personalizing educational materials. According to the survey results, in Fig. 2, 25% of survey participants consider the creation of individual programs to be a useful advantage, 18% of participants noted an improvement in student engagement through the use of AI, 6% of respondents use technology to manage the learning process [2].

Despite the significant advantages of using AI technologies in education, AI assistants face a number of disadvantages and limitations. One of the key issues is the bias that arises from the data they were trained on. This can lead to the AI giving out false information, presenting it as reliable. In addition, the lack of transparency in the learning process makes it difficult to understand the logic by which AI makes decisions. As a

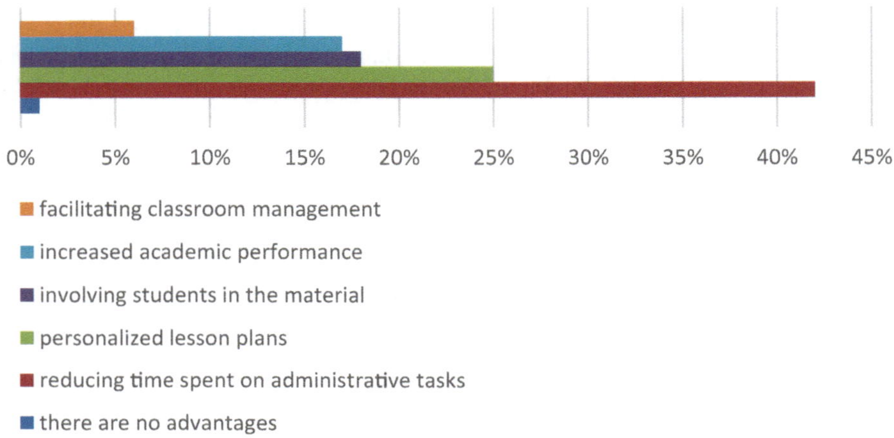

Fig. 2. Survey results: areas where artificial intelligence provides assistance.

result, algorithms based on low-quality data can harm students, affecting their learning and perception of information [4].

Using the Carnegie Learning survey, which is noted in Fig. 3, five main problems in the use of AI in education were identified that the respondents faced. About 53% of teachers say that the main problem with using AI is cheating with its help. 50% worry that they don't have enough skills to use AI at the moment. 33% say they have encountered technical difficulties. 27% had difficulty integrating AI into the educational process. 24% are afraid of leaks of personal information [2] (Fig. 3).

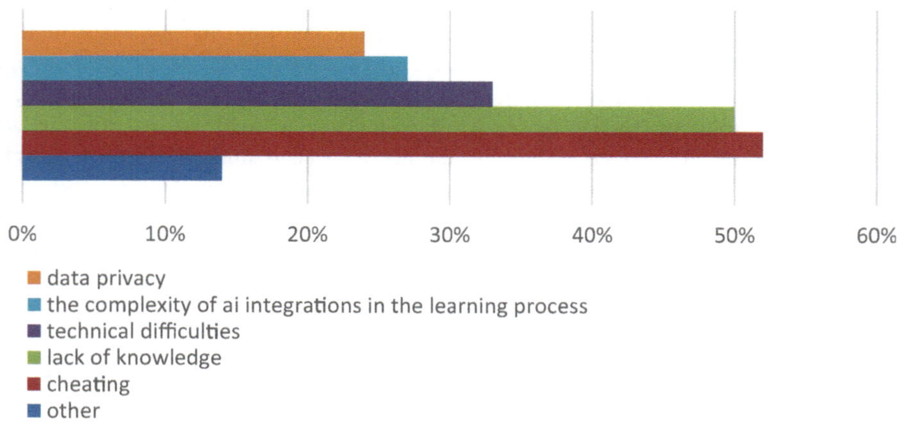

Fig. 3. Survey results: problems in using AI.

AI assistants are beginning to be actively used all over the world. For example, China has launched a smart learning environment, Smart Learning Environments [5, 6]. The main purpose of such an environment is to create an interactive and adaptive educational atmosphere using modern technologies, including artificial intelligence, big data and the Internet of Things. Smart learning environments use AI to analyze data about students, such as learning style, academic performance, and preferences. Based on this information, the systems can offer customized learning materials and assignments that meet the needs of each student. Smart learning environments collect and analyze data about the learning process, which allows teachers to track students' progress and respond to their needs in a timely manner.

The UK has launched the AI4Schools educational project, which aims to provide the knowledge and skills needed to adapt to a world governed by artificial intelligence. The project is intended for high school students, as well as for their teachers. AI4Schools is focused on creating individualized educational plans for students. AI analyzes data on students' academic performance, learning styles, and interests to offer tailored resources and assignments that meet their needs. AI4Schools offers tools that allow teachers to better understand their students' progress, identify difficulties, and provide timely support [7].

AI in Education in the USA is a rapidly developing field that uses artificial intelligence technologies to improve the educational process at all levels. The main areas include personalized learning, data analysis to assess student performance, interactive and adaptive platforms, support for the professional development of teachers, as well as the introduction of chatbots and virtual assistants. AI also promotes inclusive education and is actively researched in the context of ethics and data security [7].

The private International University of Applied Sciences IU in Germany is actively implementing generative AI models to solve auxiliary tasks. IU has developed an AI assistant named Syntea. Synthea is a chatbot that can answer questions, conduct a dialogue, and test students' knowledge in personalized tests. Sven Schütt, CEO of the IU Group, said in an interview with McKinsey, "With the help of AI, students can master the program at least 50% faster than without such personalized support." Experts also noticed that the evaluation of student papers using generative AI is more objective and fair than manual. Sven Schütt also noted that most students like to use digital assistants at different points in their studies, but up to 30% never use them [8].

Moscow City Pedagogical University has created an AI system that predicts exam results and warns students who are at risk of expulsion. The neural network was trained on the data of more than 3500 students, taking into account over 20 parameters: grades, gender, age, place of residence, form of study (budget or commerce), educational program and USE results. The AI also analyzes course selection, activity in the library and on the remote platform, as well as participation in public life. Based on this data, the system predicts the success of exams in the next session—with a high, average or satisfactory result. If a student does not achieve a satisfactory level, he may be a candidate for expulsion. This data is available to students in their personal accounts on the university's website. As a result of using the system, the number of deductions decreased from 5% to 2.5% [9, 14].

In addition, the academic council of the Moscow City Pedagogical University allowed students to use artificial intelligence technologies, including chatbots, when creating graduation papers. This information was announced in an interview with the RBC news company. The university noted that "the generated texts can serve as an excellent basis for new ideas and analysis." Students are required to process the received materials on their own. Igor Remorenko, Rector of Moscow State Pedagogical University, believes that the legalization of AI contributes to the development of new skills—working with modern tools, analyzing big data and verifying them [10].

Moscow State University uses artificial intelligence technologies to analyze student performance and personalize educational programs. MSU strives to integrate AI into educational processes and methodologies to make learning more effective and individualized. AI helps to adapt educational materials to the unique needs of each student and provides access to modern educational resources and online platforms. The University is actively developing research in the field of artificial intelligence, attracting leading experts and scientists [11, 16].

The National Research Technological University MISIS is actively developing and implementing artificial intelligence systems for online learning and student support [12, 17, 18]. These technologies help to create adaptive curricula that take into account the individual needs and level of training of each student. The AI systems at MISIS are also used to automate administrative processes, making it easier for students to access the necessary resources and information. In addition, such systems can support students in the learning process by providing recommendations for improving academic performance and offering additional materials for study. The university is conducting research to improve the interaction between students and teachers through AI, which contributes to a more efficient and personalized educational process.

Sechenov University has created a chatbot for training cardiologists. The chatbot was trained on a 37 GB data array. The data included Russian and international clinical guidelines, scientific articles, and other open datasets. The chatbot provides access to professional information about drugs, provides recommendations for diagnosis and treatment based on current clinical guidelines, helps assess certain risks associated with the patient's health, prepares standard documents such as appointment protocols and discharge epicrisis. The chatbot is useful for both novice specialists and experienced doctors, as it reduces the time required to complete routine tasks and allows doctors to pay more attention to patients. The software has already been included in the Russian software registry and fully complies with Russian information security requirements [13, 19].

An AI assistant has been created at the National Research University of the Higher School of Economics, designed for students and employees of educational offices. This project is a chatbot that provides round-the-clock access to consultations on educational, administrative and social issues. The introduction of an AI assistant has helped to reduce the burden on the staff of educational offices and provide quick responses to student requests. This project was implemented as part of the strategic initiative of the Higher School of Economics "AI Technologies for Humans" [13, 20, 24–30]. Anna Korovko, Senior Director for Basic Educational Programs at the National Research University of Higher School of Economics, said in an interview for CNew: "The new tool will allow us, in addition to unloading administrators, to collect feedback on what issues students most often worry about during different periods of the academic year. This way we can improve the information system for students. The next step in the development of the system is a chatbot for teachers. It will help you navigate through difficult rules and requirements" [14, 15, 21].

3 The Effectiveness of AI Assistants

Let's highlight the key features of AI assistants, thanks to which they are an effective teacher's tool in the educational process.

1. Creating educational materials: An AI assistant can be a good helper in creating tests, help with adapting the material to the students' level, and also help to create a lesson plan correctly.
2. Educational process support: AI is able to design simple programs using programming languages, offer code examples, and adapt hints to the student's level of knowledge.
3. Analysis of academic performance and automation of task verification: the task of checking essays, detecting plagiarism, evaluating practical work according to a given sample of students can be transferred to an AI assistant, thereby freeing up the teacher's time for other tasks.

According to the survey results (Fig. 4), teachers most often use AI to create educational materials (test tasks) – 52% (more than half of the respondents). A popular application of AI is also the analysis of academic performance – 46%. In third place is the support of the educational process – 13%.

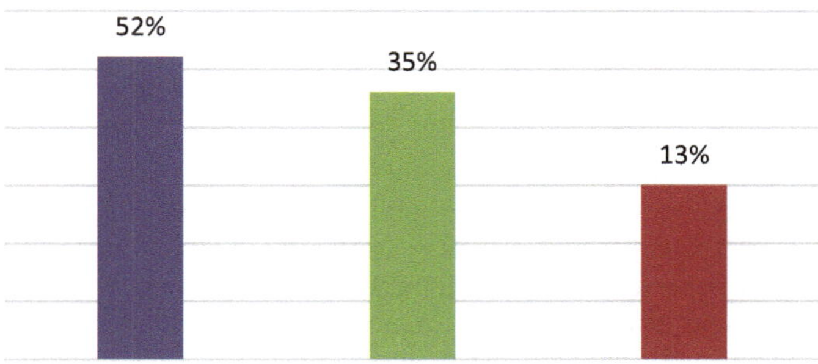

Fig. 4. Survey results: What are you using AI for?

The following methods were used to evaluate the effectiveness of AI assistants:

- Experiment (comparing task completion time with and without an AI assistant).
- Performance analysis (measuring response accuracy and query processing speed).
- Accuracy assessment (expert analysis of the correctness of responses).

This approach allowed us to identify the strengths and weaknesses of each assistant and formulate recommendations for use in real conditions.

Three chatbots with different functional capabilities were chosen as the basis for an experimental study of the effectiveness of AI assistants and the development of practical recommendations. The selection criteria were: different models of natural language processing; differences in response generation algorithms; varying degrees of integration with external data sources. As part of the study, the effectiveness of AI assistants was evaluated in comparison with the work of a teacher in two key parameters:

- preparation of test tasks;
- checking practical assignments.

Four subjects participated in the study: three modern DeepSeek AI assistants (version 2.0), GPT chat, GigaChat (Sberbank's neural network platform) and a qualified teacher with 5 years of scientific and pedagogical experience, an active associate professor of the relevant department, author of scientific publications in the field under study. The scope and nature of the experimental tasks assigned to the subjects is the development of a hundred test tasks in a given area, balanced in complexity, various types of questions and verification of practical work.

The presented research parameters – the composition of the participants and the scope of the tasks performed, were specially selected to achieve the following scientific goals:

1. Ensuring the validity of the experiment – the representativeness of the sample allows for a correct comparison between AI assistants and the teacher;
2. Creating an adequate load – the volume of one hundred test questions creates a sufficient load to identify the real performance of systems, assess the stability of work during a long task, and analyze the quality of generated content;
3. The possibility of complex analysis – a variety of task types allows you to evaluate the flexibility of algorithms, the ability to adapt to different formats, and the ability to work with context.
4. Time characteristics – to evaluate the optimization of time spent on routine typical tasks.

When analyzing the time spent on developing test materials, the teacher found that creating one hundred questions requires significant time resources. The time range was from three to 10 h. The key factors affecting the duration of the work are the complexity of the wording of the questions, the required depth of study of the content, and the need to create original assignments. The greatest time expenditure was observed when developing complex tasks, creating questions of a high level of complexity, and the need to include graphic elements. The data obtained indicate the complexity of the process of manual development of test materials and confirm the need to find ways to optimize this process.

Artificial intelligence can generate a test of 100 questions in a matter of minutes, depending on the complexity and specifics of the questions. This allows teachers to save a significant amount of time. However, it is important to understand that automation reduces the time needed to create tests, but human control remains critical for their validity and pedagogical value. The teacher may need some time to review the generated questions and make sure they are correct.

As an example, consider the DeepSeek, GPT, and GigaChat chat platforms. We will give all three AIS the task of compiling a test of one hundred questions on the topic "The use of IT technologies in education."

At the request of "Make a test of 100 questions on the topic "The use of IT technologies in education", DeepSeek provided a result in 3 min, consisting of ten sections covering key aspects of the topic, with ten questions in each section. Figure 5 shows a request to generate questions for the DeepSeek neural network. However, it should be noted that at this stage the questions were presented without answers, which is reflected in Fig. 6. It took 40 min to analyze all one hundred questions, and, overall, DeepSeek generated the correct questions. However, some of them required paraphrasing and clarification.

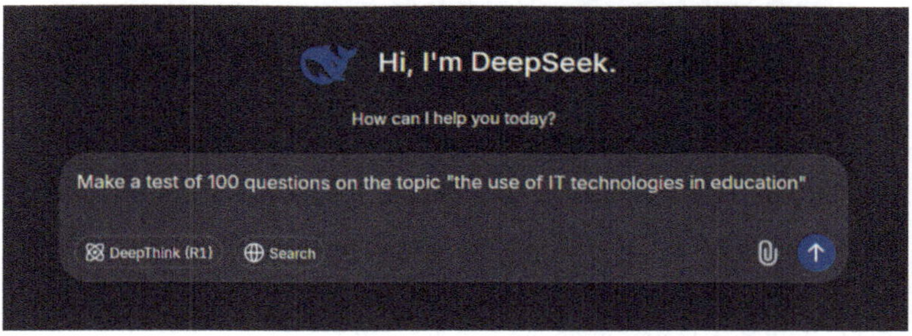

Fig. 5. Deep Seek neural network request

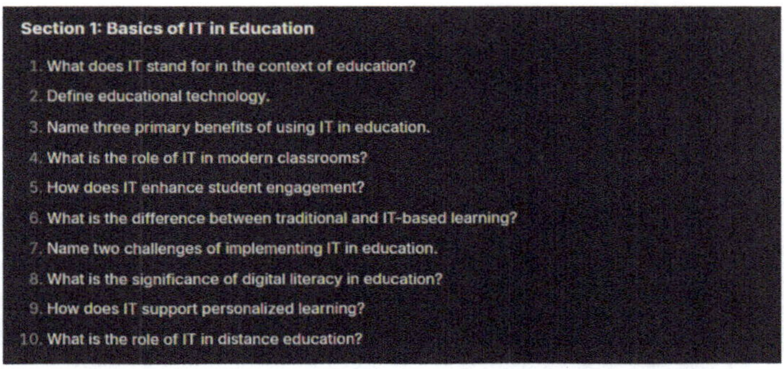

Fig. 6. The process of generating questions by the Deep Seek neural network

During the study, the functionality of the DeepSeek neural network was tested, instructing it to generate answer options for the first block of test questions, specifying the conditions beforehand.: "The answer options can be either open or closed, and can also consist of one or more answers." It took the AI only 1 min to complete this task. Of the ten questions, only one turned out to have incorrectly worded answers. However, it is worth noting that the disadvantage of the answers provided is that in multiple choice questions they are always arranged in order from top to bottom, which can make it difficult for students to choose, or vice versa simplify it. Figure 6 shows the process of generating questions by the Deep Seek.

Figure 7 shows a request for task adjustments to the DeepSeek neural network.

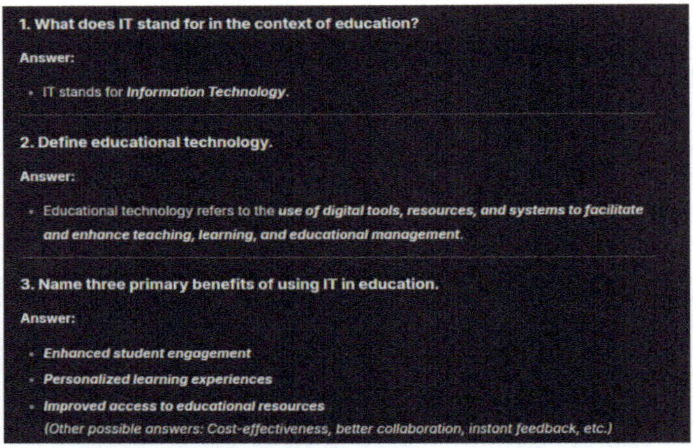

Fig. 7. Adjusting the DeepSeek neural network task нейросети DeepSeek

We spent about 8–10 min reviewing and correcting all ten issues. Thus, the total time spent on composing a test of 100 questions will be about 2 h. This suggests that using AI such as Deep Seek can significantly reduce the time required to prepare learning materials, while allowing teachers to focus on more important aspects of the educational process.

Figure 8 shows the query "Make a test of 100 questions on the topic "The use of IT technologies in education." ChatGPT was successful within 1 min. Unlike DeepSeek, ChatGPT immediately provided multiple choice questions, which significantly accelerated the process.

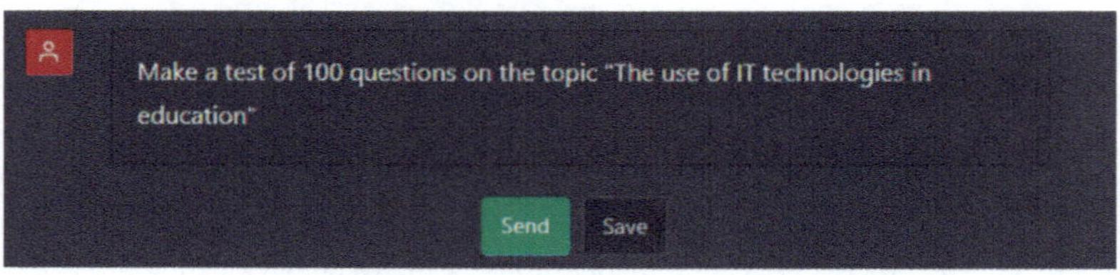

Fig. 8. Request to the ChatGPT neural network

Figure 9 shows a request for query adjustments to the ChatGPT neural network.

During the verification and correction phase, approximately 10–12 min were spent analyzing and correcting errors and inaccuracies in 10 questions. Of these, only one question required a change in wording, and one required a correction of the answers (Fig. 9). Thus, the total time required to compile and adjust a test of 100 questions can be estimated within 2 h.

Figure 10 shows the request "Make a test of 100 questions on the topic "The use of IT technologies in education." Giga Chat was completed in 2 min. Similar to Deep Seek, Giga Chat structured the questions, dividing them into 5 key sections, including

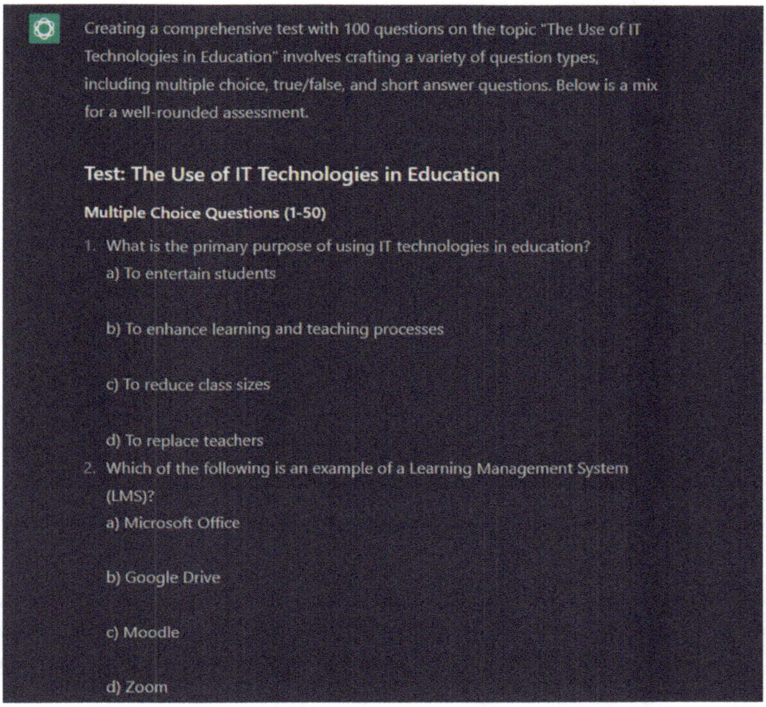

Fig. 9. Adjusting the ChatGPT neural network query

20 and 10 questions, respectively. Giga Chat also provided unanswered questions at this stage. Figure 11 shows the analysis stage of all 100 questions, which took approximately 30 min. Giga Chat demonstrated high accuracy in the formulation of questions, and no correction of their content was required.

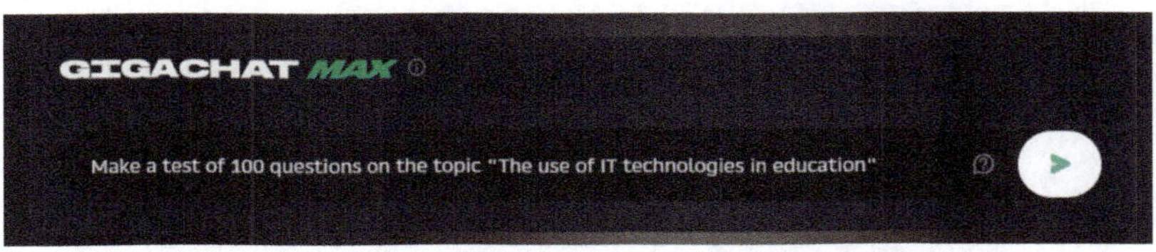

Fig. 10. Query to the Giga Chat neural network

Next, a request was made to GigaChat to compile answer options for the first 10 questions from the first section, specifying the conditions beforehand: “The answer options can be either open or closed, and can also consist of one or more answers.” Giga Chat provided answers almost instantly, but all the answers were presented in an expanded form, which is why some of the tests had to be redone manually.

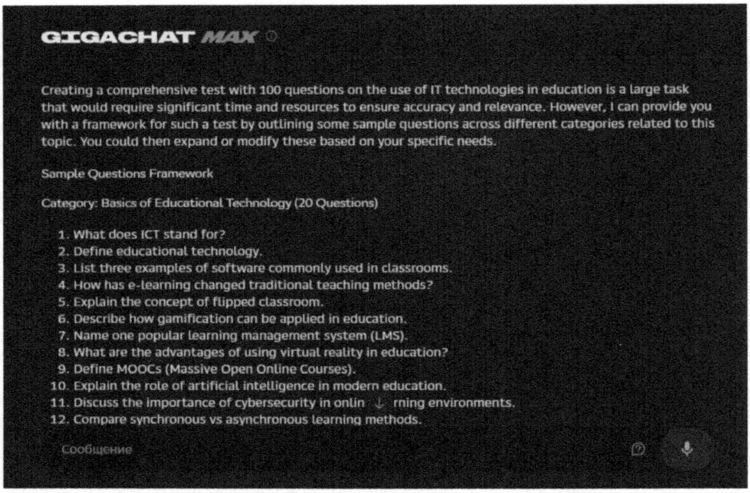

Fig. 11. The analysis stage is the compilation of questions of the GigaChat neural network

It took about 20 min to fix 10 multiple choice questions. Thus, the total time required to compile and adjust a test of 100 questions using GigaChat can be estimated within 3–4 h (Fig. 12). These results emphasize the need for additional verification and adaptation of the test responses.

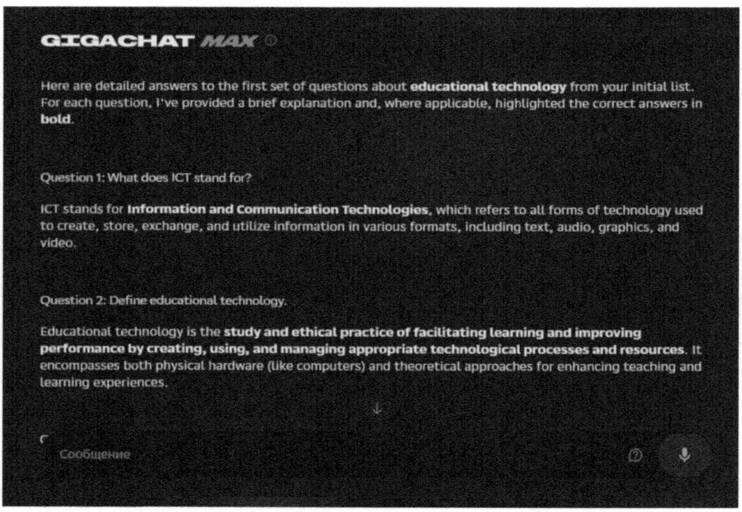

Fig. 12. The stage of adjustment of the GigaChat neural network

The conclusions of this stage of the analysis of the effectiveness of AI assistants are shown in Table 1.

Table 1. Efficiency of using AI assistants

Parameters	Without using an AI assistant	Using DeepSeek	Using GPT chat	Using GigaChat
Test preparation time (min)	300	120	120	210

The second parameter for comparing the effectiveness of AI assistants was the verification of practical work. An experimental study has shown that checking practical work on a given topic can take a long time for a teacher, depending on the complexity of the task and the amount of work. If the work contains a lot of details or requires in-depth analysis, the verification time is greatly increased. On average, a practical work check can take from 15 min to 60 min. However, AI can verify similar work in seconds or minutes, depending on the complexity of the algorithms and the amount of data that needs to be processed.

Let's consider the possibilities of the DeepSeek, GPT and GigaChat chat platforms in checking and evaluating practical tasks. To do this, we will ask you to check and rate the practical work, which contains a summary and an answer to the test. The request looks like this: "Check the correctness of the writing of the practical work and give a fair grade, where 5 is the highest score and one is the lowest."

Figure 13 shows an assessment of the practical operation of the DeepSeek neural network. The chat successfully completed the assessment of the practical work in 30 s. The model conducted a detailed analysis of each response, providing detailed explanations of the estimates. The main disadvantage of DeepSeek was the reduction of points for the lack of specifics and examples in cases where their presence was not mandatory according to the terms of the assignment.

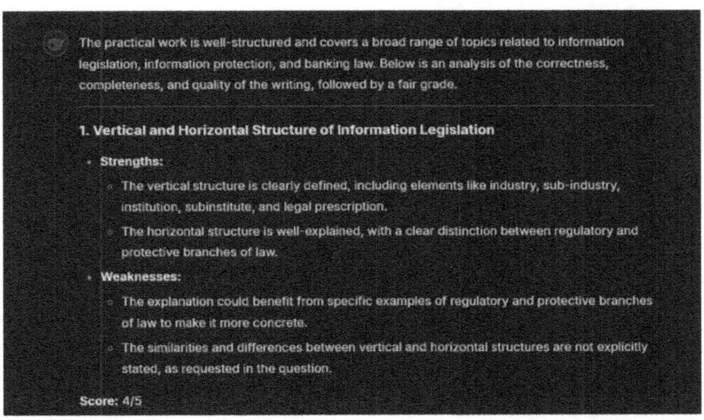

Fig. 13. Assessment of the practical work of the DeepSeek neural network

At the same time, the key advantage of the model is the availability of detailed and reasoned comments, which greatly simplifies the process of analyzing and correcting estimates. Figure 14 shows an analysis of the job evaluation.

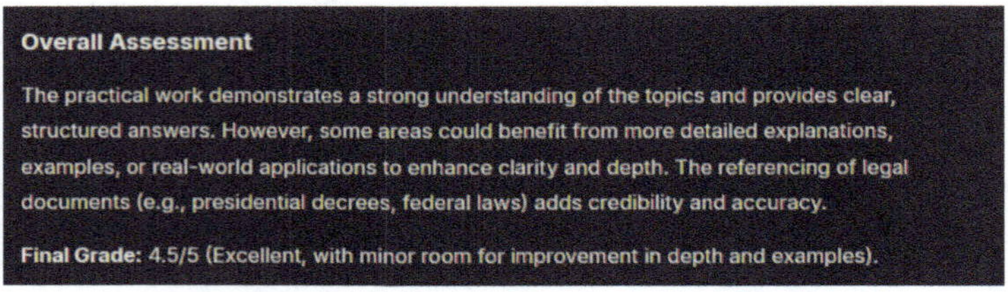
Overall Assessment

The practical work demonstrates a strong understanding of the topics and provides clear, structured answers. However, some areas could benefit from more detailed explanations, examples, or real-world applications to enhance clarity and depth. The referencing of legal documents (e.g., presidential decrees, federal laws) adds credibility and accuracy.

Final Grade: 4.5/5 (Excellent, with minor room for improvement in depth and examples).

Fig. 14. DeepSeek neural network assessment analysis

Overall, Deep Seek has demonstrated high efficiency in task verification. Thanks to the preliminary assessment using AI, the total time for checking practical work was halved from 35 min to 18 min.

ChatGPT proved to be the fastest tool, completing a practice check in 5 s. At the same time, he proved himself to be the most rigorous inspector. Figure 15 shows an assessment of the practical operation of the ChatGPT neural network.

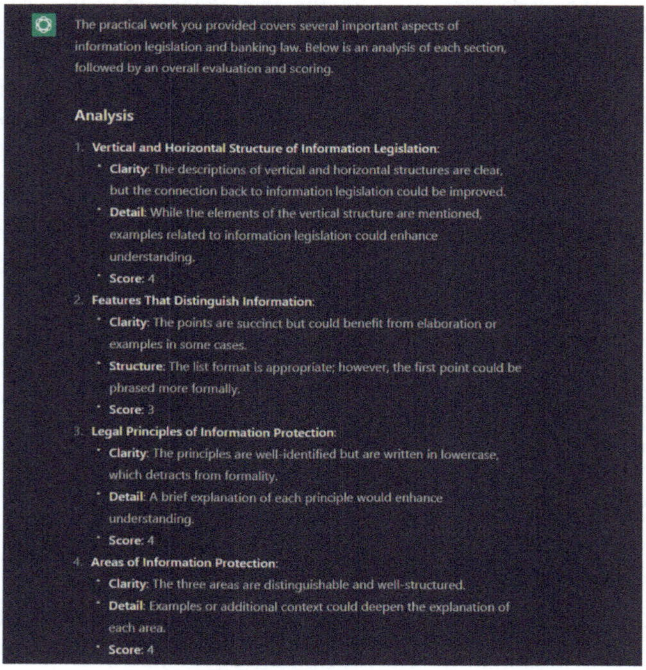
The practical work you provided covers several important aspects of information legislation and banking law. Below is an analysis of each section, followed by an overall evaluation and scoring.

Analysis

1. **Vertical and Horizontal Structure of Information Legislation:**
   - **Clarity:** The descriptions of vertical and horizontal structures are clear, but the connection back to information legislation could be improved.
   - **Detail:** While the elements of the vertical structure are mentioned, examples related to information legislation could enhance understanding.
   - **Score:** 4
2. **Features That Distinguish Information:**
   - **Clarity:** The points are succinct but could benefit from elaboration or examples in some cases.
   - **Structure:** The list format is appropriate; however, the first point could be phrased more formally.
   - **Score:** 3
3. **Legal Principles of Information Protection:**
   - **Clarity:** The principles are well-identified but are written in lowercase, which detracts from formality.
   - **Detail:** A brief explanation of each principle would enhance understanding.
   - **Score:** 4
4. **Areas of Information Protection:**
   - **Clarity:** The three areas are distinguishable and well-structured.
   - **Detail:** Examples or additional context could deepen the explanation of each area.
   - **Score:** 4

Fig. 15. Assessment of the practical work of the neural network chat GPT

Similar to DeepSeek, ChatGPT provided detailed explanations for each assessment. However, one of the disadvantages is the underestimation of points for the lack of references and direct citations, which, according to the terms of the assignment, were not required. Although the GPT chat requires an additional recheck of the assignment to give a fairer grade, thanks to the detailed explanations, the process took 25 min. Figure 16 shows an analysis of the assessment of the practical work of the ChatGPT neural network.

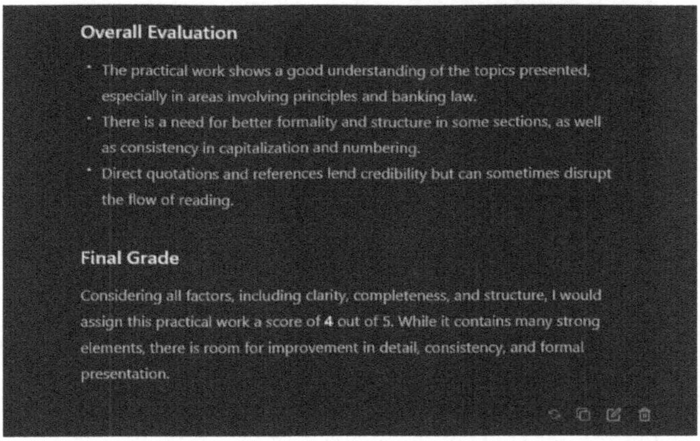

Fig. 16. Analysis of the assessment of practical work by the neural network chat GPT

GigaChat handled the practical work check in 1 min. Like DeepSeek and ChatGPT, GigaChat provided detailed ratings with explanations for each response. However, similar to other models, GigaChat lowered scores for lack of specifics and examples in cases where their presence was not mandatory. Figure 17 shows an assessment of the practical work of the GigaChat AI assistant.

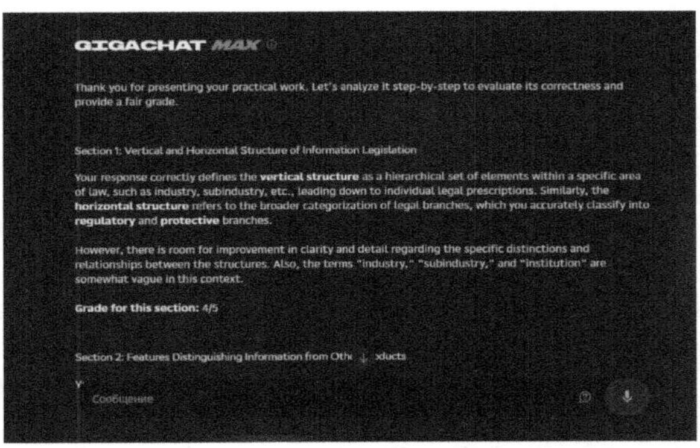

Fig. 17. Assessment of the practical work of the GigaChat neural network

Figure 18 shows an assessment of the practical operation of the Giga Chat neural network.

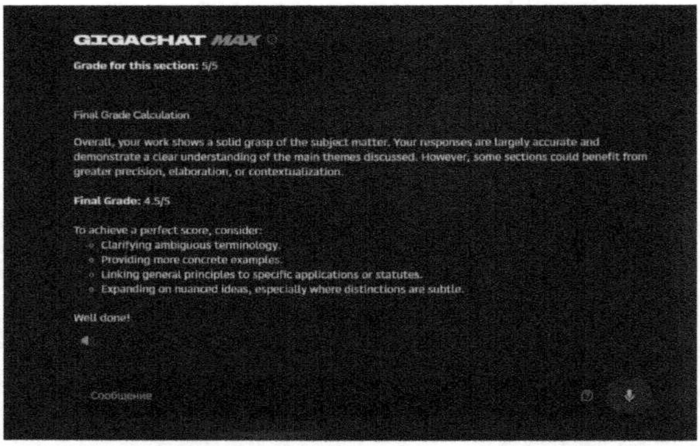

Fig. 18. Analysis of the assessment of practical work by the GigaChat neural network

In terms of the quality of verification, Giga Chat showed results comparable to DeepSeek. The use of a preliminary assessment using AI has reduced the total time for checking the practical assignment by half– from 35 min to 20 min.

The conclusions of this stage of the analysis of the effectiveness of AI assistants are shown in Table 2.

Table 2. Time spent checking practical work when using AI assistants

Parameters	Without using an AI assistant	Using DeepSeek	Using GPT chat	Using GigaChat
Work verification time (min)	35	18	25	20

Thus, all three models have demonstrated high efficiency in evaluating practical work, but require additional adjustments to take into account the specifics of the assignment and eliminate unnecessarily strict evaluation criteria. Using AI tools significantly speeds up the verification process, reducing time costs by 50%.

Table 3 provides a comparative analysis of the productivity of AI assistants and teachers.

In terms of speed and versatility, ChatGPT leads the way, providing instant and versatile responses, which is optimal for users who value promptness and a wide range of information. In terms of depth of analysis and objectivity, DeepSeeK shows the best results, which makes it the preferred choice for complex specialized queries where accuracy and consistency are critical. GigaChat, demonstrating good evaluation objectivity, is inferior to competitors in query processing speed and depth of study, which reduces its effectiveness in scenarios requiring urgent and detailed analysis. When comparing

Table 3. Comparative characteristics of work efficiency

Parameters	Teacher	DeepSeek	ChatGPT	GigaChat
Time spent	Low	Medium	High	Low
Depth of knowledge	High	High	High	Medium
The fairness of the assessment	High	High	Medium	High
Versatility	Medium	Medium	High	Medium

AI assistants and teachers, it should be noted that the speed of AI exceeds human from three to ten times. Teachers have a slight advantage in the accuracy of the compilation (98% versus 90–95% for the best AI), but for most tasks the difference is not critical. Teachers reserve an advantage in reviewing papers, since automated analysis contains 5–10% more errors than expert assessment.

4 Conclusion

The introduction of AI assistants into the educational process is becoming a trend in modern education. Artificial intelligence technologies open up wide opportunities for both teachers and students, creating an immersive learning environment. AI assistants make learning more flexible, adapted to individual educational trajectories. Simplification of routine tasks performed by the teacher leads to the fact that the teacher can focus on strategically important aspects of his work, creatively approach tasks. Examples of the use of AI assistants in the world's leading universities and Russian universities demonstrate how technologies are effectively integrated into the educational process.

Despite the obvious advantages, the use of AI in education comes with a number of limitations. One of the main challenges is choosing the AI assistant that is most effective at solving the tasks. The analysis allows us to draw the following conclusions: AI assistants provide a significant reduction in the time required to create tests and verify work compared to traditional methods. DeepSeek is the leader in processing speed and creation. It combines in-depth analysis with objective assessments, making it ideal for precise analytical tasks. ChatGPT stands out for its exceptionally fast responses and wide functionality, and is optimal for working with diverse queries, although it is somewhat inferior in the accuracy of estimates. GigaChat demonstrates good evaluation objectivity, but it is noticeably lagging behind in processing speed and depth of analysis, which reduces its applicability for urgent and complex tasks. These results indicate the significant potential of AI assistants in optimizing the learning process, while the choice of a specific tool should be determined by the teacher in accordance with the specifics of the tasks being solved. It is important to understand that the productive use of artificial intelligence in the field of education involves a harmonious combination of digital solutions with traditional pedagogical approaches. Only a systematic approach combining technological innovations with methodological support guarantees the successful digitalization of education without prejudice to its fundamental principles.

For successful integration of AI assistants into the educational process, further scientific work is required to minimize potential risks. A key area of such research may be the creation of training programs for teachers and students to work with digital assistants. The implementation of these measures will improve the quality of the educational process, ensure comfortable adaptation for all participants, and maximize the benefits of artificial intelligence.

References

1. Russell, S.: In: Russell, S., Norvig, P. (eds.) Artificial intelligence: a modern approach, 4th edn. Pearson, Hoboken, NJ (2021) ISBN 978-0134610993
2. The State of AI in Education: Artificial Intelligence in Education 2024 Report. https://olymp.hse.ru/mirror/pubs/share/986545727
3. Shirokolobova, A.G.: Artificial intelligence as a tool for optimizing the work of a higher school teacher. Pedagogy. Questions of theory and practice. **9**(2) (2024)
4. Agaltsova, D.V., Valkova, Y.E.: Artificial intelligence technologies for university teachers. https://cyberleninka.ru/article/n/tehnologii-iskusstvennogo-intellekta-dlya-prepodavatelya-vuza/viewer artificial intelligence in education: teachers' opinions on AI in the classroom. https://www.forbes.com/advisor/education/it-and-tech/artificial-intelligence-in-school/
5. Liu, Y., Chen, L., Yao, Z.. The application of artificial intelligence assistant to deep learning in teachers' teaching and students' learning processes. https://www.frontiersin.org/journals/psychology/articles/10.3389/fpsyg.2022.929175/full
6. Schools across China embrace digitalization, build smart campuses. https://www.immersivelearning.news/2024/07/12/schools-across-china-embrace-digitalization-build-smart-campuses/
7. AI in Education: Personalizing Learning for Every American Student/AI in Education: Personalizing learning for every American student. https://groupify.ai/blog/personalized-ai-education-for-american-students
8. Education for all: An interview with Dr. Sven Schütt. https://www.mckinsey.com/industries/education/our-insights/education-for-all-an-interview-with-dr-sven-schutt
9. Ilishev, I.K.: Research of the influence of artificial intelligence on modern education. https://elibrary.ru/item.asp?id=68625589
10. Building a Smart Education Ecosystem to Promote Lifelong Learning Opportunities for All. https://gse.bnu.edu.cn/202309/4910.html
11. Artificial intelligence in education. https://habr.com/ru/articles/740730/
12. NUST MISIS digitalizes educational processes using the company's Cloud platform. https://nauka.tass.ru/nauka/16503715?utm_source=yandex.ru&utm_medium=organic&utm_campaign=yandex.ru&utm_referrer=yandex.ru
13. How Russian and foreign universities are implementing AI in student education. https://www.vedomosti.ru/technology/articles/2024/11/06/1073238-kak-rossiiskie-vnedryayut. AI4Schools – AI Education for High School Students/AI4Schools – AI education for high school students. https://lamarr-institute.org/blog/ai-education-high-school-ai4schools/
14. Aithal, S., Aithal, P.S.: Optimizing the Use of Artificial Intelligence-Powered Gps as Teaching and Research Assistants by Professors in Higher Education Institutions: A Study on Smart Utilization processes. https://papers.ssrn.com/sol3/papers.cfm?abstract_id=4729191Развитие. Artificial Intelligence AI of Moscow State University. https://proza.ru/2023/12/07/101МГУ, https://proza.ru/2023/12/07/101
15. Lanchakov, A.B., Filin, S.A., Yakushev, A.Z., Stoll, A.G.: On the use of artificial intelligence in education. https://elibrary.ru/download/elibrary_70862354_82664620.pdf

16. Yashina, I.A.: Artificial intelligence in teaching programming to students of a pedagogical university. https://elibrary.ru/download/elibrary_74523714_76547982.pdf
17. Lapina, M., Tokmakova, M., Movzalevskaya, V., Karabasheva, M., Balakrishnan, D.: Analysis of modern methods and approaches using it in education and pedagogy. In: Lapina, M., Prakasha, G., S., Grigoriev, S. (eds.) SLET 2022: International Conference on Innovative Approaches to the Application of Digital Technologies in Education and Research. SLET 2022 Lecture Notes in Networks and Systems, pp. 40–53. Springer, Cham (2024). https://doi.org/10.1007/978-3-031-78776-8
18. How are AI assistants changing higher education? https://www.frontiersin.org/journals/computer-science/articles/10.3389/fcomp.2023.1208550/full
19. B The Higher School of Economics has developed an AI assistant for students and staff of educational offices. https://www.cnews.ru/news/line/2024-11-22_v_niu_vshe_razrabotali_ii-assistenta
20. News from the world of innovation. https://innovanews.ru/info/press_release/v-nizhegorodskojj-vshe-razrabotali-ii-assistenta-dlja-studentov-i-sotrudnikov-uchebnykh-ofisov/
21. Artificial Intelligence in Education: A Review. https://ieeexplore.ieee.org/abstract/document/9069875
22. Kormakova, V.N., Klepikova, A.G., Lapina, M.A., Rugelj, J.: ICT competence of a teacher in the context of digital transformation of education. In: CEUR Workshop Proceedings, vol. 2914, pp. 138–150 (2021)
23. Kormakova, V., Klepikova, A., Musaelian, E., Baybikowa, G., Lapina, M.: Formation of ICT Competencies of Postgraduate Students of Teacher Education Based on Interactive Techniques: CEUR Workshop Proceedings SLET-2019 – Proceedings of the International Scientific Conference Innovative Approaches to the Application of Digital Technologies in Education and Research, vol. 2494, pp. 11–21 (2019)
24. Lapina, M., Shiriaev, E., Babenko, M., Istamov, I.: High-speed convolution core architecture for privacy-preserving neural networks. Programm. Comput. Softw. **50**, 417–424 (2024). https://doi.org/10.1134/S0361768824700282
25. Rusanov, M., Babenko, M., Lapina, M., Sajid, M.: Identification of exploited unreliable account passwords in the information infrastructure using machine learning methods. Big Data Cogn. Comput. **8**, 159 (2024). https://doi.org/10.3390/bdcc8110159
26. Rugelj, J., Lapina, M.: Game Design Based Learning of Programming: CEUR Workshop Proceedings SLET-2019 – Proceedings of the International Scientific Conference Innovative Approaches to the Application of Digital Technologies in Education and Research, vol. 2494, pp. 29–42 (2019)
27. Sherbakov, S., Lapina, M., Lapin, V., Rugelj, J.: Methodological support of game modeling in the educational process. In: Paper presented at the SLET-2020: International Scientific Conference Innovative Approaches to the Application of Digital Technologies in Education, CEUR Workshop Proceedings, vol. 2861, pp. 351–361 (2020)
28. Basan, E., Makarevich, O., Lapina, M., Mecella, M.: Analysis of the Impact of a GPS Spoofing Attack on a UAV CEUR Workshop Proceedings. **3094**, 6–16 (2022)
29. Proshkin, N., Basan, E., Lapina, M.: Radio frequency method for emulating multiple UAVs. In: 17th International Conference on Intelligent Environments, IE 2021–Proceedings, p. 9486599 (2021)
30. Maksimova, E., Lapina, M., Lapin, V.: Synthesis of models for ensuring information security of subjects of critical information infrastructure under destructive influences. CEUR Workshop Proc. **3094**, 108–117 (2022)

Educational Perspectives of the Metaverse

Yulia Andrusenko[1], Egor Kosogov[1(✉)], Vitaliya Movzalevskaya[1], and D. Vinodha[2]

[1] North Caucasus Federal University, Stavropol 355000, Russia
egorkosogov@yandex.ru
[2] Christ University, Bangalore 560074, Karnataka, India

Abstract. The article presents a comprehensive review of the transformational potential of metaverses in the field of modern education. A multidimensional analysis of this technology is carried out, considering it not only as a technical innovation, but also as a fundamentally new educational paradigm. The paper explores in detail the unique capabilities of metaverses for creating immersive learning environments where traditional educational formats are combined with innovative approaches such as gamification and project activities in virtual space. The article pays special attention to the practical aspects of the use of metaverses in the educational process. Methodologically, the article identifies three key areas for the successful realization of the educational potential of the metaverses: technical (providing infrastructure and accessibility), organizational (developing a regulatory framework and standards) and pedagogical (creating effective didactic models). In conclusion, the article emphasizes that the full realization of the educational potential of the metaverse is possible only with a systematic approach that comprehensively takes into account technological, organizational and pedagogical aspects. Such a multi-level approach will make it possible to transform the metaverse from a promising, but still experimental technology into an effective educational tool of a new generation. Only in this case will the metaverses be able to become a catalyst for profound changes in the learning system, ensuring its adaptation to the requirements of the rapidly developing digital reality.

Keywords: educational metaverse · digital educational environment · Roblox in learning · VR/AR · digital transformation of learning · pedagogical innovation · gamification · artificial intelligence in education

1 Introduction

The relevance of this topic is due to a significant increase in interest in metaverse technologies as a new educational environment. This is due to the development of virtual and augmented reality technologies, as well as the high rate of digitalization. Metaverse technologies have gained great interest in the context of global changes, and educational institutions are faced with the need to adapt to new forms of learning [1]. The metaverse offers unique opportunities to create interactive and immersive educational spaces where students can actively participate in the learning process, interact with learning materials and each other in a virtual environment.

M. Lapina et al. (Eds.): DET 2024, LNNS 1616, pp. 32–43, 2025.
https://doi.org/10.1007/978-3-032-04365-8_3

From a technological point of view, the metaverse combines a variety of tools such as artificial intelligence, blockchain and virtual reality, which makes it possible to create adaptive and personalized educational platforms. These technologies can not only improve the quality of the educational process, but also guarantee safety and transparency in the management of learning outcomes [21, 22].

The theory of cognitive learning focuses on the cognitive activity of the student and the formation of appropriate mental models. Students learn fundamental concepts from their teacher or form learning resources, and then use logical deduction to gain new knowledge [25, 26].

Artificial intelligence methods and machine learning algorithms are currently used both in manufacturing, to improve automation efficiency, and in medicine, medicine, and financial structures for big data analysis. With the growing popularity of language models and the launch of open source GPT, artificial intelligence methods have entered even more areas of human activity [23, 24, 27].

From an educational perspective, the integration of the metaverse into the educational process represents significant methodological potential, allowing for innovative approaches to organizing educational activities based on the principles of active learning. Learning through the metaverse forms students' key competencies such as analytical skills, creative thinking, and digital skills that meet the requirements of the modern labor market. This technology platform provides a multi-level perception of educational content through immersive capabilities.

Of particular value is the ability of metaverses to transform the traditional educational process into a dynamic interactive environment where theoretical knowledge finds practical application.

2 Literature Review

An analysis of the literature shows a growing academic interest in educational metaverses, however, there is a need for further research on their long-term pedagogical impact and the development of comprehensive methods of integration into the educational process.

In the works of leading researchers of digital education, the metaverse is defined as the union of physical and digital reality, in which users can interact with each other in real time through digital avatars [1]. Research analysis shows that modern educational universes are based on virtual worlds, blockchain platforms, and digital twin technologies. Through the use of special virtual (VR) and augmented reality (AR) technologies, the user creates a sense of presence in a virtual space. Users can actively participate in events, interact with other participants, and create content [2, 7].

One of the first areas where the metaverses have demonstrated significant potential is the educational system. The transfer of the educational process to the virtual space of the metaverse is an important stage in the digitalization of education. However, a complete transition to this format remains a difficult task.

In the work of digital didactics by researchers from the University of Copenhagen, Gustav Peterson and Guido Makranski, the metaverse opens up new horizons for education, offering unique opportunities both from a technological and pedagogical point of view. Metaverse technologies allow for the creation of immersive educational environments where students can interact with educational material in 3D format. For example, students can study human anatomy through virtual models of organs, and design and test virtual models by designing and testing bridges or buildings in a safe environment. Thanks to 3D models and virtual tours, students can visit historical sites, museums or scientific laboratories without leaving the classroom [3, 4].

When teaching using the metaverse, AI assistants can be used to create adaptive learning systems that adapt to the individual needs and level of knowledge of students, providing a personalized approach to learning. Blockchain and NFT technologies provide secure storage and management of educational certificates and achievements, as well as create new forms of assessment and remuneration. The integration of social media into educational platforms allows students to share knowledge, work in groups and participate in discussions, and multiplatform makes learning more accessible and flexible.

Virtual reality as a basic element of the metaverse opens up new perspectives for learning. Immersive VR headsets that track the user's movements create the effect of total immersion in the digital environment. By complementing the system with tactile interfaces, the developers achieve an unprecedented level of realism. These technologies have already transformed professional training from flight simulators to medical simulators, demonstrating potential that goes far beyond the entertainment industry [3].

Augmented Reality (AR) is a technology that combines the real world with digital elements. Unlike VR, which creates a completely virtual environment, AR adds computer graphics to our environment: text, images, 3D models and animations. A device with a camera and a screen is enough to work with AR. Virtual objects are precisely integrated into the physical space, maintaining scale and allowing them to be viewed from different angles. Today, AR is used in a wide variety of fields, from education and design to industry [3–5].

Table 1 provides a comparative description of the two augmented reality technologies.

Augmented reality technology makes the learning process more interactive, visual, and exciting. For example, AR technologies allow you to "animate" special textbooks and manuals. In particular, Moscow State University is developing a textbook for schoolchildren with elements of augmented reality, with an emphasis on the study of biology. This initiative is aimed at improving educational outcomes and creating a more attractive learning environment [6].

Augmented reality (AR) technologies are actively used in the medical field. In particular, medical students at the University of Cambridge use the HoloScenarios medical training program in combination with the Microsoft HoloLens 2 augmented reality headset to train on holographic models of patients. The headset projects holograms of people showing various symptoms, which allows future doctors to diagnose possible diseases and develop algorithms for first aid. This approach promotes the development of clinical thinking and practical skills in a controlled and safe environment [8, 9].

The introduction of augmented and virtual reality technologies into the educational process faces a number of challenges. Firstly, the high cost of equipment limits the availability of these technologies for many educational institutions. Secondly, teachers need

Table 1. Comparative characteristics of VR and AR

Virtual Reality (VR)	Augmented Reality (AR)
A completely artificial digital environment that replaces the physical world. The environment, sounds, and what is happening around the user are modeled by the developers	Superimposing digital (virtual) elements on a real environment without total immersion
User interaction: integrates it into an artificially created world. For example, it takes a gamer into the world of a computer game	Interacts with the environment: integrates artificially created objects into the real world around the user. For example, it complements the class with an anatomical 3D model of a person
To immerse yourself in virtual reality, you need a VR helmet or special glasses and a controller	Augmented reality objects can be seen through special AR glasses, the camera of a smartphone, tablet or computer

additional training to effectively use new tools. Thirdly, the content for some academic disciplines is still limited, which reduces the effectiveness of AR/VR applications. In addition, XR technologies provide listeners with more audiovisual stimuli than they are used to receiving in real life. At the moment, there is insufficient data on how information perception works in the metaverse and which practices are most effective. It is also worth considering that prolonged exposure to immersive space can cause dizziness and nausea, so the time spent in the metaverse must be monitored [10].

Nevertheless, the process of informatization is gradually contributing to the introduction of multimedia and interactive approaches into the educational process, which makes it possible to replace traditional methods of knowledge transfer, such as rote memorization of text from textbooks, with project activities and independent research.

Every day, metaverses are being more actively introduced into the educational process. For example, the Faculty of Engineering at the University of Tokyo (Todai) initiated the creation of an educational platform called the Metaverse School of Engineering. As part of this project, the learning process is organized in a virtual space where students and teachers interact through avatars, regardless of geographical location. The main objective of the initiative is to create an inclusive educational environment accessible to all, regardless of age, gender or place of residence. It is expected that this approach will increase interest in engineering specialties and will contribute to the faster development of professional competencies in this field [11, 12].

Stanford University, a private California research university, has implemented an innovative approach to learning within the framework of the Virtual People course, completely transferring the learning process to virtual reality. The experiment involved 263 students, each of whom was provided with a set of VR headsets to access classes. The course program included interactive elements such as virtual field trips, group discussions, and practical tasks performed in a VR environment. Teachers and students interacted exclusively through avatars, which allowed them to create an immersive experience close to real learning. This experiment became part of the university's research on the use of VR technologies in education, aimed at increasing the involvement and effectiveness of the educational process. Jeremy Bailenson, founder and director of the Virtual Human Interaction Lab at Stanford University: "The Virtual People course offers students more than just a taste of VR. Here, virtual identity becomes the main means of their interaction with each other. As far as I know, no one in our educational practice has combined almost three hundred students in a single VR environment for several months. Our course allows students to try out and develop applications that their predecessors could only dream of. It's nice to see how far virtual reality has come at the beginning of the 21st century. Our class is a metaverse that demonstrates how to build an infrastructure that includes hardware, software, and content that truly work together. I am sure that this project will be brilliantly developed in the capable hands and minds of Stanford University students" [13].

The University of Hong Kong (Hong Kong University of Science and Technology, HKUST) is actively implementing virtual learning technologies: already some classes are held in virtual classrooms, and in the near future it is planned to launch a full-fledged meta-university. This project is an interactive 3D platform combining educational, communication and innovation functions. Students, faculty, and graduates from all over the world will be able to use the platform to participate in lectures, seminars, group projects, and scientific research. A special feature of the metaverse will be the gamification of learning, which includes interactive elements such as virtual labs, simulations and collaborative tasks. The platform will work as a single space, providing seamless interaction between users, which will allow not only to learn, but also to create new projects, share experiences and develop professional relationships [11, 14].

Educational initiatives related to metaverse technologies are actively developing in Russia. One of these projects is NEYMARK. MetaVerse is a virtual educational center that already allows students from different regions of the country to participate in lectures, perform laboratory work and develop digital projects. The platform makes it possible to start educational activities even before the completion of the physical campus, which speeds up the process of launching the center and attracts applicants. As part of NEYMARK, students gain access to sophisticated equipment through virtual simulations, practice public speaking skills, hold meetings, and defend projects in an interactive environment. This approach not only expands the geography of educational opportunities, but also creates conditions for practical training and the development of professional competencies in a digital environment [15].

Drone management training using augmented and virtual reality (AR/VR) technologies is also actively developing in Russia. Virtual simulators allow you to practice piloting skills without the risk of damage to real vehicles, simulating various scenarios - from extreme weather conditions to emergency situations, as implemented in the UAVProf project with its digital training ground. AR technologies are used for interactive training in drone assembly and repair, as well as in geodesy using AR terrain markings. Game formats are widely used to engage students: virtual drone racing and biathlon competitions, and multiplayer teamwork simulations. Modern systems include elements of artificial intelligence, creating adaptive VR simulators with automatic skill assessment. These technologies are already being used in leading universities, such as Tomsk Polytechnic

University, and educational projects like UAVProf and Agrimax A30, which makes it possible to accelerate the training of specialists while reducing training costs [16].

3 From Virtual Worlds to Gaming Practices: Key Mechanics

Digital metaverses create revolutionary learning opportunities by offering fundamentally new immersive and interactive educational spaces. At the same time, the true effectiveness of such technologies is determined not so much by their technical component as by the competent application of game mechanics that can transform complex academic disciplines into an exciting cognitive process. The fundamental basis of this approach is the natural human need for play activity, theoretically substantiated in the works of L.S. Vygotsky and J. Piaget.

Modern educational platforms such as Roblox and Minecraft successfully implement these principles through the creation of fully immersive digital environments, the development of interactive learning simulations, and the organization of social interaction in virtual spaces. Gamification of the educational process provides a qualitative transformation of traditional learning through several key aspects:

1. Activation of cognitive motivation (dynamic interactive environment replaces passive perception of information with active involvement in the process).
2. Democratization of complex concepts (visualization of abstract data and concepts through game mechanics greatly facilitates their understanding).
3. The applied nature of learning (direct application of knowledge in virtual environments provides instant practical benefits).
4. Formation of flexible skills (teamwork, creative thinking, the ability to solve non-standard tasks).
5. Individualization of the educational process (the level of complexity is automatically adjusted, personalized trajectories are built, individual cognitive interests are taken into account).

The key game mechanics in learning are characterized by the following features: progression and levels of advancement, rewards for completing tasks, learning is embedded in history, competitions, teamwork, feedback and instant results, open worlds and freedom of choice, role-playing mechanics [17].

Table 2 provides a comparative description of learning through game practices and traditional forms of learning.

Modern metaverses are increasingly integrating game elements as a key tool to increase user engagement. In the field of education, such gamified approaches open up new perspectives, transforming the traditional learning process into a dynamic interactive system. The use of game mechanics in educational metaverses allows achieving three fundamentally important effects: interactivity – students move from passive perception of information to active interaction with educational content through virtual simulations, quests and practical tasks.; Motivation – game design elements create sustained engagement, turning learning into an exciting process with clear goals and rewards; effectiveness – game formats promote better learning through emotional engagement, repeated repetition through practice and instant feedback. This transformation of the

Table 2. Comparative characteristics of gamification and traditional forms of education.

Comparison criteria	Game mechanics	Traditional forms of education
Engagement and motivation	They use elements of games, which increases motivation and engagement	It relies on discipline and external motivation (grades, teacher requirements)
Interactivity and active participation	They require active interaction, students make decisions, experiment, and participate in simulations	More often passive (lectures, memorization)
Feedback	Instantaneous, which helps to adjust actions	Feedback has been postponed (evaluations, work review)
Flexibility and adaptability	Allows you to adapt the difficulties to the student's level	More often they are deprived and united for all
Emotions and perception	Causes positive emotions, reduces stress	They can be associated with boredom or pressure
Application of knowledge	Real-world situations are often modeled, where knowledge is immediately applied in practice	Sometimes it is divorced from practice

educational paradigm is especially relevant in the context of digitalization, when traditional teaching methods require adaptation to new realities and needs. Metaverses with their game mechanics offer an organic way to combine educational tasks with digital interfaces and interaction formats familiar to modern students.

Let's look at the process of learning programming through game mechanics using the example of the Roblox metaverse.

Roblox is a metaverse, an online gaming platform and game creation system where users can develop their own projects and interact within virtual worlds. Roblox is available to anyone: it is completely free, has no excessive hardware requirements, and works confidently on all platforms, including smartphones [18, 20]. The Roblox metaverse is not just a platform for games, but a powerful educational tool that allows you to learn programming in an interactive, fun way. Roblox allows students to create their own games and learn how to program, develops problem solving and teamwork skills, offers ready-made lessons and training materials for teachers, and also contains educational games in various subjects.

Roblox is actively used to teach programming in Lua, develop creative thinking and teamwork. The platform provides built-in tools for 3D modeling, animation, and game design, as well as educational resources for teachers [19, 20, 23–29]. This makes it an effective tool for use in STEAM education.

The effectiveness of the Roblox game metaverse is due to a unique combination of technological and pedagogical advantages.:

- Natural game motivation – built-in gamification elements activate students' internal motivation;
- The playful way of presenting the material reduces the psychological barrier before learning programming;
- The possibility of physical testing of the created scripts;
- The ability to publish and test projects in the real community;
- Instant visualization of code results;
- Integration with professional development tools;
- Multilevel complexity system.

The advantages of using the metaverse over traditional courses are shown in table three. Please try to avoid rasterized images for line-art diagrams and schemas (Table 3). Whenever possible, use vector graphics instead (see Fig. 1).

Table 3. Comparative characteristics of the Roblox metaverse with traditional teaching methods.

Comparison criteria	Roblox	Traditional courses
Motivation	High (game form)	Often low (dry theory)
Feedback	Instant (change in game)	Requires a separate check
Practice	Creating real projects	Abstract tasks
Social interaction	Teamwork	Individual work

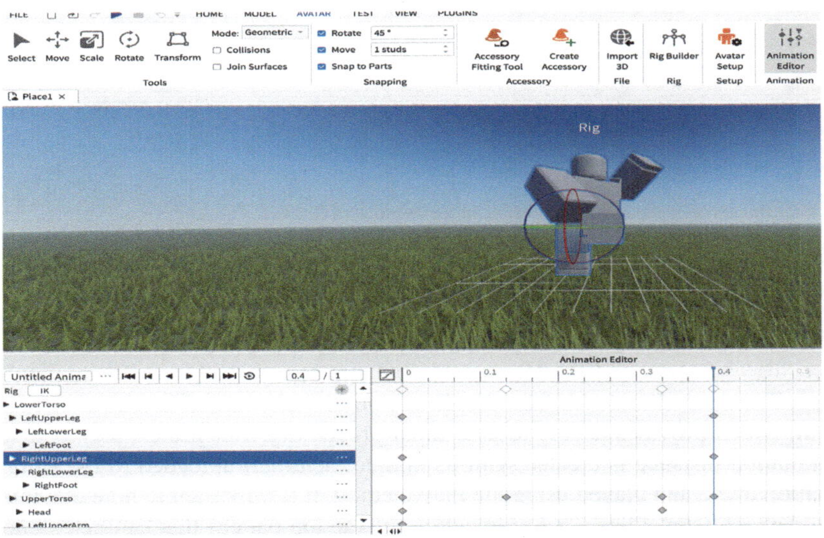

Fig. 1. Creating animations

To demonstrate the educational capabilities of the Roblox platform in the field of programming, let's look at a practical example of creating an interactive animation activated by pressing the "F" key. To do this, we will use Roblox Studio, the official development environment provided by the Roblox platform itself. Roblox Studio is a powerful development environment with a built–in script editor, visual tools for creating a 3D world, access to the Roblox API (players, physics, animations), the ability to test games locally or with friends, as well as integration with the cloud for publishing projects and storing assets. This example illustrates the basic principles of working with an animation system and processing user input, which forms the basis of interactivity in the metaverse.

The first step is to create character animations. As shown in Fig. 1 in Roblox Studio, you need to open the Avatar section, where you can use the Rig Builder tool to create a skeletal structure (Rig) of the character. Next, the Animation Editor develops a specific animation sequence, such as a hand gesture. Special attention should be paid to the smooth transitions between poses and the naturalness of movements, which is achieved by adjusting time intervals and interpolation curves.

After saving the animation in the project library, we proceed to the software implementation of its launch. The key aspect is the use of a client script (Local Script) hosted in StarterPlayerScripts, which provides input processing for each player individually. An example of using programming in the metaverse is shown in Fig. 2.

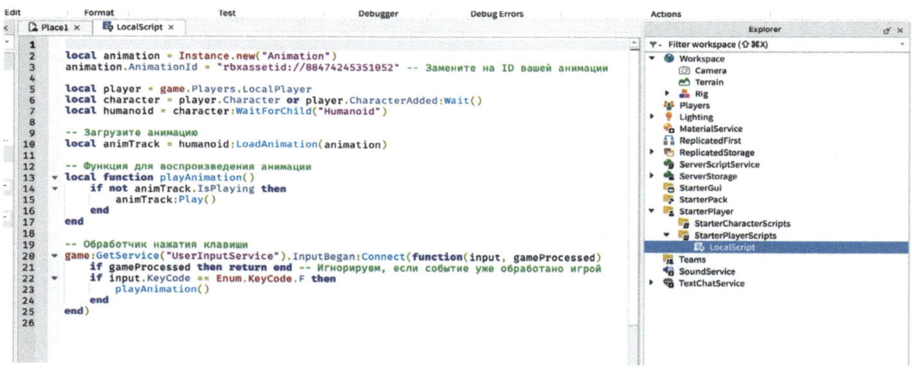

Fig. 2. A local Lua script for playing animations by pressing the F key.

The script starts by getting links to the necessary objects: an instance of the current player, his character, and the Humanoid component responsible for animation capabilities. The special feature of the implementation is the use of the UserInputService system, which provides access to user input processing. When the F key is detected, an animation instance is created with its unique identifier, uploaded to the Humanoid animation track, and played using the Play method. It is important to note that this approach demonstrates a basic action-reaction interaction pattern that can be extended to implement more complex game behavior.

The debugging process of the created solution deserves special attention. Roblox Studio provides built-in testing tools that allow you to test animation performance both in single-player mode and in a multi-user environment. This is especially important for verifying the correctness of network interaction and synchronization of animations between different clients.

The considered case of interactive animation development in the environment at Roblox Studio demonstrates the significant didactic potential of the metaverse platform in the educational process. The key advantage of the presented methodology is the implementation of an end-to-end approach from visual design to direct programming, which contributes to the formation of a student's holistic view of the process of creating game content. This training module performs an important function, laying down the basic competencies necessary for the subsequent development of complex aspects of digital technologies. The implemented sequence from simple to complex confirms the effectiveness of the Roblox Studio metaverse as a multifunctional educational platform that provides a gradual build-up of professional skills in the field of digital content development.

4 Conclusion

Metaverses represent a qualitatively new stage in the development of educational technologies, significantly different from traditional approaches to learning. Their uniqueness lies in the creation of a deeply interactive environment where the educational process organically combines quest formats with the effect of full immersion and virtual laboratory spaces that allow complex experiments to be carried out without any risks. The use of metaverses dramatically expands the boundaries of the educational space, breaking down geographical barriers and opening up opportunities for large-scale international project activities. Students from different parts of the world get a chance to work together on complex tasks in a single virtual environment, which was almost unattainable in classical educational systems. However, the implementation of these promising technologies requires careful consideration of several key factors. First of all, it is necessary to solve the issues of digital accessibility and provide the appropriate technical infrastructure. Special attention should be paid to optimizing the usage time of specialized equipment in order to avoid overloading users. No less important is the methodological aspect – each virtual activity should be clearly linked to specific educational goals, and digital formats should be harmoniously combined with traditional pedagogical approaches. The true educational value of metaverses will be revealed only with an integrated approach that takes into account both the technological potential of these platforms and the psychological and pedagogical principles of effective learning. Only in this case will the metaverse be able to go beyond the fashionable technological trend and become a truly effective tool for transforming the educational system to meet the challenges of the modern digital age.

References

1. Lin, H., Wan, S., Gan, W., Chen, J., Chao, H.-C.: Metaverse in Education: Vision, Opportunities, and Challenges (2022). https://ieeexplore.ieee.org/abstract/document/10021004

2. VR, AR, MR и XR — how do the technologies of immersion in the virtual world differ? https://education.yandex.ru/journal/vr-ar-mr
3. Ardiny, H., Khanmirza, E.: The Role of AR and VR Technologies in Education Developments: Opportunities and Challenges (2018). https://ieeexplore.ieee.org/abstract/document/8657615
4. Carmigniani, J., Furht, B., Anisetti, M., Ceravolo, P., Damiani, E.: Augmented reality technologies, systems and applications. Multimedia Tools Appl. **51**(1), 341–377 (2023). https://doi.org/10.1007/s11042-010-0660-6
5. Why are the metaverses expanding so fast? https://club.forbes.ru/practicum/pochemu-metavselennye-rasshiryajutsya-tak-bystro
6. MSU is developing school textbooks with augmented reality. https://msu.ru/press/smiaboutmsu/v-mgu-razrabatyvayut-shkolnye-uchebniki-s-dopolnennoy-realnostyu.html?tmpl=clear
7. Ar/vr in education, games and medicine: new opportunities and challenges of the digital age. https://elibrary.ru/download/elibrary_79981028_16988351.pdf
8. UK medical students training on hologram patients in world-first, https://uk.news.yahoo.com/uk-medical-students-training-hologram-153000595.html
9. Martellucci, J., Dal Mas, F., Pantalone, D., Bergamini, C., Prosperi, P.: When Surgery Meets the Metaverse (2024). https://doi.org/10.1007/978-3-031-47623-5_6
10. Metaverse 101: Where to start. https://sberuniversity.ru/edutech-club/journals/metavselennaya-modnoe-slovo-ili-obrazovatelnaya-tekhnologiya-budushchego/metavselennaya-101-s-chego-nachat/
11. Who and how integrates metaverses into learning today: current experiences and work cases. https://sberuniversity.ru/edutech-club/journals/metavselennaya-modnoe-slovo-ili-obrazovatelnaya-tekhnologiya-budushchego/kto-i-kak-vstraivaet-metavselennye-v-obuchenie-segodnya-aktualnyy-opyt-i-rabochie-keysy/
12. University of Tokyo to open Metaverse School of Engineering. https://asianews.network/university-of-tokyo-to-open-metaverse-school-of-engineering/
13. Standford Conducts the First Class Entirely in VR. https://www.avclub.pro/news/stenfordsky-universitet-perevel-obucheniye-v-vr-sredu/
14. HKUST to Launch World's First Twin Campuses in Metaverse. https://hkust.edu.hk/news/research-and-innovation/hkust-launch-worlds-first-twin-campuses-metaverse
15. The prototype of Russia's first educational metaverse was presented at the CIPR conference. https://tass.ru/nizhegorodskaya-oblast/14791689
16. Lifting force: how are drone specialists trained in the country? https://trends.rbc.ru/trends/education/cmrm/64ec4c089a79472ff541a535
17. Shamsrizi, M.: Metaverse and Gaming Potentials for Foreign Cultural and Educational Policy (2023). https://opus.bsz-bw.de/ifa/frontdoor/deliver/index/docId/957/file/ifa-2023_shamsrizi_metaverse-and-gaming_en.pdf
18. What Is the Purpose of Roblox? https://www.lifewire.com/what-is-roblox-4144781
19. Roblox education: a guide for teachers and students, https://www.progkids.com/blog/roblox-education-rukovodstvo-dlya-uchitelej-i-uchenikov
20. Metaverse or Simulacra? Roblox, Minecraft, Meta and the turn to virtual reality for education, socialisation and work. https://www.tandfonline.com/doi/full/10.1080/10494820.2022.2022899#d1e80
21. Kormakova, V.N., Klepikova, A.G., Lapina, M.A., Rugelj, J.: ICT competence of a teacher in the context of digital transformation of education. CEUR Workshop Proc. **2914**, 138–150 (2021)
22. Kormakova, V., Klepikova, A., Musaelian, E., Baybikowa, G., Lapina, M.: Formation of ICT Competencies of Postgraduate Students of Teacher Education Based on Interactive Techniques: CEUR Workshop Proceedings SLET-2019 – Proceedings of the International

Scientific Conference Innovative Approaches to the Application of Digital Technologies in Education and Research, vol. 2494, pp. 11–21 (2019)
23. Lapina, M., Shiriaev, E., Babenko, M., Istamov, I.: High-speed convolution core architecture for privacy-preserving neural networks. Program. Comput. Softw. **50**, 417–424 (2024). https://doi.org/10.1134/S0361768824700282
24. Rusanov, M., Babenko, M., Lapina, M., Sajid, M.: Identification of exploited unreliable account passwords in the information infrastructure using machine learning methods. Big Data Cogn. Comput. **8**, 159 (2024). https://doi.org/10.3390/bdcc8110159
25. Rugelj, J., Lapina, M.: Game Design Based Learning of Programming: CEUR Workshop Proceedings SLET-2019 – Proceedings of the International Scientific Conference Innovative Approaches to the Application of Digital Technologies in Education and Research, vol. 2494, pp. 29–42 (2019)
26. Sherbakov, S., Lapina, M., Lapin, V., Rugelj, J.: Methodological support of game modeling in the educational process. In: Paper presented at the SLET-2020: International Scientific Conference Innovative Approaches to the Application of Digital Technologies in Education, CEUR Workshop Proceedings, vol. 2861, pp. 351–361 (2020)
27. Basan, E., Makarevich, O., Lapina, M., Mecella, M.: Analysis of the Impact of a GPS Spoofing Attack on a UAV CEUR Workshop Proceedings, vol. 3094, pp. 6–16 (2022)
28. Proshkin, N., Basan, E., Lapina, M.: Radio frequency method for emulating multiple UAVs. In: 17th International Conference on Intelligent Environments, IE 2021 – Proceedings, p. 9486599 (2021)
29. Maksimova, E., Lapina, M., Lapin, V.: Synthesis of Models for Ensuring Information Security of Subjects of Critical Information Infrastructure under Destructive Influences CEUR Workshop Proceedings, vol. 3094, pp. 108–117 (2022)

The Impact of Interactive Distance Learning Environment on the Development of Intellectual Abilities of Students

Alevtina I. Arkhipova[1], Elena A. Pichkurenko[1], Anna G. Prigodina[2], Ekaterina A. Vladimerets[3], and Viktoriya N. Taran[4](✉)

[1] Kuban State University, Street Stavropolskaya, 149, Krasnodar 350040, Russia
[2] Kuban State Technological University, Street Moskovskaya, 2, Krasnodar 350072, Russia
[3] School No 20 named by Pavel Tyulaev, Street Sormovskaya, 171, Krasnodar 350000, Russia
[4] V.I. Vernadsky Crimean Federal University, Simferopol 295007, Russia
victoriyayalta@gmail.com

Abstract. The article discusses the development of the education system through its digitalization and the creation of an interactive learning environment. Despite the classical learning models, the use of a hermeneutic approach to mastering new disciplines, which was laid down in reflective teaching methods, will allow a deeper understanding of the material being studied and a more meaningful application of it. As part of the hermeneutic approach, the category of understanding is considered as the art of interpreting texts, so the processes of global digitalization exacerbate the relevance of solving the problems of using information technologies and computer didactics. The article also reveals the essence of the concept of "interactive learning environment" (ILE), provides its analysis based on various definitions. Various approaches to the development of the ILE structure in the form of electronic structures that have different target orientations are described. Such structures perform various functions in the general educational space depending on the infrastructure of the educational space itself. This study provides the creation of an information learning environment through software implementation in accordance with regulatory state federal documents. The ISO technology software component chain is built. A new ISO definition is given. The main methodological and pedagogical principles that form the basis of the development of ISO are formulated. Thus, the application of the principles of the hermeneutic approach and the development of an interactive learning environment, which are outlined in the work, improves the quality of education throughout the educational space, and also provides the teacher and student with the opportunity to improve their learning skills and use the knowledge they have acquired, which will allow them to use them in the future for the development of their country and to enhance its prestige in the international arena.

Keywords: Interactive Learning Environment · Distance Learning · Web Technologies · Interactive Technologies

M. Lapina et al. (Eds.): DET 2024, LNNS 1616, pp. 44–55, 2025.
https://doi.org/10.1007/978-3-032-04365-8_4

1 Introduction

The concept of "interactive learning environment" is used by authors very often, but at the present stage there is no clear definition of this concept. For the most part, this concept is used intuitively, without reference to regulatory documents. For example, in the National Standard of the Russian Federation "Information and Communication Technologies in Education. Terms and Definitions" (GOST R 52653–2006), unlike other terms, this one is not described and is not included in the list of documented concepts. In our opinion, the term "distance educational technologies" is the closest in meaning to it. Distance educational technologies (or interactive learning environment) are educational technologies implemented in direct contact between teachers and students mainly through the use of telecommunications and information technologies [5, 8, 9].

In designing the ISO, it is very important to rely on the hermeneutic potential of scientific texts, which contributes to the organization of the mental activities of students, the setting of a hermeneutic task, the development of a didactic version of educational technologies, and the creation of computer programs [1]. Bogin G.I. believed that the solution to this problem lies in addressing the hermeneutic approach. At the same time, the process of understanding the studied text should be accompanied by the transformation of knowledge into the mental experience of the student [3].

As the authors [18] note, traditional and innovative methodological support should be combined and transformed through the introduction of modern educational products and integrated into interactive educational environments. Practicing teachers can develop teaching didactics through their innovative activities and share their experience of using it.

A great contribution to the formation of professional competence of teachers was made by scientists: S.A. Beshenkov, Ya. A. Vagramenko, V.V. Laptev, E.S. Polat, I.V. Robert, E.P. Belozertsev, and others. Their research in the theory of e-learning focuses in particular on the design of content in interactive learning environments (ISO).

2 Modern Learning Technologies

2.1 Traditional Electronic Learning Tools

Computer didactics, as noted by the authors [6, 13], is widely discussed and used in the practice of world education. In particular, it is well known that currently the most popular learning tools in world universities are electronic means of providing educational content [22]. Researchers from foreign universities claim that most students prefer those electronic learning services and software platforms that contain tasks of a theoretical and practical nature. Such services and platforms have proven themselves well on foreign platforms and within the country, and are also the most widespread (see Fig. 1).

The methodology of teaching at the present stage is undergoing a new formation. New technical and educational tools are being formed that take into account the age and psychological aspects of the formation of the knowledge potential of students.

Thanks to the improvement of the Internet infrastructure, a toolkit for creating interactive video lectures with higher multimedia saturation has appeared. The interactive learning environment contributes to the differentiation of the learning process due to the

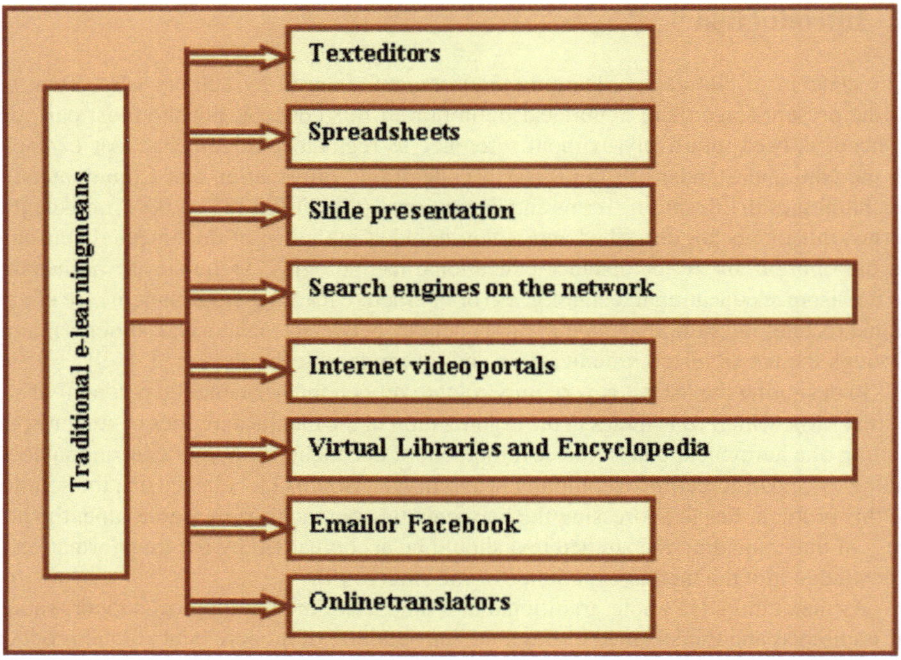

Fig. 1. Traditional e-learning.

individualization of educational content, which ultimately leads to the improvement of students' academic work and the ability to understand complex learning concepts [4, 20].

Interactive video lectures are created using some software component that facilitates the organization of interactive educational elements of the lesson using the technology of collective intelligence (CI). Collective intelligence ensures the interaction of people in the process of application and methods of information processing. Numerous network structures, cloud technologies have formed the prerequisites for the development of collective intelligence technologies, which are used, for example, in crowdsourcing projects. So, we can summarize. When considering various methods for defining the concept of "interactive learning environment", its properties and structure, the following conclusions were obtained.

- "Interactive learning environment" is a concept that still does not have a precise definition, since it is not fixed in the regulatory framework. Many authors use synonyms of this concept in their scientific works and terms that border on this concept. For example, such synonyms as digital educational environment, electronic educational environment, virtual educational environment, etc. are used. In addition, this term is often found in works on the methodology of teaching preschoolers, but preschoolers do not yet have sufficient e-learning skills.
- The authors consider the interactive learning environment in the methodological literature as a type of educational material in electronic format. For example, electronic

textbooks, problem books, virtual laboratory workshops, supplements to training courses, electronic educational resources, electronic libraries, etc. It can be noted that in one article the educational structure can very often be called differently. In addition, the teaching methodology is usually based on classical pedagogy, that is, lecture presentation of theoretical material. At the same time, the interactive educational environment is a direct interaction of the subjects of the pedagogical process.

- Some authors supplement the interactive learning environment with the use of social networks and social repositories (social bookmarks and geoservices, photo and video services), as well as collective hypertexts organized in the form of reference resources such as Wiki, social media and databases. For example, interactive assignment designers LearningApps.org, network offices and network diaries can also be considered as tools of the interactive learning environment. These electronic resources also include Blendspacelogo - a service of materials on certain topics that are integrated into the textbook. Mind maps help to visualize the processes of thinking and reasoning in the form of a graphic record using certain symbols.

Thus, all of the listed electronic resources are designed to solve various learning problems and for different target audiences, and also perform different functions in the infrastructure of the general educational space.

2.2 Structure of the ISO Software Component

Let us consider the structure of the interactive learning environment. Here, too, there is a diversity of approaches among the authors. Thus, Berezyuk S.I., Sirotina I.K. and Faley A.V. characterize the interactive learning environment for mathematics using web technologies [18].

The structure proposed in the table is typical for electronic educational environments in various disciplines and areas. Due to the lack of a single generally accepted classification of the interactive educational environment, in some definitions the concepts of "environment, space, platform" are identified.

The following Table 1 gives the components and their internal structure.

Quite often, authors in their pedagogical works consider issues related to the use of ready-made software components in the structure of an interactive learning environment. But at the same time, teachers are only required to develop educational content according to the specifics of a ready-made environment or resource, i.e. the path "from the computer to the content of training". We propose the path "from the content of training to the computer", i.e. the creation of a digital component of interactive learning environment technologies according to the standards of the content of training with the construction of a chain of training and technology for its implementation (see Fig. 2).

We have obtained a set of basic characteristics of this concept, from which it follows that the object of study is interactive learning technologies, which are structural elements of the subject interactive educational environment.

So, we can formulate the following definition of an interactive learning environment.

An interactive educational environment is an information and reference structure aimed at studying scientific theory through interactive technologies, in the process of using which reflexive methods of educational and cognitive activity prevail [5]. At the

Table 1. Components and their internal structure of the interactive learning environment for mathematics.

Component	Internet structure
Interactive content	indeed interactive texts, reference information materials, interactive seminars and interactive tests
Interactive text	animated text with elements of information visualization, capable of dialogic interaction with the reader, contains pop-up tips, hyperlinks, comments, interactive drawings and diagrams, small animated objects
Interactive workshop	workshop, consisting of a system of key problems in mathematics and including the tabs "Actualization of Knowledge" and "Decision"
Interactive online directory	presented by the systematic and structured base of theoretical material
Test component	material for independent study, which consists of two modes of operation: control and interactive

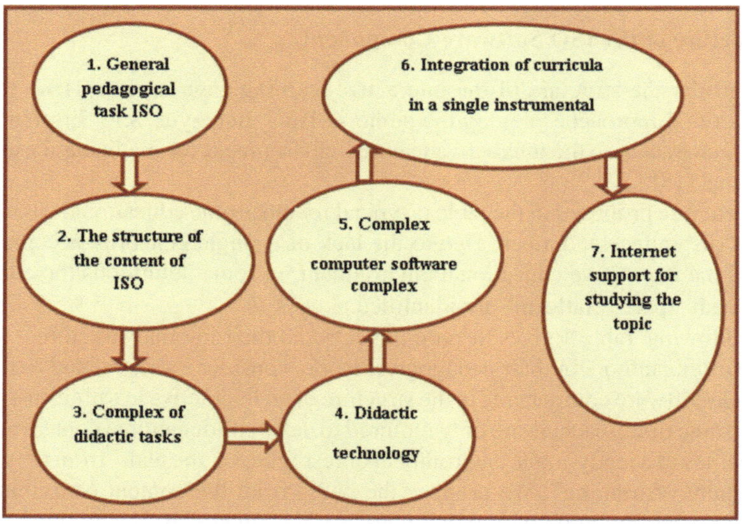

Fig. 2. Software Component ISO Technologies.

same time, the main emphasis is placed on developing the mental abilities of students, the ability to analyze, compare, look for analogies, etc., due to which the thinking experience is enriched and the culture of the student's thinking activity is formed [12]. The use of a hermeneutic approach using the intellectual and developmental potential of scientific texts with computer support to create an interactive learning environment is one of the most important factors in achieving success [11, 13, 14].

This approach promotes differentiation of the learning process with an emphasis on independent work of students. The teacher, using a creative approach to the learning process, has the opportunity to flexibly manage the learning process, control it and direct the students' activities when presenting educational information using educational, methodological and didactic support located on the electronic resources of the discipline [7, 10, 19].

Thus, the teacher does not experience inconvenience from strict regulation in planning the educational process, has the opportunity to show creativity in professional activity. A modern teacher gladly uses new tools of pedagogical activity to improve the learning process and obtain high results and finds optimal ways to include it in the learning process. The principle of pedagogical freedom implies the refusal of strict restrictions in the choice of means and methods of teaching in his pedagogical activity. In this regard, it can be stated that the tendency, when the teacher is offered a huge number of didactic manuals or lesson scenarios with a minute-by-minute progress of the educational process, can be used at the stage of formation of a young teacher in the process of developing pedagogical experience and professionalism. In general, the strict framework of a lesson or class does not allow for high results, unlike creative interactive learning.

3 Methodological and Pedagogical Principles

We formulate the basic principles that form the basis for the development of ILE.

3.1 Methodological

1. The principle of the relationship between theory and practice, in which theory acts as a vector for the development and improvement of practice, and practice as a factor for the verification of theoretical models.
2. The Gnoseological principle. This is the principle of reliance on the theory of knowledge.

   The essence of the principle is to construct the content of training based on the structure and genesis of the relevant scientific theory.
3. Philosophical and pedagogical hermeneutics, which justify replacing the target orientation of the educational process for the formation of ready-made normative knowledge for the organization setting of separate reflexive thought activity when using hermeneutic teaching methods [16].
4. Systematic. This principle requires the creation of pedagogical conditions for the formation of system knowledge appropriate to the structure of the studied scientific theory while reflecting its genesis.

3.2 Pedagogical

1. The principle of the leading role of fundamental theoretical knowledge. Fundamental knowledge corresponds to systemic thinking, includes the structure of the scientific theory being studied, conceptual apparatus, goal setting and the main methods and

models for solving problems in this area. This principle is implemented by combining technologies and learning tools in an interactive learning environment that reflect the structure of the theories being studied and generalize the content of the training.

2. The principle of maximum interactivity. The interactive learning environment operates with printed digital versions of the book containing educational information and didactic support, and is aimed at maintaining active interaction between subjects of the educational process through information and communication technologies. At the same time, it should be made as interactive as possible that students have more opportunities for self-studying material [17, 23].
3. The principle of the teacher's creative approach to learning. The role of the teacher changes. He does not simply declare a set of educational information, but is included in the learning process as a leader of the educational and cognitive process.
4. The principle of methodological inversion and iteration. The principle allows presenting the same information, considering it from different positions, different angles, repeatedly changing the form of a fragment of content due to the use of numerous techniques for working with text and its visualization. In this case, special attention is paid to the semantic transformation of definitions of scientific concepts that have a complex form, for this they also resort to the concept of logical algebra [21]. To implement this principle, a large set of didactic materials and computer forms representing the same didactic objects is necessary.
5. The principle of communicativeness. It is assumed that there is a possibility of accessing the global computer network, using electronic resources, switching to websites of interactive learning environments, remote interaction of subjects of educational activity using a computer or smartphones and tablets. In addition, for feedback it is necessary to use educational and control Internet technologies, including the use of the technology designer "Power of Knowledge" (certificate of registration in Rospatent No. 2013612181/69) [18].
6. The principle of variability of the form of content presentation. The substantive and methodological components of the environment should be maintained in the general context, but at the same time its specific content, software implementation and visual forms of information presentation can vary.
7. The principle of feedback with the professional community. In this case, the interactive learning environment acts as a tool for the professional community of teachers to exchange experience and developments in professional activities, technical, communication, etc. Due to the innovative nature of teaching tools, it is often necessary to provide advisory work to explain how to use their technologies, as well as training materials to improve the qualifications of teachers themselves and their independent creation of similar materials, including for teachers for teachers. Feedback can be obtained remotely through ICD sites: http://icdau.ru, http://ya-znau.ru, http://icdau.kubsu.ru, "http://школьные-годы.рф" [23].
8. The principle of transparency of navigation in the ISO requires maximum visualization of the trajectory of work in the ISO structure since hypertext structures are characterized by multi-way navigation, which contributes to the destruction of the integrity of the structural representation of the content.

9. The principle of an open system, which opens the directions in computer didactics and the complex thematic ISO for each teacher for creative search. On the website of the ICD "http://ya-znau.ru" the teacher can include in the materials already available on the site his products on the selected topic. This is the process of organizing collective creativity, as a result of which the ISO tools on various topics of the training course are enriched. Moreover, the ISO model acts as a tool for the individual creative activity of a teacher, acquires individual features, and at the same time, the teacher turns into a co-author for the creators of new ICD resources. Obviously, as a result of this approach, a teacher working in an interactive learning environment will constantly improve professional skills.
10. The principle of dynamic development. This principle states that the software and hardware of the educational process are constantly evolving. As a result, the educational content and methodological and substantive components are replenished, improved, modernized and accumulated. Scientific, technical and technological progress presupposes rapid and deep mastery of the material, which also changes rapidly with the change in technical or technological digital tools.
11. The principle of social activity. The essence of the principle is that their entire system in its development responds to current social demands. Educational tools and instruments contribute to solving problems arising in society. For example, in recent years' little attention has been paid to the educational component during training, patriotic and civic education. Thus, in the practice of e-learning, a direction has emerged to support educational work, which were differentiated by the areas of spiritual and moral education, intellectual and aesthetic, patriotic.
12. The principle of compliance with regulatory requirements for electronic educational resources. Standard requirements for electronic educational resources have appeared, although each educational organization has the opportunity to slightly change these requirements in accordance with the purpose of training and the direction of the training program. These properties are enshrined in the relevant regulatory documents, primarily in GOST R 52653–2006. Uniform technical requirements govern the development of resources, including ISO, intended for the e-learning system.

4 Pedagogical Experiment

In parallel with theoretical developments on the topic of research, a pedagogical experiment was held. In the cities of Armavir, Yeysk, Primorsko-Akhtarsk, project seminars were organized by the GBOU DPO "Institute for the Development of Education" of the Krasnodar Territory. Teachers were offered to test local learning technologies, as well as solve technological tasks, conduct tests of generalizing repetition, and questionnaires. During the questionnaire, the number of experts was 162 teachers (see Fig. 3).

The results of the questionnaire are shown in Fig. 3, which shows the attitude of experts towards ISO as a whole. The answers to the questions of the questionnaire were built on a scale: (+) - I rate highly in general; ($\pm$) - I rate highly, but there are shortcomings; ($\mp$) - I rate low, but there are advantages; ($-$) - generally rated low; (0) - I find it difficult to assess.

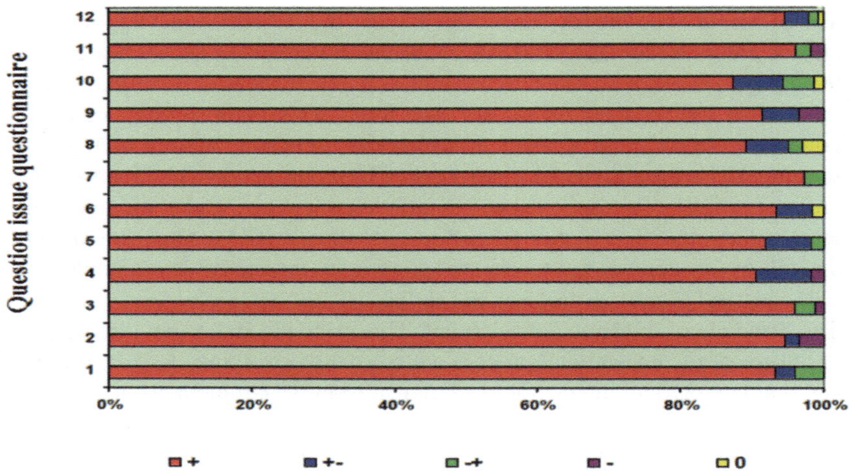

Fig. 3. The attitude of teachers to the innovation environment of education.

The questionnaire confirms the feasibility of creating an "electronic education environment."

5 Conclusions

Thus, the above principles of development and use of the interactive educational environment allow us to draw the following conclusions for work in the conditions of distance learning:

- when working with an interactive learning environment, a student receives a stable motivation to study the program material of the discipline through the use of exciting interactive content that focuses the student on productive, independent and non-violent acquisition of the knowledge necessary in professional activity;
- when working with an interactive learning environment, a teacher gets the opportunity to achieve high results in professional activity, implement a creative approach to teaching and use innovative tools that are available to schoolchildren and students from an early age;
- when implementing work with an interactive learning environment, the country receives a generation of smart, literate, highly professional specialists and patriotic members of society who will lead their country to the number of the most advanced and economically developed countries in the world.

Acknowledgments. The authors of the article express gratitude to the administration of Kuban State University, which created comfortable conditions for the scientific and pedagogical activity, as well as to the software engineer, Ph.D., R.I. Zolotarev for creating four sites.

The article was prepared with the financial support of the Kuban Scientific Foundation of scientific project № FNI-GO-20.1–53/20 "Designing software and methodological support for distance education of talented children in the paradigm of digitalization of education (using the example of physical and mathematical disciplines)."

References

1. Arkhipova, A.I., Pichkurenko, E.A.: The educational text as a fundamental didactic phenomenon and its role in the context of pedagogical hermeneutics. School days: the scientific and methodical magazine with the digital application **68**: 26-42. (2016).
2. Arkhipova, A.I., Grushevsky, S.P., Pichkurenko, E.A., Sevryugina, N.I., Shmalko, S.P.: Hermeneutical approach to the design process interactive learning environment technologies. In: Vth International Scientific and Practical Conference "Distance Learning Technologies" (DLT 2020). Yalta, Crimea, September 22–25 (2020)
3. Bogin, G.I.: Finding of ability to understand: introduction to philological hermeneutics. Published: M.: Psychol. Bus. Online. 516 (2001)
4. Konopko, E.A., Pankratova, O.P., Abdullaev, D.A., Ediev, A.M., Taran, V.N.: Digital toolkit and education an overview of distance learning resources. In: Vth International Scientific and Practical Conference "Distance Learning Technologies" (DLT 2020) (2020)
5. Zakirova, A.F.: Theoretical bases pedagogical hermeneutics and options of her realization in scientific learning practice. Sci. Educ. **6**, 19–42 (2012). https://doi.org/10.17853/1994-5639-2012-6-19-42
6. Innovative computer didactics, http://icdau.ru, date of the address 2021/08/24.
7. Konopko, E.A., Pankratova, O.P., Nersesyan, E., Abdullaev, J.A.: Training of teachers for professional activity in the digital environment of the educational space. In: International Scientific Conference "Innovative Approaches to the Application of Digital Technologies in Education and Research" (SLET-2019) (2019)
8. Markelov, K.A., Polyanskaya, E.V., Mineva, O.K., Taran, V.N.: Paradigm transformation of education system in digital reality. In: Vth International Scientific and Practical Conference "Distance Learning Technologies" (DLT 2020) (2020)
9. Gallini, N.I., Gorbunova, N.V.: Electronic educational resources as a subsystem of an integrated information and analysis area of the university. In: IVth All-Russian scientific and practical conference with international participation "Information Systems and Technologies in Modeling and Control" (ISTMC'2019) (2019)
10. Hoic-Bozic, N., Mezak, J., Tomljenovich, K.: Enhancing teachers' computational thinking skills through game-based learning. In: International Scientific Conference "Innovative Approaches to the Application of Digital Technologies in Education and Research" (SLET-2019) (2019)
11. Pichkurenko, E.A., Arkhipova, A.I.: Hermeneutical approach to the creation of training materials based on models and technologies of innovative computer didactics. The monograph about the Internet the Application. KSEI, Krasnodar (2016)
12. Pichkurenko, E.A., Arkhipova, A.I., Shaposhnikova, T.L.: The theory of the development of the highest mental functions in the context of a hermeneutical approach to educational process. Current problems of training in mathematics at school and higher education institution according to the ideas of L.S. Vygotsky. In: Materials of the III International Scientific Conference. November 17–19, 2016. M.V. Empodov, L.I. Bozhenkova - FGBOU to Moscow Pedagogical State University (MPSU) (2016)
13. Prigodina, A.G., Arkhipova, A.I., Pichkurenko, E.A., Danovich, L.M.: Use of hermeneutical receptions for the organization of reflexive activity of students of engineering higher education institution in the course of studying mathematical texts. The World of Science. Pedagogics and psychology, 2019 No. 2 (March–April). **7** (2019)

14. Prigodina, A.G.: Scientific and pedagogical approaches to didactic adaptation of students and computer support of studying scientific concepts. Mod. Probl. Sci. Educ. **2013**(6), 236 (2013)
15. Baldwin, R.: The great convergence: information technology and the new globalization. The Belknap Press of Harvard University, Cambridge, MA (2016)
16. Prakasha, G.S., Muttungal, P., Lapina, M., Taran, V.: Flipped classroom strategy in online teaching: challenges faced by higher education teachers. In: Lapina, M., Prakasha, G.S., Grigoriev, S. (eds.) International Conference on Innovative Approaches to the Application of Digital Technologies in Education and Research. SLET 2022 Lecture Notes in Networks and Systems, vol. 1222. Springer, Cham (2025). https://doi.org/10.1007/978-3-031-78776-8_24
17. Power of knowledge: http://ya-znau.ru/, date of the address 2021/08/29.
18. Sirotina, I.K., Berezyuk, S.I., Falei, A.V.: Creation of the interactive educational environment means of web technologies. In: The International Congress on Computer Science: Information Systems and Technologies. Minsk, November 4–7 (2013) http://elib.bsu.by/handle/by/handle/123456789/52259
19. Kulikova, T.A., Poddubnaya, N.A.: Training of the future teacher in applying digital tools for the organization of project activities. In: Vth International Scientific and Practical Conference "Distance Learning Technologies" (DLT 2020) (2020)
20. Timofeeva, E., Grigoryan, L., Marchenko, T., Khalatyan, K.A.: Model of mathematics distance learning in university training E-environment. In: Proceedings of the International Scientific Conference "Innovative Approaches to the Application of Digital Technologies in Education and Research" (SLET-2019) (2019)
21. Shevlyak, A.G.: The inclusion of the elements of the logic algebra in the training of mathematics at school and in the university. School Years. **2011**(36), 59–64 (2011)
22. Shevchenko, G., Brynza, S., Rybakova, A., Kochkin, D.: The concept of electronic learning with the application of digital technologies. In: Proceedings of the International Scientific Conference "Innovative Approaches to the Application of Digital Technologies in Education and Research" (SLET-2019) (2019)
23. Electronic educational resources, http://icdau.kubsu.ru/, date of the address 2021/08/28.
24. Taran, V., Savelova, L., Abdullaev, J.: Increasing learning motivation through the use of educational and technological tools. In: Lapina, M., Prakasha, G.S., Grigoriev, S. (eds.) International Conference on Innovative Approaches to the Application of Digital Technologies in Education and Research. SLET 2022 Lecture Notes in Networks and Systems, vol. 1222. Springer, Cham (2025). https://doi.org/10.1007/978-3-031-78776-8_7
25. Lutsenko, L., Testova, I., Chirkov, M., Taran, V., Mezentseva, O.: Methodical specifics of the subject distance camp organization "young cryptograph". In: Lapina, M., Prakasha, G.S., Grigoriev, S. (eds.) International Conference on Innovative Approaches to the Application of Digital Technologies in Education and Research. SLET 2022 Lecture Notes in Networks and Systems, vol. 1222. Springer, Cham (2025). https://doi.org/10.1007/978-3-031-78776-8_21
26. Lapina, M., Tokmakova, M., Movzalevskaya, V., et al.: Analysis of modern methods and approaches using it in education and pedagogy. In: International Conference on Innovative Approaches to the Application of Digital Technologies in Education and Research, 2022, vol. 1222, pp. 40–52. Springer, Cham (2025). https://doi.org/10.1007/978-3-031-78776-8_5
27. Zolotarev, V., Arkhipova, A., Lapina, M., et al.: Assessing player interaction for a social networking cooperative educational game. In: International Conference on Innovative Approaches to the Application of Digital Technologies in Education and Research, 2022, vol. 1222, pp. 114–121. Springer, Cham (2025). https://doi.org/10.1007/978-3-031-78776-8_12
28. Gerasimova, E., Konopko, E., Usmanov, M.: Online tools for organizing students' teamwork. In: International Conference on Innovative Approaches to the Application of Digital Technologies in Education and Research, 2022, vol. 1222, pp. 167–173. Springer, Cham (2025). https://doi.org/10.1007/978-3-031-78776-8_17

29. Konopko, E., Pankratova, O., Azarov, I., Guseinov, R., Hassan, M.A.: Online teaching of students of pedagogical courses by teachers from abroad. In: International Conference on Innovative Approaches to the Application of Digital Technologies in Education and Research, 2022, vol. 1222, pp. 183–193. Springer, Cham (2025). https://doi.org/10.1007/978-3-031-78776-8_19
30. Makoveichuk, K., Olifirov, A., Petrenko, S.: Integration of MOOC formats in an educational organization based on connectivism theory. In: International Conference on Innovative Approaches to the Application of Digital Technologies in Education and Research, Stavropol, Russia, 2022, vol. 1222, pp. 219–231. Springer, Cham (2025)
31. Taran, V.N., Usmanov, M.I.: Learning technologies through augmented reality projects. In: VIth International Scientific and Practical Conference "Distance Learning Technologies" (DLT 2021).Yalta, Crimea, September 20–22, vol. 3057, pp. 253–259 (2021). ceur-ws.org/Vol-3057/paper30.pdf.

Cloud Technologies of Distance Education Process in the Education System

Alevtina I. Arkhipova[1], Elena A. Pichkurenko[1], Anna G. Prigodina[2], Ekaterina A. Vladimerets[3], and Viktoriya N. Taran[4(✉)]

[1] Kuban State University, Street Stavropolskaya, 149, Krasnodar 350040, Russia
[2] Kuban State Technological University, Street Moskovskaya, 2, Krasnodar 350072, Russia
[3] School No 20 named by Pavel Tyulaev, Street Sormovskaya, 171, Krasnodar 350000, Russia
[4] V.I. Vernadsky Crimean Federal University, Simferopol 295007, Russia
victoriyayalta@gmail.com

Abstract. The article presents an interactive web-oriented textbook, which was developed based on a hermeneutic approach to the learning process. The student is intended primarily for students of the pedagogical direction, but will also be useful for practicing teachers of schools and higher educational institutions as a tool for digital technologies. The methodology and models of teaching are also presented in the article based on a competence-based and systemic approach to the educational process, as well as using a cultural and historical component. The presented textbook is located in the cloud storage, consists of 5 author's resources and is registered in the structure of Rospatent of the Russian Federation. In describing the program "Matrix of Technology of Innovative Computer Didactic", the main software components are allocated to ensure the functioning of the remote educational process. Details described the program "The model of educational planning", which presents materials and types of professional-pedagogical activities in the subject learning system. Thus, the interactive textbook for digital and distance education is designed mainly for independent work of students and pupils, which is fundamentally different from classical textbooks. The article also presents a methodology for planning the structure of the educational process. J. Carnegie wrote that the goal of a real teacher is to "make want" to learn, which is facilitated by working with this textbook.

Keywords: Digitalization of Education · Didactic Technologies · Educational Environment · Virtual Cluster of Pedagogical Innovations · Educational Technology Designer

1 Introduction

"The digital mainstream of domestic education" has become the dominant direction of its development. However, a popular Internet article about the digitalization of education calls digitalization a "headache" that is acquiring "political resonance" and "marks the prospects of its ominous transformation". (https://tsargrad.tv/articles/god-totalnoj-cifrovizacii-10-letnij-plan-unichtozhenija-obrazovanija-v-dejstvii_307691). Most media report on distance education are sharply critical, sometimes scathing, in character.

M. Lapina et al. (Eds.): DET 2024, LNNS 1616, pp. 56–71, 2025.
https://doi.org/10.1007/978-3-032-04365-8_5

But we love the old forgotten pioneer rule "When criticizing, suggest!". Thus, we will consider the "pros and cons" of digitalization of education [16] and at the same time express our observations based on pedagogical experience and propose some measures to improve the effectiveness of digital learning.

Undoubtedly, digitalization of the educational process is broader than the classical education system, which is based on a textbook and a teaching aid exclusively in printed format. The process of creating digital information services for the education system fell on the shoulders of IT specialists. The first stage was the creation of video lessons. Then video courses began to appear in combination with previously created digitized educational publications. However, for distance education, which became relevant and widespread in the early 2020s due to the pandemic, the textbook and the problem of its digital implementation came to the fore.

2 Hermeneutic Approach in Educational Technologies

2.1 Objectives of the Hermeneutic Approach in Pedagogy

Learning through a live dialogue with a teacher or independent study of the material using a book-textbook has always been the main technology of educational pedagogy, starting from Ancient Greece or the Renaissance. Digital learning leads to the rejection of paper sources of knowledge, replacing them with electronic resources, turning the dialogue between a teacher and a student into an electronic dialogue on specific platforms and with the help of appropriate tools.

Distance learning offered its own set of methods and technologies, but actually abandoned the book, thereby adopting the main drawback of classical education - the lack of independent thinking, deep understanding of educational information.

Digital technologies offer the student material in a ready-made form, where there is no incentive to look for their own non-standard solution or their own view of the situation, different from the source of information, is not welcomed. Standardized forms of control most often represent tests, the answer to which is often not difficult to guess or find the answer on the Internet.

Teaching methods have changed dramatically. What is the reason for this? In our opinion, there is a financial interest behind all this. Education is a well-funded social sphere, which covers numerous federal programs and grants. In favor of the "fashion for everything Western" in education, leading positions, including financial ones, have been taken by "high technical specialists" who have never worked in a school.

But there is a well-known truth: "Render unto Caesar the things that are Caesar's, and unto God the things that are God's".

In connection with this rule, "high technical specialists" impose their "pedagogical" ideas and principles. Practicing teachers invariably lose this battle.

Digital platforms and resources that are proposed for use in distance learning are often cumbersome, complex, resource-intensive, slow and inconvenient for a teacher who is accustomed to a live dialogue with a student. Learning to use these resources occurs with the help of instructions of several dozen pages, hundreds of drawings, many incomprehensible terms and statements. But we know that "beauty is in simplicity". The simpler the explanation, the simpler and more intuitive the work in the information field, the more interesting and effective the learning process. This is why poor students and teachers struggle with these software monsters, which sometimes work erratically.

In our opinion, there is a financial motive here. We know that the more complex and intricate the connections in the system, the higher the multi-level navigation, the more visualization channels, the more expensive the system. This is why poor students and teachers struggle with these software monsters, which sometimes work erratically.

American scientists have long concluded that the most ineffective teaching method is a lecture (see Dale's cone - Fig. 1).

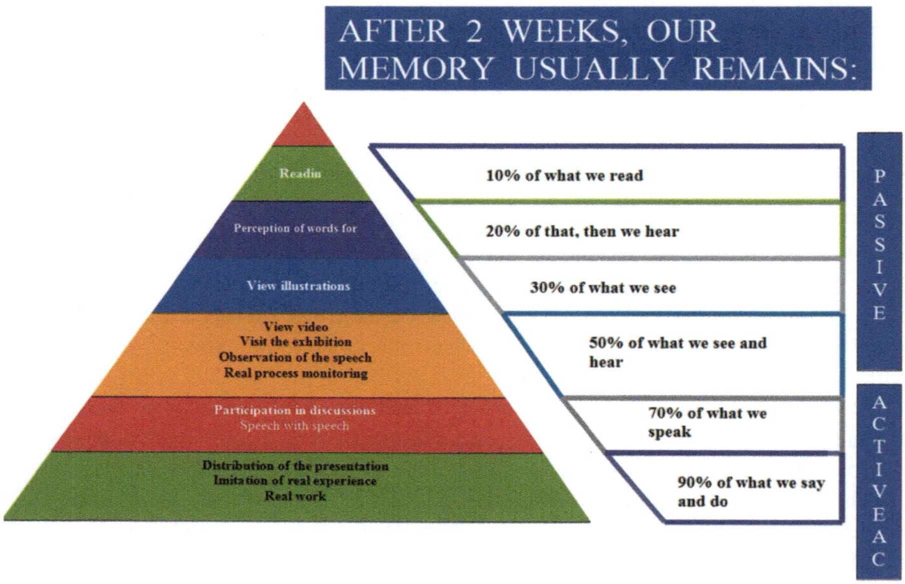

Fig. 1. The Dale Cone.

The words of the great physicist and great teacher A. Einstein: "I never teach my students, I only try to create conditions in which they can learn" should become the main basic principle of distance education.

We offer a methodology for developing training materials for distance learning, which is based on the hermeneutic approach. To study the material, a trajectory is built: "from the whole to the constituent parts and from them to the whole".

We use the methods of analysis and synthesis, for which all texts are pre-structured, divided into blocks. The text is carefully proofread and checked so that the presentation is simple and clear language, the wording is understandable and concise. Illustrations are selected so that they clearly reveal the main content of the theory being studied.

Text blocks are logically complete parts.

The first stage is the stage of targeted comprehension. This stage consists of step-by-step reading of the entire text, whereby the entire text is first read, then the first block is reread and linked to the entire text. Then the second block of text is reread and a link with the entire text is sought in it, and so on.

Let's consider what didactic and digital technologies are used at this stage.

Firstly, the technology of the "structural scheme", where all parts of the material being studied are presented as separate blocks with links, and the reading of parts of the text is recorded in the scheme,, where the connection between the parts of the proposed scheme is also highlighted.

The technology of "Blind scheme" and "Scrolling-contour" is also used in this part of the work on information.

The second stage is memorization. At this stage, individual parts of the text are learned through the use of interactive technologies. At this stage, a set of technologies is proposed, which we have combined into the so-called "Iteration Technology" (in the usual sense, "repeated repetition" using the results obtained in the previous repetition circle). We use information visualization in various forms and variants, as well as using such software technologies as: an interactive dictionary, knowledge formulas, search by algorithm, educational quests, and game technologies. A set of technologies for lexical and logical text processing is also proposed. All this helps to identify and eliminate gaps in knowledge, which leads to a higher quality of training [13].

It is obvious that the purpose of using pedagogical hermeneutics is not to offer the student to learn a standard set of facts and obtain certain knowledge in a certain section of the discipline, but to "make the student smarter" [7], to teach him to navigate a huge array of information and to identify key basic information, to distinguish it from secondary or unimportant information. This task is solved by using the "Technology of Generalization, Analysis, Synthesis and Evaluation". It is at this stage that the creative and cognitive abilities of students are developed [4].

The creation of an interactive textbook in the "UCHKOM" program implements the set tasks of distance education of students [3].

An equally important part of the textbook is its component for the teacher who will use this textbook in his professional activities. The second part of the textbook consists of descriptions of didactic tasks, the structure of software components and a description of the technologies used. The textbook on the teacher's side contains algorithms for creating and editing educational content, as well as initial software templates.

The tool shell of the textbook contains individual and complex programs, as well as methodological support for the subject, including models and technologies for distance learning, some of which are registered in the Federal State Information System of Rospatent of the Russian Federation. Table 1 provides a list of programs of disciplines that make up the methodological support for distance learning and are included in the interactive textbook developed at the Faculty of Mathematics and Computer Science of the Kuban State University [2].

In addition to registered programs, our computer didactic toolkit also includes many other small programs (we call them local programs), such as over a hundred game learning programs [8].

Table 1. List of programs included in the methodological support of distance learning.

№	The name of the program	Certificate of registration in the FSIS Rospatent of the Russian Federation	Authors
1.	UCHKOM (textbook + computer)	№ 2012610691	A.I. Arkhipova, R.I. Zolotarev, T.L. Shaposhnikova
2.	Travel by city and country	№ 2011612440	A.I. Arkhipova, R.I. Zolotarev
3.	Roads	№ 2015612432	A.I. Arkhipova, R.I. Zolotarev
4.	The power of knowledge	№ 2013615237	A.I. Arkhipova, R.I. Zolotarev
5.	Interactive model of distance tutorial	№ 2021613258	A.I. Arkhipova, R.I. Zolotarev, D.V. Yus.
6.	Matrix of technologies of innovative computer didactics	№ 2020660381	A.I. Arkhipova, V.I. Ivanov
7.	Virtual cluster of pedagogical innovations	№ 2019665694	A.I. Arkhipova, E.A. Pichkurenko, R.I. Zolotarev
8.	An innovative educational environment as the basis of a virtual cluster of pedagogical innovations	№ 2018665371	A.I. Arkhipova, R.I. Zolotarev
9.	Interactive model of planning the educational process	№ 2018663497	A.I. Arkhipova, S.P. Grushevsky, S.P. Shmalko
10.	System knowledge model in the structure of subject learning	at the registration stage	
11.	The structure of the training course "Innovative computer didactics" in the system of professional training of teachers	at the registration stage	
12.	The educational potential of information technologies of computer didactics	at the registration stage	
13.	Drug Prevention Networking Resource	at the registration stage	
14.	Model of a network pedagogical experiment	at the registration stage	

2.2 Interactive Textbook. Recursive Model

The methodology of the hermeneutic approach to distance learning is based on a systems approach in designing the text component, a cultural-historical approach in planning and technologies for managing the model of systemic knowledge and the genesis of scientific theory [14, 15, 17]. The construction of the course corresponds to the following chain of planning stages:

Contents → Teaching methods → Computer support technologies → Network technologies and information and communication technologies [18, 23].

Historically, scientific knowledge has been transformed into course content based on curricula and programs developed in the governing structures. Therefore, the author's contribution, in this case, has been insignificant. Recently, however, the authors of the course have been given greater freedom, especially in terms of the methodological support for the course, with the last generation of the FEM (Funds of Evaluation Means).

Therefore, we have found it appropriate to base the structure and content of the courses on a set of approaches: systemic, hermeneutic, competence-based, and cultural-historical.

The methodological component of the course model is based on Internet technologies for supporting independent work [1]. At the same time, the educational course should contribute to the formation of a person and a citizen.

Historically, scientific knowledge has been transformed into course content based on curricula and programs developed in the governing structures. Therefore, the author's contribution, in this case, has been insignificant. Recently, however, the authors of the course have been given greater freedom, especially in terms of the methodological support for the course, with the last generation of the FEM (Funds of Evaluation Means).

Therefore, we have found it appropriate to base the structure and content of the courses on a set of approaches: systemic, hermeneutic, competence-based, and cultural-historical.

Thus, methodological approaches synthesize all the structural elements of the distance learning process, combine them into one common system, with each participant in the educational process performing a specific function. This approach contributes to the cognitive and creative development of students, as well as improving the professional training of teachers [9, 12, 21].

The main function of the distance learning process falls on the interactive textbook due to the development of a special structure. Didactic tasks are solved in each of the shells of the textbook model.

The model of the interactive course textbook formed the basis for the design of the interactive digital environment of distance learning (see Fig. 2).

So, in the picture we see the general scheme of the study guide and the navigation buttons for the sections of the interactive textbook.

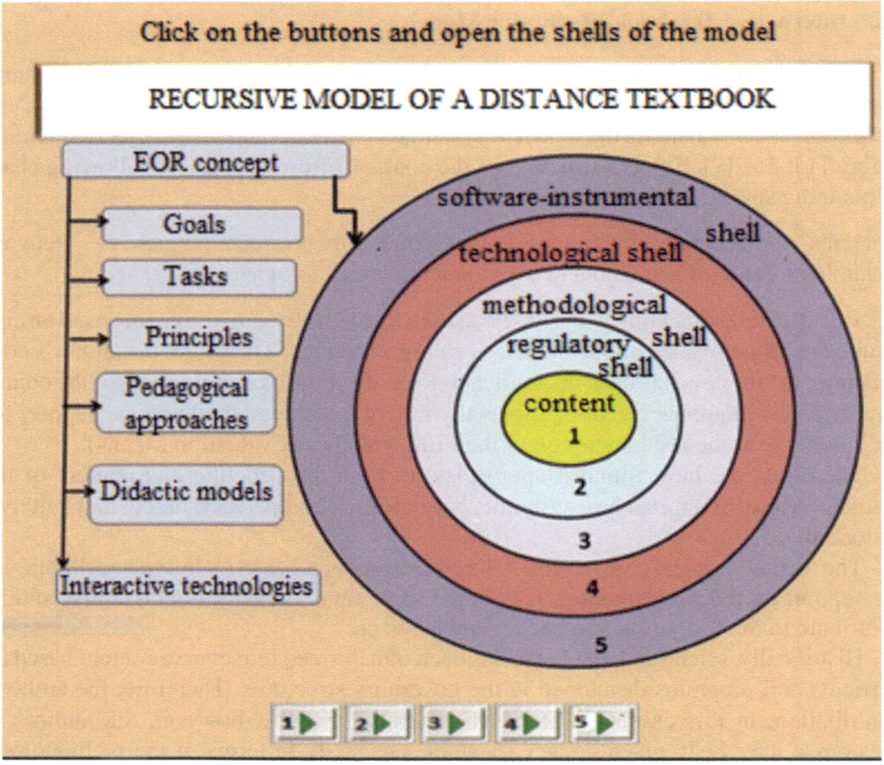

Fig. 2. The title page of the interactive.

2.3 Innovative Components of the Course Model

An interactive textbook for distance learning implements several tasks simultaneously, unlike a classic textbook.

Firstly, the textbook contains normative scientific information that must be learned by students.

Secondly, the information being studied has an additional interactive digital form, implemented through network learning technologies.

Thirdly, the textbook is a gentle form of training, without mental fatigue of students due to interactive gaming technologies.

Fourthly, monitoring of learning results as a feedback tool is also implemented when using an interactive textbook.

In addition, information and communication services create the prerequisites for mutual learning based on the principle: "learn it yourself, teach your friend". Exchange of messages in the chat and forum, computer dialogues allow for mutual assistance in the educational process. Research projects, team games, joint discussions, subject competitions and reports on current topics in science and technology contribute to collective learning, improve the quality of competencies that students receive during training. Thus, ultimately, distance learning stimulates students to be creative, for example, by creating their own educational materials using software templates, including acquiring collective features [6, 22].

Using the programs and algorithms "Matrices of Technologies of Innovative Computer Didactics", included in the parole (http://icdau.ru/matrica/web1.html), Mathematics students created interactive resources based on materials about the Great Victory in the Patriotic War of 1941–1945.

Let us consider once again the first part of the interactive textbook, which is intended for students and contains the following elements:

- training software interactive applications for subject preparation (for example, "Uchkom"),
- independent game components in the form of a generalization of the content of the topics studied (for example, "Ascent to the top of knowledge"),
- programs for current and final knowledge control (test systems),
- electronic resources that supplement the training material of the subject programs (for example, "Interactive maps of Germany, England, France") with country studies content),
- programs that include educational electronic resources (for example, resources "Educational potential of information technologies") in the structure of subject training.

The second part of the digital textbook for distance learning is intended for use by teachers and is focused on software and methodological support for their pedagogical activities, and at the same time has only a network form [5, 10]. This part of the textbook has functions that combine the following interactive technologies: resource "Interactive learning environment" (icdau.kubsu.ru), consists of three levels that correspond to different levels of training. Each level consists of several modules that open with "alternative switching" buttons and demonstrate the usual theoretical material and some interactive technologies for different levels of education [11, 19, 20].

- The first level contains texts on the problems of digital and distance learning.
- The second level interactive technologies of subject teaching.
- The third level some projects of schools and universities of the Krasnodar Territory, included in the structure of the virtual cluster of pedagogical innovations (for example, "The Immortal Regiment comes to school", stories of schoolchildren about their relatives - participants of the Great Patriotic War, "Military victories of Russia", "Electronic resources for physical education", etc.).

You can see the working version of the program "Textbook - Textbook + Computer" (topic "Calculation systems") here.

The Internet resource "Innovative Technology. Computer Didactics Matrix" offers the teacher to become familiar with the methods of using digital auxiliary resources, with the help of which to create and modify educational content.

The resource is built on the principle of a tabular structure and has more than 100 interactive cells (http://icdau.ru/matrica/web1.html).

The table consists of five columns, 31 rows and 155 cells. All cells are interactive buttons that open texts or working versions of technologies.

Figure 3 shows the first fragment of the program (see Fig. 3).

Internet technology ICD　Self-preparation technologies
Thematic technologies systems
Computer games

IKD technology	Educational themes			
Presentation of the theory on the topic	Natural numbers, discharges	Validity of natural numbers	Derivative and its applications	Indicative function, logarithm
1. Dictionary of knowledge	Theme 1	Theme 2	Theme 3	Theme 4
2. Knowledge gaps	Theme 1	Theme 2	Theme 3	Theme 4
3. Field of knowledge	Theme 1	Theme 2	Theme 3	Theme 4
4. Matrix of knowledge	Theme 1	Theme 2	Theme 3	Theme 4
5. Test knowledge	Theme 1	Theme 2	Theme 3	Theme 4
7. Face test	Theme 1	Theme 2	Theme 3	Theme 4
8. Knowledge crossword	Theme 1	Theme 2	Theme 3	Theme 4

Fig. 3. Fragment of the "Computer Didactics Matrix" technology matrix.

The figure shows a fragment of the " Computer Didactics Matrix ", which contains the subject area of mathematics. We can see three levels of educational programs: for basic school, middle school and high school.

Please try to avoid rasterized images for line-art diagrams and schemas. Whenever possible, use vector graphics instead.

Interactive technologies of the same type are highlighted by color. Thus, the first group ("Internet technologies") includes educational technologies that are created on the website "Power of Knowledge" (http://ya-znau.ru) using a special program "Technology Builder "Power of Knowledge" [18].

We see a list of the following technologies:

- keywords
- variable tests,
- movement, drawing with a mouse,
- text fields,
- support text,
- training maze,
- drop-down list,
- visual dictation,
- electronic trainer,

- electronic posters,
- blind diagram,
- interactive crossword puzzle.

The second group ("thematic systems") are programs-instrumental shells that combine some technologies from the previous category according to the overall content: Uchkom, Modules, Web presentations, Table of Contents, Roads, Travels in Cities and Countries. The third group of "Matrix" consists of programs "Educational computer games. "The global Internet on the specified resource contains materials of teachers created for teaching students. All materials were created in the technological designer "Power of Knowledge". In addition, the electronic journal "School Years" was published for more than 25 years. It published articles and studies devoted to the development of electronic resources by teachers for distance learning of students.

The interactive resource "Model of planning the educational process" is an information resource prepared to help the professional and pedagogical activity of a teacher in the distance learning system, due to the accumulation of program templates and accumulated materials in the form of the following results of the teacher's personal experience:

- the literature used,
- regulatory documents,
- selection of textbooks,
- didactic and control materials,
- development of extracurricular activities on the subject,
- Internet sources,
- assignments of the Unified State Exam and the Basic State Exam,
- subject Olympiads, etc.

As a rule, the methodological support of distance learning does not have clearly defined boundaries (see Fig. 4). Thus, the selection of training materials and tools to help the teacher in his professional activity improves the methodological training of the teacher and creates conditions for enriching his personal experience.

3 Experiment

So, this program presents the experience of the teacher from the rural school of (direction "Country Studies" and "Spiritual and Moral Education"). At a conference in Krasnodar with the participation of a delegation from Germany, a representative of the German embassy Christian Zeebodesaid: "We in Germany do not have such wonderful interactive maps of the German lands, and you have rural schoolchildren successfully work with them. (Map programs were created by the students of the Faculty of Physics at Kuban State University).

All subject network materials are combined into a single network node (see Fig. 5).

This application presents the experience of Rogovskaya village Knyazeva I.G., a teacher at a rural school in the village. Disciplines: "Regional Studies" and "Spiritual and Moral Education". This work was presented at a conference in Krasnodar, where members of the delegation from Germany participated. A participant from Germany, Christian Seebode, said: "We do not have such wonderful interactive maps of German

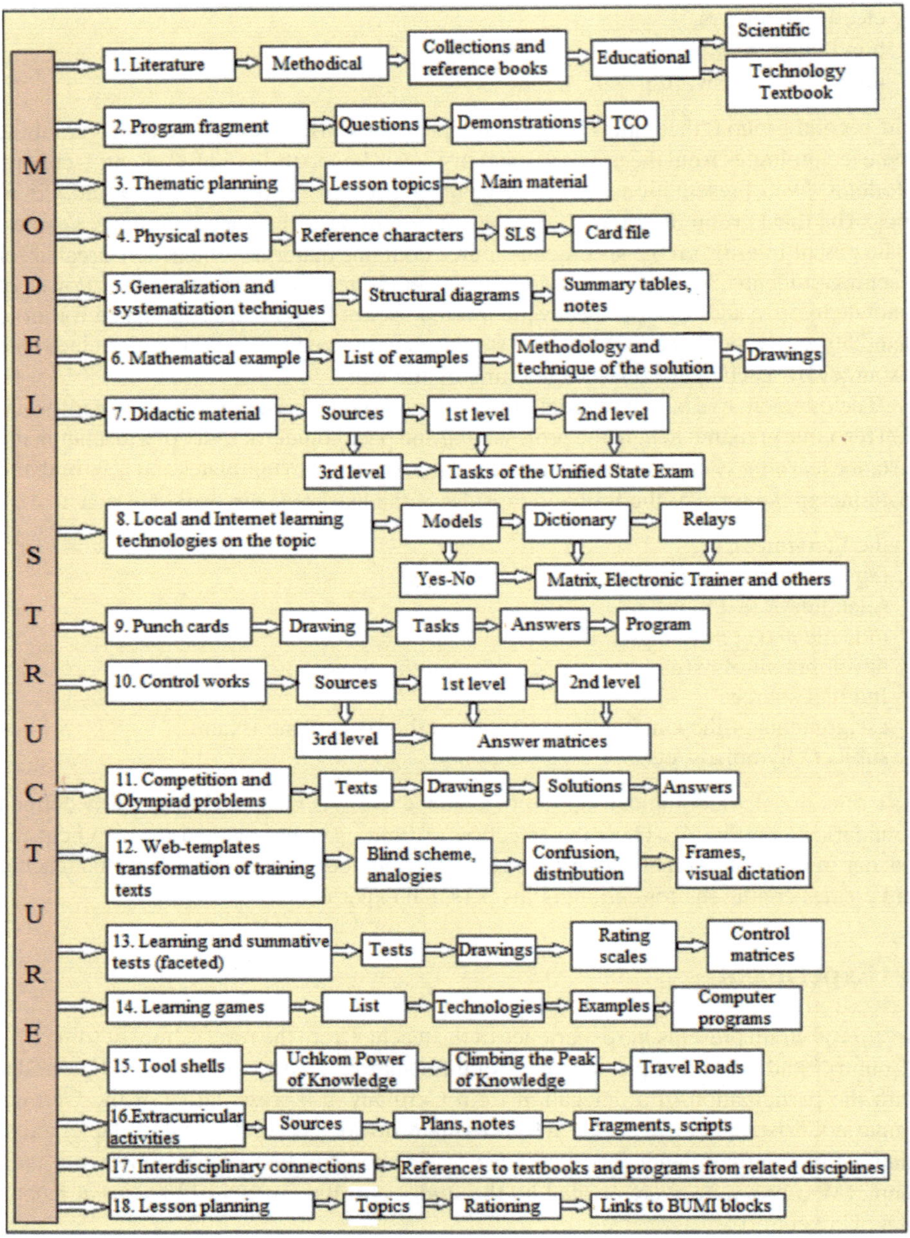

Fig. 4. Structure of the educational process planning model.

lands in Germany, but your rural schoolchildren work with them successfully. (The maps presented at the conference were created by students of the Physics Department of the Kuban State University).

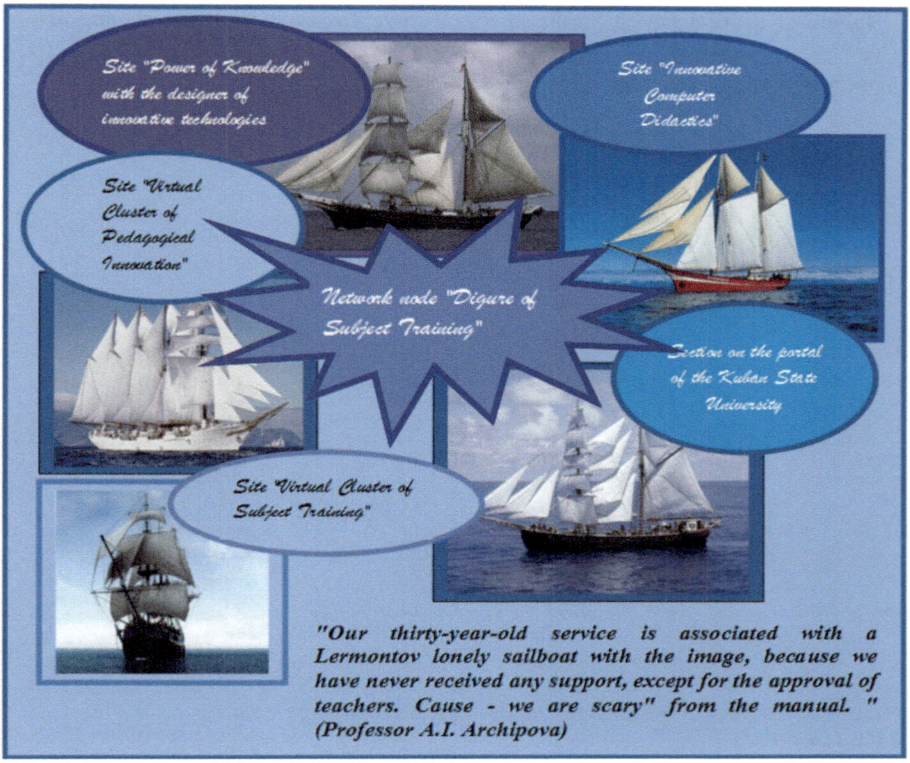

Fig. 5. Key indicators.

To carry out a quantitative and qualitative assessment of the use of a model of a training course for distance learning, a questionnaire was conducted among school teachers of the Krasnodar Territory.

It focused on identifying the expert assessment of technologies of the model of the training course for DEs, assessing the effectiveness of secondary school students using the proposed didactic computer technologies. The questionnaire statistics were obtained in the traditional way.

The number of experts was 68. The results of the questionnaire survey are listed in Table 2.

According to the results of the questionnaire from the answers of experts, we see that in general, teachers of the Krasnodar Territory highly appreciate the model of a training course for distance learning. Most of the experts think it is reasonable to use the model of the course for DE in their pedagogical activities.

The data of which are interpreted in the form of a diagram (see Fig. 6).

Table 2. Results of a questionnaire survey of school teachers in the Krasnodar territory.

Question number	Number of responses by category							
	+	%	±	%	☒	%	–	%
1.	62	91.2	4	5.8	2	3.0	0	0
2.	44	64.8	14	20.5	7	10.3	3	4.4
3.	51	75.0	12	17.6	0	0	5	7.4
4.	53	77.9	10	14.7	4	5.9	1	1.5
5.	63	92.7	4	5.8	1	1.5	0	0
6.	65	95.6	3	4.4	0	0	0	0
7.	58	85.2	7	10.3	2	3.0	1	1.5
8.	42	61.8	19	27.9	5	7.3	2	3.0
9.	61	89.7	5	7.3	0	0	2	3.0
10.	49	72.1	13	19.1	6	8.8	0	0
11.	54	79.4	8	11.7	4	5.9	2	3.0
12.	61	89.7	5	7.3	2	3.0	0	0
13.	54	79.4	12	17.6	2	3.0	0	0

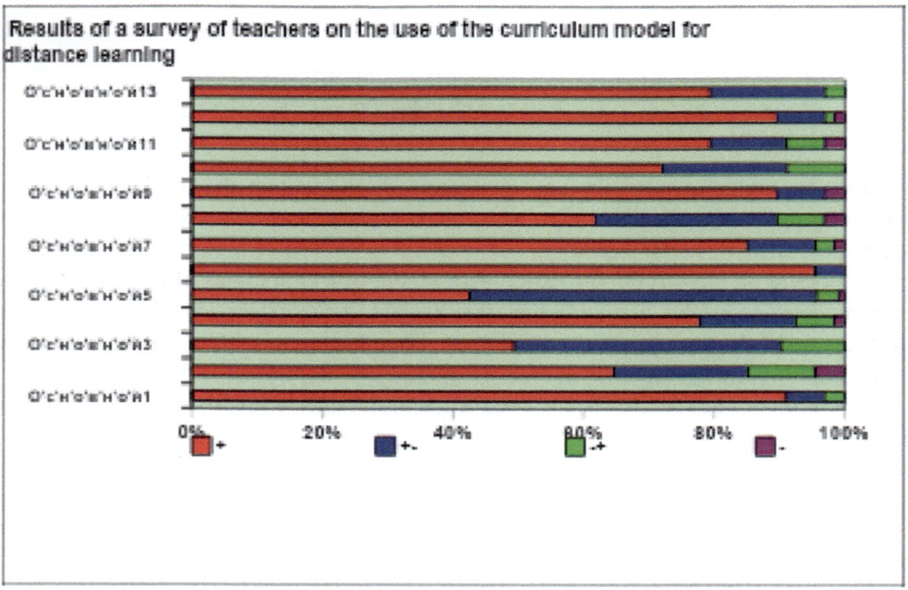

Fig. 6. Results of a questionnaire survey of school teachers of the Krasnodar Territory.

4 Conclusions

So, distance learning is a complex process that requires a comprehensive system in the form of special methodological and technological teaching techniques.

The software and methodological support for distance learning in the form of an interactive textbook is a cognitive-creative system that creates conditions for effective independent learning of pupils and students. Distance learning provides a high level of motivation for students to cognitive activity, due to the use of the enormous potential of computer technologies, is an incentive for the constant enrichment of mental experience in the future during subsequent professional activity.

Therefore, only by overcoming the conservatism of traditional approaches to the creation of new methodological support, it is possible to successfully solve the problems of general educational and professional pedagogy. A full and open examination of information training materials while abandoning the existing stereotypes of closed examination by a narrow circle of interested persons guarantees the quality of distance education. The development of interactive resources corresponding to the presented methodology is one of the ways to provide students with a high-quality educational resource. Taking into account the new realities of scientific and technological progress using information and communication technologies makes it possible to create and implement information technologies in the field of education, to improve the effectiveness of training and preparation of highly educated professionals.

Acknowledgments. The authors of the article express gratitude to the administration of Kuban State University, which created comfortable conditions for the scientific and pedagogical activity, as well as to the software engineer, Ph.D., R.I. Zolotarev for creating four sites.

The article was prepared with the financial support of the Kuban Scientific Foundation of scientific project No. FNI-GO-20.1-53/20 "Designing software and methodological support for distance education of gifted children in the paradigm of digitalization of education (using the example of physical and mathematical disciplines").

References

1. Arkhipova, A.I., Grushevsky, S.P., Pichkurenko, E.A., Sevryugina, N.I., Shmalko, S.P.: Hermeneutical approach to the design process interactive learning environment technologies. Selected Papers of the V International Scientific and Practical Conference "Distance Learning Technologies" (DLT 2020). Yalta, Crimea, September 22–25, (2020).
2. Arkhipova, A.I., Grishchenko, V.I.: Electronic Educational Resources of Innovative Computer Didactics and their Use in the Educational Work of Teachers. Monograph. KubSU, Krasnodar (2017)
3. Konopko, E.A., Pankratova, O.P., Abdullaev, D.A., Ediev, A.M., Taran, V.N.: Digital toolkit and education an overview of distance learning resources. In: Vth International Scientific and Practical Conference "Distance Learning Technologies" (DLT 2020) (2020)
4. Arkhipova, A.I., Zolotarev, R.I., Shaposhnikova, T.L., Vyazankova, V.V.: Educational and methodological set "UCHCOM" as a prototype of the textbook of the future. School Years. **37** (2011)

5. Kormakova, V.N., Klepikova, A.G., Lapina, M.A., Rugelj, J.: ICT competence of a teacher in the context of digital transformation of education. In: Selected Papers of the V International Scientific and Practical Conference "Distance Learning Technologies" (DLT 2020). Yalta, Crimea, September 22–25 (2020)
6. Zakirova, A.F.: Fundamentals of pedagogical hermeneutics: author's course of lectures. In: Tutorial, p. 322. Publishing House of TyumSU, Tyumen (2011)
7. Innovative computer didactics. http://icdau.ru, date of the address 2021/08/10.
8. Konopko, E., Pankratova, O., Nersesyan, E., Abdullaev, J.: Training of teachers for professional activity in the digital environment of the educational space. In: Proceedings of the International Scientific Conference "Innovative Approaches to the Application of Digital Technologies in Education and Research" (SLET-2019)
9. Markelov, K.A., Polyanskaya, E.V., Mineva, O.K., Taran, V.N.: Paradigm transformation of education system in digital reality. In: Selected Papers of the V International Scientific and Practical Conference "Distance Learning Technologies" (DLT 2020). Yalta, Crimea, September 22–25 (2020)
10. Gallini, N.I., Gorbunova, N.V.: Electronic educational resources as a subsystem of an integrated information and analysis area of the University. In: Selected Papers of the IV All-Russian scientific and practical conference with international participation "Information Systems and Technologies in Modeling and Control" (ISTMC'2019), pp. 243–251
11. Hoic-Bozic, N., Mezak, J., Tomljenovich, K.: Enhancing teachers' computational thinking skills through game-based learning. In: Proceedings of the International Scientific Conference "Innovative Approaches to the Application of Digital Technologies in Education and Research" (SLET-2019)
12. Pichkurenko, E.A., Arkhipova, A.I.: Hermeneutical approach to the creation of training materials based on models and technologies of innovative computer didactics. In: The Monograph about the Internet the Application. KSEI, Krasnodar (2016)
13. Pichkurenko, E.A., Arkhipova, A.I., Shaposhnikova, T.L.: The theory of the development of the highest mental functions in the context of the hermeneutical approach to the educational process. Current problems of training in mathematics at school and higher education institution according to the ideas of L.S. Vygotsky. In: Empodov, M.V., Bozhenkova, L.I. (eds.) IIIth International Scientific Conference. FGBOU to Moscow Pedagogical State University (MPSU) (2016)
14. Prigodina, A.G., Arkhipova, A.I., Pichkurenko, E.A., Danovich, L.M.: Use of hermeneutical receptions for the organization of reflexive activity of students of engineering higher education institution in the course of studying mathematical texts. World Sci. Pedagog. Psychol. **7**(2) (2019)
15. Baldwin, R.: The great convergence: information technology and the new globalization. The Belknap Press of Harvard University, Cambridge, MA (2016)
16. Power of knowledge. http://ya-znau.ru/, date of the address 2021/08/11.
17. Timofeeva, E., Grigoryan, L., Marchenko, T., Khalatyan, K.A.: Model of mathematics distance learning in university training E-environment. In: Proceedings of the International Scientific Conference "Innovative Approaches to the Application of Digital Technologies in Education and Research" (SLET-2019) (2019)
18. Shevlyak, A.G., Shernina, N.S., Zolotarev, R.I.: Computer support for the didactic adaptation of students to the study of mathematics in an engineering university. Dist. Virt. Learn. **12**, 98–107 (2011)
19. Shevlyak, A.G., Shernina, N.S.: The problem of implementing the principle of coherence at different levels of mathematical education. School Years. **29**, 19–27 (2010)
20. Shevchenko, G., Brynza, S., Rybakova, A., Kochkin, D.: The concept of electronic learning with the application of digital technologies. In: Proceedings of the International Scientific

Conference "Innovative Approaches to the Application of Digital Technologies in Education and Research" (SLET-2019)
21. Electronic educational resources. http://icdau.kubsu.ru/, date of the address 2021/08/14.
22. Lutsenko, L., Testova, I., Chirkov, M., Taran, V., Mezentseva, O.: Methodical specifics of the subject distance camp organization "young cryptograph". In: International Conference on Innovative Approaches to the Application of Digital Technologies in Education and Research. SLET 2022 Lecture Notes in Networks and Systems, vol. 1222. Springer, Cham (2025). https://doi.org/10.1007/978-3-031-78776-8_21
23. Lapina, M., Tokmakova, M., Movzalevskaya, V., et al.: Analysis of modern methods and approaches using it in education and pedagogy. In: International Conference on Innovative Approaches to the Application of Digital Technologies in Education and Research, 2022, vol. 1222, pp. 40–52. Springer, Cham (2025). https://doi.org/10.1007/978-3-031-78776-8_5
24. Zolotarev, V., Arkhipova, A., Lapina, M., et al.: Assessing Player Interaction for a Social Networking Cooperative Educational Game. In: International Conference on Innovative Approaches to the Application of Digital Technologies in Education and Research, 2022, vol. 1222, pp. 114–121. Springer, Cham (2025). https://doi.org/10.1007/978-3-031-78776-8_12
25. Gerasimova, E., Konopko, E., Usmanov, M.: Online tools for organizing students' teamwork. In: International Conference on Innovative Approaches to the Application of Digital Technologies in Education and Research, 2022, vol. 1222, pp. 167–173. Springer, Cham (2025). https://doi.org/10.1007/978-3-031-78776-8_17
26. Konopko, E., Pankratova, O., Azarov, I., Guseinov, R., Hassan, M.A.: Online teaching of students of pedagogical courses by teachers from abroad. In: International Conference on Innovative Approaches to the Application of Digital Technologies in Education and Research, 2022, vol. 1222, pp. 183–193. Springer, Cham (2025). https://doi.org/10.1007/978-3-031-78776-8_19
27. Makoveichuk, K., Olifirov, A., Petrenko, S.: Integration of MOOC formats in an educational organization based on connectivism theory. In: International Conference on Innovative Approaches to the Application of Digital Technologies in Education and Research, Stavropol, Russia, 2022, vol. 1222, pp. 219–231. Springer, Cham (2025)
28. Prakasha, G.S., Muttungal, P., Lapina, M., Taran, V.: Flipped classroom strategy in online teaching: challenges faced by higher education teachers. In: International Conference on Innovative Approaches to the Application of Digital Technologies in Education and Research, Stavropol, Russia, 2022, vol. 1222, pp. 241–253. Springer, Cham (2025). https://doi.org/10.1007/978-3-031-78776-8_24
29. Lutsenko, L., Testova, I., Chirkov, M., Taran, V., Mezentseva, O.: Methodical specifics of the subject distance camp organization "young cryptograph". In: Lapina, M., Prakasha, G.S., Grigoriev, S. (eds.) International Conference on Innovative Approaches to the Application of Digital Technologies in Education and Research. SLET 2022 Lecture Notes in Networks and Systems, vol. 1222. Springer, Cham (2025). https://doi.org/10.1007/978-3-031-78776-8_21
30. Taran, V.N., Usmanov, M.I.: Learning Technologies Through Augmented Reality Projects in the VI International Scientific and Practical Conference "Distance Learning Technologies" (DLT 2021).Yalta, Crimea, September 20–22. vol. 3057, pp. 253–259 (2021). ceur-ws.org/Vol-3057/paper30.pdf.

Review of Methods for Using Virtual Museums in Literature Lessons

Irina Babenko[1] (✉) and G. S. Prakasha[2]

[1] North Caucasus Federal University, 355017 Stavropol, Russia
irinababenko17488@yandex.ru

[2] School of Education, Christ University, Bangalore, India

Abstract. The article is devoted to the study of the possibilities of using virtual museums in school teaching of literature as an innovative tool capable of enriching traditional approaches to the study of literary texts. The article considers methodological aspects of integrating digital museum resources into the educational process, including various forms of work: virtual excursions, thematic quests, project assignments and comparative analysis of museum exhibits. Particular attention is paid to the didactic potential of virtual museums in forming students' holistic understanding of the cultural and historical context of literary works. Both the advantages of this approach and the existing limitations are analyzed. The article presents a classification of methods for working with virtual museums in chronological perspective, reflecting the evolution from simple excursion forms to complex interactive models. A conclusion is made about the need for a balanced combination of the technological capabilities of virtual museums with the depth of literary analysis. The material of the article is of practical interest to literature teachers, methodologists and developers of educational digital resources, offering specific recommendations for the effective use of virtual museums in school practice of teaching literature.

Keywords: Virtual Museums · Literature Teaching Methods · Interactive Learning · School Education · Virtual Excursions

1 Introduction

Modern education strives to integrate digital technologies that not only expand the teacher's methodological capabilities, but also make the learning process more accessible to students. In conditions where schoolchildren increasingly perceive information visually and interactively, traditional methods of teaching literature require supplementation with innovative tools. Virtual museums are becoming one of the key resources in this process, allowing to overcome the geographical and time limitations associated with visiting real museums [1].

The use of virtual museums in literature lessons is especially relevant in the context of the formation of meta-subject competencies [2]. They contribute to the development of critical thinking, visual literacy and digital skills, which are now among the key requirements of standards. In addition, such museums help to overcome the abstractness

M. Lapina et al. (Eds.): DET 2024, LNNS 1616, pp. 72–80, 2025.
https://doi.org/10.1007/978-3-032-04365-8_6

of the perception of literary works, allowing students to immerse themselves in the era, biography of the writer or cultural context through interactive exhibits, 3D tours and multimedia materials [3].

Another important aspect is the personalization of learning. Virtual museums give each student the opportunity to independently explore materials at a convenient pace, return to difficult moments and choose an individual learning path. This is especially valuable in the context of multi-level classes and inclusive education.

Finally, the relevance of virtual museums is growing due to the development of Virtual and Augmented Reality (VR/AR) technologies, which are gradually penetrating the educational environment [4]. Already today, projects such as virtual reconstructions of the estates of Russian classics or interactive maps of literary places make literature lessons more visual and memorable [5].

Thus, the introduction of virtual museums into the school curriculum is not just a tribute to digitalization, but an effective way to increase interest in literature, deepen understanding of texts and create conditions for a more conscious perception of cultural heritage [6].

The main objective of studying the possibilities of using virtual museums in literature lessons is to develop effective methods for integrating them into the educational process [7]. To achieve this goal, it is necessary to solve several key tasks.

Firstly, it is important to analyse existing virtual museums with a literary focus, identify their educational potential and classify them by type (personal museums of writers, thematic exhibitions, interactive projects, etc.). This will allow teachers to navigate the variety of digital resources and choose the most suitable ones for specific educational topics.

Secondly, it is necessary to determine the optimal forms and methods of working with virtual museums in literature lessons. It is necessary to consider how to organically integrate virtual excursions, interactive tasks and projects into the structure of the lesson so that they not only complement, but also deepen the understanding of literary works.

Thirdly, it is important to study the pedagogical possibilities of virtual museums in developing key competencies of students: analytical thinking, visual literacy, research skills. Particular attention should be paid to how digital exhibitions help schoolchildren master the cultural and historical context of works and establish interdisciplinary connections.

Fourthly, it is necessary to identify possible difficulties related to technical limitations, organization of access to resources and methods of conducting classes using virtual museums. This will allow developing recommendations for teachers to overcome such problems.

Finally, it is important to generalize the successful pedagogical experience of using virtual museums in teaching literature and offer specific examples of their use at different stages of the lesson (motivation, learning new material, consolidation, project activities).

The solution to these problems will not only theoretically substantiate the effectiveness of virtual museums in education, but also offer teachers practical tools for their implementation in the educational process [8].

2 Virtual Museums: Concept and Possibilities

A virtual museum is a digital platform that reproduces or supplements the exhibitions of real museums in multimedia form and creates original thematic collections that exist exclusively in the online format [1, 2]. Unlike traditional museums, virtual analogues provide access to cultural heritage without geographical and time restrictions, using modern information technologies. In an educational context, virtual museums act as a powerful didactic tool that allows transforming the process of studying literature, making it more visual, interactive and personally oriented [3].

The key feature of virtual museums is their multimedia. They can include digital copies of manuscripts, rare editions, photographs, audio recordings of writers' voices, video tours, 3D models of household items and interiors, as well as elements of virtual and augmented reality. For example, a virtual tour of Tolstoy's Yasnaya Polyana estate allows students not only to see the writer's office, but also to visit the spaces where his great works were created. Such immersion in the material environment of the era significantly enriches the perception of literary texts.

Time-based virtual museums offer a wide range of educational opportunities.

Firstly, they provide access to unique materials that are physically located in various museums around the world, from Dostoevsky's personal belongings to sketches of the sets for the first productions of Chekhov's plays.

Secondly, interactive formats (chronological feeds, thematic maps, comparative tables) help to systematize literary material.

Thirdly, game elements (quests, quizzes, creative tasks) increase students' motivation. Particularly valuable is the possibility of integrating virtual museums into project activities, when schoolchildren themselves create digital exhibitions based on the works they have studied.

The technological capabilities of virtual museums are constantly expanding. Today, there are already projects with elements of artificial intelligence, where virtual guides answer users' questions, or platforms that allow for a comparative analysis of different editions of works. A promising direction is the creation of virtual museums, the content of which is regularly updated thanks to crowdsourcing, when teachers and students from all over the country can supplement the expositions with their materials [9–11]. All these features make virtual museums not just a digital analogue of traditional museums, but a fundamentally new educational resource that opens innovative ways to study literature (see, Table 1).

The concept of "Virtual Museum" emerged in the late 1980s and early 1990s, alongside the development of computer technology and the Internet. The term was first used in academic circles to describe digital projects based on real museums. One of the first examples was the "computer museum" project, launched in 1990 in Boston, which presented an online collection of exhibits.

In the 1990s, with the spread of the World Wide Web, the concept of a virtual museum began to develop rapidly. In 1994, the Louvre created one of the first official museum websites with elements of a virtual exhibition. At the same time, the first fully digital museums appeared that had no physical analogues, such as the "Virtual Museum of Computer Art" (1995).

Table 1. Evolution of the concept of "Virtual Museum".

Period	Events and characteristics	Examples
1980–1990	The term and implementation of IT in museums	The Computer Museum (Boston, 1990)
1991–1999	Museum websites	Louvre (1994) and others
2000–2009	UNESCO term approval and development of Internet applications	UNESCO Digital Heritage 2002
2010-present	VR/AR technologies, digitalization of museum collections	Google Arts & Culture (2011)

In the 2000s, the term gained widespread recognition. In 2002, the virtual museum was officially defined as "an organization that digitally collects, preserves, interprets, and makes accessible the heritage of humanity. The development of Web 2.0 technologies made it possible to create interactive virtual exhibits with the possibility of user interaction.

In the 2010s, with the advent of virtual and augmented reality (VR/AR) technologies, the concept expanded significantly. Projects like Google Arts & Culture (launched in 2011) made virtual museums a mass phenomenon. Today, the term includes both digital copies of real museums and original online collections, immersive VR exhibits, and even cultural metaverses.

The revolution of the term reflects technological progress: from simple web galleries of the 1990s to complex interactive systems of the 2020s, which turned the virtual museum into an independent cultural phenomenon.

3 Methods of Using Virtual Museums in Literature Lessons

In modern educational practice, virtual museums are becoming an effective tool for in-depth study of literary works and biographies of writers (see, Table 2). When used, it requires a systematic approach that includes several interrelated stages [12]. At the preparatory stage, the teacher defines pedagogical goals, selects an appropriate virtual museum, checks the technical base and develops an excursion route. It is important to clearly correlate the content of the digital museum with the topic of the lesson: when studying Pushkin's work, this can be a virtual tour of Mikhailovskoye, for an analysis of "War and Peace" - an exposition of Yasnaya Polyana.

The main forms of working with virtual museums are varied and depend on the educational objectives. The most common format is a virtual tour, which can be either frontal (with collective viewing through a projector) or individual (when students independently study the exhibition). Interactive quests, where students look for answers to problematic questions by analyzing museum exhibits, are especially effective [7]. For example, when studying the biography of Dostoevsky, you can suggest comparing photographs of places in St. Petersburg with descriptions from his works. Project activities allow you to consolidate knowledge through the creation of your own digital exhibitions using platforms such as Artsteps or Google Sites.

Table 2. .

Period	Methodology	Characteristic	Technical base	Advantage	Restrictions
1990–2000	Excursion	Teacher-led virtual tours	Computer class, CD	Systematic presentation	Passivity of students
2001–2010	Research	Conducting research on your own	Web platforms	Development of logic	Development of various scenarios is required
2011–2015	Project	Create your own projects on a given topic	Internet	Development of logic	Development of various scenarios is required
2016-present	Game	Game technologies for studying objects	Smartphones or computers	Development of logic	Special software required

The most important methodological aspect is the establishment of meaningful connections between museum objects and literary text. Contextualization techniques help students see how the realities of the era were reflected in works. Virtual reality technologies create an immersive effect [4]: for example, a VR tour of Pushkin's Petersburg allows you to visualize the environment in which the heroes of "Eugene Onegin" lived. Reflective tasks (essays, discussions) complete the work, helping to comprehend the experience gained.

The assessment criteria include depth of analysis, creativity and active participation. A lesson on Chekhov's works can combine the study of his Yalta house-museum with the analysis of symbolic objects in the stories, ending with the creation of a digital collage. This method turns passive assimilation of information into research activity, developing analytical thinking, visual literacy and skills in working with digital resources in schoolchildren. At the same time, it is important to maintain a balance between technological capabilities and literary analysis itself, so that the virtual museum remains a means of understanding the literary text, and not an end [13]. The implementation of this method requires from the teacher both technological competence and a deep understanding of literary material, which together contributes to increasing the motivation of students and the quality of literary education [7].

Modern pedagogical practice offers various methods of integrating virtual museums into the educational process, each of which has its own characteristics, advantages and limitations. The traditional excursion method involves a frontal acquaintance with a virtual exposition under the guidance of a teacher. This approach ensures a systematic perception of the material but limits the independence of students. In contrast, the research method, where schoolchildren independently study museum resources along a given route, develops analytical skills, but requires careful preparation of didactic materials and clear evaluation criteria.

The project-based methodology, which involves students creating their own virtual exhibits, maximally activates creative potential and promotes deep assimilation of the material but is highly labor-intensive and time-consuming. The game-based methodology using quests and quizzes based on museum resources increases motivation and involvement, especially in middle school, but requires a careful balance between entertainment and educational components.

The comparative method, when students analyze different virtual museums of one writer or era, deserves special attention. This approach develops critical thinking but requires the teacher to have deep subject preparation. The integrative method, which combines virtual excursions with text analysis, written work, and discussions, demonstrates the greatest effectiveness in forming a holistic perception of a literary work, although its implementation is difficult in conditions of limited study time.

Technologically advanced methods using VR equipment provide the maximum effect of presence and emotional impact but face the problem of availability of technical means. More democratic web-oriented methods using standard browser solutions are easier to implement but are inferior in clarity. An important criterion for comparison is the age-specificity of the methods: for high school students, research and project approaches are preferable, while in middle school, game and excursion forms are more effective.

All the methods considered have a common goal - the use of digital museum resources for an in-depth understanding of literary works through the visualization of the historical and cultural context. The choice of a specific method should be determined by the educational objectives, technical capabilities of the educational institution and the level of training of students. The most promising seems to be a flexible combination of various approaches, allowing for the maximum use of the educational potential of virtual museums in teaching literature.

4 Advantages and Possible Difficulties of Using Virtual Museums in Literature Lessons

The introduction of virtual museums into the educational process opens new prospects for teaching literature, but along with obvious advantages, it also has several limitations that must be considered [6].

The key advantages include, first, the accessibility of cultural heritage [14]. Virtual museums allow students from any region to get acquainted with the exhibits of the world's leading literary museums without leaving the classroom. This is especially valuable when studying the works of writers whose museums are in remote places [15] (for example, Tolstoy's Yasnaya Polyana estate or Pushkin's Mikhailovskoye).

An important advantage is the clarity and interactivity of digital exhibits [16, 17]. 3D tours, zoom functions for examining exhibit details, audio guides and video materials create the effect of presence and help to better understand the historical and cultural context of the works [18–20]. For example, a virtual visit to Dostoevsky's St. Petersburg apartment allows you to see the environment in which Crime and Punishment was created.

Virtual museums provide unique opportunities for differentiated learning [21, 22]. Students can independently choose the pace and depth of study, return to difficult moments, use additional resources (digitized manuscripts, expert comments). This is especially important for children with special educational needs.

The advantage is the development of digital competencies. Working with virtual museums develops in schoolchildren the skills of critical selection of information, working with multimedia sources, which meets the requirements of the standard.

However, using this technology is also associated with difficulties. The main problem is technical limitations: the need for a stable Internet connection, modern equipment, which is not available in all schools, especially in rural areas. Some platforms require powerful computers or VR glasses, which makes them inaccessible to many educational institutions.

There is also a methodological difficulty - the risk of replacing a deep analysis of the text with a superficial viewing of the exhibition [19, 20]. To prevent the virtual museum from turning into an entertainment element, the teacher must carefully think through the system of questions and tasks that link museum objects with a literary work [7, 12].

Another limitation is the lack of quality Russian-language resources [7, 23, 24]. Many foreign platforms (for example, Google Arts & Culture) offer limited content on Russian literature, and domestic analogues are often inferior in technical execution.

In addition, there is a psychological aspect: some students (especially in the younger grades) may have difficulty independently navigating the digital space, which requires additional instructions and support from the teacher.

Thus, the maximum efficiency of virtual museums is achieved by maintaining a balance between their technological capabilities and substantive literary work. Overcoming existing limitations requires both improving the technical base of schools and developing high-quality methodological materials that help teachers competently integrate this resource into the educational process. Under these conditions, virtual museums become a powerful tool that can significantly enrich the teaching of literature.

5 Conclusion

The use of virtual museums in literature lessons is a promising direction of modern education, harmoniously combining traditional approaches to the study of literary texts with innovative digital technologies. As the analysis showed, this method has significant pedagogical potential: it not only expands educational opportunities by overcoming geographical and time limitations, but also qualitatively transforms the process of perceiving literary works. Virtual exhibitions allow students to visually become acquainted with the historical and cultural context of the era, biographical details of the lives of writers, and the peculiarities of the creative process, which contributes to a deeper and more emotionally charged understanding of literary texts.

However, the effective implementation of this technology requires a well-thought-out methodological approach. It is important to remember that a virtual museum is not an end, but a means of achieving educational results. The success of its application depends on several factors: the competent selection of museum resources in accordance with educational objectives, the development of a system of questions and tasks that link the virtual exposition with literary analysis, as well as the provision of the necessary technical base. Of particular importance is the training of teachers who can organically integrate digital resources into a traditional literature lesson.

The development prospects of this area relate to the improvement of virtual and augmented reality technologies, expansion of access to high-quality educational resources, and development of specialized methodological recommendations. Provided that the existing technical and methodological limitations are overcome, virtual museums can become an important component of modern literary education, contributing to the formation of students' holistic perception of a work of art as part of the cultural and historical process, the development of their analytical abilities and digital literacy. Thus, a reasonable combination of traditional and innovative approaches to teaching literature opens new opportunities for improving the quality of humanitarian education in the digital age.

References

1. Styliani, S., Fotis, L., Kostas, K., Petros, P.: Virtual museums, a survey and some issues for consideration. J. Cult. Herit. **10**(4), 520–528 (2009)
2. Carrozzino, M., Bergamasco, M.: Beyond virtual museums: experiencing immersive virtual reality in real museums. J. Cult. Herit. **11**(4), 452–458 (2010)
3. Schweibenz, W.: The virtual museum: an overview of its origins, concepts, and terminology. Mus. Rev. **4**(1), 1–29 (2019)
4. Tsichritzis, D., Gibbs, S.J.: Virtual Museums and Virtual Realities, pp. 17–25. ICHIM (1991, October)
5. Aristova, U.V., Rolich, A.Y., Staruseva-Persheeva, A.D., Rolich, A.O.: Digital museum transformation: from a collection of exhibits to a gamut of emotions. In: International Conference on Digital Transformation and Global Society, pp. 419–435. Springer International Publishing, Cham (2020, June)
6. Tserklevych, V., Prokopenko, O., Goncharova, O., Horbenko, I., Fedorenko, O., Romanyuk, Y.: Virtual museum space as the innovative tool for the student research practice. Int. J. Emerg. Technol. Learn. (iJET). **16**(14), 213–231 (2021)
7. Domansky, V.A., Mironova, N.A., Popova, N.A.: The digital environment in the field of teaching classic Russian literature. Educ. Pedag. J. **1**(5), 5–24 (2023)
8. Hirsch, F.: Getting to know "the peoples of the USSR": ethnographic exhibits as soviet virtual tourism, 1923–1934. Slav. Rev. **62**(4), 683–709 (2003)
9. Anton, M., Nicolae, G., Moldoveanu, A., Balan, O.: Virtual museums-technologies, opportunities and perspectives. Romanian J. Hum. Comput. Interact. **11**(2) (2018)
10. Bentkowska-Kafel, A.: Virtual Museum–The concept and transformation. Display: consume: respond–digital engagement with art. In: The 28th Annual Conference of Computers and the History of Art (CHArt) held on, vol. 15, pp. 1–25 (2016)
11. Pujol, L., Lorente, A.: The virtual museum: a quest for the standard definition. Archaeol. Dig. Era, 40 (2012)
12. Veltman, K.H.: Developments in virtual museums. Published in Italian as: La crescita nel settore dei musei virtuali. In: Museo contro Museo. Le strategie, gli strumenti, i resultati. Giunti, Florence, pp. 263–286 (2001)
13. Mussayeva, S.T.: Virtual museums as a sociocultural phenomenon and their functional specificity. In: Industrial Technologies and Engineering ICITE-2020, pp. 106–109 (2020)
14. Laurìa, A.: Accessibility to cultural heritage as a design challenge. In: Accessibility As a Key Enabling Knowledge for Enhancement of Cultural Heritage, pp. 90–104 (2016)

15. Clifford, J.: Museums as contact zones. In: Representing the Nation: A Reader: Histories, Heritage and Museums, pp. 435–457 (1999)
16. Foo, S.: Online virtual exhibitions: concepts and design considerations. DESIDOC J. Libr. Inform. Technol. **28**(4), 22–34 (2008)
17. Abubakar, J.A., Mahmud, M., Jahnkassim, P.S.: User perception on interactive digital exhibits and learning experience. In: The 5th International Conference on Information and Communication Technology for the Muslim World (ICT4M), pp. 1–6. IEEE (2014, November)
18. Bruno, F., Bruno, S., De Sensi, G., Luchi, M.L., Mancuso, S., Muzzupappa, M.: From 3D reconstruction to virtual reality: a complete methodology for digital archaeological exhibition. J. Cult. Herit. **11**(1), 42–49 (2010)
19. Kersten, T.P.: 3D models and virtual tours for a museum exhibition of vietnamese cultural heritage exhibits and sites. In: Digital Heritage. Progress in Cultural Heritage: Documentation, Preservation, and Protection: 7th International Conference, EuroMed 2018, Nicosia, Cyprus, October 29–November 3, 2018, Proceedings, Part I 7, pp. 528–538. Springer International Publishing (2018)
20. Zidianakis, E., Partarakis, N., Ntoa, S., Dimopoulos, A., Kopidaki, S., Ntagianta, A., et al.: The invisible museum: a user-centric platform for creating virtual 3D exhibitions with VR support. Electronics. **10**(3), 363 (2021)
21. Freed, T.: Immersive virtual environments as unique learning spaces in museum education (Doctoral dissertation, Concordia University) (2008).
22. Sylaiou, S., Mania, K., Paliokas, I., Pujol-Tost, L., Killintzis, V., Liarokapis, F.: Exploring the educational impact of diverse technologies in online virtual museums. Int. J. Arts Technol. **10**(1), 58–84 (2017)
23. Kormakova, V.N., Klepikova, A.G., Lapina, M.A., Rugelj, J.: ICT competence of a teacher in the context of digital transformation of education. CEUR Workshop Proc. **2914**, 138–150 (2021)
24. Kormakova, V., Klepikova, A., Musaelian, E., Baybikowa, G., Lapina, M.: Formation of ICT Competencies of Postgraduate Students of Teacher Education Based on Interactive Techniques: CEUR Workshop Proceedings SLET-2019 – Proceedings of the International Scientific Conference Innovative Approaches to the Application of Digital Technologies in Education and Research, vol. 2494, pp. 11–21 (2019)

Methods of Literary Text Analysis with the Help of Artificial Intelligence

Irina Babenko[1(✉)] and Vijay Anant Athavale[2]

[1] North Caucasus Federal University, 355017 Stavropol, Russia
irinababenko17488@yandex.ru
[2] Walchand Institute of Technology, Solapur, India

Abstract. This paper examines modern methods of computer analysis of literary texts using artificial intelligence and natural language processing technologies. The study demonstrates how language models BERT, GPT, specialized libraries spaCy, Gensim and computing platforms can be adapted to solve traditional literary problems - from stylistic and thematic analysis to the study of narrative structures and intertextual connections. Particular attention is paid to the possibilities and limitations of using machine learning in the humanities, including problems of interpretability of results and the need to combine quantitative methods with qualitative philological analysis. The work systematizes existing digital tools for literary studies Voyant Tools, Mallet, Stylo, offering practical recommendations for their use in research practice. Particular emphasis is placed on the methodological aspects of integrating computer technologies into humanities research, where formal methods serve not as a replacement, but as a powerful addition to traditional approaches. The results of the study show that modern NLP technologies open up new prospects for analyzing large text corpora, identifying hidden patterns and posing new research questions in literary studies, while simultaneously requiring critical understanding and meaningful interpretation of the data obtained. The work is of interest to researchers at the intersection of digital humanities and computational linguistics, offering a comprehensive overview of current methods and their practical application in the analysis of literary texts.

Keywords: computer analysis of literary texts · artificial intelligence in literary studies · NLP · BERT · GPT · Digital Humanities

1 Introduction

Modern Artificial Intelligence (AI) technologies are actively transforming the methods of working with text information, opening new opportunities for the analysis of literary works [1]. Traditional literary criticism has long relied on the subjective interpretations of researchers, but with the development of Natural Language Processing (NLP) and machine learning, tools have emerged that allow us to identify hidden patterns, stylistic features, and semantic structures in large amounts of text [2]. This makes the analysis more objective, scalable, and accessible. However, despite the rapid development of NLP methods, their application to the analysis of literary works remains understudied, which makes this problem particularly relevant [3].

M. Lapina et al. (Eds.): DET 2024, LNNS 1616, pp. 81–95, 2025.
https://doi.org/10.1007/978-3-032-04365-8_7

First, traditional methods of literary analysis are often subjective, depending on the interpretation of a particular researcher. The use of AI allows us to minimize this subjectivity by offering quantitative and statistically valid conclusions. For example, machine learning algorithms can identify stylistic features, the frequency of certain lexical constructions, or the emotional tone of a text, which provides a more objective basis for scientific conclusions.

Secondly, the volume of literary heritage available in digital format is constantly growing, and manual analysis of such data sets is becoming almost impossible. AI tools are able to process thousands of texts in a short time, classifying them by genre, topic or author's style, which is especially valuable for comparative studies and identifying historical and literary trends.

Thirdly, the integration of AI into literary studies promotes the development of interdisciplinary research, combining the humanities with computer technologies. This not only expands the methodological base of philology, but also raises new questions related to the interpretation of results, the ethics of using algorithms, and preserving the cultural context in automated analysis.

The aim of this work is to study modern methods of literary text analysis using artificial intelligence, as well as to evaluate their effectiveness in comparison with classical approaches. The study will consider key NLP technologies, such as word vector representations Word2Vec [4], GloVe [5], transformer models Bidirectional Encoder Representations from Transformers (BERT) [6], Generative Pre-trained Transformer (GPT) [4], methods of classification and clustering of texts. The practical part of the work involves the application of these methods to the analysis of specific literary works with subsequent interpretation of the results.

The significance of the study is that it not only demonstrates the capabilities of AI in philology, but also outlines the boundaries of its application, including problems of data interpretability, ethical aspects, and possible distortions of meaning during automatic text processing. The results of the work can be useful to literary scholars, digital humanities, NLP system developers, and specialists working at the intersection of computer technology and the humanities [8].

Finally, the relevance of the problem is emphasized by its practical significance. The results of such studies can be applied not only in the academic environment, but also in education, in publishing and even in digital humanities [9], where AI is used to preserve and study cultural heritage.

2 Analysis of Literary Text: Traditional Methods

2.1 Linguistic Analysis of Literary Text

Linguistic analysis of a literary text is a comprehensive approach to studying the linguistic features of a work, which allows us to reveal its deep meanings, stylistic originality and artistic value [10]. Unlike a regular text, a literary work has a complex structure, where each linguistic element - from phonetics to syntax - carries an aesthetic and semantic load.

One of the key features of linguistic analysis is its multi-level nature [11]. It includes the study of phonetic structure, vocabulary, morphology, syntax, and level structures such as composition and narrative. For example, the analysis of repetitions of certain sounds or words can reveal hidden emotional accents, and the study of syntactic constructions can reveal the author's style or the intentional complication/simplification of speech to achieve artistic effect.

Another important feature is attention to context. Linguistic analysis is not limited to an isolated examination of linguistic units, but takes into account their role within the entire work, its genre, era, and the author's intent [12]. Thus, the use of archaisms or neologisms can serve not only as a stylistic device, but also as a way of creating historical coloring or expressing the author's position.

Of particular complexity is the analysis of figurative means of language - metaphors, comparisons, symbols, which require not only a linguistic but also a literary approach [13, 14]. For example, a metaphorical series in poetry can be systemic and reflect the key themes of the work, and symbols can have cultural or philosophical connotations that go beyond the direct meaning of words.

In modern literary studies, linguistic analysis is increasingly combined with digital methods, which allows for the processing of large volumes of texts and the identification of patterns that are invisible in traditional analysis. However, despite the development of technology, the interpretation of results remains a creative process that requires a deep understanding of both linguistic mechanisms and the artistic specificity of literature.

Thus, linguistic analysis of a literary text is not simply a study of its linguistic structure, but a way of revealing the unique unity of form and content that makes a work of art of words. Its features - multi-level, contextuality and attention to imagery - make it an indispensable tool in the hands of a researcher seeking to understand the depth and originality of a literary text.

2.2 Stylistic Analysis of Literary Text

Stylistic analysis is the most important tool for interpreting a literary text, allowing one to reveal the aesthetic value of a work through the prism of linguistic expression [15]. Unlike linguistic analysis, which studies the language system in a work, stylistic analysis focuses on the artistic function of linguistic means, their role in creating imagery and emotional impact.

The key feature of stylistic analysis is its focus on the individual author's style [16]. Each writer creates a unique system of linguistic devices - from the choice of vocabulary to the organization of the narrative - which forms his recognizable style. For example, Hemingway's laconic telegraphic style, Pasternak's metaphor-rich language, or Dostoevsky's complex syntactic constructions are all objects of stylistic study.

The most important aspect of the analysis is the study of tropes and stylistic figures. Metaphors, comparisons, epithets, hyperboles, gradations and other expressive means are not considered in isolation, but in the system of their interaction, creating a special artistic reality [17]. In this case, not only their presence is analyzed, but also their originality, frequency, and distribution in the text. For example, through metaphors can become structural elements of the entire work.

Particular attention is paid to the analysis of language levels:

- The lexical-phraseological level examines the selection of words, the use of synonyms, archaisms, neologisms, terms, dialectisms;

- The syntactic level analyzes the construction of phrases - rhythm, inversions, parallelisms, repetitions;
- The phonetic level deals with sound organization, which is especially important in poetry;
- The compositional level studies the style of narration, changes in points of view, and speech characteristics of characters.

Modern stylistic analysis increasingly uses digital methods that allow us to identify statistical patterns in the use of linguistic means. However, high-quality interpretation always requires an understanding of the cultural and historical context and the author's intent.

The most important task of stylistic analysis remains the disclosure of how the form of a work becomes content, how linguistic means are transformed into artistic values. This study allows not only a deeper understanding of the text, but also an assessment of the skill of the writer, who managed to express complex ideas and emotions through the fabric of language.

Thus, stylistic analysis is a bridge between linguistics and literary studies, revealing the secrets of artistic creativity through a thorough study of the linguistic form of a work. Its importance is especially great in the study of individual authorial styles and various literary trends.

2.3 Structural and Narrative Analysis of Literary Text

Structural and narrative analysis of a literary text is a profound study of the internal organization of a work of art and the methods of constructing a narrative [18]. This approach considers the text as a complex system, where each element – from composition to speech characteristics – performs a certain function in creating a holistic artistic image. Structural analysis is based on the identification of stable models and principles of text construction. It examines the architectonics of a work, including its division into parts, chapters or stanzas, analyzes the spatio-temporal organization, identifies the system of characters and their interrelations, studies leitmotifs and recurring elements, and analyzes plot schemes, using, for example, V. Propp's theory of character functions or C.G. Jung's concept of archetypes. Particular attention is paid to binary oppositions, which often become the structural basis of the artistic world of a work.

Narrative analysis focuses on the methods of organizing the narrative and includes the study of the narrative instance, the analysis of points of view and the ways of expressing them, the study of types of narration, the consideration of speech characteristics of characters and the analysis of the temporal organization of the text [19]. The most important category here is the concept of focalization, developed by J. Genette, which describes the methods of limiting and presenting information to the reader [20]. Modern narratology pays special attention to the problem of the "unreliable narrator", when the narrative is deliberately built on the distortion or limitation of information. Structural and narrative analysis, when combined, allow us to consider a work as a complex sign system, where formal elements become carriers of deep meanings. These methods are especially productive in the comparative study of works, helping to identify both stable narrative models characteristic of certain eras and genres, and the unique features of a particular text [21]. This approach demonstrates that the methods of narration themselves are the content dominant of the work, and the analysis of the formal structure allows us to understand the mechanisms of the artistic impact of literature on the reader [22].

Modern structural-narrative analysis actively uses the achievements of semiotics and communication theory, considering the literary text as a complex sign system. Particular attention is paid to the problem of the "unreliable narrator", when the narration is conducted from a deliberately distorted position.

Methods of structural and narrative analysis are especially productive in comparative studies of works, allowing to identify stable narrative models characteristic of certain eras, genres or authors. At the same time, they help to reveal the unique features of a specific text, its innovative features in the field of narrative technique.

These approaches demonstrate how the formal elements of a work become carriers of meaning, and the methods of narration become the dominant content [23]. Such analysis allows us to understand not only "what" is being told, but also "how" it is done, thereby revealing the mechanisms of the artistic impact of literature.

3 Artificial Intelligence in Text Processing

3.1 Natural Language Processing Methods and Their Properties

NLP methods offer a variety of approaches to analyzing literary texts, each with unique capabilities and limitations. Traditional statistical methods such as Term Frequency—Inverse Document Frequency and Latent Semantic Analysis (LSA) provide a basic understanding of the thematic content of texts through word frequency characteristics [24]. However, these approaches do not take into account the semantic relationships between words and contextual meanings, which is especially important when working with fiction, where words often acquire special, non-literal meanings.

More advanced word embedding methods such as Word2Vec [4] and GloVe [5] have made a breakthrough in text analysis by capturing semantic relationships between words through their distributed representations in vector space. These algorithms demonstrate impressive results in detecting stylistic features and thematic clusters, but are still limited in understanding more complex language structures. For example, they do not take into account the polysemy of words, which is especially typical for literary texts, where the same word in different contexts can carry completely different semantic shades.

Transformer architectures such as BERT [6], GPT [7] and their variants represent a new stage in the development of NLP, offering contextual understanding of text. These models are able to analyze words in their full context, taking into account complex syntactic structures and semantic nuances, which makes them especially valuable for literary analysis. BERT [6], with its bidirectional approach, copes well with the tasks of text classification and detection of hidden semantic structures, while GPT [7], due to its ability to generate coherent text, can be used to analyze narrative strategies and stylistic features.

Comparing these methods, it can be noted that traditional statistical approaches remain useful for the initial, superficial analysis of large text arrays, while modern neural network models open up new possibilities for the in-depth study of literary texts.

However, it is important to consider that complex models require significant computing resources and large amounts of data for training, and their "black box" nature complicates the interpretation of the results. In literary studies, the issue of balance between quantitative methods of analysis and the traditional qualitative approach, where understanding the cultural and historical context and the author's intent plays a key role, is especially acute. Thus, the optimal approach to the analysis of literary texts using NLP should combine the strengths of different methods, taking into account both the technical capabilities of the algorithms and the specifics of literary texts.

3.2 Machine Learning and Neural Networks in Text Analysis

Modern methods of machine learning and neural networks have revolutionized computer analysis of texts, opening up new perspectives for literary studies. These technologies make it possible to identify complex patterns in works of art that remain invisible in traditional approaches to analysis.

Machine learning in text analysis can be divided into two main areas: classical algorithms and deep learning. Classical methods such as the naive Bayes classifier, Support Vector Machine (SVM) [25], and random forests have found application in author attribution, genre classification, and determining the emotional tone of text [1]. These algorithms work with pre-selected features - word frequency, n-grams, stylistic markers. Their advantage is the relative simplicity and interpretability of the results, but they are limited in their ability to understand context and semantic nuances.

Deep learning, and neural networks in particular, overcome these limitations by automatically extracting hierarchical representations from text data. Recurrent Neural Networks (RNNs), especially their Long Short-Term Memory (LSTM) and Guided Recurrent Units (GRU) variants, have demonstrated effectiveness in processing sequences, which is important for analyzing narrative structures [2]. These models are able to account for the dependencies between widely spaced elements of text, which is critical for understanding literary works with their complex composition.

A real breakthrough in text analysis is associated with the emergence of the transformer architecture and models based on it, such as BERT [6], GPT [7] and their numerous modifications. These models use an attention mechanism that allows analyzing the relationships between all words in the text regardless of their positional distance. The ability of such models to:

- to capture the subtle stylistic features of the author's writing;
- identify intertextual connections and allusions;
- analyze narrative strategies;
- discover hidden thematic layers;
- model the emotional dynamics of the text.

The application of neural network approaches in literary studies faces a number of methodological challenges. The "black box" problem complicates the interpretation of results, which contradicts the traditional humanities focus on the validity and transparency of analysis. In addition, most modern models require huge amounts of data for training, while many literary studies focus on individual works or the work of a specific author.

A promising direction is the development of specialized architectures for the analysis of literary texts, taking into account their specificity - metaphor, symbolism, complex narrative structures. A compromise solution may be a combination of neural network methods with traditional approaches, when neural networks are used for primary processing and pattern detection, and their results are then interpreted by humanities researchers.

It is important to note that machine learning and neural networks do not replace traditional literary studies, but offer new tools for working with texts, allowing one to process large amounts of data and discover non-obvious patterns. These methods are especially valuable for comparative studies, analysis of the evolution of authorial style, and the study of historical and literary processes in large corpora of texts.

4 Artificial Intelligence Tools for Literary Text Analysis

4.1 Language Models and NLP Frameworks for Literary Text Analysis

Modern language models and NLP frameworks open up new horizons in the study of literary texts, offering philologists powerful tools for deep analysis of works of art. Among the most promising technologies, transformer architectures should be highlighted, which have made a real revolution in natural language processing. The BERT model [6] and its specialized modifications, such as literary-oriented versions, allow for a comprehensive analysis of texts at various levels - from identifying stylistic features to determining complex intertextual connections. Of particular value is BERT's ability to analyze context in both directions, which is critical for understanding literary texts with their complex syntactic organization and polysemy of linguistic units. RuBERT [26] and other pre-trained models adapted to the morphological features of the Russian language are especially effective for working with Russian-language literature.

Generative models of the GPT series [7] find unique application in literary studies. Their ability to analyze long text sequences makes them indispensable in studying narrative structures and compositional features of works. GPT [7] allows modeling various scenarios of plot development, analyzing the features of the author's style through the tasks of text continuation, and identifying characteristic lexical-syntactic patterns. However, when working with these models, it is necessary to take into account their tendency to generate plausible, but actually unreliable interpretations, which requires special caution when analyzing the results.

For more traditional text analysis tasks, tools such as FastText [27] are widely used, which is especially effective when working with rare words and morphological variants, which is relevant for the analysis of poetic texts and works with authorial neologisms. Libraries Natural Language Toolkit (NLTK) [28] and spaCy [29] provide a robust foundation for text preprocessing, including tokenization, lemmatization, and syntactic analysis, which is especially important when working with classical works where punctuation and graphics in the text can carry semantic meaning.

A special place is occupied by specialized frameworks such as Hugging Face Transformers [30], which combine many pre-trained models and provide convenient interfaces for their adaptation to specific literary studies tasks. These tools allow researchers to focus on interpreting the results, rather than on the technical aspects of implementing the models [31]. It is important to emphasize that the effective use of these technologies in literary studies requires careful tuning and, often, additional training on specialized literary corpora, since standard models are optimized for working with modern news and scientific texts, and not with fiction and poetry.

The combination of these technologies creates a powerful analytical toolkit that allows us to study literary texts at a qualitatively new level - from the automatic detection of stylistic features and thematic transitions to complex analysis of intertextual connections and the transformation of artistic techniques in a historical-literary perspective. However, it is important to remember that these tools should not replace traditional methods of philological analysis, but rather complement them, providing researchers with new data for meaningful interpretation.

4.2 Natural Language Processing Methods and Their Properties

For professional analysis of literary texts using computer methods, a number of specialized libraries for natural language processing have been developed, each of which solves a specific range of problems. The NLTK library is a fundamental toolkit that provides a wide range of functions for basic text processing - from tokenization and stemming to syntactic analysis and named entity extraction [28]. Its modular structure and extensive collection of corpora make NLTK especially useful at the initial stages of working with texts, when primary cleaning and normalization of data is required. However, for working with fiction texts, where the accuracy of morphological analysis is important, it is often preferable to use the spaCy library, which is characterized by high performance and accuracy, especially in the tasks of identifying parts of speech and syntactic dependencies [29]. SpaCy supports several languages and offers ready-made models for processing Russian-language texts, which is critically important when analyzing Russian literature.

For more complex topic modeling tasks and working with distributed word representations, the Gensim library becomes indispensable. Its implementation of the Latent algorithms Dirichlet Allocation (LDA) [32] and Latent Semantics Indexing (LSI) [33] allows to reveal hidden thematic structures in large text corpora, and built-in tools for working with Word2Vec [4] and Doc2Vec [34] make it possible to analyze semantic connections between words and documents. A special feature of Gensim is its efficiency when working with large volumes of data without loss of quality of results, which is important when analyzing complete works or comparative studies of the works of several authors.

Specific tasks of stylometric analysis often require the use of specialized packages, such as Stylo for R or native Python implementations. These tools provide ready-made methods for calculating text frequency characteristics, analyzing lexical diversity and other parameters important for studying the author's style. Special libraries are developed for working with poetic texts, capable of analyzing the rhythmic structure, rhymes and other formal features of the verse.

Libraries for visualizing text analysis results deserve special attention, such as pyLDAvis [35], which provides interactive tools for exploring the results of topic modeling, or Matplotlib and Seaborn, which allow you to create informative graphs and diagrams to present the identified patterns. These tools are especially important in the final stages of research, when it is necessary to clearly present the obtained results.

Modern researchers increasingly use a combination of these libraries to create complex analysis pipelines where each technology is used at the stage where it is most effective. For example, text preprocessing can be done using spaCy [29], topic modeling can be done using Gensim, and results can be visualized using pyLDAvis [35]. This integrated approach allows for the full use of the strengths of each library, ensuring the depth and reliability of literary analysis.

4.3 Computational Platforms and Infrastructure for Literary Analysis

Modern studies of literary texts using artificial intelligence methods require a special computing infrastructure capable of processing large volumes of text data and supporting resource-intensive algorithms. Different platforms are used for different stages of analysis, each with unique characteristics and advantages.

Cloud platforms such as Google Colab [36] and Kaggle Notebooks [37] are a convenient solution for the initial stages of research and experimentation with small and medium-sized texts. These services provide free access to GPU and TPU accelerators, which significantly reduces the time it takes to train models. Google Colab Pro offers advanced features, including more powerful video cards and increased task execution time, making it a suitable option for working with medium-sized models. A special feature of these platforms is their interactivity - researchers can combine code execution with Markdown descriptions, creating full-fledged computational notebooks that are convenient for documenting the progress of research [38].

For larger projects that require processing complete works or analyzing large literary corpora, professional cloud solutions are needed. Amazon SageMaker [39] and Google Cloud AI Platform [40] provide a complete environment for deploying complex ML pipelines, including tools for data preprocessing, model training, and subsequent deployment. These platforms support distributed computing, which is critical when working with transformer architectures such as BERT- large [41] or GPT [7], which require significant computing resources. Microsoft Azure also offers specialized services for natural language processing that can be adapted for literary studies.

When working with particularly large amounts of data or when complex multi-task computations are required, distributed data processing frameworks such as Apache Spark [42] or Dask [43] are used. These tools allow the computational load to be efficiently distributed across multiple nodes, which speeds up the processing of large text collections. Ray is a modern alternative for distributed execution of machine learning tasks, particularly effective in hyperparameter tuning and scaling experiments.

Specialized database management systems are used to store and organize literary texts in digital form. Relational DBMS PostgreSQL, MySQL are used for structured data, while document-oriented databases MongoDB, Elasticsearch are better suited for storing full texts of works with metadata. Elasticsearch additionally provides powerful full-text search tools, which are useful for analyzing intertextual connections [44].

Of particular importance in literary studies is the infrastructure for data labeling. Platforms such as Prodigy [45] or Label Studio [46] allow philological experts to participate in the creation of labeled datasets, which is necessary for training specialized models on literary texts. These tools support complex labeling schemes that take into account the specifics of literary works - from identifying characters and settings to marking stylistic devices and rhetorical figures.

Modern computing platforms for literary analysis increasingly include specialized components that take into account the specifics of humanities research. For example, there are pre-configured virtual machines with installed NLP tools and examples of literary text analysis, which significantly reduces the entry threshold for philological researchers.

4.4 Specialized Tools for Literary Analysis

Modern digital literary analysis has a whole arsenal of specialized tools developed taking into account the specifics of literary texts. Among them, a special place is occupied by Voyant Tools – a web platform that provides a set of visualizations for analyzing text corpora. This tool allows one to study the frequency of words, their distribution in the text, identify key terms through tag clouds, and analyze lexical connections through the Cirrus tool [47]. A special feature of Voyant is the ability to work with multilingual texts, which makes it useful for comparative analysis of works of different literary traditions.

For advanced topic modeling, professional researchers often turn to MAchine Learning for Language Toolkit (MALLET) [48]. Developed at the University of Massachusetts, this package offers advanced topic modeling algorithms, including improved versions of LDA. MALLET is especially valuable when working with large historical-literary corpora, where it is necessary to identify hidden thematic structures and their evolution over time. The tool supports working with metadata, which allows one to take into account time periods, genre affiliation, and other parameters during analysis.

Stylometric studies rely on specialized packages such as Stylo for R or Stylometry for Python. These tools provide ready-made methods for:

- quantitative analysis of the author's style
- determining the authorship of disputed texts
- revealing the chronological evolution of the writer's style
- comparison of stylistic features of different authors

AntConc is a specialized concordance, indispensable in linguo-stylistic analysis [49]. This tool allows:

- build frequency dictionaries
- identify collocations
- analyze keywords
- explore the use of specific lexical units in context

For the analysis of poetic texts, special tools have been developed, such as PoetryNLP and Scandroid, which specialize in [50]:

- analysis of metrics and rhythm
- identifying rhyme structures
- study of the sound organization of the text
- visualization of poetic forms

Digital humanities scholars are increasingly using the Text Encoding Initiative (TEI), a system for tagging literary texts that allows them to be preserved and analyzed [51]:

- structural elements of works
- stylistic features
- historical and cultural context
- intertextual connections

Contemporary research projects often use Zotero and Omeka, systems for organizing and presenting literary data that allow [52]:

- create annotated bibliographies
- build digital exhibitions
- visualize literary networks
- publish interactive research

The development of specialized software for literary studies is moving towards the creation of complex platforms that combine analytical functions with visualization and collaboration capabilities. Tools such as Lexos [53] and TAPoR [54] offer integrated environments for digital literary analysis, including:

- text preprocessing
- statistical analysis
- stylometric studies
- interactive visualization of results

A special direction is represented by tools for the analysis of intertextuality, such as Tesserae, which specialize in [55–57]:

- identifying allusions and reminiscences
- comparison of parallel places
- analysis of the transformation of motives
- study of literary influences

Modern specialized tools increasingly include elements of artificial intelligence, allowing automation of:

- attribution of texts
- genre classification
- analysis of narrative structures
- identifying emotional tone

These tools do not replace traditional methods of philological analysis, but they significantly expand research capabilities, allowing for the processing of large arrays of texts and the identification of patterns that are inaccessible with "manual" analysis. Their development is moving towards greater specialization for specific literary tasks and better integration with traditional methodologies of the humanities.

5 Conclusion

Modern methods of artificial intelligence and digital technologies open up new prospects for literary studies, significantly expanding traditional approaches to the analysis of literary texts. The use of NLP models such as BERT and GPT, specialized text processing libraries and computing platforms allows us to study literary works at a qualitatively new level - from the automatic detection of stylistic features and thematic structures to complex analysis of narrative techniques and intertextual connections. However, the effective use of these technologies requires careful methodological justification and critical understanding of the results, since formal algorithms cannot completely replace the depth of traditional philological analysis. The most important advantage of computer methods is their ability to process large arrays of texts, revealing patterns that are invisible during "manual" analysis, which is especially valuable for comparative studies and the study of historical and literary processes. At the same time, the interpretation of the obtained data should always take into account the cultural and historical context and artistic specificity of the works. The optimal approach seems to be the integration of quantitative and qualitative methods, where digital tools are used for primary analysis and pattern identification, and their results are subject to meaningful interpretation by humanities researchers. The prospects for the development of digital literary studies are associated with the creation of more specialized tools that take into account the features of literary texts, and the development of methodological approaches that allow for an organic combination of computer analysis with traditional philological methods. Thus, artificial intelligence does not replace, but significantly enriches literary studies, opening up new opportunities for understanding the complex nature of a literary text and the mechanisms of its impact on the reader.

References

1. Horáková, T., Houška, M., Dömeová, L.: Classification of the educational texts styles with the methods of artificial intelligence. J. Baltic Sci. Educ. **16**(3), 324–336 (2017)
2. Samarakou, M., Tsaganou, G., Papadakis, A.: An e-learning system for extracting text comprehension and learning style characteristics. J. Educ. Technol. Soc. **21**(1), 126–136 (2018)
3. Kanchon, M.K.H., et al.: Enhancing personalized learning: AI-driven identification of learning styles and content modification strategies. Int. J. Cogn. Comput. Eng. **5**, 269–278 (2024)
4. Church, K.W.: Word2Vec. Nat. Lang. Eng. **23**(1), 155–162 (2017)
5. Pennington, J., Socher, R., Manning, C.D.: Glove: global vectors for word representation. In: Proceedings of the 2014 conference on empirical methods in natural language processing (EMNLP), pp. 1532–1543 (2014)
6. Koroteev, M.V: BERT: a review of applications in natural language processing and understanding. arXiv preprint arXiv:2103.11943 (2021).
7. Achiam J. et al. Gpt-4 technical report. arXiv preprint arXiv:2303.08774 (2023).
8. Zhou, M., et al.: Progress in neural NLP: modeling, learning, and reasoning. Engineering. **6**(3), 275–290 (2020)
9. Berry, D.M.: Introduction: Understanding the digital humanities. Understanding digital humanities, pp. 1–20. Palgrave Macmillan UK, London (2012)
10. Fries, C.C.: Meaning and linguistic analysis. Language. **30**(1), 57–68 (1954)

11. Heine, B., Narrog, H. (eds.): The Oxford Handbook of Linguistic Analysis. Oxford University Press (2015)
12. Crossley, S.A., et al.: A linguistic analysis of simplified and authentic texts. Mod Lang. J. **91**(1), 15–30 (2007)
13. Brubaker, B.H.: Health promotion: A linguistic analysis. Adv. Nurs. Sci. **5**(3), 1–14 (1983)
14. Takahashi H. A Cognitive Linguistic Analysis of the English Imperative. (2012)
15. Carter, R.: Methodologies for stylistic analysis: Practices and pedagogies. Language and style. **43**(1), 55–68 (2010)
16. Craig, H.: Stylistic analysis and authorship studies. A companion to digital humanities. pp. 271–288 (2004)
17. Short, M., Semino, E.: Evaluation and Stylistic Analysis. The Quality of Literature: Linguistic Studies in Literary Evaluation, pp. 117–137. John Benjamins Publishing Company (2008)
18. Barthes, R., Duisit, L.: An introduction to the structural analysis of narrative. New Liter. Hist. **6**(2), 237–272 (1975)
19. Barthes, R.: Introduction to the structural analysis of narratives. Narrative theory: Critical concepts in literary and cultural studies. pp. 65–116 (2004)
20. Edmiston, W.F.: Focalization and the first-person narrator: a revision of the theory. Poetics Today. **10**(4), 729–744 (1989)
21. Barthes, R.: Introduction to the Structural Analysis of the Narrative (1966).
22. Barthes, R., et al.: Structural Analysis of Narrative. Criticism: The Major Statements (1975).
23. Esin, C.: Narrative analysis approaches. Qualitative research methods in psychology: Combining core approaches., 92–117 (2011)
24. Pascual, E.Z.: Learning generators: NLP and learning styles in English text books. Revista de Estilos de Aprendizaje. **2**(3) (2009)
25. Quan, Z., Pu, L.: An improved accurate classification method for online education resources based on support vector machine (SVM): Algorithm and experiment. Educ. Inform. Technol. **28**(7), 8097–8111 (2023)
26. Andreev, A., et al.: Text processing using LLM for automatic creation of agricultural crops knowledge bases. BIO Web Conf. EDP Sci. **130**, 01029 (2024)
27. Shukla, S.M., Magoo, C., Garg, P.: Comparing fine tuned-LMs for detecting LLM-generated text. In: 2024 3rd Edition of IEEE Delhi Section Flagship Conference (DELCON), pp. 1–8. IEEE (2024)
28. Mehta, H., Bharti, S.K., Doshi, N.: Comparative analysis of part of speech (POS) tagger for Gujarati language using deep learning and pre-trained LLM. In: 2024 3rd International Conference for Innovation in Technology (INOCON), pp. 1–3. IEEE (2024)
29. Ali, N.F., et al.: Automated literature review using NLP techniques and LLM-based retrieval-augmented generation. arXiv preprint arXiv:2411.18583 (2024)
30. Jain, S.M., Hugging face.: Introduction to transformers for NLP: With the hugging face library and models to solve problems, pp. 51–67. Apress, Berkeley, CA (2022)
31. Agatha, V., Setyawan, I.: Web chat-based application with large language model and transformers from hugging face for self-learning on storytelling skills. In: 2024 International Electronics Symposium (IES), pp. 614–618. IEEE (2024)
32. Tan, L.: Integration of pre-trained large language model and knowledge graph embedding into topic modeling with Latent Dirichlet Allocation (2024). https://summit.sfu.ca/_flysystem/fedora/2025–02/etd23286.pdf
33. Kim, S., Lee, H., Lee, Y.: HEISIR: hierarchical expansion of inverted semantic indexing for training-free retrieval of conversational data using LLMs. arXiv preprint arXiv:2503.04141 (2025).
34. Dogru, H.B., et al.: Deep learning-based classification of news texts using doc2vec model. In: 2021 1st International Conference on Artificial Intelligence and Data Analytics (CAIDA), pp. 91–96. IEEE (2021)

35. Onah, D.F.O., Pang, E.L.L.: MOOC design principles: topic modelling-PyLDavis visualization & summarisation of learners' engagement. In: Online Conference (2021)
36. Bisong, E., Google colaboratory.: Building machine learning and deep learning models on google cloud platform: a comprehensive guide for beginners, pp. 59–64. Apress, Berkeley, CA (2019)
37. Wang, A.Y., et al.: What makes a well-documented notebook? a case study of data scientists' documentation practices in kaggle. In: Extended Abstracts of the 2021 CHI Conference on Human Factors in Computing Systems, pp. 1–7 (2021)
38. Baumer, B., Udwin, D.: R markdown. Wiley Interdisc. Rev. Computat. Stat. **7**(3), 167–177 (2015)
39. Liberty, E., et al.: Elastic machine learning algorithms in amazon sagemaker. In: Proceedings of the 2020 ACM SIGMOD International Conference on Management of Data, pp. 731–737 (2020)
40. Borra, P.: A survey of google cloud platform (GCP): features, services, and applications. Int. J. Adv. Res. Sci. Communic. Technol. (IJARSCT), 4 (2024)
41. Wang, S., et al.: Smiles-bert: large scale unsupervised pre-training for molecular property prediction. In: Proceedings of the 10th ACM International Conference on Bioinformatics, Computational Biology and Health Informatics, pp. 429–436 (2019)
42. Salloum, S., et al.: Big data analytics on Apache Spark. Int. J. Data Sci. Analyt. **1**(3), 145–164 (2016)
43. Dugré, M., Hayot-Sasson, V., Glatard, T.: A performance comparison of dask and apache spark for data-intensive neuroimaging pipelines. In: 2019 IEEE/ACM Workflows in Support of Large-Scale Science (WORKS), pp. 40–49. IEEE (2019)
44. Davoudian, A., Chen, L., Liu, M.: A survey on NoSQL stores. ACM Comput. Surv. (CSUR). **51**(2), 1–43 (2018)
45. McLean J. R. Prodigy studios: A study of the relationhip between timbre and light : Thesis. 1997.
46. Urbieta, I., et al.: Weblabel: openlabel-compliant multi-sensor labelling. Multim. Tools Appl. **83**(9), 26505–26524 (2024)
47. Sampsel, L.J.: Voyant tools. Music Ref. Serv. Quart. **21**(3), 153–157 (2018)
48. King, D.E.: Dlib-ml: A machine learning toolkit. J. Mach. Learn. Res. **10**, 1755–1758 (2009)
49. Anthony, L.: AntConc: design and development of a freeware corpus analysis toolkit for the technical writing classroom. In: IPCC 2005. Proceedings. International Professional Communication Conference, 2005, pp. 729–737. IEEE (2005)
50. Esposito, S.: Named Entity Recognition for Historical Italian Texts: Overcoming Data Limitations through Strategic Annotation and Model Adaptation. https://amslaurea.unibo.it/id/eprint/35312/1/AI_Thesis-8.pdf
51. Zhang, Y. et al.: Text-like encoding of collaborative information in large language models for recommendation. arXiv preprint arXiv:2406.03210. (2024)
52. Idri, N.: Zotero software: a means of bibliographic research and data organisation; teaching bibliographic research. Arab World English J. (AWEJ) Spec. Iss. CALL. (2) (2015)
53. Zhang, C., et al.: Lexos 2017: building reliable software in python. J. Comput. Sci. Coll. **33**(6), 124–134 (2018)
54. Rockwell, G., Sinclair, S., Radzikowska, M.: Text Analysis Portal for Research (TAPoR) Archive (2020). https://era.library.ualberta.ca/items/f2da0666-f523-44d4-a83c-fa06351a1e94
55. Mattingly, S.M., et al.: The tesserae project: Large-scale, longitudinal, in situ, multimodal sensing of information workers. In: Extended Abstracts of the 2019 CHI Conference on Human Factors in Computing Systems, pp. 1–8 (2019)

56. Kormakova, V.N., Klepikova, A.G., Lapina, M.A., Rugelj, J.: ICT competence of a teacher in the context of digital transformation of education. CEUR Workshop Proc. **2914**, 138–150 (2021)
57. Kormakova, V., Klepikova, A., Musaelian, E., Baybikowa, G., Lapina, M.: Formation of ICT Competencies of Postgraduate Students of Teacher Education Based on Interactive Techniques: CEUR Workshop Proceedings SLET-2019 – Proceedings of the International Scientific Conference Innovative Approaches to the Application of Digital Technologies in Education and Research, vol. 2494, pp. 11–21 (2019)

IT-Technologies as a Backbone Element of the Educational Process

Margarita E. Beglaryan[1,2](✉), Natalia Y. Dobrovolskaia[3], and Viktoriya N. Taran[4]

[1] North Caucasus Branch of the Russian State University of Justice, st. Krasnih Partizan, 234, 350020 Krasnodar, Russia
rita_beg@mail.ru

[2] Krasnodar Cooperative Institute (branch) Russian University of Cooperation, st. Mitrofana Sedina, 168/1, 350015 Krasnodar, Russia

[3] Kuban State University, st. Stavropolskaya, 149, 350040 Krasnodar, Russia

[4] V.I. Vernadsky Crimean Federal University, Simferopol 295007, Russia

Abstract. Modern education involves the use of information and communication technologies at different stages of the educational process. IT-technologies not only make it possible to automate the routine work of organizing the educational process, for example, scheduling, but also to improve the quality of education through the use of innovative intelligent digital technologies. A teacher who does not have specialized skills in programming and computer science sometimes loses the advantages of computer technologies due to a lack of understanding of their capabilities and place in their professional activity. The typology of IT technologies, in the context of their application in education, is designed to reveal the advantages of using computer applications and services, both in the activities of an individual teacher and for organizing the educational process as a whole. The relevance of the study is determined by the contradiction between the requirements of the state educational standard of teaching, indicating the maximum use of the capabilities of IT technologies in the educational process and the lack of pedagogical research on typologies and didactic properties of computer technologies and services. We will single out individual areas of pedagogical activity and, in accordance with them, classify various information and communication technologies and computer services. In our opinion, it is possible to define three types of application of IT-technologies in education: individual professional activity of a teacher, within the framework of which the teacher carries out the solution of their daily educational tasks; support of pedagogical activities, including technologies and services that ensure the organization of educational activities, both for the teacher and the institution as a whole; analysis of learning outcomes, designed to correct, predict and improve the quality of pedagogical influences.

Keywords: Information Technology · Digitalization of Education · Online Learning · Interaction in the Digital Space

M. Lapina et al. (Eds.): DET 2024, LNNS 1616, pp. 96–103, 2025.
https://doi.org/10.1007/978-3-032-04365-8_8

1 Introduction

In accordance with the main functions of the teacher, it is possible to distinguish the types of pedagogical activity, each of which must be supported in modern education using one or another computer technology [1, 2].

Constructive activity involves the selection of educational material, the design of educational tasks, the design of the activities of students, and, therefore, the teacher must have the skills of interacting with information retrieval systems, electronic libraries, macros for constructing tasks [3, 4].

The procedure for the exchange of information between the teacher and students is the basis of communication activities. Here one needs skills in working with text editors, spreadsheets, mail servers, instant messengers. Communication activities within educational institutions are additionally provided by specialized applications: electronic journals in schools, websites of universities and colleges. The preparation and implementation of various types of activities of students and teachers themselves are included in the organizational activities of the teacher.

This includes educational information environments of educational institutions existing at various quality levels, which include not only automation of load distribution and scheduling of classes, but also various remote and cloud services and technologies.

2 Educational Technologies of the Teacher

The assessment activity of a teacher in modern conditions is not effective without computer tests of various levels, systems and technologies for visualization and analysis of learning outcomes [6]. We will combine IT-technologies and services that provide the listed types of teacher's activities into the first type of the proposed classification - individual pedagogical activity (see Fig. 1).

All of these technologies do not require specialized skills from the teacher and make it possible to automate his professional activities and bring them to a higher quality level, immerse the teacher in a new worldview and a new philosophy of teaching [5].

The use of automatic cloud services in professional pedagogical activity not only facilitates quality control of training, but also makes the learning process interesting, varied, which increases the motivation for learning among students and schoolchildren. Another important aspect of the proposed technologies is the diversity of approaches for special learners - people with health developmental needs, which helps to improve the quality of their training and opportunities for inclusive learning process. The interest in technological personal growth is increasing, with technology making everyone equal in such a learning space.

3 IT-Technologies of Pedagogical Activity

The second type of information and communication technologies and services provides support for pedagogical activities. Basically, these are ready-made applications that a teacher or student can use in their activities (see Fig. 2). Here we can highlight the technologies of augmented reality, which make it possible to increase the motivation of

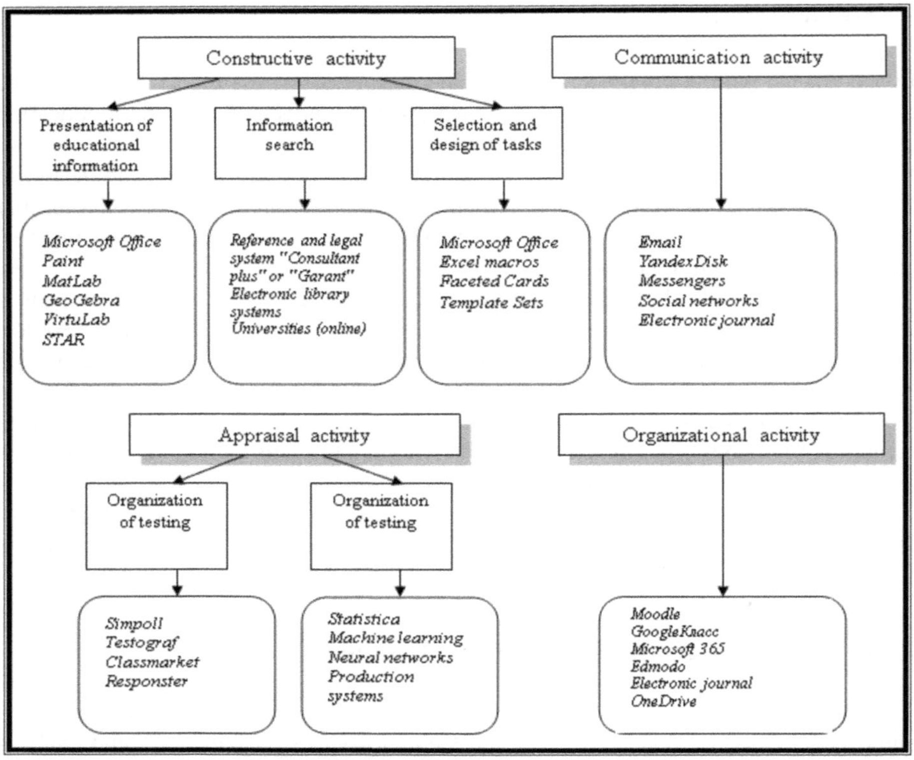

Fig. 1. Type 1. Individual pedagogical activity.

students to learn new things, immersing them in the subject area. Science disciplines are supported by virtual laboratories where learners can simulate various processes and phenomena [7]. The motivation of students is increased by 3D technologies that develop spatial thinking.

A fairly popular didactic technology that uses computer capabilities is gamification technology - learning through play. The developers offer many training programs where the study of new material and the formation of new skills is carried out using gaming technologies. Web-technology is quite popular in education. The development of various sites with educational information, remote testing systems, distance learning courses is in demand and is necessary to gain new knowledge [8]. These technologies are at the heart of the applications that the teacher can include in the educational process.

On the other hand, the teacher's professional activity is provided by a variety of programs and services that automate the organization of the educational process. These include e-journals, workload and scheduling programs, databases of teaching materials, and information retrieval systems. These are ready-made applications that improve the quality of the teacher's work.

Cloud technologies and services must be specified separately in this type of technology. Clouds are not only high-quality storage of information (YandexDisk, OneDrive),

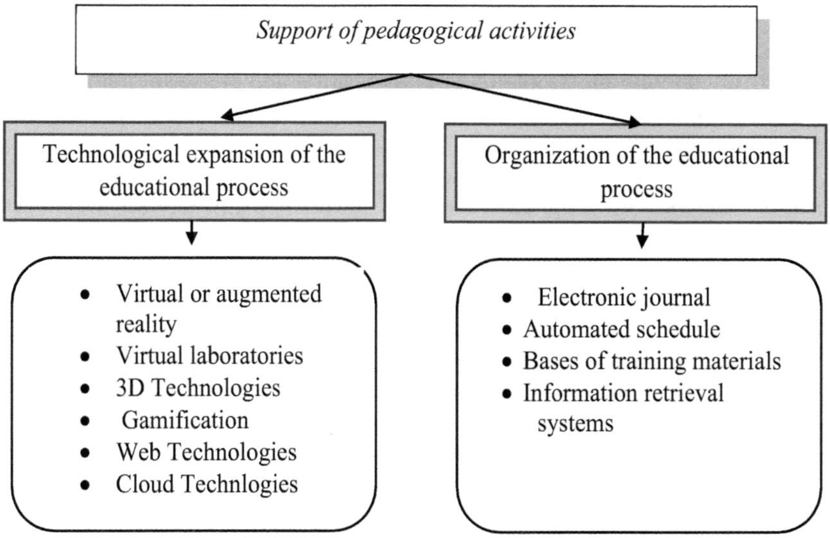

Fig. 2. Type 2. Support of pedagogical activity.

a means of data exchange (mail services, instant messengers), but also provide cloud options for typical applications: text editors, spreadsheets, graphic editors, planners, translators, geolocation programs, etc. ... Cloud technologies and services have quite powerful didactic properties: they implement distance learning, act as a repository of educational materials, allow you to create and modify educational materials of various formats, provide the ability to develop tests [9, 10]. To incorporate cloud technologies into practice, the teacher does not need programming skills; confident computer skills are enough (see Fig. 2).

4 Technologies for Assessing Knowledge

The third type of IT technologies used in teaching is aimed at analyzing learning outcomes. Here you can highlight the tools available to the teacher who does not have special knowledge in mathematical disciplines, for example, the Statistica package or visualization tools and a set of statistical formulas in Microsoft Excel spreadsheets [11, 12]. Using these tools, the teacher can calculate the main quantitative indicators of learning outcomes: the highest, lowest, average score, visualize the results in the form of diagrams and graphs. The Statistica package offers broader opportunities, for example, dividing the analyzed data into clusters of similar characteristics. Moreover, the analyzed objects can include different types of characteristics: gender, year of study, grades in a number of subjects [13].

However, for a better analysis of learning outcomes, a large sample of data and, therefore, special mathematical methods of analysis are required. In this case, Data Mining technology is used. This technology involves data mining using machine learning methods for analysis. The technology is based on basic statistical methods of analysis:

cluster analysis, regression, discriminant analysis. To improve the quality of the methods, the technology of neural networks is used, which allows training on the existing data and then effectively solving the problems of classification and clustering on the new data. The combination of these methods - statistics and machine learning - defines the Data Mining technology.

Currently, a separate area is highlighted - the intellectual analysis of educational data - Education Data Mining, designed to determine the relationship between the educational process and the efficiency of knowledge assimilation. It not only uses a set of statistical methods and machine learning technologies, but also uses psychometric algorithms. The trainees are divided into groups according to the level of assimilation of information, which makes it possible to select adapted tasks and forms of presentation of the material, and, therefore, to fulfill the principle of student-centered learning.

The use of Education Data Mining methods is necessary in the development of high-quality distance courses, when there is no possibility of direct contact with the audience, which means that there is no way to adjust the course at the time of training for a specific level of perception of the new material. The second direction in the development of Data Mining technology in the field of education is considered to be Learning Analytics. Collecting and analyzing data on learning outcomes allows predicting the level of further training of a particular student, and, therefore, adjusting or optimizing learning impacts.

When processing large volumes of unstructured data in education, as in other areas, Big Data technologies are used. To build a learner's profile, data is collected from social networks; adapting a distance course or curriculum requires a large amount of data on learning outcomes.

Big Data technologies involve the stage of data cleaning (removal of various outliers that distort further analysis), the stage of generating variables for building data models and the stage of actually building an analytical model. The data that require structuring and analysis include: personal data of the student, information on the interaction of students with remote systems, data on the effectiveness of training materials, administrative data. A general list of technologies that provide the analysis of learning outcomes (see Fig. 3).

All types of the specified classification of IT technologies have partial intersections. This is due to the fact that the applications and services under consideration perform many pedagogical tasks, for example, the university website (school website, electronic journal) solves the problems of the teacher's communicative activity, on the other hand, contributes to the organization of the educational process of the entire educational institution.

5 Materials and Methods of Classification

The proposed classification is basic and considering highly specialized education, for example, legal it is necessary to introduce the fourth type – specialized software. This includes the following technologies and services [14, 15]:

- Automation of legal practice (management of a law firm, invoicing, interaction with clients, workflow, setting tasks);
- document scanning technologies;

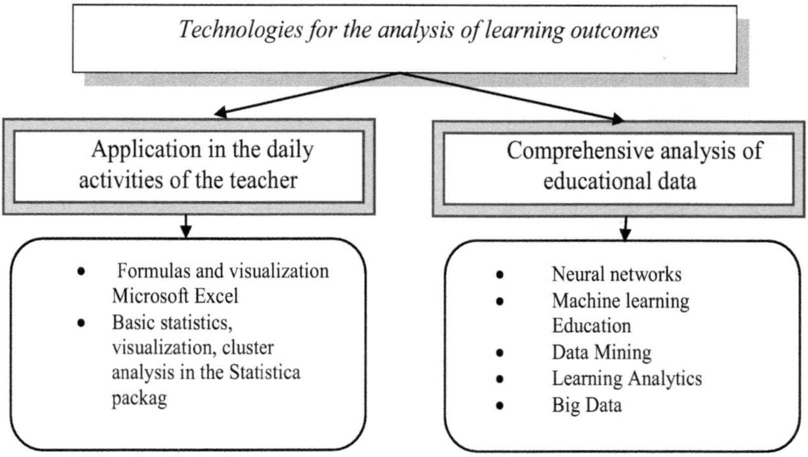

Fig. 3. Type 3. Analysis of learning outcomes.

- monitoring of registration data of the court register to protect against raider seizures and control of counterparties;
- search and text-analytical technologies for law firms (Ayfie);
- conducting judicial reporting (Epiq Systems);
- analytics based on the global patent database (Minesoft);
- intellectual base of legal documents (Rocket Lawyer Incorporated);
- Online billing and legal practice management (Rocket Matter);
- solving controversial issues with real estate (Smokeball);
- solutions that help lawyers to automate legal documentation and organize document search. (WealthCounsel, LLC).

6 Results and Conclusions

The use of the above applications, technologies and programs presupposes not only knowledge and ability to use them in practice, but also the observance of copyright or license rights for their operation. That is necessary to prepare a new galaxy of teachers, engineer-teachers who will master all the above teaching methods [16]. Today you can be an advanced user, tomorrow this is not enough for a teacher.

Law and copyright observance are reaching a new level in a new era. In the Civil Code of the Russian Federation, Art. 1274. "Free use of a work for informational, scientific, educational or cultural purposes" [17], allows you to use the product for educational purposes without purchasing a license. Does the modern software manufacturer agree with this state of affairs? If it is a demo version, then it is possible. But if it is a University or an Institute that uses this article, then a situation will arise when the software manufacturer will have to go to court or to the competent authorities, as his rights will be violated.

Financial availability of applications and technologies for education remains unresolved. Many companies understand that today a student uses the program, and tomorrow he becomes a leader and buys the same program, because he is confident in its reliability and usefulness. A striking example of such work with universities is the reference and legal giants of Russia, firms Garant and Consultant. They are doing serious work with universities, in particular, with the Russian State University of Justice. By providing legal frameworks to students, companies are secured by clients in the future.

The proposed classification reveals the didactic capabilities of IT-technologies, expands the scope of their application. The introduction of information technology into everyday pedagogical activity improves the quality of education and contributes to the training of highly qualified specialists.

Security remains the main problem of the open information society and the educational space as part of it [18–20]. That is also necessary to keep in mind the issues of collective and personal security when developing platforms and technologies for learning, excluding the possibility of pressure and manipulation of personal and other data.

References

1. Martin, F., Sun, T., Westine, C.D.: A systematic review of research on online teaching and learning from 2009 to 2018. Comput. Educ. **159**, 104009 (2020)
2. Mazzucato, A., Argiulo, A.: Enhancing students' cultural competences through embodied knowing. In: Proceedings of First Workshop on Technology Enhanced Learning Environments for Blended Education Conference - The Italian e-Learning Conference 2021. Foggia, Italy, January 21–22 (2021)
3. Markelov, K.A., Polyanskaya, E.V., Mineva, O.K., Taran, V.N.: Paradigm transformation of education system in digital reality. In: Vth International Scientific and Practical Conference "Distance Learning Technologies" (DLT-2020), pp. 186–198 (2020) http://ceur-ws.org/Vol-2914/
4. Beglaryan, M.E., Dobrovolskaia, N.Y.: Integration of learning forms as a new paradigm of the educational information space. In: Vth International Scientific and Practical Conference "Distance Learning Technologies" (DLT 2020), pp. 48–61 (2020) http://ceur-ws.org/Vol-2914/
5. Garrick, B., Pendergast, D., Geelan, D.: Introduction to the philosophical arguments underpinning personalised education. In: Theorising Personalised Education, pp. 1–16. Springer (2017)
6. Taran, V., Savelova, L., Abdullaev, J.: Increasing learning motivation through the use of educational and technological tools. In: Lapina, M., Prakasha, G.S., Grigoriev, S. (eds.) International Conference on Innovative Approaches to the Application of Digital Technologies in Education and Research. SLET 2022 Lecture Notes in Networks and Systems, vol. 1222. Springer, Cham (2025). https://doi.org/10.1007/978-3-031-78776-8_7
7. Gerasimova, E., Konopko, E., Usmanov, M.: Online tools for organizing students' teamwork. In: International Conference on Innovative Approaches to the Application of Digital Technologies in Education and Research, 2022., vol. 1222, pp. 167–173. Springer, Cham (2025). https://doi.org/10.1007/978-3-031-78776-8_17
8. Bogomolov, A.I., Nevezhin, V.P., Piskun, E.I., Larionova, M.A.: Systems of artificial intelligence in the educational process. In: Vth International Scientific and Practical Conference "Distance Learning Technologies" (DLT 2020), pp. 75–84 (2020) http://ceur-ws.org/Vol-2914/

9. Dobrovol'skaya, N.Y., Beglaryan, M.E.: Primer integracii form obucheniya kak realizaciya novoj paradigmy informacionno-obrazovatel'nogo prostranstva. In: Informacionnye sistemy i tekhnologii v modelirovanii i upravlenii: sbornik trudov V Mezhdunarodnoj nauchno-prakticheskoj konferencii. Otv. redaktor K.A. Makovejchuk, Simferopol', 2020, pp. 398–401 (2020) EDN MPTEBY
10. Sysoev, A.S., Dobrovol'skaya, N.Y.U.: Raspredelennaya sistema sbora znanij v zadachah obucheniya iskusstvennogo intellekta. In: II vserossijskoj nauchno-prakticheskoj konferencii "Prikladnaya matematika: sovremennye problemy matematiki, informatiki i modelirovaniya". Krasnodar. FGBU "Rossijskoe energeticheskoe agentstvo" Minenergo Rossii Krasnodarskij CNTI-filial FGBU "REA" Minenergo Rossii, pp. 401–405 (2020).
11. Lutsenko, L., Testova, I., Chirkov, M., Taran, V., Mezentseva, O.: Methodical specifics of the subject distance camp organization "young cryptograph". In: Lapina, M., Prakasha, G.S., Grigoriev, S. (eds.) International Conference on Innovative Approaches to the Application of Digital Technologies in Education and Research. SLET 2022 Lecture Notes in Networks and Systems, vol. 1222. Springer, Cham (2025). https://doi.org/10.1007/978-3-031-78776-8_21
12. Peters, C.A.: Statistics for Analysis of Experimental Data (2024). https://www.princeton.edu/~cap/AEESP_Statchap_Peters.pdf, last accessed 2024/10/20.
13. Saire-Peralta, E.: Predictive model for assigning exercises to students in spreadsheet functions using artificial neural networks. In: Proceedings of the International Congress on Educational and Technology in Sciences Chiclayo, Peru, November 16–18 (2021) http://ceur-ws.org/Vol-3037/paper4.pdf
14. Polishchuk, T.V., Voznosymenko, D.A.: Use of digital instruments during processing of results of pedagogical research. In: Proceedings of the Second Workshop on Technology Enhanced Learning Environments for Blended Education, October 5–6, 2021, Foggia, Italy, http://ceur-ws.org/Vol-3025/paper2.pdf
15. Rocha, R.V., Valle, P.H., Maldonado, J.C., Bittencourt, I.I., Isotani, S.S.: AIMED: agile, integrative and open method for open educational resources development. Advanced Learning Technologies (ICALT), (2017) IEEE 17th International Conference on IEEE, PP. 163–167. Retrieved from https://ieeexplore.ieee.org/document/8001747.
16. Lapina, M., Tokmakova, M., Movzalevskaya, V., et al.: Analysis of modern methods and approaches using it in education and pedagogy. In: International Conference on Innovative Approaches to the Application of Digital Technologies in Education and Research, 2022, vol. 1222, pp. 40–52. Springer, Cham (2025). https://doi.org/10.1007/978-3-031-78776-8_5
17. Civil Code of the Russian Federation: Parts 1–4, from 18.12.2006 N 230-FZ (as amended on 22.07.2024) (as amended and supplemented, entered into force on 08.08.2024). http://www.consultant.ru/document/cons_doc_LAW_64629/84bbd636598a59112a4fe972432343dd4f51da1d, last accessed 2024/10/20.
18. Zegzhda, D.P., Vasilev, Y.S., Poltavtseva, M.A., Kefeli, I.F., Borovkov, A.I.: Advanced production technologies security in the era of digital transformation. Voprosy kiberbezopasnosti [Cybersecurity issues]. **2**(26), 2–15 (2018). https://doi.org/10.21681/2311-3456-2018-2-2-15. (In Russ.).
19. Anosov, R., Anosov, S., Shakhalov, I.: Conceptual model of information technology security. Risk Anal. [Cybersecurity issues]. **2**(36), 2–10 (2020). https://doi.org/10.21681/2311-3456-2020-2-02-10
20. Butusov, I., Romanov, A.: Methodology of security assessment automated systems as objects critical information infrastructure. Voprosy kiberbezopasnosti [Cybersecurity issues]. **1**(25), 2–10 (2018). https://doi.org/10.21681/2311-3456-2018-1-2-10

Machine Learning Approaches for Detection of Cyberbullying in Virtual Space

D. Vinodha[1], E. A. Mary Anita[1], J. Jenefa[1], and Maria Lapina[2](✉)

[1] Christ University, Bangalore, India
[2] North-Caucasus Federal University, Stavropol, Russia
mlapina@ncfu.ru

Abstract. Cyberbullying, hostile behavior of a group or an individual to defame or harass the victim mentally with the help of social media and other e-communication platforms, has the potential to create a lifelong negative impact on mental health with the power of inducing suicidal thoughts. It is on the rise among the early adolescents of the age group from 8 to 16. Hence it is vital to detect Cyberbullying at an early stage to safeguard the victims at the high risk of developing depression, anxiety, and suicidal ideas. It also helps to mitigate psychological, academic, and social consequences. Existing cyberbullying detection approaches primarily depend on static monolingual questionnaires and are not personalised. With the developments in Artificial Intelligence, many neural network-based approaches are explored to detect cyberbullying. This study discusses and provides comparative analysis of various machine learning approaches for detecting cyberbullying victimization among school students highlighting their effectiveness and limitations.

Keywords: Cyberbullying · Cognitive AI Chatbot · Natural Language Processing · Deep Learning

1 Introduction

Cyberbullying and victimization are on the rise due to increased access to digital worlds like social media, online gaming platforms, and mobile phones. It has become a significant issue and major concern of mental health, particularly among adolescents, teen adults, and school children, as it leads to depression, anxiety, self-isolation, feelings of shame, learning difficulties, violent behavior, and the feelings of unsafe and suicidal thoughts or attempts. The Times of India reported that Cyberbullying cases have increased by 31% in Mumbai, India, between 2021 and 2022. According to a study by McAfee, 85% of Indian children admit that they have been bullied on online platforms. Another survey by UNICEF reports that over one-third of Indian students aged 13–15, approximately 36%, are being victimized by harassment and bullying at school campuses. The numbers are likely to be higher as the majority of the victims don't come forward to register a case. These alarming statistics highlight the urgent need for collective action to safeguard our children. The Indian government has recognized Cyberbullying as an issue of public concern and has taken the initiative by providing a helpline and portal to encourage victims to anonymously raise complaints.

M. Lapina et al. (Eds.): DET 2024, LNNS 1616, pp. 104–111, 2025.
https://doi.org/10.1007/978-3-032-04365-8_9

The major factors contributing to the rise of cyberbullying victimization include the increased usage of the internet, lack of awareness, nuclear family, lack of parental care, and lack of parental and professional guidance. Though schools and colleges have taken the initiative in the form of mentoring and counseling sessions, most of the victimization goes unnoticed due to fear, disgrace, and lack of trust in the conventional reporting system. Hence, it is essential to develop systems for detecting cyberbullying at an early stage by addressing the challenges of traditional methodologies that fail to capture evolving patterns of cyberbullying and lack personalization. the rise of machine learning based approaches has opened new avenues for enhancing the accuracy by analysing large-scale data collected in the form of text, audio, and video. These systems face significant challenges in handling non-transitional communication forms adopted by the Gen Z and Gen Alpha kids and their behaviour change.

This study provides a comprehensive review and analysis of existing approaches for detecting cyberbullying victimization among school students It highlights the research gap by examining the strengths and limitations and contributes to the development of more effective strategies for detecting and preventing cyberbullying at an early stage.

2 Literature Review

The study [1] proposes a technique using a convolutional neural network with a bidirectional long short-term memory layer, trained on a pre-processed dataset of tweets using GloVe word embeddings and the focal loss function. Syed et al. [2] created a labeled dataset of Bengali text to detect different forms of cyberbullying, which include flaming, threats, pull-a-pig, and racism. The dataset is manually collected by extracting the bullying text from the comment sections of three major OSNs, including Facebook, Instagram, and YouTube. Only the characters from Bengali are considered. Also, sentences with fewer than 3 words are ignored. A vector of size 1065 is used for extracting features using TF-ID, Bag of words. The semantic relationship is captured by employing the Keras embedding mechanism. The performance of various learning models of both machine and deep learning was evaluated and compared using precision, recall, F1-score and accuracy. The experiment is also extended to three transformer models namely BERT, BanglaBERT, and XLM-RoBERTa by fine-tuning the epoch size, batch size, learning rate, and dropout rate. From the results, it is shown that XLM-RoBERTa outperforms compared to all the other learning models for the given dataset.

Nath et al. [3] proposed a bidirectional Long Short Term Memory-based deep learning model with two layers for identifying Cyberbullying using Bengali text. The dataset is created by merging the data from Kaggle and Mendeley with manually added Bengali comments. The comments are labeled with two broad categories as toxic and non-toxic. The toxic comments include the cyberbullying text category of threat, racism, obscenity, and insult. As part of the cleaning process, emojis were removed. To avoid issues arising from sentences of varied length, all the sentences are converted to a length of 80 using padding. The authors proposed a model of 3 layers: embedding, two bidirectional LSTMs, and two dense layers. Embedding layer is used to reduce the dimension of input and helps the model to understand the relationship between the inputs by converting every word into a vector of size 200. Two LASTM layers are made up of 64 and 32 neurons. The first dense layer is made up of 64 neurons of the ReLU activation function to inactive the neurons present in the previous layer, and the second dense layer has only one neuron with a sigmoid function for binary classification. Three different optimizers are employed to improve the performance of the model by varying the learning rate and weight of each layer.

Andrea Pereraa et al. [4] proposed a supervised learning model by employing a support vector machine, Naïve Bayes, and Logistic regression models for detecting cyberbullying of categories including race, physical, politics, and sexuality. The sentences are broken down using conjunction words during the preprocessing phase to improve accuracy. The model employs lexicon-based sentiment analysis to categorize English words as positive and negative. Only the negative words are given as input to the learning model.

Batista et al. [5] proposed an approach to detect cyberbullying in an improved way by fine-tuning the sentence transformer language model. The authors attempted to generate meaningful sentence embeddings by creating many training datasets. Each dataset comprises a pair of sentences of bullying and nonbullying types. This helps the model to recognize the multiple layers of meaning related to cyberbullying that require careful consideration or analysis. The model is subjected to automatic and semantic evaluation. In addition, impact of preprocessing on the performance of the model is analyzed.

The authors in [6] proposed a deep learning model based on CNN-BiGRU for detecting cyberbullying text in Tamil and Tanglish-based social media communication. CNN-BiGRU, a hybrid neural network learning model, captures the local features and long-range dependencies by combining the strengths of both CNN and BiGRU. The CNN applies a convolution to extract local patterns and reduce dimensionality for BiGRU, which captures the temporal features by reducing vanishing gradient issues.

The authors in [7] proposed a system with three components to detect, protect, and analyze cyberbullying. The first component is designed using a machine learning model to classify the message as bullying or not bullying, and the evaluated results of nine different models show that the Stochastic Gradient Descent model outperforms the other models. The second component is to develop an education adopter by leveraging the benefits of chatbots to provide education on cyberbullying prevention mechanisms. The third component is developed using a hierarchical clustering approach using the BERTopic embedding mechanism to cluster bullying text of different themes and assist the chatbot in providing more personalized answers.

Roehrer et al. in [8] highlighted the co-design process for developing a chatbot that provides support for students to deal with online harm. The process has focused on both students and pastoral care teams and incorporated diverse perspectives to identify key findings for the chatbot design. Some of the key findings include privacy, emotional support, and personalized engagement valued by the students. Building trust and interest among students in chatbots and enhancing the role of chatbots in assisting adolescent victims are identified as major challenges.

The authors in [9] proposed a work to recognize 7 kinds of emotions from the facial expressions of persons using machine learning models. The authors created a dataset to train the models by considering only the top 10 enhanced frames extracted at every other second. During the preprocessing process, Haar cascade face detection techniques are used and faces are aligned to improve the feature extraction in later phases. The dataset is used to train CNN,RNN, SVM and KNN. The results show that the accuracy of each model is 92%,95%, 905 and 87% respectively.

The authors in [10] proposed ResNet-based learning multi-modal for recognizing of nine different emotions by fusing audio-video attention models. The multi-head visual spatial attention model is employed to select the frame region that carries the salient feature of emotion recognition. Following this, a multi-head visual channel attention model is developed to select a meaningful feature map. Also, to discriminate the vital frames from unimportant ones, the temporal attention model is employed. Also, a novel loss function that considers the semantic relationship between emotions is formulated to guide the attention models. The proposed model exhibits an accuracy level of 89.25% and 84.57% while testing over the data set of RAVDESS and CREMA-D.

Salas-Cáceres et al. in [11] proposed an LSTM-based multimodal facial emotion recognizer by fusing visual and audio embeddings using a Multimodal concatenation approach. To ensure effective fusion, the two audio and video modalities are correlated using the EmbraceNet module consisting of the docking and embracement layer. These two layers ensure the embedding vectors are the same in size and combine them into a single vector. The model has classified the prediction with accuracy 88.1%, 86.8%, and 80.3% while testing against the three popular datasets RAVDESS, CREMA-D and SAVEE respectively.

Das et al. [12] proposed another multimodal emotion recognizer for the Bengali language. The authors employed three different hybrid mechanisms for extracting features from textual, visual, and audio modalities. To facilitate the interaction among these modalities, two forms of cross-attention methods are created and finally concatenated to fuse the features together. The performance of the work is showcased using F1 score with a value of 0.64

Another multimodal emotion recognizer is proposed by Zhang et al. [13] by extracting information from audio and textual content using LSTM and BERT-based learning models. To enhance the recognizer to capture the salient features from both modalities, the authors employed three levels of attention mechanisms. In the first level, local salient features are learned separately from audio and text sequences. In the second level, joined learning is done using a cross-modal attention method. This is followed by the global modal attention to extract the final information about the emotion. The model is experimented on over the two data sets IEMOCAP and MELD.

The survey on emotion recognition from facial expression and body expression by the authors in [14] highlighted that more emotions are revealed by facial expressions than the body gesture. Two algorithms, including supervised least squares multiset kernel canonical correlation analysis (SLAM KCCA) and sparse supervised least squares multiset kernel canonical correlation analysis (SSL SMACCA), are presented in [15] to carry out efficient facial expression, speech, and body gesture feature fusion.

3 Analysis

Comparative analysis of the existing systems based on the methodology adopted and types of cyberbullying identified is presented in Tables 1 and 2.

Table 1. Comparative Analysis.

Existing Research works	Machine Learning Models adopted	Dataset	Types of cyberbullying
Koshiry et al. [1]	CNN with Bi-LSTM	Kaggle, Twitter	Binary
Syed et al. [2]	Convolutional Neural Network, Long Short-Term Memory, Bidirectional LSTM, m-BERT, BanglaBERT, and XLM-RoBERTa	Labeled Bengali Text (Size)	Flaming, Pull-a-pig, Threat and Racism
Nath et al. [3]	Bi-LSTM	12282 versatile comments	Binary
Andrea Pereraa et al. [4]	SVM, Naïve Bayes, Logistic regression	Author-generated dataset	Racism, physical, politics and sexuality
Batista et al. [5]	Sentence BERT	Bullying 3.0, Myspace, Hate Speech	Binay (bullying,not bullying)
Lian et al. [7]	Regression Analysis	Author-generated dataset	Significant relationship between the COVID-19 pandemic's influence and cyberbullying
Roehrer et al. [8]	SVM, Naïve Bayes, Logistic regression	Author-generated dataset	Aggression, Victimization
Mocanu [10]	3DCNN(3D Resnet), 2D(ResNet) CNN integrated with spatial, channel and temporal attention mechanism Audio-video cross-attention fusion	RAVDESS,CREMA-D	Emotion classification
Salas-Cáceres et al. [11]	LSTM	RAVDESS,CREMA-D, SAVEE	Emotional categories such as anger, fear, joy, and sadness
Das et al. [12]	Video: ResNet-50, ViT, DeepFace, FaceNet, and OpenFace Audio: CNN, YAMNet, VGGish, DeepSpectrum Text: mBERT, XLM-R, Bangla BERT	MAViTE-Bangla dataset	Emotion classification
Zhang et al. [13]	Audio BiLSTM Text BERT Cross model attention	IEMOCAP and MELD	Emotion classification

Existing studies have investigated the prevalence of cyberbullying and associated social influencing risk factors by conducting a self-administered questionnaire with a static set of questions. They are not personalized. It is also monolingual and available only in text mode. The students of the current IT era may not find it convenient and secure to

Table 2. Comparative Analysis – Strength and Limitations.

Existing Research works	Strength	Limitation
Koshiry et al. [1]	Use of Focal Loss helps to mitigate class imbalance, ensuring that the model does not become biased towards the majority class	Low recall value- potential false negatives, imbalance class
Syed et al. [2]	Effective cyberbullying detection on Social media platforms like Facebook, Youtube & Instagram	Only for Bengali language, Removal of emoji
Nath et al. [3]	High accuracy	Only for Bengali language, Removal of emoji
Andrea Pereraa et al. [4]	Used real-world data, Identifies Additional Influences	Limited Sample Size, Biased Responses, Context-Specific Findings
Batista et al. [5]	Explores Social Factors - social dominance orientation and perceived social injustice	Limited to Workplace Context
Lian et al. [7]	Real-world Data, Regional Contribution	Limited to Malaysia, Sample Size Constraints
Roehrer et al. [8]	Focus on Cyber Gossip, real-world data	Self-Report Bias
Mocanu [10]	high accuracy, temporal sensitivity	potential overfitting
Salas-Cáceres et al. [11]	novel loss functions that takes into account the relationships between discrete emotion classes	potential overfitting and computational complexity
Das et al. [12]	The multimodal approach with hybrid feature extraction results enhancement in the robustness of emotion recognition	moderate F1 score and computational complexity
Zhang et al. [13]	highlights the challenges related to feature learning and fusion	Limited Discussion on Generalization

share their bullying experience. Consequently, students may be reluctant to disclose their experiences, fearing a lack of confidentiality. Also, the static set of questions focuses only on a fixed set of influencing factors and may fail to detect emerging dynamic influencing factors in the rapidly evolving digital era. The target sample size in these studies is often limited, further constraining the generalizability and depth of insights gained.

Body postures convey emotional intensity more effectively than facial expressions or words, especially in common settings like classrooms, playgrounds, or mentoring sessions. Thus, to accurately recognize emotions linked to cyberbullying victimization, it is essential to consider facial expressions, body postures, and tone of voice together.

However, the survey reveals that research on body gesture emotion recognition is limited. Also, these gestures are significantly influenced by the variation in culture, environment, and social context that affects non-verbal communication. Existing emotion-recognizing AI systems lack contextual understanding and fail to recognize emotions correctly when used in multilingual and culturally diverse regions like India, as they are trained using biased Western datasets. There is also a lack of consideration for non-standard forms of communication prevalent among early adolescents, such as memes, GIFs, and emojis, which are often used in cyberbullying contexts. Hence it is important to design cross-cultural emotion recognition systems trained using non-standard dataset forms.

4 Conclusion

The study analysed the collection of existing machine learning approaches for detecting cyberbullying using text, audio, and video inputs. The comparative analysis of existing approaches based on the machine learning model and the data sets used for classifying victimization is tabularized. The major research gaps including a lack of personalization, underutilization of body posture for analyzing the emotional behavior changes and limitations in providing personalized mentoring plans are highlighted.

Most of the existing research emphasizes only detection rather than intervention and the provision of actionable insights and personalized mentoring plans to tailor individual emotional states and contexts. Addressing these gaps will significantly enhance the effectiveness of emotion recognition systems in combating cyberbullying.

References

1. El Koshiry, A.M., Eliwa, E.H.I., Abd El-Hafeez, T., Khairy, M.: Detecting cyberbullying using deep learning techniques: a pre-trained glove and focal loss technique. PeerJ Comput. Sci. **27**(10), e1961 (2024 Mar). https://doi.org/10.7717/peerj-cs.1961. PMID: 38660150; PMCID: PMC11042001
2. Sihab-Us-Sakib, S., Rahman, M.R., Forhad, M.S., Aziz, M.A.: Cyberbullying detection of resource constrained language from social media using transformer-based approach. Nat. Lang. Process. J. **9**, 100104., ISSN 2949-7191 (2024). https://doi.org/10.1016/j.nlp.2024.100104
3. Nath, S.S., Karim, R., Miraz, M.H.: Deep learning based cyberbullying detection in Bangla language. Ann. Emerg. Technol. Comput. (AETiC). **8**(1), 50–65 (2024)
4. Pereraa, A., Fernandob, P.: Cyberbullying detection system on social media using supervised machine learning. Procedia Comp. Sci. **239**, 506–516 (2024)
5. Gutiérrez-Batista, K., Gómez-Sánchez, J., Fernandez-Basso, C.: Improving automatic cyberbullying detection in social network environments by fine-tuning a pre-trained sentence transformer language model. Soc. Netw. Ana. Min. **14**, 136 (2024). https://doi.org/10.1007/s13278-024-01291-0
6. Gulecha, R.S., Neelamegam Rajaram Subramanian, S.K., Abirami, S.: Offensive text detection for Tamil language. In: Chakravarthi, B.R., et al. (eds.) Speech and Language Technologies for Low-Resource Languages. SPELLL 2023. Communications in Computer and Information Science, vol. 2046. Springer, Cham (2024). https://doi.org/10.1007/978-3-031-58495-4_16

7. Lian, A.T., Costilla Reyes, A., Hu, X.: CAPTAIN: an AI-based Chatbot for cyberbullying prevention and intervention. In: Degen, H., Ntoa, S. (eds.) Artificial Intelligence in HCI. HCII 2023 Lecture Notes in Computer Science, vol. 14051. Springer, Cham (2023). https://doi.org/10.1007/978-3-031-35894-4_7
8. Roehrer, E., Pokawinkoon, P., Watters, P., Sauer, J.D., Scanlan, J.: Adolescent-centric design of an online safety Chatbot. J. Comput. Inf. Syst., 1–14 (2024). https://doi.org/10.1080/08874417.2024.2401991
9. Prathwini, P.: DeepEmoVision: unveiling emotion dynamics in video through deep learning algorithms. Int. J. Adv. Comput. Sci. Appl. (IJACSA). **15**(3) (2024)
10. Mocanu, B., Tapu, R., Zaharia, T.: Multimodal emotion recognition using cross-modal audio-video fusion with attention and deep metric learning. Image Vis. Comput. **133**, 104676 (2023)
11. Salas-Cáceres, J., Lorenzo-Navarro, J., Freire-Obregón, D., et al.: Multimodal emotion recognition based on a fusion of audiovisual information with temporal dynamics. Multimed. Tools Appl. (2024). https://doi.org/10.1007/s11042-024-20227-6
12. Das, A., Sarma, M.S., Hoque, M.M., Siddique, N., Dewan, M.A.A.: AVaTER: fusing audio, visual, and textual modalities using cross-modal attention for emotion recognition. Sensors. **24**(18), 5862 (2024). https://doi.org/10.3390/s24185862
13. Zhang, S., Yang, Y., Chen, C., Liu, R., Tao, X., Guo, W., Yicheng, X., Zhao, X.: Multimodal emotion recognition based on audio and text by using hybrid attention networks. Biomed. Sig. Process. Control. **85**, 105052., ISSN 746-8094 (2023). https://doi.org/10.1016/j.bspc.2023.105052
14. Leong, S.C., Tang, Y.M., Lai, C.H., Lee, C.K.M.: Facial expression and body gesture emotion recognition: a systematic review on the use of visual data in affective computing. Comput. Sci. Rev. **48**, 100545., ISSN 1574-0137 (2023). https://doi.org/10.1016/j.cosrev.2023.100545
15. Yan, J., Li, P., Du, C., Zhu, K., Zhou, X., Liu, Y., Wei, J.: Multimodal emotion recognition based on facial expressions, speech, and body gestures. Electronics. **13**, 3756 (2024). https://doi.org/10.3390/electronics13183756

Digital Competencies in the Structure of Training of Advertising and Public Relations Specialists

Elena Ezhova[1(✉)], Alisa Kushchenko[1], E. A. Mary Anita[2], and Elena Semikovennova[1]

[1] North Caucasus Federal University, Stavropol, Russia
eezhova@ncfu.ru

[2] CHRIST (Deemed to be University), Bangalore, India

Abstract. The article examines the problem of the formation of digital competencies in the training of specialists in the field of media communications. An extensive review of the literature on the issue under study is given. The article examines the trends in the transformation of the profession of a communicator in the context of the intensive development of the latest technologies in the field of big data, gamification, artificial intelligence, etc. The article presents the research results on the basis of which an innovative educational program for training specialists in the field of advertising and public relations has been developed. The program includes modules related to general digital literacy, artificial intelligence resources, management in the digital environment, Internet marketing, technologies for working on various media platforms, as well as project activities using digital technologies.

Keywords: Higher Education · Digitalization · Educational Program · Media Communications · Neural network · Gamification · Project-based learning

1 Introduction

Today, the mission of educational programs in the field of advertising and public relations is to train specialists with flexible competencies related to the ability to work in a team and make informed effective decisions, and the ability to adapt to the rapidly changing conditions of the innovative post-industrial economy. At the same time, the most important technological competence of a specialist in the communications industry is the possession of skills for the full cycle of production and broadcasting of a communication product in various media environments.

Artificial intelligence methods and machine learning algorithms are currently used both in manufacturing, to improve automation efficiency, and in medicine, medicine, and financial structures for big data analysis. With the growing popularity of language models and the launch of open source GPT, artificial intelligence methods have entered even more areas of human activity [24, 25, 27].

M. Lapina et al. (Eds.): DET 2024, LNNS 1616, pp. 112–120, 2025.
https://doi.org/10.1007/978-3-032-04365-8_10

The transformation of the profession of a communicator today is primarily related to the processes of digitalization, including the introduction of gamification and neural network technologies, big data analysis technologies, etc. The dynamics of these transformations is very intense, which necessitates the constant updating of educational programs for training specialists in the media communication industry.

The theory of cognitive learning focuses on the cognitive activity of the student and the formation of appropriate mental models. Students learn fundamental concepts from their teacher or form learning resources, and then use logical deduction to gain new knowledge [26].

2 The History of the Issue

The problem of digitalization of higher education has been the focus of scientific understanding in recent years. Thus, M. E. Weindorf-Sysoeva and M. L. Subocheva (2023) [1], P.A. Petryakov, M.N. Pevzner, A.S. Shustrov (2024) [2] devoted their research to the study of trends in the development of the didactics of the digital age. A large number of scientific studies in the post-covid period are devoted to the issues of distance and blended learning. In the works of U. S. Zakharova, K. A. Vilkova, R. N. Abramov (2023) [3], A. D. Melnik (2023) [4], distance education is considered as a resource for students to adapt to instability and rapidly changing living conditions in the future. The impact of digitalization of higher education on the development of a person's creative potential is considered in the works of N. B. Kirillova (2023) [5], N. A. Simchenko, G. V. Astratova and V. V. Klimuk (2023) [6], E. N. Shutenko, A. I. Shutenko (2023) [7]. O. L. Podlinyaev and S. V. Mindeeva (2022) [8] analyze the problem of the importance of media competence of young scientists in research activities. Part of the work is devoted to the use of digital technologies in practice-oriented and project-based activities of students, for example, these researches: T. A. Boronenko, A. V. Kaisina and V. S. Fedotova (2024) [9], N. I. Isupova, E. V. Soboleva, G. A. Kobeleva and O. Yu. Zaslavskaya (2023) [10], M. Lapina, N. Gorbunova, M. Tokmakova, V. Movzalevskaya and W. H. Elashmawi (2025) [11].

In the last few years, scientists have been reflecting on the problem of using artificial intelligence in the higher education system, which is the subject of the works of M. Ararat-Isaeva, M. Ararat-Isaev, M. Tokmakova, and V. Lapin (2025) [12], P. V. Sysoev (2023) [13].

Of particular interest to our research are works that are devoted to the problem of using digital technologies in the system of training specialists in the communications industry. Let us note the research of D. V. Zubko, S. E. Kalyuzhnaya, Yu. V. Klyuev and T. S. Tarakanova (2023) [14], E. A. Popova, L. P. Plekhanova и A. A. Lyulushin (2024) [15], Ju. V. Artemova, A. A. Dyakina, N. V. Osipova и L. N. Shcherbatykh (2022) [16] in this regard. The works of A. I. Chishevich (2023) [17], V. V. Ilyicheva, K. E. Kuzmenkova and E. B. Gromova (2022) [18], A. B. Menzhurova and B. Yu. Erdyneev (2023) [19] are devoted to the analysis of digital competencies of media specialists, which are most in demand in modern conditions.

3 The Demand for an Innovative Training Program for Specialists in the Field of Advertising and PR

In order to meet global trends in the innovative economy, we have developed and implemented an educational program on advertising and public relations with the profile "PR and advertising on the Internet", since it is the Internet environment that the main range of communications is now moving to.

The need to develop a new educational program focused on training public relations and advertising specialists for innovation in the digital space of mass media communication was identified based on the analysis and formulation of requests at various levels:

1. a government order formulated in state educational standards and other regulatory documents;
2. a regional order aimed at eliminating the imbalance of supply and demand in replenishing human resources;
3. employers' requests for personnel with a certain set of competencies;
4. requests from applicants and their parents.

Online surveys, as well as constant contacts with representatives of business communities, ministries and departments, leading media in the region as key partners of the educational program revealed their interest and need for graduates of the educational program who possess the skills to use innovative digital technologies in the field of advertising and public relations.

The Atlas of New Professions, developed by the Moscow School of Management Skolkovo, presents professions that correspond to the content of the educational program and are included in the structure of training for the new educational program "PR and Advertising on the Internet" in terms of their competencies [22, 23]. Among the professions that will be in high demand in the medium and long term, such as "corporate social responsibility consultant", "smart advertising designer", "semantic field producer", "private content maker", "online sales management manager", "personal brand manager", "corporate anthropologist", "moderator of user communities", "developer of media programs", "editor of content aggregators", "brand manager of spaces" [20].

We have previously described the results of a study based on the content analysis of the Head Hunter online recruitment site and open data from the social media API in order to analyze the most sought-after competencies of specialists in the field of media communications and identify the most typical offers in terms of working conditions [21]. As a result of a study of large databases (more than 6.5 thousand vacancy units), the following conclusions were drawn:

1. the hypothesis has been confirmed that professions related to the use of digital technologies are becoming particularly important in the communication sphere today: smm manager, copywriter, brand manager, targetologist;
2. among the highly professional skills (Hard Skills), such competencies as text creation (copywriting), writing articles, brand promotion, and marketing research turned out to be the most in demand;

3. among the highly professional digital skills (IT Skills), such competencies as smm, working with social networks, targeted advertising and other types of online advertising, content creation, working in Adobe Photoshop, web design turned out to be the most in demand;
4. among the narrow professional skills (Soft Skills), such competencies as competent oral and written speech, knowledge of English, creativity, teamwork, and business communication skills turned out to be in high demand;
5. the most sought-after skills of working in various social networks. social networks and messengers turned out to have skills in Telegram, VKontakte;
6. it was found that the most sought-after positions in the communications industry were: smm manager – 31.2%, copywriter – 21.1%, brand manager – 15.4%. At the same time, the most highly paid professions belong to the "External PR" group: creative director, art director, creator" [21 , p. 58];
7. about 70% of vacancies involve full–time employment, 24.6% - remote work. However, the average salary does not change significantly. The largest number of vacancies with remote work conditions are in the Advertising and Digital Communication segments, which are directly related to the use of Internet technologies.

4 Competitive Advantages of an Innovative Educational Program

The course of study "Advertising and public Relations" is currently being implemented in 192 universities in Russia, many of which have digital profiles. The field of training "Advertising and public relations" is also being implemented in foreign universities. For example, in the UK, the University of Greenwich (London) is studying for the profile "Advertising and Digital Marketing Communications BA"; in Italy, The International University of Languages and Media (IULM) (Milan) is implementing a bachelor's degree program in the profile "Communication, Media and Advertising"; in the USA at Iowa State University (Moscow The profile "B.S. in Public Relations" is being implemented.

The main advantages of the proposed educational program "PR and advertising on the Internet" are:

1. focus on the development of fundamental, applied and research competencies in the field of public relations, advertising in the digital environment;
2. a symbiosis of classical (traditional) and project-oriented learning models: using gamification, neural network, and other technologies, etc.;
3. the program focuses on the needs of the region in human resources for the media industry.

In order to update the content of the educational program, the following "digital" modules have been integrated into the curriculum (see **Ошибка! Источник ссылки не найден.**):

1. **Module related to general digital literacy:**
   (a) Digital literacy and data processing
   (b) Information technologies
   (c) Personal cybersecurity
2. **Module related to the study of artificial intelligence resources:**

   (a) Artificial intelligence systems
   (b) Artificial intelligence in the professional field
3. **Module related to management in the digital environment:**
   (a) Organization and conduct of communication campaigns
   (b) Reputation management and crisis management communications
4. **Module related to Internet marketing:**
   (a) Integrated marketing communications
   (b) Online Marketing
   (c) Web analytics in advertising and public relations
   (d) Website design and SEO promotion
   (e) Online advertising
   (f) Targeting
   (g) SMM and Community Management
   (h) Typology of promotional content
   (i) Digital copywriting
   (j) Brand promotion technologies in the digital environment
5. **Module related to project activities using digital technologies:**
   (a) Fundamentals of graphic design in advertising and public relations
   (b) The latest technologies for the production of audiovisual content
   (c) Multimedia directing
   (d) Podcasting
6. **Module related to project activities using digital technologies:**
   (a) Fundamentals of project activities
   (b) Teamwork and effective communication
   (c) Design workshop on advertising photography
   (d) Design workshop for photo and video production

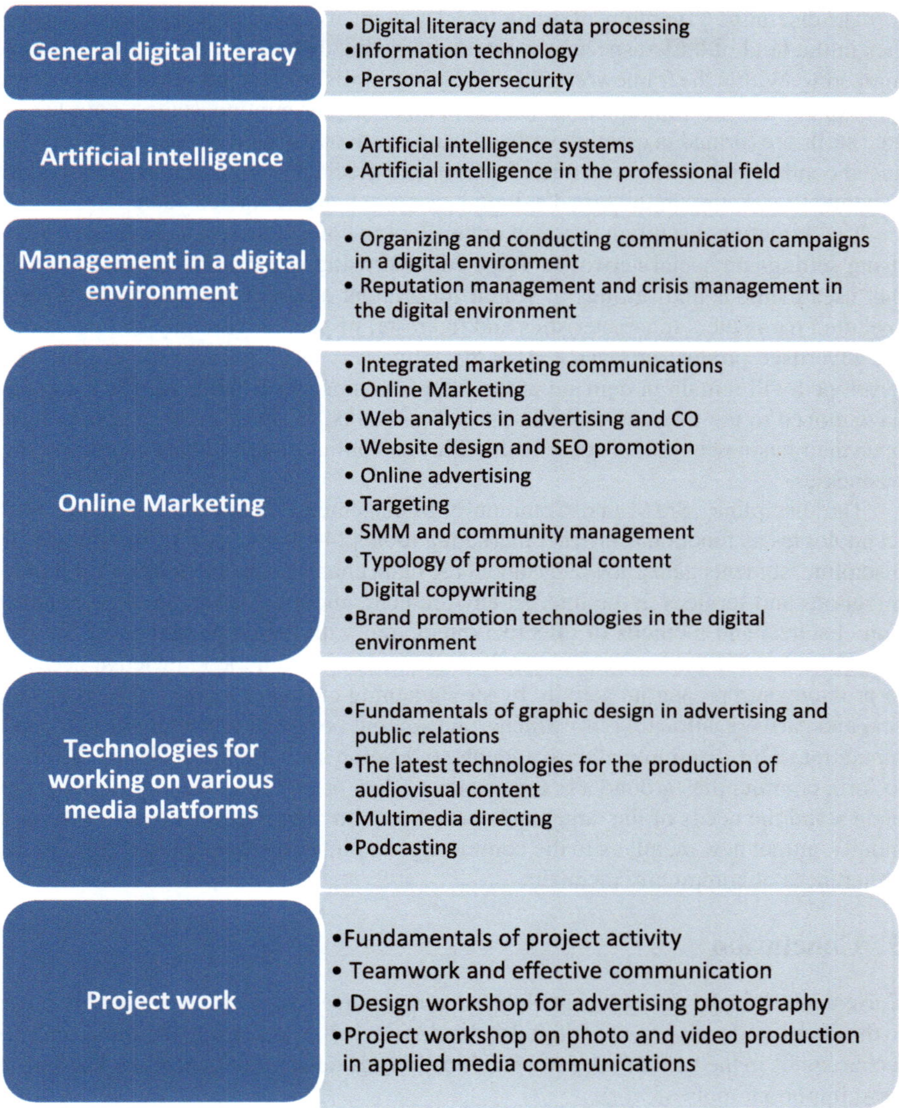

Ошибка! Источник ссылки не найден. Modules of the curriculum of the educational program on advertising and public relations

Let's look at the content of some of the disciplines.

Thus, the discipline "Website Design and SEO Promotion" is aimed at mastering the basic principles of Web technologies for creating and publishing static and dynamic websites on the Internet. Within the framework of this discipline, students study the basics of website design; acquire practical skills in developing and formulating SEO text and its implementation into the body of the site; website promotion using SEO, etc.

The discipline "Typology of promotional content" is aimed at developing competencies in the field of basic aspects of content marketing as one of the key areas in digital marketing. Within the framework of the discipline, skills are developed to develop various types of content: informational, educational, involving, promoting, selling, entertaining, etc.; skills are formed in creating verbal, visual and acoustic content, as well as their balanced combination in the structure of a creolized media text. In addition, issues related to content marketing in the brand's digital ecosystem are being considered.

The Targeting discipline is aimed at developing competencies in the field of advertising settings on social networks. Targeted advertising is a type of online advertising that uses methods and settings to search for a target audience in accordance with the specified parameters (characteristics and interests) of people who may be interested in the advertised product or service. This discipline is relevant today, and the skill being developed will remain in demand in the future, as most advertising and PR campaigns have moved to the Internet, including social networks. In addition, most entrepreneurs open their businesses online or use both offline and online environments to promote their businesses.

The discipline "SMM and Community Management" is aimed at studying SMM technologies as functional Internet marketing tools [24–29]. As part of the study of the discipline, students gain knowledge about the main channels and techniques for promoting goods and services in the Internet environment, about the selection of communication channels and methods of effective interaction with the target audience. Studying the principles of community management allows you to gain the knowledge necessary to promote entrepreneurial activity by strengthening customer loyalty, attracting sponsors and partners, attracting investments and solving problems related to the company's reputation. This discipline allows students to develop such competencies as the ability to form communities around a brand, product, idea, or problem, identify brand values, understand the needs of the target audience, create an organizational chart of the community, attract new members to the community, and track the dynamics of requests and preferences of community members.

5 Conclusion

Current trends in the development of the telecommunications sector are primarily related to the intensive development of digital technologies. In the profession of a communicator, technologies in the field of big data, gamification, artificial intelligence, etc. become the most important tools of work.

In this situation, there is a need in the higher education system in the field of media communications to respond promptly to technological innovations and train new personnel with the competence to flexibly respond to the demands of the time.

The transformation of educational programs in the field of Advertising and Public Relations is based on the principle of combining classical (traditional) and project-oriented learning models using gamification, neural network, etc. technologies, etc. The innovative educational program included modules related not only to general digital literacy, but also artificial intelligence resources, management in the digital environment, Internet marketing, technologies for working on various media platforms, as well as project activities using digital technologies.

This transformation makes it possible to train personnel capable of adapting their skills to the new needs of the economy in conditions of dynamic technological processes and global instability.

References

1. Weindorf-Sysoeva, M.E., Subocheva, M.L.: Didactics of the digital age: some aspects of development. Bull. Tomsk State Univ. **490**, 160–168 (2023)
2. Petryakov, P.A., Pevzner, M.N., Shustrov, A.S.: Digital generation in the assessment of modern university students: myths and reality. Prospects Sci. Educ. **6**(72), 53–72 (2024)
3. Zakharova, U.S., Vilkova, K.A., Abramov, R.N.: Application of MOOE in teaching in Russian universities: why is there no growth in demand for them in the conditions of forced transition to distance learning? Higher Educ. Russia. **32**(2), 125–148 (2023)
4. Melnik, A.D.: The role of blended learning in higher education: students' "first steps" to unstable work and life. Integrat. Educ. **27**(3), 506–521 (2023)
5. Kirillova, N.B.: Impact of digital culture on shaping young people's creative activity. Perspect. Sci. Educ. **2**(62), 10–22 (2023)
6. Simchenko, N.A., Astratova, G.V., Klimuk, V.V.: Creative human capital and assessment of its manifestation in organizational behavior in the context of digitalization of higher education. Perspect. Sci. Educ. **6**(66), 647–672 (2023)
7. Shutenko, E.N., Shutenko, A.I.: Mechanisms for activating the personal potential of students in the context of digitalization of university education. Higher Educ. Russ. **32**(10), 91–114 (2023)
8. Podlinyaev, O.L., Mindeeva, S.V.: Media competence in the system of student research work. Bull. Tomsk State Univ. **485**, 177–186 (2022)
9. Boronenko, T.A., Kaisina, A.V., Fedotova, V.S.: Transdisciplinary understanding of the fundamental and practice-oriented nature of pedagogical education. Perspect. Sci. Educ. **3**(69), 27–43 (2024)
10. Isupova, N.I., Soboleva, E.V., Kobeleva, G.A., Yu, Z.O.: Using the interactive multifunctional cloud portfolio for constructing and implementing individual educational routes for students in the project activities. Perspect. Sci. Educ. **2**(62), 641–657 (2023)
11. Lapina, M., Gorbunova, N., Tokmakova, M., Movzalevskaya, V., Elashmawi, W.H.: Features of the project approach in information security training. In: International Conference on Innovative Approaches to the Application of Digital Technologies in Education and Research, pp. 313–322. Springer, Cham (2025)
12. Ararat-Isaeva, M., Ararat-Isaev, M., Tokmakova, M., Lapin, V.: Experience of using Chatbots in education: optimization of surveys and development of multidisciplinary projects. In: International Conference on Innovative Approaches to the Application of Digital Technologies in Education and Research, pp. 275–283. Springer, Cham (2025)
13. Sysoev, P.V.: Artificial intelligence in education: awareness, readiness and practice of using artificial intelligence technologies in professional activities by higher education teachers. Higher Educ. Russ. **32**(10), 9–33 (2023)
14. Zubko, D.V., Kalyuzhnaya, S.E., Klyuev, Y.V., Tarakanova, T.S.: University media education in the digital age: methods and practices of training future specialists in communication professions. Prospects Sci. Educ. **3**(63), 585–602 (2023)
15. Popova, E.A., Plekhanova, L.P., Lyulushin, A.A.: Characteristics of modern network communication in the aspect of formation of media and information literacy of students (using the example of the analysis of media texts of social networks). Perspect. Sci. Educ. **2**(68), 41–61 (2024)

16. Artemova, J.V., Dyakina, A.A., Osipova, N.V., Shcherbatykh, L.N.: Media education strategy in training future professionals for the digital economy: experience of theory and regional practice. Perspect. Sci. Educ. **5**(59), 547–565 (2022)
17. Chishevich, A.I.: Digitalization in the context of development of competencies in journalism education. In: Multilevel Language Training in the Conditions of a Multicultural Society: Proceedings of the X International Scientific and Practical Conference, 2023. Kazan, pp. 443–448 (2023)
18. Ilyicheva, V.V., Kuzmenkova, K.E., Gromova, E.B.: Peer learning and edutainment technologies in the development of professional competencies of students of media specialties. Bulletin of Peoples' Friendship University of Russia. Series: Literary criticism. Journalism. **27**(1), 158–170 (2022)
19. Menzhurova, A.B., Yu, E.B.: Professional competencies of a media specialist in the digital age. In: Digital University: Improving the Organization of Training and the Strategy for the Development of Education in the Context of Digitalization: Proceedings of the All-Russian Scientific and Methodological Conference with International Participation, 2023. Ulan-Ude, pp. 200–203 (2023)
20. Atlas of Professions of the Future and Professions of Retirees. 16/03/2021. Skolkovo-resident (2021). https://skolkovo-resident.ru/atlas-professij-budushchego-skolkovo/
21. Ezhova, E.N., Zamozhnykh, E.A.: Innovative approaches to the implementation of educational programs in the sphere of media communications in the context of the new reality. In: MEDIA Education: Digital Environment in the Context of Forced Metamorphosis: Proceedings of the VII International Scientific and Practical Conference (Chelyabinsk, 2022), pp. 56–62. Publishing house of Chelyabinsk State University, Chelyabinsk (2022)
22. Kormakova, V.N., Klepikova, A.G., Lapina, M.A., Rugelj, J.: ICT competence of a teacher in the context of digital transformation of education. CEUR Workshop Proc. **2914**, 138–150 (2021)
23. Kormakova, V., Klepikova, A., Musaelian, E., Baybikowa, G., Lapina, M.: Formation of ICT competencies of postgraduate students of teacher education based on interactive techniques. In: CEUR Workshop Proceedings SLET-2019 – Proceedings of the International Scientific Conference Innovative Approaches to the Application of Digital Technologies in Education and Research, vol. 2494, pp. 11–21 (2019)
24. Lapina, M., Shiriaev, E., Babenko, M., Istamov, I.: High-speed convolution core architecture for privacy-preserving neural networks. Program. Comput. Softw. **50**, 417–424 (2024). https://doi.org/10.1134/S0361768824700282
25. Rusanov, M., Babenko, M., Lapina, M., Sajid, M.: Identification of exploited unreliable account passwords in the information infrastructure using machine learning methods. Big Data Cogn. Comput. **8**, 159 (2024). https://doi.org/10.3390/bdcc8110159
26. Sherbakov, S., Lapina, M., Lapin, V., Rugelj, J.: Methodological support of game modeling in the educational process. In: Paper presented at the SLET-2020: International Scientific Conference Innovative Approaches to the Application of Digital Technologies in Education, CEUR Workshop Proceedings, vol. 2861, pp. 351–361 (2020)
27. Basan, E., Makarevich, O., Lapina, M., Mecella, M.: Analysis of the impact of a GPS spoofing attack on a UAV. CEUR Workshop Proc. **3094**, 6–16 (2022)
28. Proshkin, N., Basan, E., Lapina, M.: Radio frequency method for emulating multiple UAVs. In: 17th International Conference on Intelligent Environments, IE 2021 – Proceedings, p. 9486599 (2021)
29. Maksimova, E., Lapina, M., Lapin, V.: Synthesis of models for ensuring information security of subjects of critical information infrastructure under destructive influences. CEUR Workshop Proc. **3094**, 108–117 (2022)

Using the REN'AI Program for Developing Scientific Thinking in Youth Through Visual Novels

Nadezhda Gallini[1(✉)], Andrey Saveliev[1], and Igor Grishin[2]

[1] Vernadsky Crimean Federal University, Humanities and Education Science (Branch), 2a, Sevastopolskaya Street, Yalta 298635, Russia
gallini.nadi@yandex.ru

[2] Knowledge Engineering Laboratory, Lomonosov Moscow State University, 1, Leninskie Gory, Moscow 119991, Russia

Abstract. The article examines the development and conceptual analysis of the Ren'AI program as a tool for fostering scientific thinking in youth while studying courses on artificial intelligence technologies in a distance learning environment. The program offers an interactive approach by using visual novels to engage students in the process of analyzing and solving scientific problems. The study includes an analysis of the structure of Ren'AI, its algorithms, and its educational potential. A literature review confirms the effectiveness of gamified methods and interactive learning in developing critical thinking. The developed algorithm for creating visual novels provides a structured approach to forming educational content and can be adapted for various academic disciplines. In the future, testing of the program among applied informatics students is planned, which will allow for an empirical evaluation of its impact on material comprehension and student engagement.

Keywords: scientific thinking · visual novels · interactive learning · gamification · Ren'AI · digital education · critical thinking · distance learning · artificial intelligence technologies

1 Introduction

The rapid development of artificial intelligence (AI) technologies is reshaping various domains, including science, economics, and education. In today's world, acquiring critical analytical skills, formulating hypotheses, and providing logical justifications are fundamental aspects of AI-related education. Scientific thinking serves as the foundation for understanding and developing AI technologies, encompassing the ability to analyze complex algorithms, recognize patterns, test hypotheses, and make informed decisions. However, AI education remains a challenging endeavor due to its abstract nature and the necessity of a strong mathematical background. This demands well-developed critical thinking abilities and the capacity to adapt to constantly evolving scientific paradigms.

M. Lapina et al. (Eds.): DET 2024, LNNS 1616, pp. 121–126, 2025.
https://doi.org/10.1007/978-3-032-04365-8_11

One of the major challenges in AI education is the inherent complexity of its algorithms, which require an intuitive understanding of machine learning and neural networks. Mastery of these subjects calls for an interdisciplinary approach that integrates mathematics, programming, philosophy, ethics, and cognitive sciences. Effective learning requires students to cultivate analytical and critical thinking skills, enabling them to comprehend algorithmic principles, detect errors, interpret results, and refine models. Traditional distance learning methods often fail to provide adequate engagement, prompting the need for innovative educational approaches.

Interactive teaching methods, particularly gamification and game-based learning, have gained widespread recognition in education, especially in technical fields. Research indicates that gamified learning formats enhance student motivation and facilitate deeper comprehension of complex subjects. Among the most promising tools for engaging students in learning is the use of visual novels-interactive narrative-based scenarios where users make decisions that influence the storyline. This approach offers several advantages: an engaging narrative fosters motivation, interactive modeling integrates real scientific problems into the storyline, and decision-making tasks help develop cognitive and analytical skills.

The Ren'AI program is an innovative educational platform that leverages the visual novel format to teach AI-related topics. Unlike conventional text-based lectures and video courses, Ren'AI provides an interactive learning experience by simulating various AI scenarios. Key features of the platform include flexibility, visualization of scientific concepts, and an analytical system that monitors student progress, adapting content to individual needs.

This study explores the potential of Ren'AI in enhancing scientific thinking among students learning AI technologies. It examines how the interactive mechanics embedded in the platform support students in grasping complex AI-related concepts and fostering critical analysis skills. While digital technologies are actively being integrated into education, many AI learning platforms still rely on traditional methods. In contrast, Ren'AI introduces a new learning format that blends gamification, visual storytelling, and interactive modeling. The program's adaptive content, real-world AI problem simulations, and accessibility for remote learning make it a promising tool for digital education. This approach not only simplifies the comprehension of intricate topics but also encourages the development of analytical skills, which are crucial in today's AI-driven world.

2 Materials and Methods

The development of the Ren'AI educational platform involved an extensive review of existing research in relevant fields. This analysis helped identify effective approaches and technologies suitable for integration into the platform.

Contemporary research highlights the significance of interactive methodologies in education. For instance, Fan Yuanyuan and Sun Yu examined the distribution of scientific and technological resources in research institutions, emphasizing the importance of incorporating engineering technologies into education [2]. Similarly, Chandra et al. investigated interactive learning material design for future educational initiatives, underscoring the role of interactivity in effective learning [5].

Visual novels have been recognized as a valuable educational tool. Kovalevskaya and Shishkina explored the editorial processes involved in preparing visual novels as an interactive digital publication, reinforcing their potential in educational contexts [13].

Li, Tang, and Huang analyzed STEAM curriculum design and practical studies aimed at developing advanced scientific thinking skills among students [4]. Their findings underscore the necessity of integrating science, technology, engineering, arts, and mathematics into educational frameworks to promote scientific reasoning. Meanwhile, Octaviani et al. demonstrated that online scientific collaboration enhances students' professional scientific thinking, suggesting its applicability to interactive learning platforms [3].

The application of interactive systems and deep learning technologies in education continues to expand. Hong, for example, demonstrated how digital resources and deep learning can enhance English language acquisition, illustrating the potential of AI-driven interactive learning [8]. Shi et al. proposed an interactive reinforcement learning strategy that can be adapted for educational purposes, particularly in online learning platforms [12].

Furthermore, research on immersive technologies has revealed their promising educational applications. Kwon, John, and Kim designed a framework for classifying visual motion imagery using a 3D virtual brain-computer interface, emphasizing the educational potential of such systems [1]. Wang explored the application of data mining techniques in visual communication, highlighting their potential use in adaptive educational environments [9].

Additionally, Chen and Chen investigated the integration of ChatGPT and reinforcement learning in localized energy trading, demonstrating AI's capacity for autonomous decision-making and system adaptation [7]. Arif et al. studied visual error management through advanced Jacobian control, which could be relevant for developing adaptive learning systems [10]. Meanwhile, Fang examined interactive learning systems designed for art education, further supporting the educational value of interactive digital methods [11].

Zhang, Wang, and Ma proposed a novel HEVC bitrate control system designed for visual analysis, which may contribute to optimizing multimedia content delivery in interactive educational environments, including those utilizing visual novels [6].

The evaluation of these studies affirmed the effectiveness of interactive methodologies, gamification, and visual novels in modern education. These insights served as the basis for developing the Ren'AI platform, aimed at fostering scientific thinking through interactive AI-based learning.

3 Results and Discussion

The Ren'AI program was developed to create an engaging and interactive digital learning environment that uses visual novels to help students develop scientific thinking and learn about artificial intelligence technologies. The goal is to make complex technical concepts more approachable by presenting them in a game-like format, where students make decisions and see the outcomes of their choices.

The program is built on the Ren'Py game engine, allowing it to integrate text-based dialogues, multiple-choice interactions, and multimedia content. This setup ensures an immersive and adaptive learning experience. The way students interact with Ren'AI is based on dialogue-driven scenarios—users choose different responses, which not only shape the story but also influence the learning material they encounter.

The process of creating an educational visual novel follows a structured approach. First, the developer builds a knowledge base that includes scripts, texts, images, and videos. Then, course parameters like genre, themes, and presentation style are set. After that, the system adapts materials based on student choices, allowing different learning paths to emerge. As students progress, they make decisions that shape both the storyline and their learning experience. At the final stage, the program tracks their progress, identifies difficult topics, and suggests additional materials for further study.

Figure 1 illustrates the Ren'AI interaction process. The top section of the diagram outlines the steps developers take to set up educational content, while the lower section shows how users engage with the platform—from selecting a topic to creating interactive materials like text, illustrations, and videos.

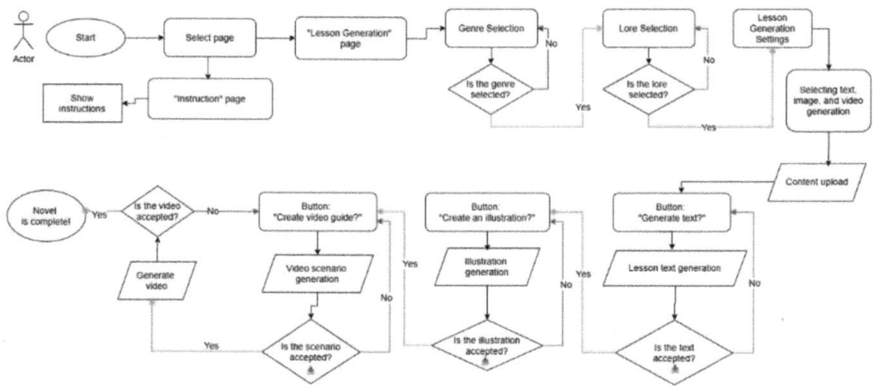

Fig. 1. Novel Development Algorithm

The platform includes a user-friendly interface that makes navigation smooth. It also features a scenario generation system, giving students the ability to choose different learning paths. Interactive elements such as dialogue windows, decision-making options, and built-in tests keep students engaged. Additionally, an adaptive learning mechanism tracks their progress and adjusts the difficulty of scenarios accordingly.

An analysis of Ren'AI's structure and capabilities suggests that the platform can be effectively used in various educational settings, from online courses to university programs. Future improvements will focus on incorporating machine learning to personalize courses based on students' skill levels and learning styles.

The framework behind Ren'AI demonstrates its flexibility for different educational goals. The use of interactive decision-making and automated content generation makes it an effective tool for AI education, helping students develop analytical and critical thinking skills. In the next phase of development, Ren'AI will be tested with real students to evaluate its effectiveness and identify areas for improvement.

4 Conclusions

The development and conceptual analysis of the Ren'AI platform highlight the potential of visual novels as an effective means for fostering scientific thinking and AI literacy. The interactive, dialogue-based approach enhances student engagement, while adaptive content mechanisms support deeper exploration of complex subjects.

A review of existing studies confirmed the educational benefits of gamification and interactive learning methods, which informed the creation of Ren'AI. The platform's structured novel development workflow demonstrates its adaptability for diverse courses and proficiency levels.

Future research will involve testing Ren'AI among students to empirically evaluate its impact on knowledge retention and critical thinking development. The platform presents a promising direction for digital education by integrating interactive storytelling, gamification, and adaptive learning. Further advancements will focus on incorporating machine learning technologies and expanding analytical capabilities to optimize personalized learning experiences.

References

1. Kwon, B.-H., Jeong, J.-H., Kim, D.-J.: A novel framework for visual motion imagery classification using 3D virtual BCI platform. In: 2020 8th International Winter Conference on Brain-Computer Interface (BCI), Gangwon, Korea (South), pp. 1–5 (2020). https://doi.org/10.1109/BCI48061.2020.9061621
2. Yuanyuan, F., Yu, S.: Allocation of scientific and technological resources in scientific research institutes: current situation analysis and enlightenment: - Thinking based on the progress of engineering technology. In: 2021 International Conference on Management Science and Software Engineering (ICMSSE), Chengdu, China, pp. 290–295 (2021). https://doi.org/10.1109/ICMSSE53595.2021.00068
3. Oktaviani, H.I., Yuliati, L., Herdiani, A., Purwodani, D.L., Putra, W.P.: Online scientific collaboration model to support the professionality of learners in developing scientific thinking abilities. In: 2020 6th International Conference on Education and Technology (ICET), Malang, Indonesia, pp. 136–141 (2020). https://doi.org/10.1109/ICET51153.2020.9276614
4. Li, H., Tang, Y., Huang, G.: STEAM curriculum design and practical research: training students' scientific advanced thinking ability. In: 2022 10th International Conference on Information and Education Technology (ICIET), Matsue, Japan, pp. 244–249 (2022). https://doi.org/10.1109/ICIET55102.2022.9779031
5. Chandra, M.A., Dhanendra, N.W.D.A., Renato, R., Fredyan, R., Pranoto, H.: Developing interactive learning media design for futuristic learning activities. In: 2023 15th International Congress on Advanced Applied Informatics Winter (IIAI-AAI-Winter), Bali, Indonesia, pp. 135–138 (2023). https://doi.org/10.1109/IIAI-AAI-Winter61682.2023.00033
6. Zhang, Q., Wang, S., Ma, S.: A novel visual analysis oriented rate control scheme for HEVC. In: 2020 IEEE International Conference on Visual Communications and Image Processing (VCIP), Macau, China, pp. 491–494 (2020). https://doi.org/10.1109/VCIP49819.2020.9301817
7. Chen, Y., Chen, G.: Interactive Learning - Implementation of ChatGPT and Reinforcement Learning in Local Energy Trading. In: 2024 IEEE 34th Australasian Universities Power Engineering Conference (AUPEC), Sydney, Australia, pp. 1–6 (2024). https://doi.org/10.1109/AUPEC62273.2024.10807616

8. Hong, Y.: Enhanced english learning through integrated interactive systems and deep learning utilizing digital resources. In: 2024 International Conference on Interactive Intelligent Systems and Techniques (IIST), Bhubaneswar, India, pp. 58–63 (2024). https://doi.org/10.1109/IIST62526.2024.00056
9. Wang, Y.: Visual communication data mining: novel approach for open project under the background of internet on enhancing the professional practical ability. In: 2021 5th International Conference on Electronics, Communication and Aerospace Technology (ICECA), Coimbatore, India, vol. 2021, pp. 1175–1178. https://doi.org/10.1109/ICECA52323.2021.9675909
10. Arif, Z., Fu, Y., Siddiqui, M.K., Zhang, F.: Visual error constraint free visual servoing using novel switched part Jacobian control. IEEE Access. **10**, 103669–103693 (2022). https://doi.org/10.1109/ACCESS.2022.3203734
11. Fang, Z.: Design and implementation of interactive learning system for art teaching. In: 2023 International Conference on Ambient Intelligence, Knowledge Informatics and Industrial Electronics (AIKIIE), Ballari, India, vol. 2023, pp. 1–5. https://doi.org/10.1109/AIKIIE60097.2023.10390424
12. Shi, Z., Ma, W., Yin, S., Zhang, H., Zhao, X.: Interactive reinforcement learning strategy. In: 2021 IEEE SmartWorld, Ubiquitous Intelligence & Computing, Advanced & Trusted Computing, Scalable Computing & Communications, Internet of People and Smart City Innovation (SmartWorld/SCALCOM/UIC/ATC/IOP/SCI), Atlanta, GA, vol. 2021, pp. 507–512. https://doi.org/10.1109/SWC50871.2021.00075
13. Kovalevskaya, N.I.: Visual novel as a form of interactive digital publication: features of editorial preparation. Kovalevskaya, N.I., Shishkina, N.I. Proc. BSTU. Series 4: Print and Media Technologies. 1(279): 37–44 (2024). https://doi.org/10.52065/2520-6729-2024-279-5. EDN WQQJPP.

Digital Footprints in the Moodle System and Analysis of the Temporal Dynamics of Students' Learning Activity

Ilshat Garafiev(✉) and Gullshat Garafieva

Kazan National Research Technological University, Kazan, Russia
GarafievIZ@corp.knrtu.ru

Abstract. This paper proposes a methodology for analyzing digital footprints in the Moodle Learning Management System (LMS) to identify procrastination patterns and classify students based on temporal activity dynamics. The study introduces novel metrics, including last_week_intensity (activity concentration in the final week), late_activity_ratio (proportion of activity in the last 25% of the course), and activity_trend (linear regression slope of weekly engagement), to quantify behaviors. Using data from 19 students enrolled in a 17-week Human Resource Management course, the authors apply the K-Means clustering algorithm to categorize learners into three distinct groups: procrastinators, achievers, and diligent underachievers. The results reveal significant correlations between procrastination metrics, engagement duration, with procrastinators demonstrating short-term participation (5 weeks) and high end-of-course activity (88% intensity). The study highlights the potential of LMS data for early identification of at-risk students and designing adaptive pedagogical interventions.

Keywords: Academic Performance · Behavioral Clustering · Digital Learning Analytics · Educational Data Mining · K-Means Algorithm · Learning Management Systems · Moodle Platform · Procrastination Patterns · Student Engagement

1 Introduction

Modern educational environment is actively transforming under the influence of digital technologies, forming new formats of interaction between participants of the educational process. One of the key elements of this transformation is students' digital footprints - data generated as a result of their activity in e-learning platforms, social networks and other online environments [1–5]. Learning Management Systems (LMS) such as Moodle are becoming an integral part of the educational process. These platforms not only provide access to learning materials, but also generate significant amounts of data on student behavior, opening up new possibilities for analyzing learning strategies, predicting outcomes, and personalizing learning. However, a key challenge lies in the interpretation of this data: despite its volume, many studies are limited to superficial metrics, ignoring dynamic patterns of activity such as procrastination, uneven workload distribution, and decreased engagement.

M. Lapina et al. (Eds.): DET 2024, LNNS 1616, pp. 127–138, 2025.
https://doi.org/10.1007/978-3-032-04365-8_12

Procrastination, manifested in the concentration of activity before deadlines, remains one of the most common problems negatively affecting academic performance. Traditional assessment methods based on aggregate measures (e.g., total number of logins) fail to identify temporal anomalies in student behavior, thus reducing the effectiveness of pedagogical interventions. In addition, existing approaches rarely consider relative time frames (e.g., activity in the last 25% of a course) and do not offer tools for clustering students based on their unique behavioral profiles [6–9].

The aim of this paper is to propose a methodology for analyzing digital footprints in the LMS Moodle to identify procrastination behaviors and cluster students based on their academic activity:

1. Develop dynamic metrics such as last_week_intensity (activity intensity in the last week), late_activity_ratio (proportion of activities in the last 25% of the course), and activity_trend (activity trend) that reflect the temporal distribution of student effort.
2. Clustering students based on their behavior in the LMS using the K-Means algorithm and interpreting the identified groups through the lens of academic performance.
3. Analyze the relationship between the identified patterns of activity and risks of academic failure.

The empirical basis of the study was data from 19 students of a 17-week Human Resource Management course. The results demonstrate that the proposed metrics identify three stable clusters of students: "procrastinators", "achievers" and "diligent underachievers", each characterized by unique behavioral profiles and requiring differentiated pedagogical strategies.

The work contributes to the development of adaptive learning by offering practical tools for early identification of at-risk students and designing personalized educational trajectories. The results obtained are relevant for teachers, administrators of educational programs and developers of LMS-solutions focused on improving the academic resilience of students.

2 Background

Modern higher education is actively integrating digital technologies, such as learning management systems (LMS), which not only facilitate access to learning materials but also generate voluminous data on student behavior. This data, including the timing, frequency, and nature of course interactions, has become a valuable resource for analyzing learning strategies, predicting academic outcomes, and personalizing the educational process [10–12]. However, a key challenge is transforming raw data into meaningful insights that identify risks such as procrastination, uneven workloads, and decreased engagement [13–15].

Procrastination, defined as the voluntary postponement of tasks despite realizing the negative consequences [16], remains one of the most common problems in academia. Studies show that up to 70% of students regularly postpone tasks, which correlates with increased stress, reduced quality of work, and risk of academic failure [17–19]. In the LMS context, procrastination manifests itself in a surge of activity before deadlines, which can be quantified through dynamic metrics such as the proportion of activities in the last week of a course or the distribution of effort over time [20, 21].

Despite a growing body of work analyzing educational data [22–25] most studies are limited to aggregate measures (e.g., total number of logins), ignoring temporal patterns of activity. Furthermore, existing methods rarely consider relative time frames (e.g., activity in the last 25% of a course) and do not offer tools for clustering students based on their unique behavioral profiles. This creates a gap in understanding how different learning strategies are related to academic outcomes and psychological factors. The present paper aims to address these issues by identifying patterns of procrastination through the development of new dynamic metrics such as: last_week_intensity (activity intensity in the last week), late_activity_ratio (proportion of activities in the last 25% of the course) and clustering students' learning activity based on their activity in the LMS (Moodle).

3 Methodology

The Moodle system allows to generate course reporting for each user, taking into account his number of course visits per week and the number of changes - activities that he carried out in this week. This article presents the results of analyzing student activity in a Human Resource Management course for 19 students over the course of a 17-week academic semester. The difficulty was that each student had a different end of course deadline, when he or she had completed the activity in the course, several indicators were calculated relative to this week. First, *duration_weeks*, the duration of the activity in weeks, was determined:

$$duration\_weeks = \left\lceil \frac{\Delta_{days} + 1}{7} \right\rceil LR = \frac{likes}{views} \tag{1}$$

where is $\Delta_{days} = (last\ \_\ date\ -first\ \_\ date).\ day$ - the difference in days between the last and first activity, 1 day is added to include the start date in the count. The result is divided by 7 (days in a week) and rounded up to a whole number. Assuming that students had different completion dates, the sum of activities (posting replies, comments) in the last week of the course for each student is calculated, last_week_changes – changes in the last week:

$$last_{week_{changes}} = \sum changes\ for\ the\ week\ corresponding\ to\ the\ student's\ last\ activity\ date\ LR = \frac{likes}{views} \tag{2}$$

Then the late_activity indicator was calculated – activity in the last 25% of time:

$$late_{activity} = \sum changes\ over\ the\ last \lceil duration_{weeks} \times 0.25 \rceil weeks\ TR = \frac{comments}{views} \tag{3}$$

where is duration_weeks - duration of activity in weeks, calculated as the sum of activities in the last 25% of the student's total activity time.

The indicator last_week_intensity was introduced - the intensity of changes in the last week::

$$last\_week\_intensity = \frac{last\_week\_changes}{total\_changes + \epsilon} ReRate = \frac{reports}{views} \tag{4}$$

where $\epsilon = 1e - 1$ – to avoid division by 0, *last_week_changes* - changes in the last week, *total_changes total* number of changes, calculated as the proportion of changes occurring in the last week relative to the total number of changes.

The late_activity_ratio indicator was calculated – the proportion of the latest activity:

$$late\_activity\_ratio = \frac{late\_activity}{total\_changes + \epsilon}\ ReRate = \frac{reports}{views} \tag{5}$$

where $\epsilon = 1e - 1$ – для избежания деления на 0, *total_changes – общее количество изменений*

Then the *activity_trend* indicator, the activity trend, was introduced. The linear trend is calculated through the method of least squares:

$$slope = \frac{n \cdot p\sum(x \cdot py) - \sum x \cdot p\sum y}{n \cdot p\sum(x^2) - \left(\sum x\right)^2}\ ReRate = \frac{reports}{views} \tag{6}$$

where x is the number of weeks (1, 2, ..., n), y is the change for each week, n is the number of weeks of activity. *Slope* is the slope of the linear regression line plotted for student activity by week. A positive slope value indicates an increase in activity, while a negative value indicates a decrease in activity.

Then *max_rolling_2w*- the maximum moving average over 2 weeks was determined:

$$\max\_rolling\_2w = \max\left(\frac{Changes_t + Changes_{t-1}}{2}\right) ReRate = \frac{reports}{views} \tag{7}$$

where *t*– week, calculated as the maximum average activity for any two consecutive weeks.

Then *avg_weekly_changes*- average changes per week-was calculated:

$$avg\_weekly\_changes = \max\left(\frac{total\_changes}{duration\_weeks + \epsilon}\right) ReRate = \frac{reports}{views} \tag{8}$$

where $\epsilon = 1e - 1$ – to avoid division by 0, duration_weeks - duration of activity in weeks, total_changes - total number of changes

The *stability* indicator - activity stability - was then calculated:

$$stability = 1 - \left(\frac{\underset{rolling_{2w}}{\max\ +}}{avg_{weekly_{changes}} + \epsilon}\right) ReRate = \frac{reports}{views} \tag{9}$$

where $\epsilon = 1e - 1$ – to avoid division by 0, *max_rolling_2w*- maximum moving average over 2 weeks, *avg_weekly_changes*- average changes per week. Indicates how evenly distributed the activity is, if the value is close to 1 = stable activity, value close to 0 = sudden spikes in load.

The collection, processing, and analysis of data was carried out based on the pandas, numpy, matplotlib, seaborn, sklearn libraries for the general-purpose programming language Python.

Table 1. Student course completion totals in the Moodle system.

№	total_views	total_changes	duration_weeks	last_week_changes	late_activity	last_week_intensity	late_activity_ratio	activity_trend	max_rolling_2w	avg_weekly_changes	stability
1	248	37	17	6	13	0,16	0,35	-0,03	4,00	2,18	0,89
2	279	38	17	6	8	0,16	0,21	0,03	5,50	2,24	0,86
3	161	35	17	3	7	0,09	0,20	0,03	5,50	2,06	0,84
4	254	31	17	0	7	0,00	0,23	0,00	5,00	1,82	0,84
5	232	43	17	3	8	0,07	0,19	0,08	8,00	2,53	0,81
6	217	38	15	7	0	0,18	0,00	0,39	7,50	2,53	0,80
7	134	36	17	10	11	0,28	0,31	0,13	7,50	2,12	0,79
8	299	50	15	3	10	0,06	0,20	-0,10	10,50	3,33	0,79
9	334	102	17	26	23	0,25	0,23	0,62	25,00	6,00	0,75
10	171	30	17	6	8	0,20	0,27	0,10	8,00	1,76	0,73
11	195	41	15	0	1	0,00	0,02	0,64	11,00	2,73	0,73
12	282	43	17	12	17	0,28	0,40	0,23	13,00	2,53	0,70
13	97	34	15	15	11	0,44	0,32	0,08	11,00	2,27	0,68
14	116	10	17	0	7	0,00	0,70	-0,13	3,50	0,59	0,65
15	163	28	17	16	4	0,57	0,14	0,31	10,00	1,65	0,64
16	61	22	5	14	8	0,64	0,36	-1,60	8,00	4,40	0,64
17	109	22	7	22	0	1,00	0,00	2,36	11,00	3,14	0,50
18	94	22	14	0	0	0,00	0,00	0,44	11,00	1,57	0,50
19	69	11	3	11	11	1,00	1,00	-5,50	11,00	3,67	0,00

Source: author.

4 Results

The calculated indicators for 19 students are presented in Table 1.

From the initial indicators it was necessary to select the most significant ones for further clustering. The correlation matrix is presented in in Fig. 1.

The value of the correlation coefficients of the correlation coefficients of the students' course completion rates in the Moodle system is presented in Table 2.

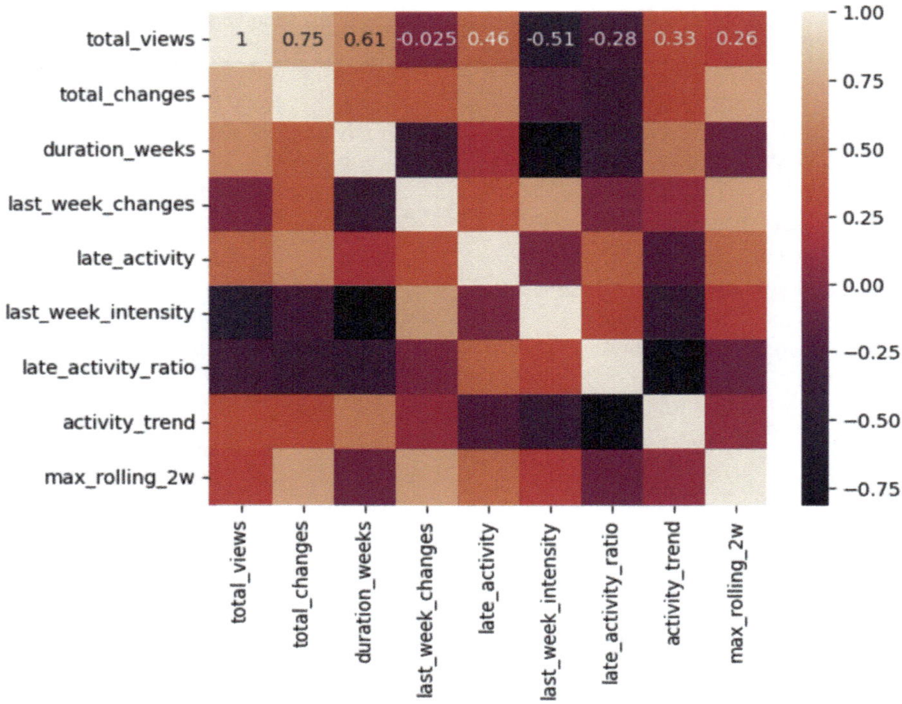

Fig. 1. Correlation matrix of students' course completion rates in Moodle system (Source: author).

Based on the data in Table 2, we can identify several strong positive correlations when the correlation coefficient value is greater than 0.6. The correlation coefficient between the parameters total_views and total_changes equal to 0.75 means that the more the student reviews the materials, the more often he/she makes changes in the course. The correlation coefficient between total_views and duration_weeks of 0.61 means that students with more views stay active longer. The correlation coefficient between the parameters last_week_changes and max_rolling_2w 0.67 means that, peak activity in 2 weeks is related to activity in the last week. The correlation coefficient between the parameters duration_weeks and activity_trend 0.55 means that, the longer a student is active, the more their activity increases over time. The correlation coefficient between parameters total_changes and late_activity 0.61 means, active students are more likely to make changes at the end of the course. The correlation coefficient between last_week_intensity and last_week_changes 0.70 means that the intensity in the last week correlates with the number of changes in that period.

We can identify several strong negative correlations when the value of correlation coefficient is less than −0.6. The correlation coefficient between duration_weeks and last_week_intensity parameters −0.79 means that students with short activity show maximum intensity at the end. The correlation coefficient between the parameters late_activity_ratio and activity_trend −0.81 means that the more activity is concentrated at the end, the more the overall trend decreases. It should be noted that the growth

Table 2. The value of correlation coefficients of students' course completion rates in the Moodle system.

	total_views	total_changes	duration_weeks	last_week_changes	late_activity	last_week_intensity	late_activity_ratio	activity_trend	max_rolling_2w
total_views	1.00	0.75	0.61	-0.02	0.46	-0.51	-0.28	0.33	0.26
total_changes	0.75	1.00	0.42	0.41	0.61	-0.28	-0.32	0.32	0.71
duration_weeks	0.61	0.42	1.00	-0.33	0.19	-0.81	-0.35	0.55	-0.10
last_week_changes	-0.02	0.41	-0.33	1.00	0.36	0.69	-0.03	0.10	0.71
late_activity	0.46	0.61	0.19	0.36	1.00	-0.03	0.45	-0.23	0.47
last_week_intensity	-0.51	-0.28	-0.81	0.69	-0.03	1.00	0.30	-0.36	0.24
late_activity_ratio	-0.28	-0.32	-0.35	-0.03	0.45	0.30	1.00	-0.81	-0.11
activity_trend	0.33	0.32	0.55	0.10	-0.23	-0.36	-0.81	1.00	0.07
max_rolling_2w	0.26	0.71	-0.10	0.71	0.47	0.24	-0.11	0.07	1.00

Source: author.

Source: author.

of activity is weakly related to the total number of views as the correlation coefficient between activity_trend and total_views is 0.33.

The obtained correlation matrix allows us to draw several conclusions. First, students with high total activity (high values of total_views, total_changes) are more likely to distribute the load evenly (duration_weeks) and are less prone to procrastination (last_week_intensity = −0.28). Second, procrastinators (last_week_intensity, late_activity_ratio) have short activity (duration_weeks = −0.79) and show a sharp decline in overall activity trend (activity_trend = −0.81). Third, students with peak activity (max_rolling_2w) are active in the last week (last_week_changes = 0.67) and are more likely to show high overall change (total_changes = 0.71).

For further clustering of students to avoid duplicate information, we leave one indicator from each strongly correlated pair. From the pair of indicators total_views and total_changes, we leave the indicator total_changes with a correlation coefficient of 0.75, since the indicator total_views is less informative for action. Of the pair of last_week_changes and max_rolling_2w, we leave max_rolling_2w with a correlation coefficient of 0.67 because it shows peak load, not just the last week. From the pair of activity_trend and late_activity_ratio we leave late_activity_ratio with a correlation coefficient of −0.81, as it more clearly reflects procrastination. In total we have

5 indicators for clustering of students. The total_changes indicator shows total activity (a key indicator of engagement), moderately correlated only with duration_weeks, which is acceptable. The duration_weeks indicator highlights students with long-term engagement and is strongly negatively correlated with last_week_intensity (the shorter the course, the higher the activity at the end). Last_week_intensity is an indicator of "rush" work and helps identify procrastinators. The late_activity_ratio indicator shows how much a student puts off tasks until the end, independent of other traits (correlations <0.45). The max_rolling_2w indicator identifies periods of overload and does not duplicate information about the last week. The K-Means algorithm was used to cluster the data: relevant features were selected from the original dataset and then standardized using StandardScaler to eliminate scale differences. The quality of separation was assessed by analyzing the mean values of key features within clusters and the silhouette coefficient (0.52), which indicates moderately good separability of clusters while maintaining the homogeneity of their structure.

Average values of indicators by clusters are presented in Table 3.

Table 3. Average values of indicators by clusters.

cluster	total_changes	duration_weeks	last_week_intensity	late_activity_ratio	max_rolling_2w	number of students	total_changes	duration_weeks	last_week_intensity
0	18.33	5.00	0.88	0.45	10.00	3	18.33	5.00	0.88
1	34.40	16.27	0.17	0.24	8.07	15	34.40	16.27	0.17
2	102.00	17.00	0.25	0.23	25.00	1	102.00	17.00	0.25

Source: author.

Source: author.

Next, we used the PCA (Principal Component Analysis) method to reduce the dimensionality of the data; it allowed us to transform the set of our correlated features into a smaller number of uncorrelated variables (principal components) while maintaining the maximum variance of the data. The PCA visualization of the clusters is presented in Fig. 2.

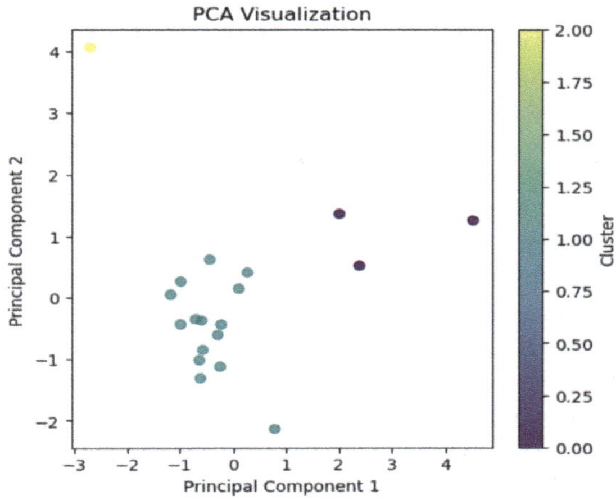

Fig. 2. PCA visualization of clusters (Source: author).

Visualization of the clusters showed a clear separation of three clusters, Cluster 0: "Procrastinators" - three students were included. This cluster is characterized by low total activity (Total Changes = 18.33) and short duration of course participation (Duration Weeks = 5.0). Most activity (88% of Changes, Last Week Intensity = 0.88) was concentrated in the final week, with 45% of activities (Late Activity Ratio = 0.45) completed late in the course. Moderate peak workload (Max Rolling 2w = 10.0) indicates a lack of systematic work. The profile corresponds to students who joined the course late or delayed completing assignments until deadlines, resulting in a spike in activity at the end of the course with minimal overall contribution. Cluster 1: Successful Students - 15 students were included in this cluster. Students in this cluster showed moderate overall activity (Total Changes = 34.4) and long term participation (Duration Weeks = 16.27). Only 17% of changes (Last Week Intensity = 0.17) and 24% of total activity (Late Activity Ratio = 0.24) occur in the final stages of the course, with no pronounced peaks in workload (Max Rolling 2w = 8.07). This indicates high engagement at the beginning of the course followed by a decrease in activity, which may be due to the fact that they did not have to redo the work handed in. Cluster 2: "Diligent underachieving students" 1 student was included in this cluster. The cluster contains students with maximum total activity (Total Changes = 102.0) and prolonged participation (Duration Weeks = 17.0). Activity is evenly distributed (Last Week Intensity = 0.25, Late Activity Ratio = 0.23), with high peak activity (Max Rolling 2w = 25.0), reflecting systematic work and adaptation to task volume. The profile corresponds to the most engaged participants, but whose work was not accepted on the first attempt and they had to redo it.

5 Discussion

The conducted analysis allowed us to identify three stable clusters of students differing in patterns of activity in the Moodle system. The most significant result is the identification of a link between procrastination and a decrease in overall academic performance. Students in cluster 0 ("Procrastinators") show a highly uneven distribution of activity with a peak in the last week (Last Week Intensity = 0.88), which is consistent with research linking procrastination to increased stress and decreased quality of work [26, 27]. Interestingly, the short duration of participation (Duration Weeks = 5.0) in this cluster may indicate both late joining the course and deliberate procrastination of assignments, which requires further verification.

Cluster 1 ("Successful students"), which includes 78.9% of the sample, reflects typical behavior of students with moderate even activity. Low values of Last Week Intensity (0.17) and Late Activity Ratio (0.24) indicate a developed time management skill, which correlates with data on the positive impact of early engagement on academic performance [12]. However, the decrease in engagement towards the end of the course may be due to the lack of need to finalize assignments, which was not accounted for in the metrics. The anomaly is cluster 2 ("Diligent Underachievers"), represented by one student with extremely high total engagement (Total Changes = 102). The paradox of combining maximum engagement (Duration Weeks = 17) with low final success (inferred from the label "Diligent Failers") may be due to ineffective task strategies or technical difficulties requiring multiple retakes. This case highlights the limitations of behavioral metrics without considering contextual data and investigating regional characteristics of educational practices [28–30]. The proposed metrics system and clustering method demonstrated the potential of digital footprint analysis to diagnose learning strategies. However, the transition from pattern description to performance management requires the integration of behavioral data with pedagogical context. The results provide a basis for the development of adaptive student support systems in LMS platforms.

6 Conclusion

The conducted research allowed to identify the key patterns of students' academic activity in the Moodle system on the example of the course "Human Resource Management". Analysis of data from 19 students over 17 weeks showed that procrastination is a systemic problem. Students with high activity intensity in the last week (cluster 0, last_week_intensity = 0.88) show short engagement (5 weeks) and low overall engagement (total_changes = 18.33). This supports the hypothesis that task procrastination negatively affects the depth of mastery of the material. Uniform activity is the key to sustainability, most students (cluster 1, 78.9%) with moderate performance (total_changes = 34.4) and long engagement (duration_weeks = 16.27) confirm the effectiveness of early engagement strategies, which is consistent with current pedagogical approaches. The paradox of high engagement in cluster 2 (total_changes = 102) with extremes of engagement with no direct correlation to success emphasizes the need to integrate quantitative metrics and describe the number of retakes of a particular student. The developed metrics (last_week_intensity, late_activity_ratio, max_rolling_2w) proved effective in identifying patterns of procrastination and overload. The small sample size and lack of grade data does not allow unequivocally linking activity to academic performance, future studies would benefit from a larger sample. Analyzing digital footprints in LMS systems opens new possibilities for predicting learning strategies and early intervention. However, moving from diagnosis to performance management requires synthesizing quantitative metrics, pedagogical context, and feedback from teachers. The results of the study can provide a basis for the design of adaptive educational environments that minimize the risks of procrastination and academic failure.

Acknowledgments. The authors would like to thank the administration of the Kazan National Research Technological University for providing access to the Moodle platform and technical support in data extraction. We would like to express our gratitude to the students of the Human Resource Management course, whose activity data, anonymized and aggregated in accordance with ethical principles, made this analysis possible.

Conflicts of Interest. The authors declare no conflicts of interest.

References

1. Holden, V. M., Jones, R. F., Bartlett, J., Terry, K.: Digital Footprints: Freedom on the Move at the University of Kentucky (2024)
2. Garafiev, I.: Network Communities and Groups in the Preparation for the Unified State Exam, pp. 221–227 (2024)
3. Cheremisina, E., Kirpicheva, E., Tokareva, N., Potemkina, S., Tyatyushkina, O., Kreider, O., Nechaevskiy, A., Korenkov, V., Streltsova, O., Zhatkina, K.: How digital footprints shape IT-education and drive digital transformation. Phys. Part. Nucl. **55**(3), 317–320 (2024)
4. Appiah-Kubi, P., Zouhri, K., Basile, E., McCabe, M.: Analysis of engineering technology students' digital footprints in synchronous and asynchronous blended courses. Int. J. Eng. Pedagog. **12**(1) (2022)
5. Garafiev, I.: Level of Engagement Based on Analysis of Community Posts on VKontakte Network, pp. 46–57 (2024)
6. Benati, K., Fischer, J.: Beyond human capital: student preparation for graduate life. Educ. Train. **63**(1), 151–163 (2020). https://doi.org/10.1108/ET-10-2019-0244
7. Garafiev, I., Vafina, Y., Gatina, L.: Clustering of students by type of relationship to invest in a special human capital. WSEAS Trans. Bus. Econ. **16**(51), 463–470 (2019)
8. Campbell, A.C., Neff, E.: A systematic review of international higher education scholarships for students from the global south. Rev. Educ. Res. **90**(6), 824–861 (2020)
9. Burukuna, O.: Strategic and competitive advantages of Finnish and Russian universities' international student recruitment practices [fi=Ylempi AMK-opinnäytetyö|sv=Högre YH-examensarbete|en=master's thesis|] (2020). http://www.theseus.fi/handle/10024/344788
10. Eraković, B.R., Topalov, J.P.: Teaching and learning through Moodle, Google doc and zoom: fostering student engagement in (a) synchronous learning environments. Inovacije u nastavi-časopis za savremenu nastavu. **34**(4), 122–136 (2021)
11. Liu, D.Y.-T., Atif, A., Froissard, J.-C., Richards, D.: An enhanced learning analytics plugin for Moodle: student engagement and personalised intervention. In: ASCILITE 2015-Australasian Society for Computers in Learning and Tertiary Education, Conference Proceedings (2019)

12. Poondej, C., Lerdpornkulrat, T.: Gamification in e-learning: a Moodle implementation and its effect on student engagement and performance. Interact. Technol. Smart Educ. **17**(1), 56–66 (2020)
13. Elmoazen, R.: Using Learning Analytics to Capture Social and Temporal Dimensions of Collaborative Learning (2024)
14. Saint, J., Fan, Y., Gašević, D., Pardo, A.: Temporally-focused analytics of self-regulated learning: a systematic review of literature. Comput. Educ. Artif. Intell. **3**, 100060 (2022)
15. Salonen, A.O., Tapani, A., Suhonen, S.: Student online activity in blended learning: a learning analytics perspective of professional teacher education studies in Finland. SAGE Open. **11**(4), 21582440211056612 (2021)
16. Steel, P.: The nature of procrastination: a meta-analytic and theoretical review of quintessential self-regulatory failure. Psychol. Bull. **133**(1), 65–94 (2007)
17. Melgaard, J., Monir, R., Lasrado, L.A., Fagerstrøm, A.: Academic procrastination and online learning during the COVID-19 pandemic. Procedia Comput. Sci. **196**, 117–124 (2022)
18. Rakes, G.C., Dunn, K.E.: The impact of online graduate students' motivation and self-regulation on academic procrastination. J. Interact. Online Learn. **9**(1) (2010)
19. Ucar, H., Bozkurt, A., Zawackı-rıchter, O.: Academic procrastination and performance in distance education: a causal-comparative study in an online learning environment. Turk. Online J. Dist. Educ. **22**(4), 13–23 (2021)
20. Cerezo, R., Esteban, M., Vallejo, G., Sanchez-Santillan, M., Nuñez, J.C.: Differential efficacy of an intelligent tutoring system for university students: a case study with learning disabilities. Sustainability. **12**(21), 9184 (2020)
21. You, J.W.: Examining the effect of academic procrastination on achievement using LMS data in e-learning. J. Educ. Technol. Soc. **18**(3), 64–74 (2015)
22. Chaparro-Peláez, J., Iglesias-Pradas, S., Rodríguez-Sedano, F.J., Acquila-Natale, E.: Extraction, processing and visualization of peer assessment data in moodle. Appl. Sci. **10**(1), 163 (2019)
23. Charitopoulos, A., Rangoussi, M., Koulouriotis, D.: Educational Data Mining and Data Analysis for Optimal Learning Content Management: Applied in Moodle for Undergraduate Engineering Studies, pp. 990–998 (2017)
24. Romero, C., Ventura, S., García, E.: Data mining in course management systems: Moodle case study and tutorial. Comput. Educ. **51**(1), 368–384 (2008)
25. Ueda, H., Nakamura, M.: Data analysis for evaluation on course design and improvement of "cyberethics" moodle online courses. Procedia Comput. Sci. **112**, 2345–2353 (2017)
26. Klingsieck, K.B.: Procrastination in different life-domains: is procrastination domain specific? Curr. Psychol. **32**, 175–185 (2013)
27. Garafiev, I., Garafieva, G., Idiatullina, A., Spirchagova, E.: Assessment of online education problems during the COVID-19 pandemic in Russia. Indones. J. Electr. Eng. Comput. Sci. **23**, 1689 (2021). https://doi.org/10.11591/ijeecs.v23.i3.pp1689-1698
28. Nisar, N., Khalid, M.M., Dogar, A.H.: Determinants of students' academic achievement at secondary school level. Bull. Educ. Res. **39**(1), 145–158 (2017)
29. Garafiev, I., Garafieva, G.: Territorial features of value orientations and satisfaction with the studies of student youth in the republic of Tatarstan. E3S Web Conf. **210** (2020). https://doi.org/10.1051/e3sconf/202021018134
30. Ali, W.: Online and remote learning in higher education institutes: a necessity in light of COVID-19 pandemic. High. Educ. Stud. **10**(3), 16–25 (2020). https://doi.org/10.5539/hes.v10n3p16

Incorporating Python into the Classroom to Teach Students How to Process Data from Multiple Excel Files

Ilshat Garafiev(✉) and Gullshat Garafieva

Kazan National Research Technological University, Kazan, Russia
GarafievIZ@corp.knrtu.ru

Abstract. This article explores the integration of Python into educational environments as a tool for teaching students how to process data from multiple Excel files. With the growing importance of data literacy in a data-driven world, the study demonstrates how Python, with its powerful libraries such as Pandas and NumPy, can improve students' ability to extract, clean, and analyze complex data sets. The authors propose a structured methodology that educators can replicate to effectively teach data science skills. Using real-world statistical reporting data, the study illustrates the practical application of Python in an educational context. The results highlight the advantages of Python over traditional methods such as Excel and underscore the importance of equipping students with data analysis skills to meet future challenges.

Keywords: Data Analysis · Data Processing · Educational Technology · Excel · NumPy · Pandas · Python · Statistical Reporting · Teaching Methodology

1 Introduction

The current stage of development of society in general and the economy in particular does not have a clear definition for designation, but whatever concept we would not adhere to ("post-industrial society", "digital economy", "new technological order"), they all state the exponential growth of accumulation and use of data. To achieve professional success in industry, in the civil service and even in the field of culture, an employee must be able to process and analyze data. Organizations and enterprises themselves generate more and more data, some of which are accumulated in the .xls format, which is the basic one for Excel files. Unfortunately, the tools for working with data within spreadsheets, which seemed advanced several years ago, do not allow automating the formation of complex data sets. This must be considered when forming a culture of working with data among students; the use of digital technologies in education should evolve from using programs similar to MS Excel to writing scripts in programming languages, the most accessible of which is Python.

The main advantage of Python over other programming languages is its versatility, the ability to use it to solve a wide range of problems, including data processing and

M. Lapina et al. (Eds.): DET 2024, LNNS 1616, pp. 139–149, 2025.
https://doi.org/10.1007/978-3-032-04365-8_13

analysis. There are many studies that have considered the use of Python for educational purposes, but no algorithms for teaching university students to extract, clean and process data simultaneously from several Excel files of official statistical reporting have been described. This article addresses this gap by proposing a structured approach that can be easily integrated into existing curricula.

By utilizing real-world statistical reporting data, this study illustrates the practical applications of Python for data extraction, cleaning, and analysis. The proposed methodology is tested on a dataset comprising multiple Excel files, demonstrating how Python can streamline complex data processing tasks and enhance students' understanding of data management.

The purpose of the article is to reveal the effectiveness Python as an accessible tool for university students to process and analyze data from multiple Excel files. The authors are ready to offer university teachers a simple and reproducible algorithm for integrating Python into everyday educational practice for working with data in seminars. It should be emphasized that the development of interdisciplinary competence of students in the field of working with data is particularly relevant as a key skill necessary for successful professional self-realization.

Through this exploration, the article aims to contribute to the ongoing discourse on the role of technology in education by providing a practical framework for integrating Python into the classroom. By equipping students with the skills to process and analyze data efficiently, this study underscores the importance of data literacy in preparing the next generation for the challenges of a data-centric society.

2 Background

The integration of Python into educational settings has emerged as a highly effective tool for teaching students how to handle data from multiple Excel files. It provides a strong alternative to conventional spreadsheet-based techniques. This method not only enhances students' data analysis abilities but also equips them with programming skills that are increasingly sought after across various industries. In this section, the topic is explored in depth, highlighting the benefits of Python over traditional tools like Excel. We attempt to summarize existing research on the specifics of using Python programming in teaching various academic subjects. Research articles and systematic literature reviews that have addressed various data cleaning mechanisms confirm the ability of Python scripts to automate data processing tasks and be more efficient than operations in the Excel interface [1–7].

The combination of Python and Excel has been explored to leverage the strengths of both tools. Python's pandas library is particularly effective for cleaning, formatting, and analyzing data from Excel files. A hands-on guide for non-programmers demonstrates how Python can automate data wrangling tasks, such as scraping websites, cleaning datasets, and creating visualizations [8]. Research has proposed frameworks that integrate Excel with Python using tools like Django, enabling end users to perform data extraction, analysis, and visualization within a familiar Excel interface. This approach has been tested across various domains, showcasing its versatility and effectiveness.

Python can be incorporated into existing courses without significant restructuring. For example, using Jupyter Notebooks allows students to learn Python through instructor-led exercises, facilitating a smooth transition from traditional Excel tasks to Python-based data analysis [9].

The use of Python for processing Excel data has been applied in various academic subjects. A textbook for undergraduate chemistry students teaches basic Python. This resource has been successfully used to process NMR data and other scientific datasets [10] In a macroeconomics course, students use Python to analyze and simulate economic models, gaining familiarity with libraries like NumPy and Matplotlib. This approach has been well-received by students, who find it valuable for connecting theory to practice [11]. The researchers emphasize that real-life examples and case studies help students understand the practical applications of Python. For example, a study on big financial data analysis used business scenarios to teach Python-based data processing [12]. A study of the problems of code modification for Excel operations in the Python programming learning assistant system [13].

One study described how students used the Python programming language and a spreadsheet to simulate particle trajectories affected by electric and magnetic fields during online learning [14]. Other authors described the process of implementing the Pandas (Phyton) library to manage data obtained from Excel files that contained student report cards [15]. The lower resource intensity and therefore higher efficiency of Python compared to MS Excel is a well-known fact, but teachers need to be trained to optimize code for large data sets [1]. Students often encounter difficulties when they try to switch from Excel to Python programming, especially when it comes to writing scripts for data processing. However, previous studies of the process of switching from Excel to Python programming have found that if teachers use practical exercises and interactive tools in their classes, then students perceive this transition more positively [8, 16]. At the end of the literature review, I would like to mention a number of our studies where we used Excel and Python integration methods in our scientific research, and their preliminary results were discussed with students in economics and management courses [17–23].

The literature emphasizes the practical importance and relevance of including Python in the educational process when teaching academic subjects at universities. Python's popularity in developing data science skills is determined by its simplicity and versatility.

3 Methodology

The article uses files of statistical reporting form No. VPO-1 "Information about the organization carrying out educational activities under higher education programs - bachelor's, specialist, master's degree" as an example of processing several Excel files. Form VPO-1 is publicly available on the website of the Ministry of Higher Education and Science of the Russian Federation at: https://minobrnauki.gov.ru/action/stat/highed. For each year, form VPO-1 is available for downloading as an archive with the extension .zip. You need to download the files for each year, unpack the zip archives in any archiver, after which folders containing individual Excel files will be created on the user's computer. The folders created on the user's computer under the following names: "VPO-1 Codes 2022" and "VPO-1 Codes 2021", contain two folders: "State" and "Non-state". Each of these folders contains files with the extension "xls". Each of these folders contains files with the extension "xls". We need to move the following files from the folder named "State" to the folder "Admission": files in the name of which there is an indication of the form of study: "_full-time", files in the name of which there is an indication of the form of study: "_correspondence", files in the name of which there is an indication of the form of study: "_full-time and part-time". Python libraries such as Pandas and NumPy were used to process the data. Pandas provided convenient tools for data manipulation, including reading Excel files, merging dataframes, and cleaning data, while NumPy was used to perform numerical operations and process arrays of data. The data processing process began with extracting information from the Excel files using the pd.read_excel() function.

This skipped the first 11 rows and limited the number of rows and columns to improve processing efficiency. We used the pd.concat() function to combine all the received dataframes into one. At the data cleaning stage, unnecessary columns were removed, the remaining ones were renamed to make them more informative, and new columns were added containing additional information, such as year and region. To ensure correct interpretation of numerical data, data types were converted to float64, which allowed avoiding errors in further calculations and analysis. After preparing and cleaning the data, the resulting data frame was ready for further analysis.

4 Results

This section provides a detailed description of the code used to process and analyze the data of the statistical reporting form No. VPO-1 for 2022. The goal was to prepare a data frame for further analysis of student admission data. Next we tried to explain the code we wrote to extract data from xls files. from the "P2_1_1" sheet, divided for clarity into two main parts (Figs. 1 and 2).

The code starts by importing the necessary libraries: pandas for data manipulation and numpy for numerical operations. Data source: the process begins with the identification of a specific directory containing multiple Excel files (*.xls) related to the research topic. The path to this directory is defined in the code. File handling: a list is created to store dataframes, which will hold the data extracted from each Excel file.

This step is crucial for organizing the data before further analysis. Data extraction: the code iterates through each Excel file found in the specified directory. For each file the pd.read_excel() function is used to read a specific sheet ('P2_1_1 (2)') while skipping the first 11 rows and limiting the data to the first 1000 rows and columns A to K. This ensures that only relevant data is processed. A new column, 'file_name', is added to each dataframe to keep track of the source file for each piece of data. Data consolidation: all individual dataframes are combined into a single dataframe (df22) using pd.concat(). This step consolidates the data from all the files into one comprehensive dataset.

Figure 2 shows a snippet of code for cleaning and transforming the data. A data frame named df22 was created by retrieving data from multiple Excel files. Our code contains a data pruning routine that removes columns with indices 4–7, 9–15, and 17. After these columns, we have a data set that contains only the data we need.

The figure highlights the renaming of several columns to improve clarity and facilitate understanding. Each original index is replaced with descriptive names.

```
import pandas as pd
import numpy as np

# Defining the path to the folder with 2022 data
path22 = r'/Users/ilshatgarafiev/Downloads/Своды ВПО-1 2022 г/Прием'

# Listing all .xls files in the folder
files22 = Path(path22).glob('*.xls')

# Creating an empty list to store dataframes
dfs22 = list()

# Reading each file and appending it to the list
for f in files22:
    data22 = pd.read_excel(f, sheet_name='P2_1_1', skiprows=12, nrows=1000, usecols='A:Q')
    data22['file_name'] = f.stem  # Adding the filename as a column
    dfs22.append(data22)

# Combining all dataframes into one
df22 = pd.concat(dfs22, ignore_index=True)
```

Fig. 1. Loading and Combining Data sheet «P2_1_1» (Source: author).

```
# Dropping unnecessary columns
df22.drop([4, 5, 6, 7, 9, 10, 11, 12, 13, 14, 15, '17'], axis=1, inplace=True)

# Renaming columns
df22.rename(columns={
    1: 'napravlenie_podgotovki',  # Field of study
    2: 'code_stroki',  # Row code
    3: 'code_napravlenie_podgotovki',  # Field of study code
    8: 'priem_itogo_all_budget_vnebudget',  # Total admission (budget + non-budget)
    16: 'priem_vnebudget'  # Non-budget admission
}, inplace=True)

# Splitting the filename into region, university type, and form of study
df22[['region', 'vuz_type', 'forma_obuchenia']] = df22['file_name'].str.split('_', 2, expand=Tr
ue)

# Dropping the file_name column
df22.drop(['file_name'], axis=1, inplace=True)

# Adding a year column with the value 2022
df22.insert(8, 'year', '2022')
```

Fig. 2. Cleaning and transforming data sheet «P2_1_1» (Source: author).

Column 1 is renamed to 'napravlenie_podgotovki' (Field of study).
Column 2 is renamed to 'code_stroki' (Row code).
Column 3 is renamed to 'code_napravlenie_podgotovki' (Field of study code).
Column 8 is renamed to 'priem_itogo_all_budget_vnebudget' (Total admission, including budget and non-budget).

Column 16 is renamed to 'priem_vnebudget' (Non-budget admission).

Splitting filename for additional attributes: the figure illustrates the extraction of additional attributes from the 'file_name' column. The filename is split into three new columns: 'region', 'vuz_type' (university type), and 'forma_obuchenia' (form of study). This transformation enriches the dataset with contextual information. Splitting filename for additional attributes: the figure illustrates the extraction of additional attributes from the 'file_name' column. The filename is split into three new columns: 'region', 'vuz_type' (university type), and 'forma_obuchenia' (form of study). This transformation enriches the dataset with contextual information.

Dropping the original filename column: after extracting the necessary information, the original 'file_name' column is removed from the dataframe to reduce redundancy. Adding a year column: finally, a new column named 'year' is added to the dataframe, with a constant value of '2022'. This addition provides temporal context to the data, facilitating future analyses that may require year-based filtering or comparisons.

Next we tried to explain the code we wrote to extract data from xls files. from the "P2_1_" sheet, divided for clarity into two main parts (Figs. 3 and 4).

```
import pandas as pd
import numpy as np

# Defining the path to the folder with 2022 data
path22 = r'/Users/ilshatgarafiev/Downloads/Своды ВПО-1 2022 г/Прием'

# Listing all .xls files in the folder
files22 = Path(path22).glob('*.xls')

# Creating an empty list to store dataframes
dfs22 = list()

# Reading each file and appending it to the list
for f in files22:
    data22 = pd.read_excel(f, sheet_name='P2_1_1 (2)', skiprows=11, nrows=1000, usecols='A:K')
    data22['file_name'] = f.stem  # Adding the filename as a column
    dfs22.append(data22)

# Combining all dataframes into one
df22 = pd.concat(dfs22, ignore_index=True)
```

Fig. 3. Loading and Combining Data sheet «P2_1_1 (2)» (Source: author).

The code for loading and combining the passport "P2_1_1(2)" is identical to the code for loading and combining the passport "P2_1_1" described earlier.

The code uses the pandas renaming method to give the columns in df22 more meaningful names.

We renamed the columns in our dataset. The columns had numeric indices, and they were renamed based on the names of the indicators they reflect. The names of the indicators were written in Russian, but transliterated. So column 18 was renamed

```python
# Renaming columns
df22.rename(columns={
    1: 'napravlenie_podgotovki',  # Field of study
    2: 'code_stroki',  # Row code
    3: 'code_napravlenie_podgotovki',  # Field of study code
    18: 'priem_itogo_all_budget_vnebudget_obrazovanie_drugoi_region',  # Total admission from other regions
    19: 'priem_budget_obrazovanie_drugoi_region',  # Budget admission from other regions
    20: 'priem_itogo_all_budget_vnebudget_obrazovanie_drugoe_gosudarstvo',  # Total admission from other states
    21: 'priem_budget_obrazovanie_drugoe_gosudarstvo',  # Budget admission from other states
    22: 'priem_itogo_all_budget_vnebudget_inostrancy',  # Total admission of foreign students
    23: 'priem_itogo_vnebudget_inostrancy',  # Non-budget admission of foreign students
    24: 'priem_itogo_all_budget_vnebudget_inostrancy_SNG',  # Total admission of students from CIS countries
    25: 'priem_vnebudget_inostrancy_SNG'  # Non-budget admission of students from CIS countries
}, inplace=True)

# Splitting the filename into region, university type, and form of study
df22[['region', 'vuz_type', 'forma_obuchenia']] = df22['file_name'].str.split('_', 2, expand=True)

# Dropping the file_name column
df22.drop(['file_name'], axis=1, inplace=True)

# Adding a year column with the value 2022
df22.insert(14, 'year', '2022')
```

Fig. 4. Cleaning and transforming data sheet «P2_1_1 (2)» (Source: author).

"priem_itogo_all_budget_vnebudget_obrazovanie_drugoi_region", since this column contains data on the total number of students admitted from other regions, both at the expense of the Russian state budget and on the basis of contracts for the provision of paid educational services.

Column 19 was renamed "priem_budget_obrazovanie_drugoi_region", since this column contains data on the total number of students admitted from other regions, only at the expense of the Russian state budget. Column 20 has been renamed to 'priem_itogo_all_budget_vnebudget_obrazovanie_drugoe_gosudarstvo', since this column contains data on the total number of students admitted from other countries, both at the expense of the Russian state budget and on the basis of contracts for the provision of paid educational services. Column 21 has been renamed 'priem_budget_obrazovanie_drugoe_gosudarstvo', since this column contains data on the total number of students admitted from other countries at the expense of the Russian state budget. Column 22 has been renamed to 'priem_itogo_all_budget_vnebudget_inostrancy', since this column contains data on the total number of foreign students admitted at the expense of the Russian state budget and on the basis of contracts for the provision of paid educational services.

Column 24 was renamed "priem_itogo_all_budget_vnebudget_inostrancy_SNG", since it contains data on the total number of students from the CIS countries accepted at the expense of the state budget of the Russian Federation and on the basis of agreements on the provision of paid educational services. Column 25 was renamed "priem_vnebudget_inostrancy_SNG", since it contains data on the total number of students from the CIS countries accepted on the basis of agreements on the provision of paid educational services.

Then we need to combine the data, remember that we got it from two sheets, into one data set. The code converts the data types of specific columns in the dataframe df to float64 (Fig. 5). This ensures that numerical columns are correctly interpreted for calculations and analysis.The pd.to_numeric() function is used to perform the conversion.

```
# Changing the data type of 'priem_itogo_all_budget_vnebudget' to float64
df['priem_itogo_all_budget_vnebudget'] = pd.to_numeric(df['priem_itogo_all_budget_vnebudget'], downcast='float', errors='coerce')

# Changing the data type of 'priem_vnebudget' to float64
df['priem_vnebudget'] = pd.to_numeric(df['priem_vnebudget'], downcast='float', errors='coerce')

# Changing the data type of 'priem_itogo_all_budget_vnebudget_obrazovanie_drugoi_region' to float64
df['priem_itogo_all_budget_vnebudget_obrazovanie_drugoi_region'] = pd.to_numeric(df['priem_itogo_all_budget_vnebudget_obrazovanie_drugoi_region'], downcast='float', errors='coerce')

# Changing the data type of 'priem_budget_obrazovanie_drugoi_region' to float64
df['priem_budget_obrazovanie_drugoi_region'] = pd.to_numeric(df['priem_budget_obrazovanie_drugoi_region'], downcast='float', errors='coerce')

# Changing the data type of 'priem_itogo_all_budget_vnebudget_obrazovanie_drugoe_gosudarstvo' to float64
df['priem_itogo_all_budget_vnebudget_obrazovanie_drugoe_gosudarstvo'] = pd.to_numeric(df['priem_itogo_all_budget_vnebudget_obrazovanie_drugoe_gosudarstvo'], downcast='float', errors='coerce')

# Changing the data type of 'priem_budget_obrazovanie_drugoe_gosudarstvo' to float64
df['priem_budget_obrazovanie_drugoe_gosudarstvo'] = pd.to_numeric(df['priem_budget_obrazovanie_drugoe_gosudarstvo'], downcast='float', errors='coerce')

# Changing the data type of 'priem_itogo_all_budget_vnebudget_inostrancy' to float64
df['priem_itogo_all_budget_vnebudget_inostrancy'] = pd.to_numeric(df['priem_itogo_all_budget_vnebudget_inostrancy'], downcast='float', errors='coerce')

# Changing the data type of 'priem_itogo_vnebudget_inostrancy' to float64
df['priem_itogo_vnebudget_inostrancy'] = pd.to_numeric(df['priem_itogo_vnebudget_inostrancy'], downcast='float', errors='coerce')

# Changing the data type of 'priem_itogo_all_budget_vnebudget_inostrancy_SNG' to float64
df['priem_itogo_all_budget_vnebudget_inostrancy_SNG'] = pd.to_numeric(df['priem_itogo_all_budget_vnebudget_inostrancy_SNG'], downcast='float', errors='coerce')

# Changing the data type of 'priem_vnebudget_inostrancy_SNG' to float64
df['priem_vnebudget_inostrancy_SNG'] = pd.to_numeric(df['priem_vnebudget_inostrancy_SNG'], downcast='float', errors='coerce')
```

Fig. 5. Converting column data types to float64 (Source: author).

The downcast = 'float' argument ensures that the data is stored as efficiently as possible, while errors = 'coerce' ensures that any non-numeric values are converted to NaN (Not a Number) to avoid errors.

The results of this data science process are a well-structured and enriched database containing comprehensive information on the 2022 student intake.

5 Discussion

Our study showed that introducing Python programming elements into academic courses in economics and management contributes to the development of data science skills. Python's accessibility and ease of use make it a de facto choice for starting courses on data cleaning and merging, even for students with no prior programming experience.

The proposed algorithm for processing data from multiple Excel files complements the existing literature and fills a gap in it. Microsoft Excel and Python are software that are often used for data processing. Microsoft Excel has limitations in terms of speed and smoothness of data processing. Therefore, the Python programming language can be used as an alternative.

The practical significance of the work is that, firstly, students were able to overcome common data cleaning problems, such as handling missing values and inconsistencies, and secondly, the process of renaming columns and creating new attributes from file names not only optimized the data set but also showed students the need to label the data. Despite these advantages of using Python as an alternative to Microsoft Excel, confirmed in several previous studies [1, 9, 11], switching from Excel to Python is not without its challenges. Initially, students will have a hard time understanding Python syntax, but if the teaching practice uses interactive tools such as Jupyter Notebooks and the assignments themselves have a hands-on approach, then this transition will be much easier and less painful.

Compared to previous studies, our approach stands out in that it focuses on integrating Python into existing courses without significant restructuring. This is becoming increasingly popular among a certain type of university professors. These professors can be called supporters of traditional analog universities who do not integrate digital technologies into education, citing a lack of time and other resources.

6 Conclusion

This study highlights the significant potential of integrating Python into educational programs for teaching data science skills. The proposed algorithm for processing data from multiple Excel files simultaneously leverages Python libraries like Pandas and NumPy, requiring no specialized training. When students work with real datasets, they not only strengthen their technical abilities but also gain a clearer understanding of the full data lifecycle, which includes extracting, cleaning, and analyzing information.

The authors would like to emphasize once again that modern society functions and develops on the basis of data that is generated in huge quantities. In this regard, it is especially relevant to develop students' skills in working with data based on Python.

The authors' algorithm, based on Python, can be implemented in any university; supercomputers are not required for this. Future researchers will be able to study new pedagogical methods for teaching Python and how gaining knowledge in data science impacts students' employment opportunities after graduation.

To summarize, we can say that this study confirmed the importance and relevance of introducing Python into everyday educational practices when teaching economic and management academic subjects at the university. By cultivating a culture of inquiry and critical thinking, Python-based data science education can empower students to become informed and capable participants in an increasingly data-driven world.

Acknowledgments. The authors express their gratitude to the administration of the Kazan National Research Technological University for support in registering the described codes as part of the author's computer programs.

We have no conflicts of interests to disclose.

References

1. Kelvin, K., Wahab, W., Wulandari, M.: Computer resource utilization analysis for Microsoft excel and python in data processing. Eng. Math. Comput. Sci. J. (EMACS). **6**(2), 137–142 (2024). https://doi.org/10.21512/emacsjournal.v6i2.11736
2. Makarov, A., Namiot, D.: Overview of data cleaning methods for machine learning. Int. J. Open Inf. Technol. **11**(10), 70–78 (2023)
3. Hosseinzadeh, M., Azhir, E., Ahmed, O.H., Ghafour, M.Y., Ahmed, S.H., Rahmani, A.M., Vo, B.: Data cleansing mechanisms and approaches for big data analytics: a systematic study. J. Ambient. Intell. Humaniz. Comput. **14**, 1–13 (2023)
4. Ridzuan, F., Zainon, W.M.N.W.: A review on data cleansing methods for big data. Procedia Comput. Sci. **161**, 731–738 (2019)
5. Fatima, A., Nazir, N., Khan, M.G.: Data cleaning in data warehouse: a survey of data pre-processing techniques and tools. Int. J. Inf. Technol. Comput. Sci. **9**(3), 50–61 (2017)
6. Srivastava, P., Kaur, N.: An Overview on Data Cleaning on Real World Data. Authorea Preprints (2023)
7. Alotaibi, O., Pardede, E., Tomy, S.: Cleaning big data streams: a systematic literature review. Technologies. **11**(4), 101 (2023)
8. Kazil, J., Jarmul, K.: Data Wrangling with Python. O'Reilly (2016)
9. Mansour, M.F., Aly, T., Gheith, M.: Python based end user computing framework to empowering excel efficiency. Int. J. Res. Appl. Sci. Eng. Technol. (2024). https://doi.org/10.22214/ijraset.2024.60097
10. Weiss, C.J.: A creative commons textbook for teaching scientific computing to chemistry students with python and Jupyter notebooks. J. Chem. Educ. **98**(2), 489–494 (2021). https://doi.org/10.1021/ACS.JCHEMED.0C01071
11. Jenkins, B.C.: A python-based undergraduate course in computational macroeconomics. J. Econ. Educ. **53**(2), 126–140 (2022). https://doi.org/10.1080/00220485.2022.2038322
12. Shi, R., Wang, X.: Research on innovative teaching models for the course of financial big data analysis using python based on the concept of outcome-based. Education. **1**(3), 208–213 (2024). https://doi.org/10.62517/jhet.202415334
13. Funabiki, N., Wai, K.H., Aung, S.L., Kao, W.-C.: A Study of Code Modification Problems for Excel Operations in Python Programming Learning Assistant System, pp. 209–213 (2022)

14. Purnama, A.Y., Istiyono, E., Putranta, H., Rani, S.A., Wijayanti, A., Saputri, R.: Particle trajectory simulation using python and spreadsheet as an online learning alternative. Rev. Mexicana de Física E. **20**(2), 020202 (2023)
15. Irsan, M., Husain, A.: Implementasi Aplikasi Pandas (Phyton) Dalam Mengelola Data Excel Sebagai Media Persiapan Pelaporan Nilai Raport Siswa. Jurnal Pengabdian Masyarakat Bangsa. **2**(4), 1243–1249 (2024)
16. A Study of Code Modification Problems for Excel Operations in Python Programming Learning Assistant System. (2022). https://doi.org/10.1109/iciet55102.2022.9779025
17. Garafiev, I.: Network Communities and Groups in the Preparation for the Unified State Exam, pp. 221–227 (2024)
18. Garafiev, I.: Level of Engagement Based on Analysis of Community Posts on VKontakte Network, pp. 46–57 (2024)
19. Garafiev, I., Garafieva, G., Idiatullina, A., Spirchagova, E.: Assessment of online education problems during the COVID-19 pandemic in Russia. Indones. J. Electr. Eng. Comput. Sci. **23**, 1689 (2021). https://doi.org/10.11591/ijeecs.v23.i3.pp1689-1698
20. Garafiev, I., Garafieva, G.: Territorial features of value orientations and satisfaction with the studies of student youth in the republic of Tatarstan. E3S Web Conf. **210**, 18134 (2020). https://doi.org/10.1051/e3sconf/202021018134
21. Garafiev, I.: Regional Features of Research and Development Cooperation Between Innovative Organizations and Partners, pp. 265–274 (2024)
22. Garafieva, G.: The Role of Capital Assets in the Sustainable Development of Industrial Production, pp. 96–104 (2024)
23. Garafiev, I., Garafieva, G.: Training specialists in water resources in Russia: typology of educational programs. WSEAS Trans. Environ. Dev. **20**, 910–1923 (2024). https://doi.org/10.37394/232015.2024.20.85

Model for the Development of Technical University Teachers Digital Competencies

Elvira Gayazova[(✉)] and Tatyana Nikitina

Kazan National Research Technological University Kazan 68, K. Marx str., Kazan, Russia
klabpnr@gmail.com

Abstract. The paper is devoted to the problem of creating a unified model of digital competencies of higher education teachers. Focusing on universal digital competencies leaves aside the training of specialists for specific enterprises with their own digital resources. The models presented in the scientific literature define as constituent elements the educational ones directly related to the activities of an educational institution. The paper discusses the aspects of improving the presented models related to the needs of the labor market and specific industry in specialists. The authors of the paper have conducted a study aimed at expanding the understanding of the model of teachers' digital competence development at an engineering university based on the opinions of experts from education, digital technologies and employers. The results of the study confirmed the hypothesis: the focus on the universalization of digital competencies and weak attention to the need of enterprises in specific digital skills in employees, leads to the lack of specific digital competencies in future specialists due to the specifics of digitalization of the enterprise. The inclusion of specific digital technologies in the practical training, as well as the accompaniment of teachers by tutors from the enterprise will further create the basis for improving the quality of training for specific enterprises or industry needs.

Keywords: digitalization of education · digital environment · digital competencies · labor market · professional training of engineering personnel · tutoring institute

1 Introduction

Quality training of engineering specialists includes the formation of digital competencies, readiness to work in the conditions of digital restructuring of technological processes at production facilities. The main task is to introduce into the learning process the means and methods that ensure the acquisition of digital skills. This can be achieved by fostering a unique digital work environment within the university. In today's educational landscape, educators must continuously upgrade their digital skills and participate in professional development programs. This ensures they meet the evolving needs of the job market and effectively teach digital technologies to their students.

Ultimately, in modern conditions, the effectiveness of the educational process depends on the mastery of digital competencies by students. At the same time, the

M. Lapina et al. (Eds.): DET 2024, LNNS 1616, pp. 150–160, 2025.
https://doi.org/10.1007/978-3-032-04365-8_15

need to take into account the specifics of the industry and digitalization of specific enterprises can become the basis for the economy. A specialist with his or her own digital potential is the goal of modern digital education. This places new demands on the digital competence model of teachers, which should also include the specific needs of the enterprise. The universal digital competencies possessed by a teacher should be supplemented by specialized, industry-specific competencies. This aspect should be included in the model of digital competencies of teachers. The study of digital technologies of a particular enterprise is possible when introducing the institute of tutoring in the program of teachers' professional development. The work discusses the topical issues of formation of an innovative model for the development of digital competencies of teachers in the higher education system in accordance with the demands of the labor market and the needs of enterprises in qualified personnel.

The research aims to enhance the current model for developing digital skills among faculty members at technical universities. By examining educational theories, conducting studies, and drawing from teaching experiences, the study seeks to improve the existing framework. Additionally, the research intends to demonstrate how the updated model can effectively address the growing need for digital capabilities in higher education.

Artificial intelligence methods and machine learning algorithms are currently used both in manufacturing, to improve automation efficiency, and in medicine, medicine, and financial structures for big data analysis. With the growing popularity of language models and the launch of open source GPT, artificial intelligence methods have entered even more areas of human activity [22, 23].

Improving digital competencies in modern education is a prerequisite for teacher competence. A modern teacher is obliged to possess digital skills, first, to conduct training sessions at a high professional level. However, the existing barriers do not allow the teacher to apply them in practical pedagogical activity.

Let us consider the reasons that reduce the effectiveness of pedagogical labor. First, the reasons of organizational nature. They include insufficiently developed information and communication environment and infrastructure in the university, insufficient equipment of classrooms with computer equipment.

Second, low or below average levels of digital competencies among teaching and administrative staff. Completing additional training does not mean that the teacher can freely use digital technologies or that it takes a large number of hours.

Thirdly, there are no detailed administrative guidelines, instructions or legal acts on the mandatory use of digital technologies in the teaching process. All the above-mentioned reasons hinder the introduction of digital technologies in the educational process.

Let us note a number of other reasons, which are not obvious in the modern system of higher education. Perhaps they are the main ones and determine the quality of the educational process. Such reasons are the lack of tutors or IT specialists to support the learning process. Lack of a unified approach to the formation of digital competencies in different academic disciplines. Lack of criteria for evaluating the use of digital technologies. Motivation of Russian teachers to implement digital technologies is a topical issue.

Despite the obvious reasons that hinder digitalization in HEIs, it is necessary to talk about the further development of digitalization of the educational process and look to the future. Practical training in higher education consists in mastering competencies directly at enterprises or organizations that have their own digital potential and digital technologies. This means that the training process should also include the mastering of those digital processes that are in demand in the company in question. In addition to universal digital competencies, the teacher should also have specific digital competencies that are specific to a particular enterprise. This can be expressed in the form of case tasks offered for students to solve in practical classes. Professional development of teachers in the field of digital competencies can in this case take place at enterprises as internships. For this purpose, it is necessary to envisage the institute of tutoring in the field of studying specific digital technologies. This will facilitate the integration of specific digital tools into the educational process.

2 Methodology

The methodology of our study is based on the competence approach. It is used to determine the key digital competencies of university teachers, which are acquired in the process of professional development in accordance with scientific interests and professional interests.

Expert interviewing of representatives of enterprises, whose responsibilities include managing the digitalization of the enterprise, and university employees were used as research methods. The modeling method becomes the leading one in the development of mechanisms for implementing digital technologies in the educational process at the university.

The classical model is based on the unification of digital competence components. It includes diagnostics of the level of digital competence possession among teachers. It implies the creation of a common normative basis and common criteria of digital indicators for all digital components.

Digital competencies of university teachers are defined as readiness to effectively use information and communication technologies in the teaching process. These technologies can be applied during the preparation of lecture and practical materials, in the organization of various types of practice. It is a system of knowledge, skills, responsibility and motivation, which are acquired in the process of professional development, taking into account the professional orientation.

The model contains organizational and pedagogical conditions for the application of digital competencies by teachers. They include the content, forms and methods of the process of applying digital competencies in the practice of teaching a particular discipline or area of training.

The model for integrating digital competencies into the educational process of higher education institutions is created by considering several key factors. These include the implementation of innovative practices, social relevance, the advancement of higher education in the country, and the development of cutting-edge pedagogical technologies and digital tools for professional use [1, 21].

In the conditions of transformation of the education system in our country, associated with the departure from the Bologna system, the model of application of digital competencies will allow creating individual trajectories of professional and personal development of university teachers, to realize the need for competence improvement.

The need to include a new component in the unified model of digital competencies improvement is determined by the activity of a particular enterprise and its needs for specialists with specialized digital knowledge. Let us call this component "specialized digital competencies". Specialized digital competencies are the ability of teachers to create digital learning content in accordance with the digital processes of a particular enterprise - the place of future industrial practice. To implement the component it is necessary to retrain teachers, organize professional development and internships in the field of digitalization of a particular enterprise. For this purpose it is necessary to organize as new types of interactions the institute of tutoring in providing the educational process with digital technologies included in the activities of the enterprise.

3 Literature Review

The model of development of digital competencies of university teachers is based on sources from different scientific directions. This indicates not only the relevance of the study of the problem of digitalization of education in different scientific directions, but also points to the difficulties in considering various aspects of the model and its content. First, the scientific literature pays much attention to the informatization of education, the development of conceptual provisions for the introduction of information and communication technologies in the educational process. A. A. Strokov considers digitalization from the position of digital culture of modern Russian society in the dynamics of its impact on the modernization of the education sphere. The author proves that the analysis of the problems of digitalization of education in the conditions of the development of modern society and the introduction of new information and communication technologies is possible only when the philosophical and cultural approach is included in the methodology. According to the author, the risks of digitalization for society lie in the dehumanization of educational and social relations, deepening the crisis of people's intellectual culture [17].

A.YU. Uvarov considers the digital transformation of education as a systemic renewal of the basic components of the educational process. The author describes the SAMR model (The Substitution-Augmentation-Modification-Redefinition Model) as a priority in the development of a digital school. The model includes substitution of traditional pedagogical tools, improvement of traditional pedagogical tools, change of pedagogical practices, transformation of pedagogical practices [20].

V.M. Minina defines the peculiarities of creating digital schools and the basic components of the educational process, which include the results of educational work, the content of education, the organization of the educational process, and the assessment of its results [13].

Based on the theory of P. Gilster, the digital competence of a teacher consists in achieving the following criteria: skills of searching for the necessary information and tools for working with it, the ability to quickly master these tools (information competence); skills of communication with other users (communicative competence); skills of producing information in its various forms and formats (creative competence) [8].

The scientific community is developing different variants of models for implementing digital competencies in the educational process. In 2017, a model of digital competence for teachers (DigCompEdu) was developed for the first time. It consists of groups of competencies, there are 21 of them in total. Eight levels of competence expression are identified, organized into four groups, which depend on the complexity and comprehensiveness of the task; on the readiness to solve them independently and the dominant cognitive domain (memorization, understanding, application and creation). The Digital Competence of Educators (DigCompEdu) teacher digital competence profile, includes six areas of digital competence of educators focused on [2]:

- on the use of digital technologies in the professional teaching environment;
- developing professional skills in searching, creating and sharing digital educational resources;
- developing the necessary skills to use digital tools in teaching;
- mastery of digital tools for assessing learning outcomes;
- use of digital tools to expand students' educational opportunities;
- supporting the process of developing students' digital competence.

Osipov P.N., Gayazova, E.B., Girfanova, E.Y. have developed a model of professional training of financial managers. The model contains normative, substantive, procedural-methodological and resultant-evaluative blocks and consists of interrelated structural and functional elements that reveal the regulatory and legal aspects, the implementation of the competence approach in the process of training financial managers [9, 15].

S.A. Naumchenko developed a model for improving digital competencies of teaching staff. The model defines the following components: target component; methodological component; diagnostic component of the model as a systematic process of determining the level of digital competencies of teaching staff; organizational and management component in the form of creating special conditions based on educational and management technologies and aimed at solving the set goals and objectives; the educational component involves the organization of the process of improving digital competencies of teaching staff with the help of tutors and mentors. The model provides support for digitalization by mentors [14]. These are experts from educational communities with advanced digital competencies who will provide expert, consulting and psychological support to educators who improve their digital competencies. In the future, based on the diagnostics of the level of formed competencies, the mentor creates an individual educational trajectory of the teacher depending on the results of diagnostics and personal needs. The last component is the result component. It includes criteria for evaluating the effectiveness and the result of improving digital competencies of pedagogical staff.

The program "Digital Economy in the Russian Federation" (2019) defines the basic model of digital economy competencies as a normative document that establishes a system of uniform requirements for the formation, continuous and successive build-up of digital economy competencies throughout a person's life.

N.V. Maksimenko and T.A. Chekalina conducted a comparative analysis of approaches to the models of digital competencies of teachers: DigCompEdu, TRACK. The authors developed their own generalizing model of digital competencies of a teacher. This model includes five areas of digital competencies of a teacher:

1. information competencies - the teacher's ability to work with information resources (apply, find, store, transform, and perform their critical and methodological analysis);
2. communication competencies - the teacher's ability to use various forms of communication (e-mail, chats, blogs, forums, social networks, etc.) and work with cloud technologies;
3. media competencies - the teacher's ability to create information objects using digital resources (text, image, audio and video) and actively apply them in the organization of the educational process;
4. technological competencies - a teacher's ability to effectively and safely use technical and software tools to solve various tasks, including the use of computer networks, cloud services, etc..;
5. information security competencies - the teacher's ability to assess the risks of working in the digital space, knowledge of measures to ensure the security of personal data, as well as understanding of the negative impact that digital devices have on the environment, physical and mental health [12].

The Technological Pedagogical Content Knowledge (TPACK) model assumes three groups of competencies: technological, pedagogical and content knowledge [18]. Technological competencies are presented (see Fig. 1).

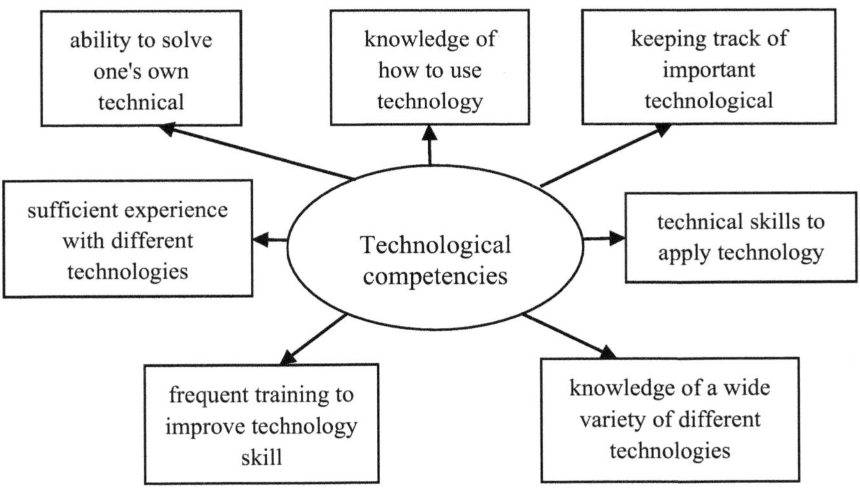

Fig. 1. Technological competencies in the Technological Pedagogical Content Knowledge model.

Foreign and domestic studies pay much attention to the risks associated with the introduction of digital technologies in the educational process.

E. V. Semenov, D. V. Sokolov study the opportunities and risks associated with the digitalization of higher education. The accessibility and flexibility of education are noted. Risks, according to the authors, are associated with market demands, reduction in the content and fundamentality of education, fragmentation of knowledge and replacement of full-fledged socialization by simple training of workforce [16].

Y. Y. Eleneva examines the state and key areas of digital transformation necessary to achieve the level of digital maturity of an educational organization of higher education. Special attention in this study is paid to the review of features, problems and risks arising in the course of digital transformation [3].

A study conducted by a team of authors within the framework of a three-year Erasmus + project has identified the future needs of teachers in relation to digital literacy in the school context [10]. Falloon G. presents a conceptual framework that expands the view of teachers' digital competencies. The author writes about the complex knowledge and skills that young people need to work in a variety of digital environments [4]. Fernández-Batanero, J. M., Montenegro-Rueda, M., Fernández-Cerero, J., García-Martínez, I. conclude in their study that teachers can acquire the necessary knowledge, skills, and enhance digital competencies to compete successfully in a technology-based world [5].

4 Results

At the preliminary stage of the study, a survey was conducted to determine the level of implementation of information and communication technologies in the educational process, in the development and implementation of lecture and practical materials. 78 employees of the engineering university took part in the survey. The results of the survey showed that the preparation of lecture material and practical tasks is determined by the level of computer technology proficiency. At the same time, young specialists track innovations in this area more often than employees with more than 30 years of experience in the university. Next figure shows the distribution of respondents' answers by basic competence in the digital environment (see Fig. 2).

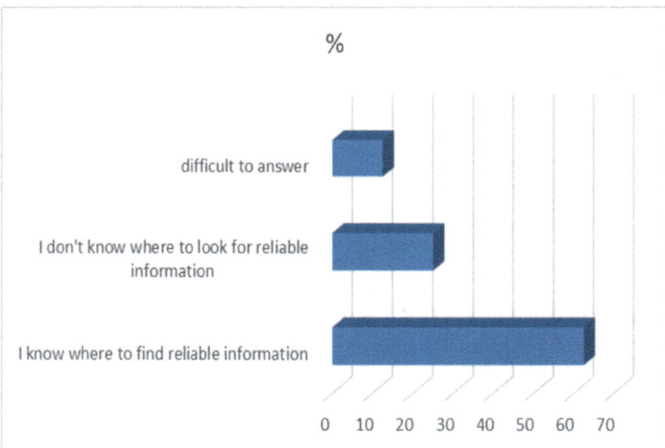

Fig. 2. Distribution of answers to the statement "I know where to find reliable information".

The study revealed the methods and tools most frequently used by the staff, the skills acquired and, accordingly, the level of digital competence formation. Despite the fact that all employees have undergone advanced training and mastered information technologies, their active use is hindered by both technological factors, which consist in the lack of equipping classrooms with information technology tools, and the low readiness of students to work in conditions of individual assignments. It was noted that there was a lack of time to develop practical interactive lessons [7].

The second set of questions focused experts' opinions on the content of the model of digital competencies development. Most agree that digitalization of higher education should involve two areas:

- Providing all participants of the educational process with universal digital competencies, uniform in their basis;

-content in the curriculum of special components defined by the digitalization of enterprises and industry for which the higher education institution provides training [6].

Summarizing the experts' answers, the authors included several indicators in the model of digital competence development. The following Table 1 gives a summary of these competencies (Table 1).

Table 1. Digital competence development model.

Competence	Summarizing
Professional motivation and reflection	It involves the use of digital technologies and resources for professional communications. This area also includes self-reflection, critical evaluation of professional activities and continuous professional development of engineering university teachers. Diagnostics of the level of digital competencies development is included
Universal digital resources	It includes the ability to work with digital content (use ready-made, modify, create own), select digital resources depending on educational and professional goals, manage them and ensure their safe use
Specialized Digital Resources	Includes readiness to work in the digital environment in connection with the activities of the enterprise. Preparation of tasks of different levels of complexity in connection with the digitalization of technological processes. Working with tutors from the enterprise in the field of digitalization of the educational process. Participation in professional internships for the development of digital competencies due to the digital processes of the particular enterprise [11]
Training	Consists of the use of digital technologies in pedagogical activities, mentoring, organizing the interaction of learners and supporting their autonomy in the learning process

(continued)

Table 1. *(continued)*

Competence	Summarizing
Personal component	Provides psychological comfort in the digital environment during professional interactions, as well as professional development and individualization of the educational process
Cognitive component	Responsible for analyzing information, critical thinking, creative use of digital technologies for education and production of new knowledge
Communicative component	Provides professional involvement in interaction with all participants of the educational process through digital technologies [19]

While it is not imperative for all faculty members within a department to specialize in digital technologies, particularly when their respective disciplines do not inherently incorporate such technologies, the pedagogical imperative of educational continuity must be upheld across the curriculum. Consequently, it is incumbent upon educators to employ a unified and comprehensible terminological framework.

5 Conclusion

The introduction of digital technologies in the activities of engineering university teachers is based on the integration of competence-based, personality-oriented and activity-based approaches. The modeling method allowed to reveal new components in the model of digital competencies development related to modern labor market requirements. Based on the expert interview, conclusions were drawn about the relevance of the model formation in the direction of universal and specialized digital competencies. The differentiation will allow to define the scope of application, create an algorithm of practice-oriented training, and improve the quality of educational material. The peculiarities of the educational environment of an engineering university on the implementation of digital technologies in the learning process are considered. The scientific novelty of the study lies in the allocation of a specialized component in the model of digital competence development. The introduction of the parameter - the activity of the tutoring institute - will allow to create a basis for harmonious implementation of digital competencies in the activity of a teacher in accordance with the needs of the industry. Training in specific digital technologies will allow future specialists to be immediately involved in the production process.

The improved model includes: digital technologies applied at the enterprise; organization of teacher internships and study of digital technologies at enterprises; a new level of digital competencies related to production processes at a particular enterprise; provision of the educational process with practice-oriented materials, training cases related to the activities of a particular enterprise; educational and professional trajectory in connection with a particular enterprise.

References

1. Asadullin R. M., Dorofeev A. V., Levina I. R.: Diagnostics of digital competencies of a teacher. Pedagogy enlightenment **1**, 1–17 (2022). https://doi.org/10.7256/2454-0676.2022.1.37153
2. Carretero, S., Vuorikari, R., Punie, Y.: DigComp 2.1: the digital competence framework for citizens. With eight proficiency levels and examples of use. Publications Office of the European Union. (2017). https://publications.jrc.ec.europa.eu/repository/handle/JRC106281. https://doi.org/10.2760/38842
3. Eleneva, Y.Y., Mozharovskaya, A.A., Demushkin, D.I.: Digital transformation of educational organizations of higher education: current state, tasks, risks. Econ. Entrepr. Law. T. **14**(4), 1149–1170 (2024). https://doi.org/10.18334/epp.14.4.120670
4. Falloon, G.: From digital literacy to digital competence: the teacher digital competency (TDC) framework. Educ. Technol. Res. Dev. **68**, 2449–2472 (2020). https://doi.org/10.1007/s11423-020-09767-4
5. Fernández-Batanero, J.M., Montenegro-Rueda, M., Fernández-Cerero, J., García-Martínez, I.: Digital competences for teacher professional development. Systematic review. Eur. J. Teach. Educ. **45**(4), 513–531 (2020). https://doi.org/10.1080/02619768.2020.1827389
6. Garafiev, I., Garafieva, G.: Territorial features of value orientations and satisfaction with the studies of student youth in the Republic of Tatarstan. In: Rudoy, D., Olshevskaya, A., Kankhva, V. (eds.) Innovative Technologies in Science and Education (ITSE-2020), E3S Web of Conferences, vol. 210, p. 18134 (2020). https://doi.org/10.1051/e3sconf/202021018134
7. Gayazova, E.B., Girfanova, E.Y.: Aspects of students' professional orientation in technological university in the digital age. In: Mantulenko, V. (ed.) Proceedings of the 3rd International Conference Engineering Innovations and Sustainable Development. CEISD 2024 Lecture Notes in Civil Engineering, vol. 540. Springer, Cham (2024). https://doi.org/10.1007/978-3-031-67372-6_45
8. Gilster, P.: Digital Literacy, p. 276. Wiley, New York, NY (1997)
9. Girfanova, E.Y., Gayazova, E.B.: Model of professional training of financial managers and conditions of its realization. In: Mantulenko, V. (ed.) Proceedings of the 3rd International Conference Engineering Innovations and Sustainable Development. CEISD 2024 Lecture Notes in Civil Engineering, vol. 540. Springer, Cham (2024). https://doi.org/10.1007/978-3-031-67372-6_47
10. Gouseti, A., Lakkala, M., Raffaghelli, J., Ranieri, M., Roffi, A., Ilomäki, L.: Exploring teachers' perceptions of critical digital literacy and how it manifests in their teaching practice. Educ. Rev. **76**(7), 1751–1785 (2023). https://doi.org/10.1080/00131911.2022.2159933
11. Knyazeva, N., Mikhaylova, I., Shindina, T., Usmanova, N.: Mechanism for Managing the Development of University Teachers Digital Competencies in the Conditions of Digital Transformation, pp. 1–5 (2024). https://doi.org/10.1109/Inforino60363.2024.10551937
12. Maksimenko, N.V.V., Chekalina, T.A.: Review of models of digital competencies of the preservice teacher in the conditions of transformation of the educational process. Profess. Educ. Russia Abroad. **2**(46), 41–50 (2022)
13. Minina, V.N.: Digitalization of higher education and its social results // Vestnik of St. Petersburg University. Sociology. Sociol. Educ. **13**(1), 22–37 (2020). https://doi.org/10.21638/spbu12.2020
14. Naumchenko, S.A.: Improvement of digital competencies of pedagogical staff of professional educational organizations with the help of ERP-systems. Pedagog. J. **12**(6A), 1008–1016 (2022). https://doi.org/10.31862/2073-9613-2022-2-54-63
15. Osipov, P., Girfanova, E.: Professional development of financial managers. In: Auer, M.E., Rüütmann, T. (eds.) Educating Engineers for Future Industrial Revolutions: Proceedings of

the 23rd International Conference on Interactive Collaborative Learning (ICL2020), vol. 2, pp. 261–268. Springer International Publishing, Cham (2021). https://doi.org/10.1007/978-3-030-68201-9_27

16. Semenov, E., Socolow, D.: Digitalization of higher education: opportunities and risks. Ideas Ideals. **14**(4–1), 137–153 (2022). https://doi.org/10.17212/2075-0862-2022-14.4.1-137-153
17. Strokov, A.A.: Digitalization of education: problems and prospects. Vestnik Mininn Univ. **2**(31) (2020). https://doi.org/10.26795/2307-1281-2020-8-2-15
18. The TPACK Framework [Electronic resource]. http://tpack.org/
19. Toktarova, V.I., Rebko, O.V., Semenova, D.A.: Comparative analysis of models of educators' digital competencies in the context of digital transformation of education. Sci. Educ. Today. T. **13**(5), 79–104 (2023). https://doi.org/10.15293/2658-6762.2305.04
20. Uvarov, A.Y.: On the model of digital school. Scientific Education: Collection of Papers of Participants of the Symposium on the Problems of Development of Giftedness of Children and Youth in Education, Yakutsk, July 08–15, 2018. Under the general editorship of A.S. Obukhov and Y.I. Semenov. Yakutsk: Journal "Researcher/Researcher". ppp. 81–98. (2018).
21. Alexandrova, L., Mozhaeva, G., Karabasheva, M., Lapina, M., Rugelj, J.: Use of IT for student-centered learning approach to education at modern university. In: Paper Presented at the CEUR Workshop Proceedings SLET-2020: International Scientific Conference Innovative Approaches to the Application of Digital Technologies in Education, vol. 2861, pp. 32–40 (2020)
22. Lapina, M., Shiriaev, E., Babenko, M., Istamov, I.: High-speed convolution core architecture for privacy-preserving neural networks. Program. Comput. Softw. **50**, 417–424 (2024). https://doi.org/10.1134/S0361768824700282
23. Rusanov, M., Babenko, M., Lapina, M., Sajid, M.: Identification of exploited unreliable account passwords in the information infrastructure using machine learning methods. Big Data Cogn. Comput. **8**, 159 (2024). https://doi.org/10.3390/bdcc8110159
24. Maksimova, E., Lapina, M., Lapin, V.: Synthesis of models for ensuring information security of subjects of critical information infrastructure under destructive influences. CEUR Workshop Proc. **3094**, 108–117 (2022)

Integrating Digital Technologies into Financial Manager Training Programs

Elena Girfanova and Elvira Gayazova(✉)

Kazan National Research Technological University, Kazan, Russia
klabpnr@gmail.com

Abstract. The use of modern digital technologies in the process of training financial managers enables future graduates to be more competitive in the labor market, to meet the demands of employers and improves the quality of the educational program implemented at the university. The ability to apply digital technologies in the process of future activities of financial managers enables them to quickly adapt to changes in the internal environment. And the introduction of digital components in the realized competences gives the opportunity to master new digital technologies, to constantly improve them and then successfully apply them in their professional activity.

The paper considers the information resources used in the process of training financial managers on the example of Kazan National Research Technological University. The application of digitalization elements in the educational process allows students to master new digital platforms independently, since their primary skills were formed already in the first courses, and not only in mastering professional, but also universal and general professional competencies. Based on the experience gained in the implementation of digital components, it is proposed to introduce them in all other disciplines of the curriculum, or to fix them to the competencies implemented in the main educational program. This will enable graduates - financial managers to freely use the acquired skills and competencies in practical activities, and will also lead to the need for teachers to constantly improve their qualifications in this area.

Keywords: digital technologies · information and communication technologies · training of financial managers · competencies · trainees

1 Introduction

Training of qualified financial managers becomes impossible without the use of information technologies. Therefore, the quality of the educational process can be improved by making certain adjustments and adding digital components to most disciplines of the curriculum.

Artificial intelligence methods and machine learning algorithms are currently used both in manufacturing, to improve automation efficiency, and in medicine, medicine, and financial structures for big data analysis. With the growing popularity of language

M. Lapina et al. (Eds.): DET 2024, LNNS 1616, pp. 161–169, 2025.
https://doi.org/10.1007/978-3-032-04365-8_16

models and the launch of open source GPT, artificial intelligence methods have entered even more areas of human activity [22, 23].

The purpose of the paper is to analyze the existing digital elements in the teaching process and to determine where they should be added.

Since at present there is still no single standard of training of specialists studying in the profile (program) of training "Financial Management", each educational organization has the right to independently make changes related to the digitalization of the educational process [1, 18, 21].

The government transforms the system of higher education through the legitimization of changes. Society, on the one hand, influencing the process of legitimization of norms and rules, carries out the reverse transaction - implements legislative acts [3]. At the same time, both subjects of education have their own preferences and needs. Society needs to receive a quality product - a well-trained bachelor, specialist, master - the vitality and viability of society and its development depend on it [5]. The government funds the education of each individual for four, 5 or 6 years depending on the qualification and degree. Any government also needs to get quality graduates, as graduates are the basis for socio-economic, political (which is relevant in the context of globalization) and even demographic development of the society [6, 7].

Therefore, digitalization of education in Russian universities has become a trend in recent years. Higher education implies different options for using digital technologies based on the direction of training. For the humanities, these are technologies for automating messages, working with big data, and monitoring social media. For socio-economic specialists, different technologies are required. These technologies should be applied in enterprises and organizations [12]. This is necessary for practice-oriented digital training of students.

2 Methodology

A set of research methods was used in the course of the study. The study and analytical analysis of sources became the basis for the formation of methodological foundations of the research. Special attention was paid to the systematization of the collected material in accordance with the research topic. In addition, the forecasting method was used to assess possible results and trends in the researched area. A study was conducted among the students of the program "Financial Management" in the field of application of MOODLE system.

3 Literature Review

A. Marey sees it as a new way of communication between people and society [10]. But researchers like E. L. Vartanova, M. I. Makeenko think it's more than just turning information into a digital format. They believe digitalization also affects infrastructure, management, behavior, and culture [19].

N. P. Petrova points out that moving to interactive ways of working together is important. This helps develop the key skills needed in the 21st century, known as the four Cs: creativity, critical thinking, cooperation, and communication. With these skills, students can learn on their own and create new knowledge [14].

To date, researchers have identified important dimensions of digital content engagement.

- information literacy;
- communication literacy;
- digital content creation [4];
- digital security;
- digital competence [2].

According to some experts, 39% of higher education institutions treat digital transformation as a special project and launch separate activities by special order [16]. Thus, the perception of digital transformation is very limited, there is no unified digital strategy, and, accordingly, the effect of the conducted, disparate activities is very mosaic [17].

4 Results

Since a financial manager has an important role in business development, such a specialist should possess not only professional but also digital competencies related to digital data processing methods, artificial intelligence, as well as softskills, which include the ability to work in a team, creativity, leadership and systems thinking [11].

The trend today is digitalization, which has affected the social and economic sciences. Teaching such disciplines also requires knowledge of digital technologies and the ability to apply them in practice [20]. Despite the computer literacy acquired during the pandemic, many teachers still have difficulties with online technologies and fall victim to cyber-fraud [9].

In this regard, it is necessary to creatively rethink almost all aspects of training a modern financier. In the block of management and financial disciplines of the university curriculum of any orientation in the study of financial and economic activity of an economic entity in modern conditions it is advisable to solve problems using the capabilities of information technologies and their tools, such as, for example, the program "Excel", "1C Enterprise" and course management system MOODLE [13].

To determine how well the virtual system works for teaching future financial managers, we did a study. We picked two student groups studying Management with a focus on Financial Management. One group had regular classes with lectures and practice sessions, while the other used the MOODLE online learning platform. They both took a 72-h Financial Management course. At the end, both groups took a test. The students who learned traditionally scored an average of 62%, but those who used the virtual platform did better with 79%. Figure 1 shows the results of the final test of the control and experimental group during the study of the discipline "Financial Management".

The discipline "Financial Management" is a discipline of the variable block and forms professional competencies. As the second MOODLE classroom we considered the classroom for the discipline "Team Building and Self-Organization". This room includes the same structural elements as the previous one. In the process of studying this discipline, students master universal competencies. Despite the different categories of

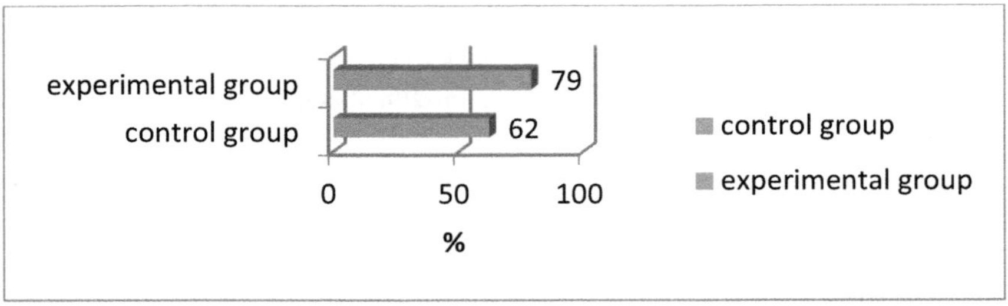

Fig. 1. Results of the final test of the control and experimental groups during the study of the discipline "Financial Management"

competencies, the efficiency of their realization turned out to be quite high. The students noted the possibility to study and consolidate the acquired competencies in this discipline at a convenient time for them, the possibility to take the final test up to 3 times and to choose the maximum obtained result [15]. Figure 2 shows a screenshot of an example of a discipline structure in MOODLE. Topics are structured logically and intuitively. This makes the possibility of using MOODLE digital environment optimal for students as the main source of information and for independent work.

This raises the need to train future financial managers in a new way. In the disciplines of the variative block and in the part formed by the participants of educational relations of the curriculum of financial managers training it is necessary to add digital components or information components everywhere. It is necessary to use 1C software products. Their number is so great nowadays that it is possible to study separate modules of 1C: Enterprise program in different disciplines. Modules as a rule do not repeat each other. In order to save time, you can use the completed directories of this program on the subsequent discipline. This program product has this possibility. In addition, you can use cloud solutions of this product, for example, the discipline "Accounting and Financial Reporting" of the curriculum for training financial managers at the undergraduate level in Kazan National Research Technological University uses a software product of the company 1C "Managing Our Firm", the program involves the use of cloud solutions. It gives an opportunity to the students who are behind to catch up the material, which they could not cope with during classroom training, at home in their free time.

In addition, many universities use the MOODLE educational platform. Kazan National Research Technological University also uses it. Teachers create classrooms for their disciplines. For each discipline for each discipline. A group of students studying this discipline is enrolled there. In order to expand horizons, students can enroll in other courses they like, which are available on the educational platform [8].

On the platform. Each course has its own structure, fixed by the regulations of the university. of the university. Each course should include lectures, questions for independent study, practical assignments. There is also a chat room for individual communication with the teacher.

The software package "1C: Management of our company" is used for practical training in the discipline "Financial Operations of Organizations" at the Masters of Financial Managers, which is fixed in the curriculum in the fourth semester. At this moment the

Fig. 2. Example of building the structure of the discipline in MOODLE (source: https://moodle.kstu.ru/).

students already possess universal, general professional and some professional competences and have skills of independent business management. The program can be easily adjusted to the specifics of the organization's activities. Each student has his/her own account, which is stored in the cloud. He can perform certain tasks not only during practical classes, but also in his free time, from any personal computer, if the organization has a contract with the company 1C. The possibilities of this configuration are very wide, but during practical classes it is enough to study financial operations on settlements with counterparties and preparation of regulated reporting in order to master the competences fixed in the curriculum [7].

The disciplines directly related to the professional activity of financial managers at Kazan National Research Technological University also use elements of gamification. For example, in the discipline "Financial Risk Management", envisaged by the curriculum of the Master's degree program "Financial Management", the business game involves simulation of the process of a production enterprise that produces several types of products. Students choose the enterprise independently. They are divided into groups of 3–5 people. Each group receives a task. For example, the first group should determine the necessary regulatory and legal documentation on the regulation and calculation of financial risks. For this purpose it is necessary to select from the system "Consultant Plus" the necessary laws and regulatory documents governing the processes of risk management, analyze them and determine the appropriate.

The second group finds on the Internet accounting statements of the enterprise for several years (usually for 3 years), on the basis of which the calculations will be carried out. The calculations are carried out either using Excel or in any previously learned software product.

The third group of students should assess the financial risks, based on the calculations presented and propose measures to optimize them. To do this, they need to study the methods of risk assessment and norms in accordance with which the assessment is carried out and the risk group is determined, where the analyzed enterprise falls. The choice of calculation methodology is chosen in advance by students and agreed with the teacher. Both Russian and foreign methods of financial risk assessment can be used. The teacher gives the opportunity to use the Internet to independently select favorite sources, recommends electronic library systems to which the university has a subscription and its own electronic teaching materials, or similar materials of their colleagues [7].

Each practical exercise is evaluated by a group of experts, which consists of the teacher and students who have been released from work. Thus, all students are involved to some extent in the entire process of calculating, analyzing and managing financial risks. Upon completion of the group assignment, students are given a similar assignment for consolidation, but already on the enterprise of small or medium-sized businesses, which they must perform independently, and the results to present in the form of a report and presentation to the audience.

Thus, this business game gives students the opportunity to work in a team, listen to colleagues in solving financial issues related to future professional activities, the opportunity to make independent decisions and be responsible for them. In addition, it provides an opportunity not only to learn new professional competencies, but also to consolidate those already acquired earlier in the learning process.

Digital transformation of the educational process changes the teacher's position and approach to student learning. The teacher's tasks include mastering the skills of processing information, presenting it in various formats, creating and designing the electronic content of the educational discipline, forms of control, which implies the study of e-course materials by students. Accordingly, the role of the teacher changes, he acts not as the main source of information, but as a tutor, who can be referred to if necessary.

5 Conclusion

Thus, we can conclude that the use of digital components in the process of training financial managers, regardless of the level of education, improves the quality of teaching as a separate discipline, and the implementation of its educational program. The use of information and educational environment does not remove the teacher from personal communication with students, and gives the latter the opportunity to take a more serious approach to mastering the discipline [10]. In his/her free time, the student can independently study the material offered by the teacher several times, understand it, ask questions in the chat room and get specific answers, possibly interested only in a particular student. At the same time, the teacher and the student are not "driven" into time limits. Everyone works at a time that is convenient for him or her. Digitalization is becoming a tool for innovative development in the education system, not replacing traditional teaching methods, but only making them more effective and graduates more competitive.

References

1. Bali, S., Liu, M.C.: Students' perceptions toward online learning and face-to-face learning courses in. J. Phys. Conf. Ser. **1108**(1), 012094 (2018). https://doi.org/10.1088/1742-6596/1108/1/012094
2. Garafiev, I.: Regional features of research and development cooperation between innovative organizations and partners. In: Mantulenko, V. (ed.) Proceedings of the 3rd International Conference Engineering Innovations and Sustainable Development. CEISD 2024 Lecture Notes in Civil Engineering, vol. 540. Springer, Cham (2024). https://doi.org/10.1007/978-3-031-67372-6_34
3. Garafiev, I., Garafieva, G., Idiatullina, A., Spirchagova, E.: The management of Russian universities on the basis of international education quality standards. Indonesian J. Electr. Eng. Comput. Sci. **23**, 1689–1698 (2021)
4. Garafiev, I., Garafieva, G.: Territorial features of value orientations and satisfaction with the studies of student youth in the Republic of Tatarstan. In: Rudoy, D., Olshevskaya, A., Kankhva, V. (eds.) Innovative Technologies in Science and Education (ITSE-2020), E3S Web of Conferences, vol. 210, p. 18134 (2020). https://doi.org/10.1051/e3sconf/202021018134
5. Garafieva, G.I.: The role of capital assets in the sustainable development of industrial production. In: Mantulenko, V. (ed.) Proceedings of the 3rd International Conference Engineering Innovations and Sustainable Development. CEISD 2024 Lecture Notes in Civil Engineering, vol. 540. Springer, Cham (2024). https://doi.org/10.1007/978-3-031-67372-6_12
6. Gayazova, E.B., Girfanova, E.Y.: Aspects of students' professional orientation in technological university in the digital age. In: Mantulenko, V. (ed.) Proceedings of the 3rd International Conference Engineering Innovations and Sustainable Development. CEISD 2024 Lecture Notes in Civil Engineering, vol. 540. Springer, Cham (2024). https://doi.org/10.1007/978-3-031-67372-6_45
7. Girfanova, E.Y., Gayazova, E.B.: Model of professional training of financial managers and conditions of its realization. In: Mantulenko, V. (ed.) Proceedings of the 3rd International Conference Engineering Innovations and Sustainable Development. CEISD 2024 Lecture Notes in Civil Engineering, vol. 540. Springer, Cham (2024). https://doi.org/10.1007/978-3-031-67372-6_47
8. Kadeeva, Z.K., Kraysman, N.V., Kadeeva, E.N.: Using Business Games to Build Engineering Competencies in Technological University Students. In: Auer, M.E., Hortsch, H., Michler, O., Köhler, T. (eds.) Mobility for Smart Cities and Regional Development - Challenges for Higher Education. ICL 2021 Lecture Notes in Networks and Systems, vol. 390. Springer, Cham (2022). https://doi.org/10.1007/978-3-030-93907-6_73
9. Kart, M., Şimşek, H.: Defining competencies in curriculum and instruction and developing a new competency model. Hum. Soc. Sci. Commun. **11**, 1496 (2024). https://doi.org/10.1057/s41599-024-03917-2
10. Marey, A.: Digitalization as a paradigm shift [Electronic resource]. Access mode: https://www.bcg.com/about/bcg-review/digitalization (date of reference: 01.02.2025)

11. Osipov, P., Girfanova, E., Ziyatdinova, J.: Educational innovations in financial management degree programs. In: Auer, M.E., Hortsch, H., Michler, O., Köhler, T. (eds.) Mobility for Smart Cities and Regional Development - Challenges for Higher Education. ICL 2021 Lecture Notes in Networks and Systems, vol. 389. Springer, Cham (2022). https://doi.org/10.1007/978-3-030-93904-5_61
12. Osipov, P., Girfanova, E., Ziyatdinova, J., Kraysman, N.V.: Integrating higher education, science and industry: towards a partnership for training competitive financial managers. In: Auer, M.E., Pachatz, W., Rüütmann, T. (eds.) Learning in the Age of Digital and Green Transition. ICL 2022 Lecture Notes in Networks and Systems, vol. 634. Springer, Cham (2023). https://doi.org/10.1007/978-3-031-26190-9_95
13. Osipov, P., Girfanova, E.: Professional development of financial managers. In: Auer, M.E., Rüütmann, T. (eds.) Educating Engineers for Future Industrial Revolutions: Proceedings of the 23rd International Conference on Interactive Collaborative Learning (ICL2020), vol. 2, pp. 261–268. Springer International Publishing, Cham (2021). https://doi.org/10.1007/978-3-030-68201-9_27
14. Petrova, N.P., Bondarev, G.A.: Digitalization and digital technologies in education. World Sci. Cult. Educ. **78**, 353–355 (2019)
15. Shageeva, F., Kraysman, N.: Development of the ability for professional interaction in future engineers at a research university. In: Auer, M.E., Rüütmann, T. (eds.) Educating Engineers for Future Industrial Revolutions: Proceedings of the 23rd International Conference on Interactive Collaborative Learning (ICL2020), vol. 2, pp. 118–128. Springer International Publishing, Cham (2021). https://doi.org/10.1007/978-3-030-68201-9_12
16. Starshinova, T.A., Kraysman, N.V.: Integrative psychological and pedagogical disciplines at an engineering university. In: Auer, M.E., Hortsch, H., Michler, O., Köhler, T. (eds.) Mobility for Smart Cities and Regional Development - Challenges for Higher Education. ICL 2021 Lecture Notes in Networks and Systems, vol. 390. Springer, Cham (2022). https://doi.org/10.1007/978-3-030-93907-6_55
17. Tabachuk, N.P., Malykhina, O.A.: Information competence of university students: historical aspects and modern trends of development. Mod. Probl. Sci. Educ. **2** (2021). https://doi.org/10.17513/spno.30550
18. Turner, E.Y., Nikitina, T.N., Kraysman, N.V., Myshko, F.G., Kaybiyaynen, A.A.: The study of subjective content and conditions for generating the professional development path in university environment. In: Auer, M.E., Hortsch, H., Michler, O., Köhler, T. (eds.) Mobility for Smart Cities and Regional Development - Challenges for Higher Education. ICL 2021 Lecture Notes in Networks and Systems, vol. 390. Springer, Cham (2022). https://doi.org/10.1007/978-3-030-93907-6_57
19. Vartanova, E.L., Vyrkovsky, A.V., Zagidullina, M.V.: Media text in the era of digital platforms: opportunities and threats. Bulletin of Moscow University. Series 10. Journalism. **49**(3), 3–13 (2024). https://doi.org/10.30547/vestnik.journ.3.2024.313
20. Zaky, H.: Sense of belonging and students' engagement in higher education: investigating the factors shaping medical students' inclusion. SN Soc. Sci. **4**, 154 (2024). https://doi.org/10.1007/s43545-024-00944-3
21. Zolotarev, V., Arkhipova, A., Lapina, M., Prakasha, G.S., Movzalevskaya, V.: Assessing player interaction for a social networking cooperative educational game. In: Lapina, M., Prakasha, G., S., Grigoriev, S. (eds.) SLET 2022: International Conference on Innovative Approaches to the Application of Digital Technologies in Education and Research. SLET 2022 Lecture Notes in Networks and Systems, pp. 144–125. Springer, Cham (2024)

22. Lapina, M., Shiriaev, E., Babenko, M., Istamov, I.: High-speed convolution core architecture for privacy-preserving neural networks. Program. Comput. Softw. **50**, 417–424 (2024). https://doi.org/10.1134/S0361768824700282
23. Rusanov, M., Babenko, M., Lapina, M., Sajid, M.: Identification of exploited unreliable account passwords in the information infrastructure using machine learning methods. Big Data Cogn. Comput. **8**, 159 (2024). https://doi.org/10.3390/bdcc8110159

Integration of Digital Solutions into the University Educational Process

Inna I. Golovanova[1(✉)], Tatiana Yu. Gorskaya[2], and Diliara R. Gainanova[2]

[1] Kazan (Volga Region) Federal University, Institute of Psychology and Education, Kazan, Russia
ginnag@mail.ru

[2] Kazan State University of Architecture and Engineering, Kazan, Russia

Abstract. The relevance of the study is due to the constant transformation processes occurring in higher education. The modern high-tech digital world places demands on the improvement of educational activities. The implementation of digital solutions is becoming not just a trend, but a necessity to ensure the competitiveness of universities and training of specialists who meet the requirements of the digital economy. In this regard, a study aimed at identifying trends used by university teachers, their attitude to the use of digital technologies and their readiness for innovation in this area is of particular relevance. The leading method of the presented study was a survey, in which 103 teachers from two leading Kazan universities took part. Structural and comparative analysis of the survey results allowed us to obtain an objective picture of the current state of integration of digital solutions into the university educational process, identify problem areas and determine prospects for further development. As a result of the study, trends and priorities in the use of digital trends were identified: artificial intelligence has become the leader in teachers' preferences in the last 3 years, the use of mobile devices is becoming less relevant, and there is little interest among teachers participating in the study of universities in the inclusion of gamification in the educational process. The data obtained will form the basis for the development of effective strategies for implementing digital solutions, advanced training of teachers and improvement of educational programs, which ultimately will contribute to improving the quality of higher education and training in-demand specialists.

Keywords: Digital trends · digitalization of higher education · digital competences of the teacher

1 Introduction

1.1 The Relevance of the Problem

Modern higher education is in the process of constant transformation and this is due to both the political and economic changes taking place in public life, as well as new demands caused by technological progress. The new requirements for student training are related to both the changing conditions of the labor market and the changing demands

M. Lapina et al. (Eds.): DET 2024, LNNS 1616, pp. 170–180, 2025.
https://doi.org/10.1007/978-3-032-04365-8_17

for the developing competencies of university graduate. The rapid digitalization of all spheres of life undoubtedly has an impact on the organization of educational activities at the university. The implementation of digital solutions is becoming not just a trend, but a necessity to ensure the competitiveness of universities and the training of specialists who meet the requirements of the digital economy. The digital transformation of society dictates the need to train specialists who have the competencies to work with modern technologies and are ready for innovation, while the modern generation of students, the so-called digital generation, has grown up in an environment saturated with technology and expects an appropriate approach to learning.

Artificial intelligence methods and machine learning algorithms are currently used both in manufacturing, to improve automation efficiency, and in medicine, medicine, and financial structures for big data analysis (Lapina et al. 2024). With the growing popularity of language models and the launch of open source GPT, artificial intelligence methods have entered even more areas of human activity (Rusanov et al. 2024).

The creation, improvement and support of a digital educational environment allows for a significant increase in the efficiency of the learning process, ensuring its personalization, expanding access to educational resources and improving communication between teachers and students. The COVID-19 pandemic has demonstrated the low readiness of higher education to quickly transition to distance and blended learning formats, which has highlighted the importance of developing digital competencies among teachers to use digital solutions in the educational process. The breakthrough in digitalization of education that occurred during the pandemic also had an impact on the further expansion of digital tools for improving the educational process. Over the past three to four years, we have seen how traditional teaching methods are increasingly giving way to interactive online platforms, electronic textbooks, virtual laboratories and other digital tools. These innovations open up new opportunities for both personalizing learning and increasing student engagement and developing their digital literacy. However, the effectiveness of the integration of digital solutions largely depends on the ability and readiness of teachers to use new technologies, their understanding of the advantages and limitations, as well as on the development of the appropriate infrastructure for this and the provision of technical and didactic support for teachers. Previously, we conducted a study related to the transformation and development of the digital environment of universities, which received an active impetus during the COVID-19 pandemic and continues to the present (Golovanova et al. 2022). At that time, we were interested in the components of the digital educational environment and the activity in its use. Today, the improvement of digital tools by teachers and their integration of existing digital trends into the educational process are becoming relevant. Among the digital trends, personalization of learning supported by artificial intelligence is of particular interest, which allows taking into account the individual characteristics and pace of mastering the material by each student. The use of collaboration and social learning enhanced by digital tools and aimed at developing teamwork and critical thinking skills is also gaining an important focus in modern realities. Among the current digital trends, it is necessary to touch upon the competitive nature of learning and gamification, which can increase motivation and engagement, as well as digital storytelling, aimed at enhancing perception and activity in mastering educational material. Teachers' use of microlearning and learning using mobile devices, which can provide flexibility and accessibility of educational content, also characterizes the integration of digital solutions into educational process. And the digital educational resources available at universities provide ample opportunities for using the Flipped classroom technology, which transforms the traditional learning model by transferring lecture material to an online environment and freeing up time in the classroom for practical work and discussions. The use of these digital trends helps to increase the effectiveness of the educational process and train competitive specialists.

1.2 Analysis of Domestic and Foreign Literature on the Research Problem

The range of research problems in the field of integration of digital solutions into the educational process of higher education covers all the trends of interest to us. At the same time, both approaches to solving educational problems and digital tools are considered.

A sufficient number of theoretical, research and methodological publications have already been devoted to the use of artificial intelligence in higher education. Thus, a group of researchers from Università Telematica Pegaso, Naples, Italy, based on a literature review, identified the potential of AI to improve the effectiveness of higher education processes such as personalization, motivation, support for interaction with the teacher, increasing student engagement in educational activities, expanding the information framework for working with educational material (Triberti et al. 2024). And researchers from Indonesia presented the results of a study on students' attitudes towards the use of AI in writing text-based work such as academic essays, noting both their positive attitudes and the positive impact on the development of students' writing abilities, while expressing concern about the decline in the development of creative and critical thinking (Rinaldy et al. 2023). Of course, the effectiveness of learning using artificial intelligence raises many questions in the teaching community. Colleagues from China, when studying the impact of AI chatbots on students' learning outcomes, concluded that short-term use of AI improves learning outcomes but with long-term use this effect disappears (Wu and Yu 2023). The integration of AI (GAI) into higher education at 40 universities from six regions of the world was studied by scientists from Monash University. They found that higher education activity in implementing GAI in educational activities is increasing and focuses on the objectivity of assessment and improvement of teaching methods, while the study identified weaknesses in addressing data privacy issues and ensuring equal access to GAI tools (Jin et al. 2024). Researchers from various Australian universities conducted an experiment to determine how useful AI is in regulating students' self-learning, whether they use AI to simplify their actions or actually benefit from it in their own development (Darvishi et al. 2024). The results of this experiment did not confirm the positive impact of AI on learning. All researchers of the issues of integration of artificial intelligence into higher education agree on the need to continue research, since AI is only just entering the pedagogical toolkit and needs experimental research from various aspects of educational activity.

Another area in the use of digital solutions that undoubtedly deserves active inclusion in the university educational process is collaboration and social learning. This trend has been particularly actively developed during the COVID-19 pandemic, using various communication platforms. The authors' (Wong et al. 2021) analysis of the experience of organizing discussions of mini-cases by student teams in online discussion groups shows how learning can be effective by changing the role of the teacher in such an educational process. Marc Aurel Schnabel и Jeremy J Ham also emphasize that the use of a blended social learning environment, the Social Network Learning Cloud (SNLC), enhances students' experience and deepens their knowledge (Schnabel and Ham 2014). Scientists from Hong Kong and Sydney also agree with them, describing their experience of organizing interactions between university students from different countries, which allowed them to draw a conclusion about the possibilities of social networks in professional exchange and the development of collective intelligence, while they define the possibility of synchronous augmented communication as an important condition (Howe and Schnabel 2011).

Research on digitalization in higher education also touches on visualization, the logic of which, according to the authors, influences both the development of mental activity and the formation of educational and professional skills (Bayne 2008). The inclusion of gamification as the use of game elements in a non-game context is increasingly being used by university teachers, as evidenced by the literature review on gamification in higher education presented by Amina Khaldi et al. It notes that gamification elements such as ratings, points, levels and feedback are the most frequently used elements in digital educational resources of universities, while they note a tendency towards more complex game elements, but at the same time weak work on studying the effectiveness of the educational content itself (Khaldi et al. 2023). Many researchers note the potential of gamification (Ghai and Tandon 2022), but at the same time they draw attention to the need to improve the pedagogical competencies of teachers using gamification (Kabilan et al. 2023). Chinese scientists (Li et al. 2023) talk about the ambiguity of the results of the experiment on obtaining students' learning outcomes using gamification in educational activities. The use of gamification in the university environment undoubtedly requires the continuation of pedagogical experimental and research activities, especially improvement in obtaining learning outcomes.

When considering innovative digital trends in higher education, it is necessary to note a number of works related to the use of storytelling methods. This is how Peter Bryant describes the experience of using storytelling to describe and evaluate the experience of students, applied in a joint student project (Bryant 2023). Russian scientists conducted a review of articles on the use of storytelling in university practice and identified positive aspects of using the method, including the development of competencies and the impact on the motivation of students (Zaitseva et al. 2024).

The flipped classroom technology, where lectures are transferred to digital content, has found wide application in higher education, especially after the COVID-19 pandemic, and along with this technology, the microlearning method has also become popular, since it is easier for students to work with short meaningful video fragments, which was reflected in an experimental study confirming the effectiveness of this combination (Fidan 2023). At the same time, the authors who studied the effectiveness of microlearning emphasize the insufficiency of didactic grounds for its successful use (Monib et al. 2024), and the researchers also note the importance of further research on students' perception of microlearning in an online format (McNeill and Fitch 2022). The conducted literature review shows the high interest of the scientific community from all over the world in the study of the integration of digital solutions into higher education and confirms the prospects and high need for continuing research activities in this area.

2 Methodological Basis

2.1 Purpose and Objectives of the Research

The purpose of our research is to identify the digital trends used by university teachers, their attitude towards the use of digital technologies and their readiness for innovation in this area.

The identification of digital trends, development of a questionnaire, creation of a questionnaire using Google Forms, organization of a survey of university teachers in digital format, analysis and interpretation of the results, obtained data based on the methods of mathematical statistics – the solution of the presented set of tasks made it possible to achieve the purpose in view.

2.2 Research Methods and Techniques

The study used a questionnaire survey of the teaching staff, for which a questionnaire "Teaching in the Digital Age" was developed, consisting of 35 questions of both open and closed types. Structural and comparative analysis was used in the analysis of the data. The data were divided into several categories: the use of digital trends, attitudes towards digitalization, and readiness to use digital tools and develop in this. A comparison was also carried out by the majors of the subjects taught by the study participants, related to the natural-technical and humanitarian-social cycle. The comparison was also carried out by positions and length of teaching experience. The temporal dynamics of the use of digital tools in the educational process of the university was made.

2.3 Research Base

The study involved 103 teachers from two Kazan universities: 62 teachers from Kazan (Volga Region) Federal University and 41 teachers from Kazan State University of Architecture and Engineering, who teach classes in humanities, natural sciences and technical disciplines. For convenience, we combined the majors taught by the respondents into two groups: natural-technical and humanitarian-social. The choice of these two large groups is due to the fact that the first group is dominated by technical and fundamental natural-scientific disciplines, and the second group by humanitarian and social disciplines. Among the teachers of the first group are the teachers of mathematics, physics, chemistry, biology, structural mechanics, and strength of materials. The second group includes teachers of foreign and Russian languages, history, philosophy, philology, economics, etc.

The numerical ratio of these groups is approximately equal: 51 in the natural sciences and 52 in the humanities. In addition, 21 professors (20% of respondents), 41 associate professors, and 41 assistants and senior lecturers (40% of respondents each) participated in the survey. The last group is combined based on the fact that the teachers of the third group conduct seminars and laboratory classes, while the first mainly give lectures, and the second group conducts both lectures and seminars. This distribution is significant, since in most cases lectures are conducted in the traditional form - the presentation of the educational material is conducted by the lecturer, while the students mainly only take notes on the material, thus, interaction is carried out only by the lecturer, and feedback is weak. This nuance must be taken into account when forming the survey conclusions. Data on the distribution of teachers in the above groups are shown in Table 1.

Table 1. The composition of respondents participating in the study.

University	Majors	Number of professors	Number of associate professors	Number of senior lecturers and assistants	Total
Kazan (Volga Region) Federal University	technical	5	10	10	25
	humanitarian	10	11	16	37
Kazan State University of Architecture and Engineering	technical	4	12	10	26
	humanitarian	2	8	5	15
Total		21	41	41	103

The average length of service for professors is 45 years, for associate professors – 16 years, for senior professors and assistants – 6 years.

3 Results

Based on the analysis of the obtained data, it can be stated that teachers mostly use cloud services for storing and sharing files and social networks for educational purposes, while professors use any digital tools much less than their colleagues. This is due to the fact that the professorial corps is older in age and focused on traditional teaching methods. According to the data presented in Table 2, it is clear that teachers who conduct mainly practical and seminar classes more often use digital tools in their teaching activities, this may also be due to the fact that their average experience is about 10 years, thus, they have not yet developed their teaching habits and methods, but are able to transform along with modern trends.

A study of the use of a recently emerging and increasingly popular digital trend such as artificial intelligence has shown that teachers are actively interested in this topic, but do not use it so actively yet. In this part of the study, it is not advisable to single out separate groups of teachers, since AI is most often used in the educational process for research activities, conducting seminars, and professors, whose average work experience is 45 years, mostly do without AI when giving lectures. Therefore, this question mainly

Table 2. Distribution of digital tools usage by teaching staff positions.

Digital tools	Professors	Associate professors	Senior lecturers and assistants
Electronic educational platforms (Moodle, Blackboard, etc.)	12	30	33
Video conferencing (Zoom, Microsoft Teams, etc.)	17	25	28
Cloud services for storing and sharing files (Google Drive, Dropbox, etc.)	8	38	40
Interactive tools (Kahoot, Quizlet. Etc.)	0	10	15
Social networks for educational purposes	10	40	41

applies to 80% of respondents who interact with students, receiving feedback from them. It should be noted that AI is also used to model some physical processes, for example, to generate test tasks or to create graphic images, for example, in the training of architects, AI is used in the classes to create elements of architectural facades, interiors, etc. Thus, 25 respondents, which is slightly more than 24% of all respondents, actively use AI, 27 (which is slightly more than 26%) use AI sometimes, 51 teachers do not use AI in their teaching activities, but 21 of them plan to use it. Consequently, it can be stated that approximately half of the teaching staff of the universities where the study was conducted use AI in the teaching process.

The priorities in the use of digital trends by teachers are presented in Table 3. It is obvious that in choosing digital tools, teachers most often give preference to social networks, while noting that social networks are used both for transmitting educational information and for organizing educational discussions and completing group projects. Flipped classroom technologies are also popular. Teachers of Kazan Federal University, especially in social and humanitarian disciplines, note that the mandatory inclusion of e-learning in the bachelor's degree programs motivates them to use the flipped classroom in organizing the mastery of disciplines.

A comparative analysis of the dynamics of the use of prevailing digital solutions in KFU and KSUAE universities, presented in Fig. 1, shows that after the COVID-19 pandemic from 2022 to 2024, the use of artificial intelligence increases sharply, and the use of mobile phones in the learning process tends to decrease, while social networks remain in demand. The graph shows the trend over the past three years, the data are given as a percentage of teachers using these tools of two universities of KFU and KSUAE.

WhatsApp is the leader in the use of social networks in general and in the educational process in particular. In 2023, the rate of use of social networks decreased slightly, the reasons for this, in particular, are that the use of WhatsApp was not recommended in Russia, many switched to another network Telegram, and in general, the hours of use

Table 3. The use of digital trends by the teaching staff

Trends	Professors	Associate Professors	Senior Lecturers and Assistants
Personalization of learning	10	32	24
Artificial intelligence in education	0	20	23
Collaboration and social learning			
Competitive learning (Challenge)	2	15	20
Social networks for educational purposes	15	42	42
Microlearning	10	15	8
Flipped classroom	10	20	30
Storytelling	0	0	3
Gamification	3	8	6
Learning via mobile phone	5	12	21

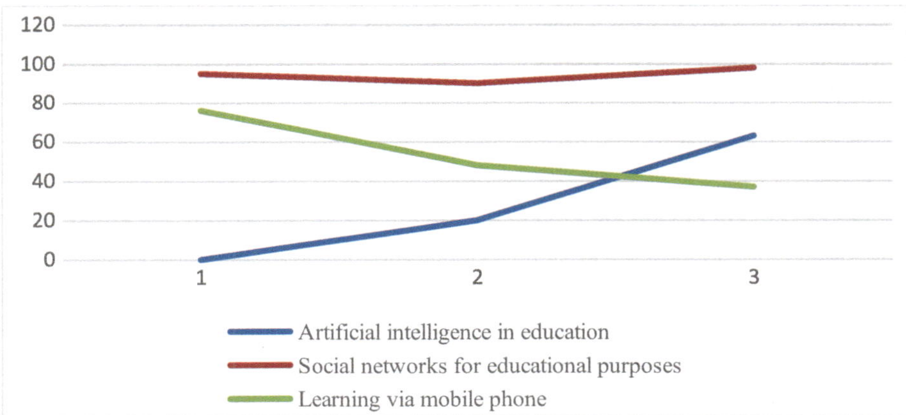

Fig. 1. Trends in the use of digital tools in teaching activities (data are given as percentages by year) 1–2022, 2–2023, 3–2024.

of these networks in the educational process decreased slightly (by about 5%). This is also due to the fact that, having gained experience in distance learning during the pandemic, when universities were not ready for a total transition to online learning, an active process of equipping classrooms with technology began that allows the use of digital resources in the educational process, and educational Internet content and digital educational resources began to be actively formed. And, as a result, the need for social networks as a source of educational communication began to decrease.

The same trend continues with the use of mobile phones, the intensity of its use in the educational process in 2022 was quite high, due to the lack of technical teaching aids in most classrooms. However, over time, classrooms began to be equipped with technology and the need for phones to organize classes began to decrease every year. When using mobile phones in class, teachers used Kahoot! Mentimeter applications, etc. So far, their use is impossible due to blocking on the territory of the Russian Federation. Teachers do not consider other options due to a loss of confidence in the availability of digital products.

On the contrary, the use of artificial intelligence is only increasing every year. If earlier teachers treated AI as an innovation in other areas, not understanding its value in the educational process, now AI has actively entered our everyday life and the capabilities of AI have begun to be applied. A tool such as ChatGPT is especially often used, through which it became possible to generate educational and scientific texts on request, such capabilities are now actively used by both students and teachers. In addition, the generation of graphic images by AI is relevant for creative areas, such as architecture. In classes using AI, building facades are generated at the request of the teacher. In addition, AI is also capable of generating case studies, both in technical and humanitarian disciplines. Thus, the use of AI has become understandable to teachers, even to those who only read lecture courses, since AI is capable of creating presentations on request of excellent quality.

It should be noted that in 2022, the use of the Internet in universities is associated primarily with the placement of educational content on YouTube, as well as on educational platforms of universities, in Moodle systems, for example, as well as for electronic educational resources and distance learning using Zoom, Microsoft Teams, etc. This trend, which began during the pandemic, when the need for electronic resources increased sharply, continues today, with an increase in consumption.

4 Conclusion

The study identified trends and priorities in the use of digital trends. It has been revealed that artificial intelligence has become a leader in teachers' preferences over the past three years, while the use of mobile devices is becoming less relevant; the interest of teachers participating in the study of universities in the inclusion of gamification in the educational process is low. It should be noted that in the universities that took part in the study, the processes of integrating digital solutions take place not only at the design and practical level of implementation, but also experimental research pedagogical activities are organized to assess the effectiveness of use in obtaining an educational result. Thus, at Kazan State University of Architecture and Engineering, a large part of the organization of the educational process is associated with work on bench and real devices that provide preparation for the implementation of the production process. And here it is important to find a balance in the combination of online forms and classroom work (Gorskaya and Zamega 2022).

Another factor is the existence of structural divisions of the university that ensure the creation and maintenance of a digital educational environment based on the university's own developments. Thus, KFU has institutes related to the training of IT specialists. These institutes organize design and research activities on the use of digital systems in the educational process, so these universities use their own software developments, apply experimental digital solutions, such as SMART laboratories (Aiupov and Golovanova 2021), VR simulators (Kugurakova et al. 2021).

The data obtained will form the basis for developing effective strategies for implementing digital solutions, advanced training of teachers and improving educational programs, which will ultimately contribute to improving the quality of higher education and training in-demand specialists.

References

Golovanova, I., et al.: Digital educational environment and online learning through the eyes of students: pros and cons. Educ. Self-Dev. **17**(3), 202–221 (2022). https://doi.org/10.26907/esd.17.3.16

Triberti, S., Di Fuccio, R., Scuotto, C., Marsico, E., Limone, P.: "Better than my professor?" How to develop artificial intelligence tools for higher education. Front. Artif. Intell. **7**, 1329605 (2024). https://doi.org/10.3389/frai.2024.1329605

Rinaldy, A., Pratiwi, Y., Andajani, K., Numertayasa, I.W., Suharti, S., Darwis, A., Marzuki: Exploring artificial intelligence in academic essay: higher education student's perspective. Int. J. Educ. Res. Open. **5**, 100296 (2023). https://doi.org/10.1016/j.ijedro.2023.100296

Wu, R., Yu, Z.: Do AI Chatbots improve students learning outcomes? Evidence from a meta-analysis. Br. J. Educ. Technol. **55**(1), 10–33 (2023). https://doi.org/10.1111/bjet.13334

Jin, Y., Yan, L., Echeverria, V., Gašević, D., Martinez-Maldonado, R.: Generative AI in higher education: a global perspective of institutional adoption policies and guidelines. Comput. Educ. Artif. Intell. **8**, 100348 (2024). https://doi.org/10.1016/j.caeai.2024.100348

Darvishi, A., Khosravi, H., Sadiq, S., Gašević, D., Siemens, G.: Impact of AI assistance on student agencyю. Comput. Educ. **210**, 104967 (2024). https://doi.org/10.1016/j.compedu.2023.104967

Wong, C., Fink, E., Bhati, A.: Future of learning and teaching in higher education post-COVID-19. In: Digital Transformation in a Post-COVID World, pp. 221–244 (2021). https://doi.org/10.1201/9781003148715-12

Schnabel, M.A., Ham, J.J.: Deakin University the social network learning cloud: architectural education for the 21st century. Int. J. Archit. Comput. **12**(3), 225–241 (2014). https://doi.org/10.1260/1478-0771.12.3.225

Howe, E.L.C., Schnabel, M.A.: The changing face of problem-based learning: social networking and interprofessional collaboration. In: Bridges, S., McGrath, C., Whitehill, T.L. (eds.) Problem-Based Learning in Clinical Education Innovation and Change in Professional Education, pp. 121–137 (2011). https://doi.org/10.1007/978-94-007-2515-7_8

Bayne, S.: Higher education as a visual practice: seeing through the virtual learning environment. Teach. High. Educ. **13**(4), 395–410 (2008). https://doi.org/10.1080/13562510802169665

Khaldi, A., Bouzidi, R., Nader, F.: Gamification of e-learning in higher education: a systematic literature review. Smart Learn. Environ. **10**(1), 10 (2023). https://doi.org/10.1186/s40561-023-00227-z

Kabilan, M.K., Annamalai, N., Chuah, K.-M.: Practices, purposes and challenges in integrating gamification-using technology: a mixed-methods study on university academics. Educ. Inf. Technol. **28**(11), 14249–14281 (2023). https://doi.org/10.1007/s10639-023-11723-7

Li, M., Ma, S., Shi, Y.: Examining the effectiveness of gamification as a tool promoting teaching and learning in educational settings: a meta-analysis. Front. Psychol. **14**, 1253549 (2023). https://doi.org/10.3389/fpsyg.2023.1253549

Bryant, P.: Student experience and digital storytelling: integrating the authentic interaction of students work, life, play and learning into the co-design of university teaching practices. Educ. Inf. Technol. **28**(11), 14051–14069 (2023). https://doi.org/10.1007/s10639-022-11566-8

Ghai, A., Tandon, U.: Integrating gamification and instructional design to enhance usability of online learning. Educ. Inf. Technol. **28**(2), 2187–2206 (2022). https://doi.org/10.1007/s10639-022-11202-5

Zaitseva, E.V., Goncharova, N.V., Daineko, L.V.: Storytelling in Higher Education: A Literature Review in International Conference on Professional Culture of the Specialist of the Future, vol. 1204, pp. 3–31. Springer Nature Switzerland, Cham (2024). https://doi.org/10.1007/978-3-031-76800-2_1

Monib, W.K., Rosyzie, A.Q., Apong, A.: Microlearning beyond boundaries: a systematic review and a novel framework for improving learning outcomes. Heliyon. **11**(2), e41413 (2024). https://doi.org/10.1016/j.heliyon.2024.e41413

McNeill, L., Fitch, D.: Microlearning through the lens of Gagne's nine events of instruction: a qualitative study. TechTrends. **67**(3), 521–533 (2022). https://doi.org/10.1007/s11528-022-00805-x

Fidan, M.: The effects of microlearning-supported flipped classroom on pre-service teachers' learning performance, motivation and engagement. Educ. Inf. Technol. **28**(10), 12687–12714 (2023). https://doi.org/10.1007/s10639-023-11639-2

Aiupov, T.A., Golovanova, I.I.: Metrological skills development of technical university students in a digital educational resource with a Smart laboratory. In: "Distance Learning Technologies", DLT 2021; Humanities and Education Science Academy (Branch) of V.I. Vernadsky Crimean Federal University, Yalta, Crimea, Sept. 20–22, vol. 3057, pp. 246–252 (2021) https://www.scopus.com/inward/record.uri?eid=2-s2.0-85122813668&partnerID=40&md5=15b5c787f6f98736c86d6d812d979ec2

Kugurakova, V.V., Golovanova, I.I., Shaidullina, A.R.: Digital solutions in educators' training: concept for implementing a virtual reality simulator. Eur. J. Math. Sci. Educ. **17**(9), 1–10 (2021). https://doi.org/10.29333/ejmste/11174

Gorskaya, T.Y., Zamega, E.N.: The prospect of children's education: pros and cons. Hum. Cap. **8**(164), 92–99 (2022). https://doi.org/10.25629/HC.2022.08.10

Alexandrova, L., Mozhaeva, G., Karabasheva, M., Lapina, M., Rugelj, J.: Use of IT for student-centered learning approach to education at modern university. In: Paper presented at the CEUR Workshop Proceedings SLET-2020: International Scientific Conference Innovative Approaches to the Application of Digital Technologies in Education, vol. 2861, pp. 32–40 (2020)

Lapina, M., Shiriaev, E., Babenko, M., Istamov, I.: High-speed convolution core architecture for privacy-preserving neural networks. Program. Comput. Softw. **50**, 417–424 (2024). https://doi.org/10.1134/S0361768824700282

Rusanov, M., Babenko, M., Lapina, M., Sajid, M.: Identification of exploited unreliable account passwords in the information infrastructure using machine learning methods. Big Data Cogn. Comput. **8**, 159 (2024). https://doi.org/10.3390/bdcc8110159

Information Technologies as a Means of Forming the Civic Position of Student Youth

Natalya Gorbunova and Ekaterina Ivanova(✉)

Humanities and Education Science Academy (Branch) of V.I. Vernadsky Crimean Federal University in Yalta, Yalta 298635, Russia

Yniver-ivanova@mail.ru

Abstract. The article discusses information resources, portals, platforms, databases and various technologies that allow students to learn the true and authentic history of the war years of our country. The study of military history events is an important aspect in the formation of a civic position, and educational activities should be based on the use of modern information technologies. The implementation of civilian information technologies helps preserve the memory of the tragic military pages of our state, outstanding soldiers and heroic deeds committed to protect the territorial integrity, independence and security of the country. Virtual tours allow you to demonstrate historical and cultural heritage, instill a sense of love, admiration and respect for native places. The authors considered such information technologies as the "ARTEFACT" platform, virtual tours, "Nuremberg. VRdict of the peoples", VR-reconstruction. "Unknown standard-bearer", GDB "Memorial", GDB "Feat of the people", portal "Memory of the people", website "No statute of limitations", and information portal "Living History".

Keywords: Civic position · student youth · military history · information technology · virtual tours · portal · platform · database · archive

1 Introduction

The history of the war years plays a special role in shaping the civic position of students, as it is a special example of the courage and self-sacrifice shown by the valiant defenders of the Motherland during the difficult years of the war. The use of information portals, platforms, electronic archives, etc. in educational work allows young people to feel not only self-confidence, but also to discover civic and patriotic guidelines such as justice, loyalty, moral commitment, and the idea of serving the Fatherland. Thus, the purpose of this article is to analyze information technologies created at the state level and aimed at shaping the civic position of students.

Knowledge of the events of the military history of the state is an integral element of the formation of the civil position of the younger generation in the modern world. Familiarization with the military exploits of our ancestors contributes to the development of civil and patriotic personality traits, as well as a sense of pride in the military past of our state. The study of military history contributes to the accumulation and transmission

M. Lapina et al. (Eds.): DET 2024, LNNS 1616, pp. 181–186, 2025.
https://doi.org/10.1007/978-3-032-04365-8_18

of military traditions, strengthening the intergenerational bond of the valiant defenders of the Motherland, creating and developing a positive image of the Armed Forces of the Russian Federation, thereby forming a willingness to perform civil and military duty with honor and dignity, conscientious labor for the benefit of the Motherland.

The formation of the civic position of the younger generation plays a special role, as it is aimed at preserving state stability, independence and security; preparing young people for life in a democratic and rule-based state with its own norms, rules, duties, etc.; solving civil issues and problems; educating an active and responsible citizen who is able to participate in the life of the state at various levels. It is worth noting some features of the formation of a civic position: taking into account individual, age and national characteristics; the influence of information flow and interpretation of information by young people; the use of modern technologies, regarding native Russian national values and emerging new modern orientations, in order to integrate individuals into civil society; participation in socially significant activities; the influence of domestic and foreign policy events on the formation of civic consciousness.

The concept of "citizenship" has been considered by different Russian scientists and has a different definition. Let's take a closer look at the definition of this definition (Table 1).

Table 1. Definition of the concept of "citizenship" in the works of Russian scientists.

Author	Definition of the concept of "citizenship"
V.T. Lisovsky and Yu.V. Berezutsky	"conscious human participation in the life of society, aimed at the realization of social values with a reasonable balance of personal and public interests" [3, 7]
T.M. Abrahamyan	"a personality characteristic that entails certain social behavior and taking responsibility for its consequences" [1]
P.A. Baranov	"the dominant sphere of personal characteristics that determines the meaning, the direction of actions, the involvement of the individual in the fate of the Fatherland, in the process as a whole" [2]
G.Ya. Grevtseva	"an integrative personality quality that includes concepts, knowledge, beliefs, behavior, and moral characteristics" [5]
N.P. Kapustin	"a set of personality traits, including social activity, following the law, and the dominance of motives of public duty" [6]

(*continued*)

Table 1. *(continued)*

Author	Definition of the concept of "citizenship"
G.N. Filonov	"the adoption of universal moral values is the humanism of human relationships, it is a manifestation of dignity, which is manifested in the communication of a person with other people, based on mutual respect and recognition of the identity of each" [8]
S.V. Gladchenkova and O.V. Gorokhovtsev	"the totality of personality qualities in relation to a citizen, society, and the state" [4]

We consider citizenship as an integrative quality of a person, aimed at realizing his belonging to the country and society, responsibility for the successful development of the state as a whole, and the formation of readiness for active action in socially significant activities in accordance with established rules and norms.

In modern society, information technologies play a special role in shaping the citizenship of the younger generation, which make it possible to use computer, multimedia and telecommunications tools that ensure social cohesion and strengthen statehood.

Russian scientists in their works consider the possibilities of information technologies and their role in the formation of civil society. E.V. Mikhailova and A.P. Skorogev in the article "The role of information technologies in the development of civil society in modern Russia (Internet practice) " presents a study of the representativeness of information technologies of a civic-patriotic orientation. In the work "The role of information technology in the formation of civic identity: the potential of the third sector" (Ya.A. Siryukova) the author discusses modern technologies that are used by civic institutions. M.M. Amirov's work "Modern information technologies in the system of civic education" provides arguments about the lack of elaboration of the problem of educating civic active youth using modern technologies. It is worth noting the works of Russian authors who prove the effectiveness of information technologies in the formation of civic identity and activism, such as: R.G. Abdulatipov, V.Y. Zorin, L.N. Timofeev, V.A. Mikhailov, A.G. Bolshakov, N.M. Mukharyamov.

2 Materials and Methods

Information portals, platforms and databases allow citizens to remotely gain knowledge about the history of their native land, the country and the world as a whole, about civil rights and freedoms, study historical materials and documents, and actively interact with activists of the civil society to share experiences and documentary sources related to the protection of human rights and freedoms.

Let's take a closer look at information technologies at the state and federal levels aimed at studying the country's military history and historical and cultural heritage, the introduction of which into the educational process will allow students to form an active civic position. In our opinion, the use of information technologies such as the "ARTEFACT" platform, virtual tours, "Nuremberg. VRdict of the peoples", VR-reconstruction. "Unknown standard-bearer", GDB "Memorial", GDB "Feat of the people", portal "Memory of the people", website "Without statute of limitations", and information portal "Living History" are suitable for this purpose. Let's consider the presented information technologies in details.

The *"ARTEFACT" platform* is an innovative cultural platform that allows you to get acquainted remotely with the cultural heritage of Russia, represented in historical and museum complexes. Augmented reality technologies and audio guides are used to get acquainted with the history of the exhibit or the object as a whole and to obtain reliable and complete information. The platform is unique in that it allows users to find out information that was owned only by museum staff and was not presented to the public. The demonstration of objects and exhibits includes not only a modern look, but also the original one, which was available to restorers and museum curators. The platform contains a public website, apps for iOS and Android, as well as a platform for creating multimedia guides. Using this platform, it is possible to get acquainted with 2,190 museums, cultural sites and exhibition halls, which in total contain 30 thousand exhibits.

Virtual tours are aimed at introducing young people to the historical and cultural heritage, traditions and customs of their native country, and getting acquainted with the achievements of Russian scientists in the field of science and technology. There are many resources in Russia that allow the younger generation to form a complete picture of various historical events. The portal "Culture.RF" presents 482 historical, cultural, and museum sites, from well-known metropolitan to small provincial ones. Using virtual tours, you can see panoramas of the Moscow Kremlin squares, see the Senate and the Grand Kremlin Palace, get acquainted with the collection of the Russian Museum, etc.

"Nuremberg. VRdict of the peoples" is a historical project developed by RIA Novosti, historians and actors of the Gorky Moscow Art Theater with the support of the Russian Military Historical Society in virtual reality format. The use of this technology allows you to plunge into the events of the International Military Tribunal (Nuremberg Trial), which was held between November 20, 1945 and October 1, 1946, over a group of major war criminals of Nazi Germany. Users have the opportunity to find evidence proving guilt, to argue the charges against the defendants in the tribunal.

VR-reconstruction. "The Unknown Standard-Bearer" is an immersive project that aims to explore the historical events that took place from April 30 to May 2, 1945, namely the installation of flags and banners on the Reichstag of the victorious Red Army. The project is a virtual reconstruction of the events and includes two episodes. The first episode allows you to see the establishment of the Victory Banner and feel yourself in the role of one of the soldiers of the Red Army, who had the honor of this triumphant and victorious action. The second episode reveals the story of the establishment of the red banners in the capital of the Third Reich, which marked the victory of the USSR and the liberation of Europe from the Nazi invaders. In the epilogue, users can see a metaphor dedicated to liberation from Nazism. The project is available in the RIA.lab application, and you must use virtual reality glasses of any format to view it.

The website "No statute of Limitations" is a federal project created with the President of the Russian Federation V. V. Putin's support in order to preserve the memory of the tragedies and deaths of the civilian population of the USSR, the cruel treatment of the Nazis and their accomplices, and war crimes during the Great Patriotic War. In recent years, attempts to justify Nazism have been observed in the European Union, Canada and Japan, while it can be stated that the rehabilitation of Nazism is becoming organized and purposeful. Based on this conclusion, it is of particular importance to consider the issue of the manifestation of Nazism in the modern world and to prevent its large-scale spread. The project's activities include: search work; identification of Nazi crimes against civilians, followed by the initiation of lawsuits aimed at recognizing the criminal acts committed by Nazi Germany against civilians as genocide; joint activities of archives at various levels to make materials public.; scientific and educational activities at various levels, organization of events, etc., in order to increase the level of citizenship and patriotism of citizens of the Russian Federation.

The *information portal "Living History*" is a unique online platform aimed at introducing the younger generation to the martial and national values of our country. The activity of this platform promotes the involvement of young people in socially significant events, active participation in voluntary and creative initiatives, in order to form and develop patriotic personality traits. As part of the project's activities, online contests, forums and various events are organized, collections of methodological recommendations, scenarios of events, civil-patriotic materials, historical films and videos are created, joint work is carried out with various federal projects and programs.

The *GDB "Heroic deed of the people*" is an electronic archive of documents related to the Great Patriotic War, created on the initiative of the Ministry of Defense of the Russian Federation. This archive is considered unique and has no analogues in the world, as it is a kind of information resource on the military history of our state. The materials for the electronic database were provided by the Central Archive of the Ministry of Defense of the Russian Federation, including award documents related to the operational management of combat operations. In total, 19 million sheets were studied, processed and digitized, which contained information on 40,391,096 awards. A separate group was created for the work, which was tasked with eliminating inaccuracies due to the age of the documents.

The *GDB "Memorial*" is an electronic archive of documents containing information about dead or missing Soviet soldiers during the Great Patriotic War, as well as in the post-war period. This archive allows citizens to find information about their relatives and friends, and learn about their fate or their burial place. The database includes such archival documents as: reports, documents from hospitals and medical sanitary battalions, funerals, trophy cards and passports of military graves. The electronic archive has no analogues in the world. To create it, ten million sheets of archival documents and over 30 thousand passports of military graves were processed and digitized.

The *information portal "Memory of the people*" was created on the basis of two resources: GDB "Heroic deed of the people of the People" and GDB "Memorial", the main purpose of which is to perpetuate the memory of the fallen defenders of the Motherland. Interactive tools allow users to learn authentic documentary information about the participants of the Great Patriotic War.

3 Conclusion

Thus, Russian government agencies and civil institutions have initiated the creation of a variety of information technologies aimed at preserving and perpetuating the memory of the valiant defenders of our Motherland, accomplished military exploits and military history in general. The introduction of the presented information technologies into the educational process of the university makes it possible to form a system of knowledge about military-historical events, the cultural heritage of our country and to achieve a stable civic position among students. The knowledge of authentic military history based on available materials and documents contributes to the formation of patriotism, pride, a sense of belonging and immersiveness to national history, civic duty and the civic position of students. The remote study of historical and cultural monuments, is aimed at popularizing native Russian traditions and customs, forming traditional values, feelings of love and respect for native places.

4 Thanks

The article was prepared within the framework of the Russian Science Foundation 24-28-20453 project "Formation of an active civic position of the student youth of the Republic of Crimea".

References

1. Abrahamyan, T.M.: Preparing Future Teachers for the Civic Education of Teenagers. PhD Thesis. Moscow
2. Baranov, P.A.: Civic Education Is a Pedagogical, Social and Cultural Phenomenon. SPbAPPO, St. Petersburg (2006)
3. Berezutsky, Y.V., Stasyuk, E.V.: Features of the formation and content of youth self-identification: a theoretical analysis. Power Manag. East Russia. **4**, 106–113 (2012)
4. Gladchenkova, S.V., Gorokhovtsev, O.V.: Formation of the civil position of lawyers at the university: towards the problem statement. Bull. Voronezh Inst. Min. Intern. Aff. Russia. **1**, 167–170 (2013)
5. Grevtseva, G.Y.: On the issue of youth civic culture. In: Collection of materials of the IX Lazarev Readings International scientific conference "Faces of traditional culture in the modern cultural space: cultural memory and culture of memory", pp. 193–196 (2020)
6. Kapustin, N.P.: Pedagogical Technologies of Adaptive Schools. Akademiya, Moscow (1999)
7. Lisovsky, V.T.: The Spiritual World and Value Orientations of Russian Youth. SPbGUP, St. Petersburg (2000)
8. Filonov, G.N.: Civic education: reality and development trends. J. Pedagogy. **8**, 45–51 (1999)

Digital and Artificial Intelligence in Education

Oksana Gordienko[1(✉)], Anastasia Sokolova[1], Yurii Braun[1], Oleg Ivanov[1], Vitalii Lapin[2], and J. Jenefa[3]

[1] Moscow Pedagogical State University (MPGU), 119435 Moscow, Russia
ov.gordienko@mpgu.su
[2] North-Caucasus Federal University, Stavropol 3550017, Russia
[3] Christ University, Bangalore, India

Abstract. The article examines the relationship between digital intelligence and artificial intelligence, with an emphasis on the pedagogical aspects of this problem. The author emphasizes that the creation of AI requires not only technical literacy, but also developed digital thinking, supported by clear ethical principles. Particular attention is paid to the role of modern education in the formation of such specialists - the school should educate not just IT specialists, but people who are able to responsibly use technology. The work analyzes the various levels of digital competencies required in professional activities, and also provides an overview of scientific publications on the implementation of artificial intelligence in the educational process. The study demonstrates how important it is to combine technological progress with humanistic values in the training of future personnel.

Keywords: artificial intelligence · digital intelligence · digital competencies · digital education First Section

1 The Penetration of AI into Education

Please note that the first paragraph of a section or subsection is not indented. The first paragraphs that follows a table, figure, equation etc. does not have an indent, either subsequent paragraphs, however, are indented.

In recent years, artificial intelligence (AI) has been actively penetrating various fields, including education, which has caused heated discussions about its potential to replace humans in a number of professional areas. This topic has attracted the attention of researchers around the world, as evidenced by numerous publications in leading scientific databases such as Elibrary, Web of Science, Scopus and Google Scholar. An analysis of more than 500 papers over the past decade has shown that the study of AI in education is conducted in different directions, combining both theoretical and practical approaches. One of the key aspects of research has become the role of AI in assessing educational achievements. Scientists such as Tyack, Khorramdel and von Davier are studying how algorithms help analyze student performance and the effectiveness of educational programs [1]. Another important area is related to intelligent learning systems

M. Lapina et al. (Eds.): DET 2024, LNNS 1616, pp. 187–196, 2025.
https://doi.org/10.1007/978-3-032-04365-8_19

such as CodeCombat and Duolingo, which, according to research by Starodubtsev, Neradovskaya and Luckin, are able to increase motivation and adapt material to individual student needs [2–4].

Special attention is paid to the use of chatbots and generative models, including ChatGPT and DALL-E. The works of Gašević and Baker show how these tools change the approach to teaching, allowing for the automation of routine tasks and the creation of personalized educational materials [5]. Personalization of learning has generally become one of the most discussed topics - König, Bhutoria and Titova explore how AI helps to tailor educational trajectories to the characteristics of each student [6–8].

An equally significant layer of research is devoted to the automation of administrative processes in education. Rezaev, Tregubova and other authors consider how algorithms simplify document flow, workload distribution and course management [9–18].

Interestingly, reviews by colleagues, including the works of Zawacki-Richter, Marín and Bukina, confirm the relevance of the identified areas, demonstrating similar trends in world science. This suggests that the development of AI in education is not a local trend, but a global process that requires further in-depth study [19, 20].

2 Digital Intelligence (DQ) as an Emerging Concept

In parallel with the research on artificial intelligence, the phenomenon of digital intelligence (DQ) is actively discussed in the scientific literature. This topic is reflected in the works of authors such as Boughzala, Cocorocchia and Petsangsri, who are trying to give a clear definition of this concept, which has not yet been established in the academic environment [21–23]. Other researchers, including Singh and Stiakakis, focus on the structure of digital intelligence, highlighting its key components - from digital literacy to ethical behavior in the online environment [24–30]. Of particular interest are the studies of Jackman, Gentile and Park, devoted to methods of developing DQ, which vary from educational programs to special trainings. An important area has become the assessment of the level of digital intelligence in both students and teachers [31]. Zhao, Yang, Eshet-Alkalai and other scientists propose methods for measuring DQ, which allow us to determine how effectively a person adapts to the digital reality [32–38]. The relevance of the topic is confirmed at the political level - the Council of Europe declared 2025 the Year of Digital Citizenship, emphasizing the need to develop relevant skills.

However, despite the abundance of publications, there are virtually no studies that examine the relationship between artificial and digital intelligence. This omission is particularly strange, since effective human-AI interaction is impossible without advanced DQ. Although both concepts contain the word "intelligence," they are fundamentally different: one refers to machine algorithms, the other to human abilities. Artificial intelligence processes data, while digital intelligence allows humans to critically evaluate information, consciously use technology, and protect themselves from digital threats. Without understanding this interdependence, any discussions about the harmonious implementation of AI in education will remain incomplete.

3 Conceptualizing Digital Intelligence

When talking about human digital intelligence, we need to take into account the theory of multiple intelligences, according to which intelligence is not something single and indivisible, but consists of several blocks or levels (G. Gardner, R. Sternberg, G. Eysenck, etc.) [39–41]. Today, one of the most important ideas is the idea of distinguishing types of intelligence for different needs of everyday life. In accordance with this, there are such types of intelligence as the familiar mental intelligence, emotional intelligence (a person's ability to recognize emotions, understand the intentions, motivation and desires of other people and their own, as well as manage their emotions and the emotions of other people in order to solve practical problems), spiritual intelligence (mastery of ethical standards, awareness of the spiritual and moral foundations of life), physical intelligence (the ability to maintain and develop human physical abilities). Today, digital intelligence should be added to the types of intelligence identified, since today we live in a world where our daily life, for our convenience, is partially digital: in Russia, we have already become accustomed to the fact that to make an appointment with a doctor, submit documents, buy tickets to the theater, for transport, etc., we turn to digital resources and resolve these issues in a few minutes. Accordingly, for a harmonious life, a modern person needs developed digital intelligence.

The concept of digital intelligence was proposed in 2015 by South Korean educator P. Yuhyun and his colleagues from Nanyang Technological University. Today, it is understood as a comprehensive set of technical, cognitive, metacognitive, and socio-emotional competencies based on universal moral values that allow people to confront challenges and use the opportunities of digital life. It should be noted that it is no coincidence that the definition of digital intelligence includes socio-emotional competencies and moral values: a person must have a clear understanding of what is and is not allowed in the digital environment, and that one person's freedom ends where another's freedom begins. Only under this condition will mobbing, bullying, and other destructive phenomena in the digital environment decrease.

The concept of digital intelligence is a complex multi-level system that includes three key stages of development. At the basic level is digital citizenship - this is the fundamental ability to interact with technology while observing security and ethical standards. Here we are talking about the simplest but critically important skills: from creating strong passwords to understanding the consequences of posting personal information online. The second level, digital creativity, involves more conscious participation in the digital environment - creating content, developing applications or using technology to implement creative ideas. The highest level is digital competitiveness, when a person does not simply consume or create, but uses digital tools to solve large-scale problems, launch start-ups or transform existing business models.

These levels are closely intertwined with eight key aspects of digital intelligence. For example, digital self-identification includes understanding your role in the online space, and digital hygiene covers the habits of regularly updating software and analyzing your digital footprint. Of particular interest is the aspect of digital emotional intelligence - the ability to recognize the emotional subtext in correspondence or video calls. When these eight areas are superimposed on three levels, a matrix of 24 specific competencies is formed. Thus, the combination of digital citizenship and security gives the skill of recognizing phishing sites, and creativity plus digital communication is manifested in the ability to create viral content. It is important that this system is not static - moving up the levels requires the simultaneous development of all eight components. You can be a brilliant programmer (high creativity), but neglect digital hygiene, putting your data at risk. Or demonstrate brilliant communication skills on social networks, but do not understand your digital rights. It is the integrated development of all elements that makes digital intelligence a truly effective tool for living in the modern world (Table 1).

Table 1. Key Aspects of Digital Citizenship.

Category	Description	Components
Digital creativity	Co-authorship in the digital environment	Healthy Use of Technology, Network Security Management
Digital competitiveness	Reformism in the digital economy	Socially beneficial use of technology, cyber risk management
Digital security	Personal digital hygiene	Content Cyber Risk Management, Cyber Security at the Organizational Level
Digital communication	Payment for communication and cooperation	Online Relationship Management, Mass Communications
Digital literacy	Information Literacy	AI Literacy, Intellectual Property Management
Digital rights	Knowing and exercising your rights in the digital environment	–

4 Integrating AI and DQ into Pedagogy

Modern pedagogy faces the challenge of forming a harmonious personality, where physical, mental, emotional, spiritual and digital intelligence are developed equally. This formula for success has ceased to be an abstract theory - today it is becoming a practical guideline for school education. Moreover, the responsibility for cultivating such qualities in the younger generation lies not only with teachers of the exact sciences, but equally with teachers of the humanities. When a historian analyzes the ethical aspects of technological revolutions with his class, when a philologist analyzes the change in language in the digital age, when a social studies teacher discusses the transformation of legal norms - all of them actually participate in the formation of the digital intelligence of students [42–44].

Of particular relevance are issues related to artificial intelligence, which should become part of the educational discourse. We are not talking about the technical aspects of the algorithms, but about the humanitarian understanding of these technologies. For example, in literature classes it is appropriate to discuss whether AI can truly create, and where the line is between borrowing and plagiarism in the case of neural networks. In social studies, it is important to raise the topic of responsibility - who should be responsible for the error of a driverless car: the developer, the owner, or the algorithm itself? History teachers should draw students' attention to how AI can distort historical facts through biased training data.

The key task of a modern teacher is to prepare children for life in a world where decisions are increasingly made by algorithms. Schoolchildren must understand the risks of digital discrimination, when facial recognition systems make mistakes in determining ethnicity, or when credit algorithms do not objectively evaluate women. It is equally important to discuss the problem of the "black box" - why we often cannot understand the logic of AI decisions, and what threats this creates. These topics should not be the domain of narrow specialists - they require an interdisciplinary approach and accessible presentation within the school curriculum. This is the only way to raise a generation capable of not only using technology, but also consciously shaping the digital environment of the future.

Today, the development of artificial intelligence requires programmers not only to have technical skills, but also to develop all types of intelligence in a comprehensive manner - especially digital intelligence with its moral and ethical component. The creator of AI must have not only knowledge in the field of algorithms, but also a clear understanding of responsibility for their developments, an awareness of the boundaries that cannot be crossed even with the most impressive technological capabilities. Only such an approach can guarantee that artificial intelligence will become an instrument of progress, and not a source of threats to society.

Speaking about modern achievements of AI, it is worth recalling the famous Turing test, proposed back in 1950 [45]. Its essence is perfectly illustrated by the science fiction parable "Who is Behind the Wall?" by Stanislav Rajtburt - a film that makes you think about the nature of the mind and the criteria for its evaluation. Interestingly, for almost 30 years of the existence of the Loebner Prize (1991–2019), not a single program has been able to truly convince the judges of its "humanity" to receive the main award. Experiments in the field of creativity are especially indicative. In 2018, philologists Orekhov and Uspensky conducted an interesting experiment: they invited participants to compare two poems - one was written by the avant-garde artist Ilya Zdanevich (1917), the other was created by an algorithm [46]. This experiment clearly demonstrates how difficult it is sometimes to distinguish human creativity from machine creativity, especially when it comes to works with non-standard structure and imagery. Here are these poems:

1. "Love is a scar on the heart of time,
Where the hours drip into the void.
Your eyes are two extinguished suns,
In which I drown without hope of salvation."
2. "Oh love! You are a crack in the mirror of existence,
Where all my failures are reflected.
Your lips are two petals of a withered rose,
Falling into the abyss of my despair."

Paradoxically, it was the second poem that was the work of a human (Zdanevich), while the first was created by an algorithm. This example shows that modern AI is already capable of imitating some aspects of the creative process, although it has not yet been able to reproduce the depth and authenticity of human experiences.

Cars in the fields and here it is cheerfully lifted like the wind, and here everything will return to the field, and in the motionless twilight under the water rose in the field of sunset, and here it is on the field above the water rose in the field, and a cheerful wave cheerful and in the evening field the wave will lay. And here it returned to the field wave, and in the field the wind stirs breezes, and here under the clearing full of waves, and here under the white sky the wave rose, and in the field the wind cheerfully raised its gaze. And here it raised its cheerful voice, and the evening flight of excitement raised in the involuntary twilight, and again in the city rose. Under the breeze the waves along the fields rose in the cheerful field, and the wind wind in the field woke up, and here under the clearing a strip the explosion of the wind raised a cheerful canopy.	Cars dig rough crowds faces in swings moon boredom ore left left gangs of soldiers circles beetle arks of the wretched lay on the pavement wooden pieces of darkness dark dark dark coffee shops tables bridges jump in fright now threshing woman bustle lackeys like tangent like tangent poles stick out street braided telephone wires red hair constellations on them speak to the earth evil beeps raise outskirts poison sleepiness cities variety show fox terriers lick lick people pus destroy workers poles eaten by wormhole wires into red shreds woe woe to the gentlemen them fall shot by soot on the sidewalks traps luminaries at the ends of hair with a dead long dead moon but the earth will char, patched with rags under damage of love, burned, burned death, smoke

This experiment with determining the authorship of poems clearly demonstrates the fundamental difference between human creativity and machine text generation. Although modern algorithms have already learned to imitate the formal features of poetry - rhythm, rhyme, imagery - they remain unable to reproduce the very "humanity" that manifests itself in unexpected associations, emotional depth and personal experience of the author. Predictions that by 2026 90% of Internet content will be created by artificial intelligence simultaneously cause admiration for technological progress and anxiety for the future of creative professions. However, it is important to understand the fundamental limitation of modern AI systems - they can combine and process existing templates, but are not able to truly feel, experience or comprehend the content they create. This fundamental difference between human and artificial intelligence becomes especially obvious when we compare their emotional capabilities. A machine can recognize and reproduce a description of laughter or sadness, but will never experience these states. This is why training a new type of teacher is especially relevant - a teacher of the future must not only master digital technologies, but also preserve and develop in students those unique human qualities that cannot be algorithmized: emotional response, moral guidelines, the ability to deeply reflect. This is the paradox of modern education - we need to prepare children for life in the digital world, without allowing technology to replace the human essence itself.

5 Conclusion

In conclusion, it is worth noting that the presented study raises extremely important issues related to the integration of digital and artificial intelligence into the educational environment. The authors convincingly demonstrate that modern technologies require not only technical literacy, but also developed digital thinking based on ethical principles and social responsibility. Particular emphasis is placed on the role of teachers, who must not only possess digital competencies, but also be able to form them in students, taking into account the moral and ethical aspects of interaction with technologies.

The conducted analysis of scientific literature and practical examples, such as the use of ChatGPT, CodeCombat and other tools, shows that artificial intelligence is already actively used in education, but its development is impossible without appropriate training of personnel. An important contribution of the work is the structuring of digital intelligence, including three levels and eight areas, which allows for a more systematic approach to the formation of relevant skills in students and teachers.

The idea that future education should be oriented towards the harmonious development of all types of intelligence – mental, emotional, spiritual, physical and digital – deserves special attention. Only such a comprehensive approach will allow us to prepare specialists who are capable of not only creating and using technologies, but also doing so consciously, taking into account humanistic values.

Thus, the study not only highlights the relevance of the topic, but also suggests specific areas for further work, including the modernization of educational programs and strengthening the role of teachers in the formation of digital culture. This is especially important in the context of rapid technological development, when the quality of training of specialists determines not only the effectiveness of the educational process, but also the future of society as a whole.

References

1. Tyack, L.: Using convolutional neural networks to automatically score eight TIMSS 2019 graphical response items. In: Tyack, L., Khorramdel, L., Von Davier, M. (eds.) Computers and Education: Artificial Intelligence, vol. 6, p. 100249. Accessed 16 Mar 2025 (2024). https://doi.org/10.1016/j.caeai.2024.100249
2. Starodubcev, V.A., Neradovskaja, O.R.: Iskusstvennyj intellekt i immersiv-nye tehnologii v vysshem pedagogicheskom obrazovanii (Artificial Intelligence and Immersive Technologies in Higher Pedagogical Education). Otkrytoe obrazovanie. **28**(2), 13–23 (2024)
3. Luckin, R.: In: Luckin, R., Holmes, W., Griffiths, M., Forcier, L.B. (eds.) Intelligence Unleashed: An Argument for AI in Education, p. 60. Pearson, London (2016)
4. Karram, O.: The role of computer games in teaching object-oriented programming in high schools – Code Combat as a game approach. In: Karram, O. (ed.) WSEAS Transactions Advances in Engineering Education, vol. 18, pp. 37–46 (2021) https://wseas.com/journals/education/2021/a085102-003(2021).pdf. Accessed 15 Mar 2025

5. Gašević, D.: Empowering learners for the age of artificial intelligence. In: Gašević, D., Siemens, G., Sadiq, S. (eds.) Computers and Education: Artificial Intelligence, vol. 4, (2023). https://doi.org/10.1016/j.caeai.2023.100130. Accessed 30 Mar 2025
6. König, C.M.: Development guidelines for individual digital study assistants in higher education. Int. J. Educ. Technol. Higher Educ. **21**, 9 (2024). https://doi.org/10.1186/s41239-024-00439-4
7. Bhutoria, A.: Personalized education and Artificial Intelligence in the United States, China, and India: a systematic review using a Human-In-The-Loop model. Comput. Educ. Artif. Intell. **3**, 100068 (2022)
8. Titova, S.V.: Intellektual'nye sistemy obuchenija dlja personalizacii i adaptacii jazykovyh kursov (Intelligent learning systems for personalization and adaptation of language courses). Vestnik Moskovskogo universiteta. **27**(4), 84–99 (2024)
9. Rezaev, A.V.: Chatgpt i iskusstvennyj intellekt v universitetah: kakoe budushhee nam ozhidat'? (chatgpt and AI in Universities: What Future Can We Expect?). Vysshee obrazovanie v Rossii. **32**(6), 19–37 (2023)
10. Bogachevskij, V.M.: Ispol'zovanie chatgpt v obuchenii grammatike anglijskogo jazyka (Using chatgpt in Teaching English Grammar). Biznes. Obrazovanie. Pravo. **2**, 306–313 (2024)
11. Alana, J.A.: The Vision of Digital Intelligence for Postgraduate Education: Improving Both Student Experience and Administrative Ease (2017). https://eric.ed.gov/?Id=ED574831. Accessed 1 Apr 2025
12. Baidoo-Anu, D.: Education in the era of generative Artificial Intelligence (AI): understanding the potential benefits of Chatgpt in promoting teaching and learning. J. AI. **7**(1), 52–62 (2023)
13. Lapina, M.A., Prakasha, G.S.: Project-based learning approach to the formation of digital competencies of students of universities in Russia and India. Inform. Educ. **37**(3), 80–87 (2022)
14. Prakasha, G.S., Lapina, M., Balakrishnan, D., Sajid, M. (eds.): Digital Technologies in Modeling and Management: Insights in Education and Industry. IGI Global (2024). https://doi.org/10.4018/978-1-6684-9576-6
15. Costa, E.B.: Evaluating the effectiveness of educational data mining techniques for early prediction of students' academic failure in introductory programming courses. Comput. Hum. Behav. **73**, 247–256 (2017)
16. Du, J.: Transforming language education: a systematic review of AI-powered chatbots for English as a foreign language speaking practice. Comput. Educ. Artif. Intell. **6**, 100230 (2024)
17. Tripathi, V.V., Prakasha, G.S., Lapina, M.: Cognitive engagement scale (CES) in an online environment: construction and validation. In: Lapina, M., Prakasha, G.S., Grigoriev, S. (eds.) SLET 2022: International Conference on Innovative Approaches to the Application of Digital Technologies in Education and Research. SLET 2022. Lecture Notes in Networks and Systems, pp. 73–87. Springer, Cham (2024)
18. Lapina, M., Shiriaev, E., Babenko, M., Istamov, I.: High-speed convolution core architecture for privacy-preserving neural networks. Program. Comput. Softw. **50**, 417–424 (2024). https://doi.org/10.1134/S0361768824700282
19. Zawacki-Richter, O.: Systematic review of research on artificial intelligence applications in higher education – where are the educators? Int. J. Educ. Technol. Higher Educ. **16**(39) (2019) https://www.researchgate.net/publication/336846972_Systematic_review_of_research_on_artificial_intelligence_applications_in_higher_education_-where_are_the_educators. Accessed 11 Mar 2025
20. Bukina, T.V.: Iskusstvennyj intellekt v obrazovanii: sovremennoe sostojanie i perspektivy razvitija (Artificial Intelligence in Education: Current Status and Development Prospects). Obshhestvo: sociologija, psihologija, pedagogika. **1**, 76–83 (2025)

21. Boughzala, I.: Digital Intelligence: A Key Competence for the Future of Work. AACSB (2019). www.aacsb.edu/insights/articles/2019/03/digital-intelligence-a-key-competence-for-the-future-of-work
22. Cocorocchia, C., Forget, I.Q.: Digital intelligence will be what matters in the future. World Econ. Forum. (2018) www.weforum.org/agenda/2018/02/digital-intelligence-internet-safety-future/. Accessed 30 Mar 2025
23. Kulworatit, C.: Development of an internet risks assessment using a digital intelligence quotient and a communication-based model. Turk. J. Comput. Math. Educ. **12** (2021) https://www.researchgate.net/publication/358624508_Development_of_an_Internet_Risks_Assessment_using_a_Digital_Intelligence_Quotient_and_a_Communication-Based_Model. Accessed 15 Mar 2025
24. Singh, C.D.: The need for digital intelligence. Nature. **562** (2018) https://media.nature.com/original/magazine-assets/d41586-018-06848-6/d41586-018-06848-6.pdf. Accessed 20 Mar 2025
25. Stiakakis, E.: Developing an understanding of digital intelligence as a prerequisite of digital competence. In: Conference: The 13th Mediterranean Conference on Information Systems (MCIS), Naples, Italy (2019) https://www.researchgate.net/publication/354810638_DEVELOPING_AN_UNDERSTANDING_OF_DIGITAL_INTELLIGENCE_AS_A_PREREQUISITE_OF_DIGITAL_COMPETENCE. Accessed 16 Mar 2025
26. Adams, N.B.: Digital intelligence fostered by technology. J. Technol. Stud. **30**(2) (2004) https://scholar.lib.vt.edu/ejournals/JOTS/v30/v30n2/pdf/adams.pdf. Accessed 1 Apr 2025
27. Chassignol, M.: Artificial Intelligence trends in education: a narrative overview. Procedia Comput. Sci. **136**, 16–24 (2018)
28. Cope, B.: Artificial intelligence for education: knowledge and its assessment in AI-enabled learning ecologies. Educ. Philos. Theory. **53**(12), 1229–1245 (2021)
29. Marnewick, C.: Digital intelligence: a must-have for project managers. Project Leadersh. Soc. **2** (2021). https://doi.org/10.1016/j.plas.2021.100026
30. Wang, S.: Artificial intelligence in education: a systematic literature review. In: Expert Systems with Applications, p. 252 (2024). https://doi.org/10.1016/j.eswa.2024.124167
31. Jackman, A.: Addressing the digital skills gap for future education. Nat. Hum. Behav. **5**, 542–545 (2021)
32. Zhao, S.: A study on the framework of digital intelligence quotient measurement for vocational college students under the perspective of embodied cognition. J. Open Learn. **28**, 31–41 (2023)
33. Aviram, A.: Towards a Theory of Digital Literacy: Three Scenarios for the Next Steps. 2006. https://citeseerx.ist.psu.edu/viewdoc/summary?Doi=10.1.1.643.8589. Accessed 30 Mar 2025
34. Lin, C.-C.: Artificial intelligence in intelligent tutoring systems toward sustainable education: a systematic review. Smart Learn. Environ. **10**(1), 41 (2023). https://doi.org/10.1186/s40561-023-00260-y
35. Memarian, B.: A review of assessment for learning with artificial intelligence. Comput. Hum. Behav. Artif. Hum. **2**(1), 100040 (2024). https://doi.org/10.1016/j.chbah.2023.100040
36. Sun, J.: Issledovanie primenenija iskusstvennogo intellekta v prepodavanii anglijskogo jazyka v universitete Kitaja (Research on the Application of Artificial Intelligence in English Teaching in a University in China). Nauchno-metodicheskij jelektronnyj zhurnal «Koncept». **12**, 88–103 (2023)
37. Prakasha, G.S., Lapina, M., Balakrishnan, D., Sajid, M. (eds.): Educational Perspectives on Digital Technologies in Modeling and Management. IGI Global (2024)
38. Lapina, M., Tokmakova, M., Movzalevskaya, V., Karabasheva, M., Balakrishnan, D.: Analysis of modern methods and approaches using it in education and pedagogy. In: Lapina, M., Prakasha, G.S., Grigoriev, S. (eds.) SLET 2022: International Conference on Innovative

Approaches to the Application of Digital Technologies in Education and Research. SLET 2022. Lecture Notes in Networks and Systems, pp. 40–53. Springer, Cham (2024)
39. Gardner, H.: Multiple Intelligences: the Theory in Practice, p. 320. Basic Books (1993)
40. Sternberg, R.: Prakticheskij Intellekt = Practical Intelligence in Everyday Life (Practical Intelligence = Practical Intelligence in Everyday Life), p. 272. Piter (2002)
41. Ajzenk, G.: Intellekt: novyj vzgljad (Intelligence: A New View). Voprosy psihologii. **1**, 111–129 (1995)
42. Christine, R.: European Framework for the Digital Competence of Educators. Digcompedu Publications Office of the European Union, Belgium (2017)
43. Ma, S.: Development and validation of internet literacy scale for high school students. Educ. Inform. Technol. **29**, 1427–1454 (2024) https://www.researchgate.net/publication/369758755_Development_and_validation_of_internet_literacy_scale_for_high_school_students. Accessed 2 Mar 2025
44. Rusanov, M., Babenko, M., Lapina, M., Sajid, M.: Identification of exploited unreliable account passwords in the information infrastructure using machine learning methods. Big Data Cognit. Comput. **8**, 159 (2024). https://doi.org/10.3390/bdcc8110159
45. Turing, A.: Computing machinery and intelligence. Mind. **59**(236), 433–460 (1952)
46. Orehov, B.: Galvanizacija avtora, ili jeksperiment s nejronnoj pojeziej (Galvanization of the Author, or an Experiment with Neural Poetry). Novyj mir. **6**, 139–158 (2018)

Distribution of Students by Competence Levels from the Point of View of Statistical Physics

Sherniyaz B. Kabdushev[1], Dinara K. Matrassulova[2], and Ibragim E. Suleimenov[3](✉)

[1] National Engineering Academy of RK, Bogenbai Batyr Street 80, 050010 Almaty, Republic of Kazakhstan
[2] Almaty University of Power Engineering and Telecommunications Named by Gumarbek Daukeyev, Baytursynuli 126/1, 050013 Almaty, Republic of Kazakhstan
[3] V.I. Vernadsky Crimean Federal University, Academician Vernadsky Avenue 4, 295007 Simferopol, Republic of Crimea
esenych@yandex.kz

Abstract. The problem of finding measurable indicators reflecting the quality of education at universities is considered. It is shown that the existing methods of compiling university rankings are guided de facto by the analysis of the prerequisites for obtaining a quality education, but not by the actual result. The problem of finding a measurable indicator that reflects the level of competence of university graduates remains unresolved. It is shown that the solution to this problem can be found through finding the optimal distribution of students by competence levels, which, in turn, can be established using classical methods of statistical physics. It is shown that it is possible to find an analog of the Boltzmann distribution, reflecting the distribution of students by competence levels for given macroscopic parameters, which are the characteristics of a particular university (by analogy with such macroscopic parameters as temperature and pressure). Using the reference theoretical distribution of students by competence levels, we can offer a measurable indicator, which is a measure of the deviation of the real distribution of grades received by students at exams from the theoretical optimum. The influence of subjective factors on the result of calculating the specified measurable indicator is discussed. It is shown that even with attempts to artificially distort the result, its impact on the quality of the educational process remains positive.

Keywords: Quality of Education · University Rating · Measurable Indicators · Statistical Physics · Entropy · Stationary Systems

1 Introduction

For the post-Soviet states, the problem of independent assessment of the quality of higher education is very acute. There are various agencies whose functions directly include the compilation of university rankings based on the analysis of measurable indicators. In particular, in Kazakhstan, one of such organizations is an Independent Accreditation and Rating Agency [1], which uses the methodology of the international rating "IAAR

M. Lapina et al. (Eds.): DET 2024, LNNS 1616, pp. 197–205, 2025.
https://doi.org/10.1007/978-3-032-04365-8_20

Eurasian University Ranking (IAAR EUR)", developed by this organization taking into account the international methodological standards for ranking IREG (Observatory on Academic Ranking and Excellence).

The basic international requirements for the methodology of rating compilation are the transparency of their activities, as well as the maximum exclusion of any subjective assessments [2]. That is why such tools as Webometrics play such an important role in the development of the rating methodology [3].

In particular, the methodology used by the IAAR provides for taking into account such indicators as the total number of articles published by university teachers and indexed in Scopus, the total number of university employees represented in this database, indicators reflecting commercialization and grant funding of research projects (R&D), etc. [1]. These indicators are also widely used in international practice [2, 4].

The activities of such organizations as IAAR are very relevant for Kazakhstan, mainly due to the mass nature of higher education and the progressive decline in its quality [5], which, however, corresponds to the observed crisis trends in the world as a whole [6, 7]. This, among other things, requires further improvement of the methods of compiling ratings from the point of view of assessing the real quality of training.

We emphasize that the international expert community recognizes that the existing methods of building ratings are far from ideal [8, 9], and it is the assessment of the result, i.e. the level of training of university graduates, that is the most difficult task (at least from the point of view of using measurable indicators).

It is obvious that no other method is applicable for assessing the competence of an individual graduate, except for expert assessments; it is almost impossible to offer a measurable indicator here. However, it is possible to go a slightly different way, focusing on the analysis of the nature of the distribution of graduates according to the estimates that the university itself puts up. So, if all students, without exception, receive an excellent grade on a particular exam, this means that either excessively simple material is being taught, or the teacher has declared a kind of strike, in fact, only pretending that he is working. Similarly, if all the marks on the exam are unsatisfactory, then either the course is excessively complex, or the actions of the teacher are not adequate.

Therefore, the distribution of estimates should correspond to some optimum. The degree of complexity of the course should provide for the possibility of obtaining the highest grades by the most gifted students while mastering the key material by everyone else. This implies the importance of mathematical modeling of the learning process at the university, which was emphasized in [10]. In particular, having found the theoretical optimum (the optimal distribution of students by competence levels upon completion of training in a particular discipline), then you can set the task of finding a measurable indicator based on comparing the actual distribution of grades with the theoretical standard.

Finding a measurable indicator of this type is a non-trivial task, however, in any case, the first step on this path is to obtain some kind of analog of the Boltzmann distribution, reflecting the distribution of students by competence levels.

This problem is solved in this paper, where it is shown that under certain assumptions, the distribution of students by competence levels can be obtained from considerations similar to those based on which classical statistical physics is built.

2 Problem Statement

The simplest model of learning for each specific subject is illustrated in Fig. 1. Formally, learning is described as a process that corresponds to the transition of "particles", i.e. students from one level to another under external influence. We show that if the system under consideration is considered stationary (i.e., the flows developing to it remain unchanged in time), then it admits a correct mathematical description using the same methods that classical statistical physics uses.

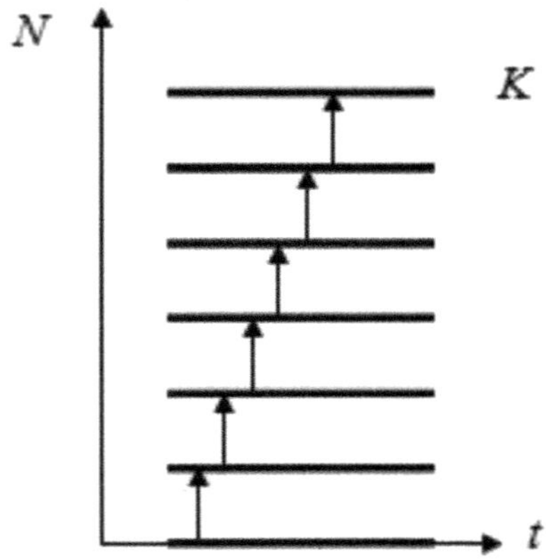

Fig. 1. Illustration of the K-level model of the learning process.

The basic assumption used in the formulation of the problem under consideration is completely analogous to the assumption underlying all statistical physics. Namely, it is assumed that under a given constraint (the constancy of the integral flow of particles "up through the levels" of the system Fig. 1) in practice, exactly the state that is most likely is realized. Quantitatively, finding such a state is reduced to finding the extremum of a function that reflects the dependence of the number of possible states of the system (the multiplicity of degeneracy) on the corresponding parameters.

Let the number of "particles" in the system of the type under consideration at level 0 be N_0. Consider the time interval Δt, assuming that n_0 particles will be transferred to level 1 during this time. Similarly, during this time, n_i particles will be transferred from level k, where there are N_i particles, to level $k+1$.

Let's calculate, following the basic idea of statistical physics, the number of ways in which such an event can be realized. It is obvious that the number of such methods coincides with good accuracy with the number of ways in which n_i particles can be selected from a set of N_i elements. This number is given by the binomial coefficient $C_{N_i}^{n_i}$

$$C_{N_i}^{n_i} = \frac{N_i!}{n_i!(N_i - n_i)!} \tag{1}$$

The application of the ideas of statistical physics is justified only if the system contains a relatively large number of particles. In this case, it is acceptable to use the Stirling formula, which leads to the following approximation of the expression for the binomial coefficient

$$C_{N_i}^{n_i} \approx \frac{(N_i)^{N_i}}{(n_i)^{n_i}(N_i - n_i)^{N_i - n_i}} \frac{\sqrt{N_i}}{\sqrt{2\pi n_i (N_i - n_i)}} \tag{2}$$

In the system as a whole, a set of flows associated with the transfer of "particles" from the previous level to the next one develops during the considered period.

The number of ways (the multiplicity of the degeneracy of the enclosing system G) by which such a set of flows can be realized is given by the product

$$G \approx \prod_{i=0}^{i=K-1} \frac{(N_i)^{N_i}}{(n_i)^{n_i}(N_i - n_i)^{N_i - n_i}} \frac{\sqrt{N_i}}{\sqrt{2\pi n_i (N_i - n_i)}}, \tag{3}$$

where K – is the number of levels, the summation goes up to $K - 1$, since the number of threads in the system is one less than the number of levels.

Formula (3) takes into account that each of the possible ways of implementing a flow from one of the system levels to the next can combine any of the possible ways of implementing all the other flows. Because of this, the total number of implementations of the entire set of threads is given by the product of the number of ways of implementing each of the threads connecting nearby levels.

Following the logic of statistical physics and taking into account that the logarithm is a strictly monotone function, instead of the multiplicity of degeneracy G, we can use its logarithm, interpreted as the entropy of the system (or its analog).

In addition, the following inequality can be considered valid for any system of the type under consideration

$$N_i \gg n_i \tag{4}$$

Inequality (4) expresses the fact that only a small part of the system elements can be moved from one level to another for a short time. (For correctness, we should talk about physically infinitesimal intervals; this implies that quite a lot of elements of the system move during the corresponding time interval, but the value Δt can be treated as a differential.)

Taking the logarithm of (3), we obtain

$$lnln\, G\left(\{N_i\}_{i=0}^{i=K-1}\right) \approx \sum\nolimits_{i=0}^{i=K-1} N_i lnln\, N_i - (N_i - n_i) lnln\, (N_i - n_i) \tag{5}$$

In the expressions under the sum sign in formula (5), the terms proportional to n_i are discarded, which can be done by (4).

Let us take into account (4) again. You can write

$$lnln\, (N_i - n_i) = lnln\, N_i + lnln\, \left(1 - \frac{n_i}{N_i}\right) \approx lnln\, N_i - \frac{n_i}{N_i} \tag{6}$$

Substituting (6) into (5), we have

$$lnln\, G\left(\{N_i\}_{i=0}^{i=K-1}\right) \approx \sum\nolimits_{i=0}^{i=K-1} n_i lnln\, N_i \tag{7}$$

Passing to the time interval Δt and assuming that we are considering a situation in which the number of transitions between levels is relatively small, relation (7) can be rewritten in the form

$$lnln\, G\left(\{N_i\}_{i=0}^{i=K-1}\right) \approx \Delta t \sum\nolimits_{i=0}^{i=K-1} J_i lnln\, N_i \tag{8}$$

Result (8), rewritten in the form

$$\frac{G}{\partial t} = \sum\nolimits_{i=0}^{i=K-1} J_i lnln\, N_i \tag{9}$$

it is permissible to interpret as the production of entropy in the system under consideration. This quantity has the dimension of frequency.

In form, relation (9) is very similar to the expression for the entropy of an equilibrium system, but it has a different meaning. First of all, it contains the dynamic characteristics of a system in a stationary state, but not in equilibrium. Let us emphasize that the production of entropy in a separate part of the encompassing system is greater, the higher the frequency of the transition (in this example, the probability of acquiring an additional "quantum of competence").

Thus, the formulation of the problem under consideration is as follows. It is required to find the maximum value of functional (9) under certain additional constraints reflecting the physical meaning of the processes under consideration.

3 Distribution of Students by Levels of Competence

To find the distribution of students (or generalized "particles") according to the levels of competence, it is necessary, first of all, to specify the type of functional (9), i.e. establish a relationship between the fluxes under consideration and the level populations.

As applied to the system under consideration (Fig. 1), in an acceptable approximation, we can write

$$\frac{dN_i}{dt} = J_{i-1} - J_i - \frac{N_i}{\tau}, \tag{10}$$

where τ - time of study at the university (lifetime of a "particle").

Ratio (10) expresses the average balance of flows to the i-th level. Students from level $i-1$ enter this level who have received an additional "quantum of competence" (the first term on the right side), students who have received another "quantum of competence" (the second term on the right side) leave, and students whose time studies at the university completed (third term).

For the level with the index «0» (the lowest level of competence), an analog of eq. (10) is written in the form

$$\frac{dN_0}{dt} = Q - J_0 - \frac{N_0}{\tau} \tag{11}$$

where Q is the inflow of students to the university, numerically equal to the ratio of the number of students entering the first year to the duration of the study.

Starting from relations (10), it is easy to show that in the stationary state

$$Q - J_i = \frac{1}{\tau}\sum\nolimits_{j=0}^{i} N_j \tag{12}$$

This ratio can be obtained, inter alia, by considering a set of levels from zero to i - th inclusive: it expresses the balance of flows to such a subsystem. The total number of students entering this system is Q; with a minus sign in formula (12) flows flowing from the considered subsystem appear.

Formula (12) allows one to express the values of all fluxes developing in the system through the values of the populations of its levels. We have

$$\frac{dlnG}{dt} = \sum\nolimits_{i=0}^{i=K-1}\left(Q - \frac{1}{\tau}\sum\nolimits_{j=0}^{i} N_j\right)\ln N_i \tag{13}$$

Following the basic logic of statistical physics, functional (13) should be supplemented with conditions expressing the effectiveness of the learning process as a whole. In the simplest case, such a condition can be written in the following form

$$\sum\nolimits_{i=0}^{i=K-1} J_i = B = const, \tag{14}$$

where B – a control parameter characterizing the average learning efficiency.

Using the method of indefinite Lagrange multipliers, we can explicitly indicate a functional, finding the extremum of which allows us to find the distribution of students by levels of performance. We have

$$S = \frac{dlnG}{dt} + \lambda B, \tag{15}$$

where λ – indefinite Lagrange multiplier.

Specifically, the equations for the values of the level populations of the system under consideration have the form

$$\frac{\partial S}{\partial N_i} = 0, i = 0, 1, \ldots, K - 1 \tag{16}$$

The population of the uppermost level is determined by their overall balance of flows in the system

$$Q - \frac{1}{\tau}\sum\nolimits_{j=0}^{K} N_j = 0 \tag{17}$$

Together with formula (14), relations (16) and (17) form a closed system of $K + 1$ equations for the values of populations and an indefinite Lagrange multiplier. This system of equations contains three control parameters Q, τ and B. However, passing to dimensionless quantities, their number can be reduced to one.

Indeed, it follows directly from relation (17) that the total number of particles (students) in the considered system C is expressed by the simple formula

$$C = \sum\nolimits_{j=0}^{K} N_j = \tau Q \tag{18}$$

The expression for the functional S can be written as

$$S = \sum_{i=0}^{i=K-1} \left(Q - \frac{1}{\tau} \sum_{j=0}^{i} N_j \right) (\ln N_i + \lambda) \tag{19}$$

This function can be considered defined up to a factor since its multiplication by an arbitrary nonzero coefficient will not change the eqs. (16). Divide (19) by Q, then

$$s = \sum_{i=0}^{i=K-1} \left(1 - \frac{1}{\tau Q} \sum_{j=0}^{i} N_j \right) (\ln N_i + \lambda) \tag{20}$$

Convenient to go to the relative populations of all levels

$$x_i = \frac{N_i}{\tau Q} \tag{21}$$

We have

$$s = \sum_{i=0}^{i=K-1} \left(1 - \sum_{j=0}^{i} x_j \right) (\ln x_i + \lambda_0), \tag{22}$$

where the reduced Lagrange multiplier is expressed as

$$\lambda_0 = lnC + \lambda \tag{23}$$

Expression (22) allows one to obtain equations for the relative populations of the levels of the system under consideration in the explicit form

$$\frac{\partial s}{\partial x_k} = \frac{1}{x_k} \left(1 - \sum_{j=0}^{k} x_j \right) - \sum_{i=0}^{i=K-1} \left\{ (\ln x_i + \lambda_0) \frac{\partial}{\partial x_k} \sum_{j=0}^{i} x_j \right\} = 0 \tag{24}$$

At the same time

$$\frac{\partial}{\partial x_k} \sum_{j=0}^{i} x_j = \left\{ 1, k \leq i \; 0, k > i \right. \tag{25}$$

Supplementing the relation (24) with expressions following from (17) and (14)

$$\sum_{j=0}^{K} x_j = 1 \tag{26}$$

$$b = \frac{B}{Q} = \sum_{i=0}^{i=K-1} \left(1 - \sum_{j=0}^{i} x_j \right), \tag{27}$$

we obtain a closed system of $K + 1$ equations for the values of the reduced populations and an indefinite Lagrange multiplier.

A special case is of interest when the number of levels in the system is equal to three.

Indeed, current practice shows that the number of students who receive unsatisfactory grades on exams at Kazakh universities is negligible, including due to the reluctance of the management to "spoil the indicators". When analyzing statistics, as a result, it is

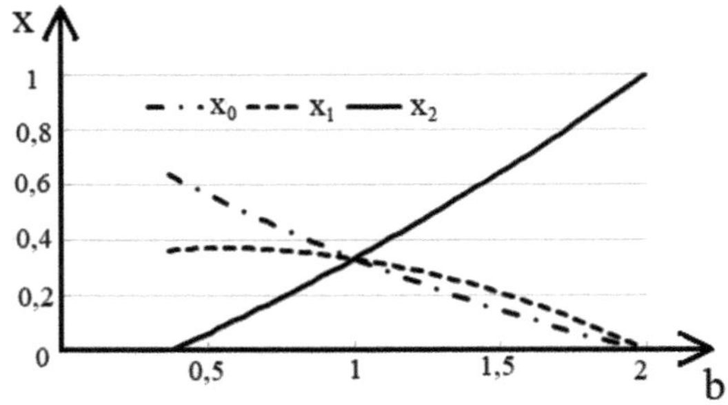

Fig. 2. The dependence of the population of the three-level system on the indicator characterizing the average rate of competence improvement.

enough to take into account only three levels that formally correspond to the ratings "satisfactory", " good "and "excellent".

The results of numerical calculations using Eqs. (24)–(27) for this case are presented in Fig. 2, where the population of all three levels of the system under consideration is deferred from parameter *b*, which characterizes the integral efforts of the teaching staff and the quality of the university environment. This parameter, we emphasize once again, is the average rate of competence improvement by students.

It can be seen that the dependence of the population of the highest level on the parameter *b* is close to linear, but the tangent of the slope angle significantly exceeds one. This suggests that the real costs of acquiring the highest competence by all students are de facto not justified. More precisely, based on the obtained solution, it is advisable to find an optimum in the future, taking into account, among other things, economic indicators.

However, even at this stage of research, the very question of the existence of such an optimum is of interest. Indeed, in the current practice, when teachers of many universities overestimate exam grades for some reason, the factor of competition between students is excluded. In particular, this applies to the assessments issued for graduation papers, master's theses, etc. Just raising the question of comparing the distribution of estimates with the optimum will ensure the creation of a competitive environment and reduce the factor of corruption risks.

4 Conclusion

Thus, it is possible to obtain a theoretical distribution of students by competence levels based on the same methods that classical statistical physics uses. This distribution can be used as a comparison object to develop a criterion that allows independent rating agencies to evaluate the quality of education at a particular university. It is the assessment of the quality of training that is the most difficult task, especially if you use publicly available information. The proposed approach creates prerequisites for solving such a problem, including through cross-comparison with other measurable indicators (the observed distribution of grades should correlate with the potential of the teaching staff, etc.).

References

1. National Agency for Accreditation and Rating of the Republic of Kazakhstan. https://iaar.agency. Accessed 15 Mar 2025
2. Luque-Martínez, T., Faraoni, N.: Meta-ranking to position world universities. Stud. High. Educ. **45**(4), 819–833 (2020)
3. Yakymenko, I., Kazymyr, V., Lytvyn, S.: Webometrics ranking analysis, and possible ways to improve the position of the University. In: Proceedings of the 2020 IEEE 11th International Conference on Dependable Systems, Services and Technologies (DESSERT), 14–18 May 2020, pp. 1–5. IEEE, Piscataway. NJ (2020). https://doi.org/10.1109/DESSERT50317.2020.9124999
4. Fernández-Cano, A., Curiel-Marin, E., Torralbo-Rodríguez, M., Vallejo-Ruiz, M.: Questioning the Shanghai ranking methodology as a tool for the evaluation of universities: an integrative review. Scientometrics. **116**, 2069–2083 (2018)
5. Obukhova, P.V., Guichard, J.P., Baikenov, A.S., Suleimenov, I.E.: Influence of mass consciousness on quality of the higher education in Kazakhstan. Procedia. Soc. Behav. Sci. **185**, 172–178 (2015)
6. Tight, M.: Mass higher education and Massification. High Educ. Pol. **32**, 93–108 (2019)
7. Cardoso, S., Rosa, M.J., Stensaker, B.: Why is quality in higher education not achieved? The view of academics. Assess. Eval. High. Educ. **41**(6), 950–965 (2016)
8. Johnes, J.: University rankings: what do they show? Scientometrics. **115**, 585–606 (2018)
9. Moed, H.F.: A critical comparative analysis of five world university rankings. Scientometrics. **110**, 967–990 (2017)
10. Suleimenov, I., Bakirov, A., Niyazova, G., Shaltykova, D.: University as an analog of the neural network. In: Proceedings of the E3S Web of Conferences, vol. 258, p. 07056. EDP Sciences, Les Ulis (2021)

Higher Education in the Context of Digitalization Necessity and Sufficiency

Ekaterina Kargapolova[1(✉)], Andrey Koshkin[1], Andrey Novikov[1], B. Deepanraj[2], and Vitalii Lapin[3]

[1] Plekhanov Russian University of Economics, Moscow, Russia
k474671@list.ru

[2] Prince Mohammad bin Fahd University, 617, Al Jawharah, 34754 Dhahran, Saudi Arabia

[3] North Caucasus Federal University, Stavropol, Russia

Abstract. The digital environment is rapidly changing contemporary socio-cultural space, including higher education. The aim of this research is to analyze the opinions of residents of the capital megapolis of Russia – the city of Moscow – regarding the necessity and sufficiency of higher education; to identify factors that influence these opinions. The results of the study fully showed that almost all respondents consider higher education to be important and necessary, but not complete and high-quality enough in the era of digitalization, which requires constant, continuous and flexible learning. The gender and current level of education of the respondents do not influence the opinion about the need for higher education, younger survey participants are more likely to doubt its necessity, while the older generation tends to consider higher education both necessary and sufficient. It has been determined that respondents' perceptions of the necessity of higher education are statistically significantly connected with defining a university as a practice-oriented platform with a qualified teaching stuff and various forms of extracurricular activities and student leisure. The connection has also been identified between the perception of a university as the source of corruption threats and the opinion about the lack of necessity for higher education.

Keywords: Era of Digitalization · Higher Specialized Education · Educational Environment · Information Technology · Transformation of Modern Education

1 Introduction

The challenges of digital digitalization require higher education to maintain its function of quality knowledge transmission to an increasingly broad audience, that includes not only youth but also people of other age groups. On the other hand, while preserving the foundation of classical universities, the transition into a new digital environment as a vanguard and flagship, higher education institutions need a set of innovations and transfer (Nešković 2024), capable of helping to establish reasonable requirements for the organization of the educational process, as well as requirements for students and teachers, administrative workers and students (Burner et al. 2025, Díaz-García et al. 2022, Frølich

M. Lapina et al. (Eds.): DET 2024, LNNS 1616, pp. 206–214, 2025.
https://doi.org/10.1007/978-3-032-04365-8_21

et al. 2025, Prakasha et al. 2025a, Prakasha et al. 2025b). As T.V. Vorotilina notes, that modern information and digital technologies can help achieve an effective transformation of educational processes and the education system as a whole. Education is entrusted with a crucial mission – preparing specialists capable of effectively performing professional tasks in the future, anticipating and forecasting that future".

Artificial intelligence methods and machine learning algorithms are currently used both in manufacturing, to improve automation efficiency, and in medicine, medicine, and financial structures for big data analysis (Lapina et al. 2024). With the growing popularity of language models and the launch of open source GPT, artificial intelligence methods have entered even more areas of human activity (Rusanov et al., 2024).

The increasing role of the technological environment is emphasized, shaping a fundamentally new type of communication in mastering digital culture (Emikh, Fomina, 2023, Tam 2024). This approach promotes a more flexible, open and interactive educational process, encourages student independence, develops critical and creative thinking (Aflalo et al., 2024; Bali et al. 2024; Clayton et al., 2024; Nieminen et al. 2025), and improves interaction between universities and accelerates innovation. However, significant social risks of this process manifest in situations of uncertainty regarding changes in "the configuration of relationships between subjects of the new educational reality" (Karpov and Karpova 2022), as well as risks of technological dependence (Zhang & Xu 2025).

2 Research Methods

In April 2024, a research group from the Department of Political Analysis and Socio-Psychological Processes at Plekhanov Russian University of Economics, led by Doctor of Political Sciences, Professor and head of the Department A. P. Koshkin, and Doctor of Sciences, Professor E.V.Kargapolova, with the participation of second and third-year students in the "Sociology" program, conducted a sociological study titled "Higher Education in Russia" among the residents of Moscow (N = 750). Among the research questions, there was a section dedicated to the mission of higher education in the context of digitalization.

The method of collecting primary sociological information was online surveying. The research group developed and tested an algorithm for implementing online surveys targeting the desired group. Interviewers were tasked with surveying respondents from the target group, and quotas were established based on gender, age, and employment status. In the Google form, respondents indicated the name of the interviewer and their contact information, that allowed for monitoring the interviewers' work based on the quota characteristics. Control procedures also included excluding respondents from the database who did not meet the parameters of the target audience based on territorial residency and removing surveys that displayed signs of halo effect. As a result, the obtained ratio of men to women corresponds to the gender structure of the general population (among those surveyed, 44,6% were men and 55,4% were women). The sample included 47,8% youth, 27,7% adults and 24,5% elderly individuals. This distribution allows for certain conclusions to be drawn about age groups. When interpreting the results, it is important to note that the majority (62,6%) of respondents have higher education, 5,4% have postgraduate education (graduate school, dissertation), 13,9% have secondary technical education (technical school, college), 13,5% have general secondary education (high school), 4,3% have initial vocational education (vocational school), and 0,3% have primary or incomplete secondary education.

3 Results

Respondents were asked the question, "Do you consider obtaining higher education necessary and sufficient in the age of digitalization?" The overwhelming majority (70,7%) of the residents of Moscow surveyed expressed that higher education is necessary but digitalization requires continuous skill improvement. Therefore, in their opinion, it is essential to constantly upgrade qualifications in their professional field. 15.4% of Muscovites believe that a person should receive higher education, while 13.9% of respondents believe that there is absolutely no need for it. They believe that in order to succeed in life, it is rather enough to improve their skills in a certain professional field.

The homogeneity of variance criterion (Levene's statistic) with a significance level $p < 0{,}01$indicates that the variances for groups (formed by categories of the variable "age groups") differ statistically significantly; that is, their variances are different and not equal (Tables 1 and 2). Therefore, when analyzing differences between groups, the Tamhane T2 test should be used, which assumes unequal variances (Table 4). There is a statistically significant difference between the mean values of the dependent variable ($p = 0{,}004$) (Table 3). The average attitudes towards higher education among different age groups differ statistically significantly. Thus, age groups differ statistically significantly in their attitudes towards education in the age of digitalization.

Table 1. Descriptive statistics (Do you consider obtaining higher education necessary and sufficient in the age of digitalization?).

		N	Average	Standart deviation	Average mistake	95% confidence interval for the average		Minimum	Maximum	Inter-component variance
						Lower limit	Upper limit			
Youth		364	**2,019**	5865	0307	1,959	2,080	1,0	3,0	
Adults		211	**2,028**	4772	0329	1,964	2,093	1,0	3,0	
Elderly people		187	**1,872**	5023	0367	1,799	1,944	1,0	3,0	
Total		762	1,986	5411	0196	1,947	2,024	1,0	3,0	
Model	Fixed effects			5379	0195	1,947	2,024			
	Random effects				0488	1,776	2,195			0055

Table 4 Multiple comparisons presents the results of the Tamhane T2 test. This criterion calculates the differences between the average values for each of the three groups. Of the three groups, only Group 3 differs significantly from the groups 1 and 2

Table 2. The homogeneity of variance criterion (Do you consider obtaining higher education necessary and sufficient in the age of digitalization?).

	Levene's statistics	1 df	2 df	value
Based on the average	**4,514**	2	759	**011**
Based on the median	**4,758**	2	759	**009**
Based on the median and with adjusted df	**4,758**	2	750,937	**009**
Based on a truncated average	4,664	2	759	010

Table 3. ANOVA (Do you consider obtaining higher education necessary and sufficient in the age of digitalization?).

	Sum of squares	df	Average square	F	value
Between groups	3,227	2	1,613	5,576	004
Inside groups	219,615	759	289		
Total	222,841	761			

regarding the importance of higher education. Groups 1 and 2 do not differ significantly from each other. This allows us to conclude that young people and adults do not differ in their attitudes toward education. They believe that digitalization requires continuous skill improvement. However, older people believe that while higher education is necessary today, the acquired knowledge is sufficient for adaptation in the labor market.

Table 4. Multiple Comparisons (Dependent variable: Do you consider obtaining higher education necessary and sufficient in the age of digitalization?).

(I) newage2	(J) newage2	Average difference(I-J)	Average mistake	value	95% confidence interval	
					Lower limit	Upper limit
Youth	2,00	−0092	0450	996	−117	099
	3,00	**1476**[a]	0479	**007**	033	262
Adults	1,00	0092	0450	996	−099	117
	3,00	**1568**[a]	**0493**	**005**	039	275
Elderly people	**1,00**	**-1476**[a]	0479	**007**	−262	−033
	2,00	**-1568**[a]	0493	**005**	−275	−039

[a]The average difference is significant on the level 0.05

The average value of the opinion indicator on the expediency and completeness of higher education in the era of digitalization is shown in Fig. 1.

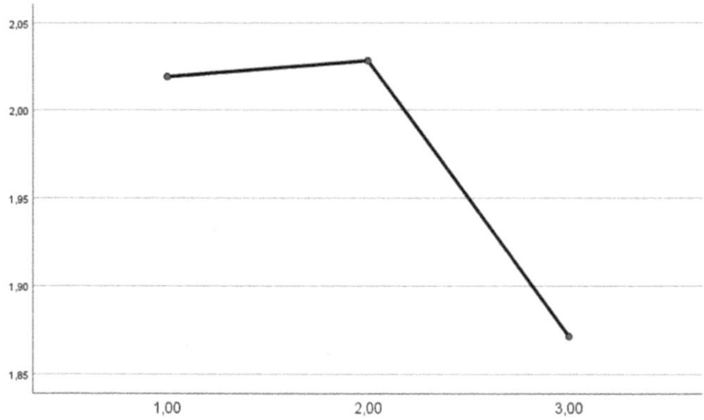

Fig. 1. The average value of the opinion indicator on the expediency and completeness of higher education in the era of digitalization.

A one-way analysis of variance (ANOVA) showed no significant differences in attitudes toward the necessity and sufficiency of higher education in the context of digitalization among respondents with different levels of education (higher, incomplete higher, secondary vocational, primary vocational) ($p = 0.496$) (Table 5).

Table 5. ANOVA (Do you consider obtaining higher education necessary and sufficient in the age of digitalization?).

			Sum of squares	df	Average square	F	Value
Between groups	(in common)		1,285	5	257	877	**496**
	linear term	Unweighted	027	1	027	091	763
		Weighted	473	1	473	1,613	204
		Deviations	812	4	203	693	597
Inside groups			221,556	756	293		
Total			222,841	761			

No statistically significant differences were found between the opinions of men and women on this issue (Tables 6 and 7).

A linear regression model was constructed using the Ordinary Least Squares (OLS) method for the dependent variable "Do you consider obtaining higher education necessary and sufficient in the age of digitalization?". The model has a low coefficient of determination (R2=,082). However, the model shows no multicollinearity (VIF < 4) or

Table 6. Conjugacy "Do you consider obtaining higher education necessary and sufficient in the age of digitalization?" * "About yourself. Your gender" (% of respondents).

Yes, higher education is necessary, and the knowledge gained is enough to adapt to the labor market	16,2	14,7	15,4
Yes, higher education is necessary, but digitalization requires constant improvement of skills. Therefore, it is necessary to constantly improve professional skills	68,8	72,3	70,7
No, today, in order to get a good job in life, it is not at all necessary to get a higher education, it is enough only to improve skills in a certain professional field	15,0	13,0	13,9
Total	100,0	100,0	100,0

Table 7. Chi-square test.

	Value	df	Asymptotic significance (2-sided)
Pearson's chi-square	1,111[a]	2	574
Likelihood ratios	1,109	2	574
Linear-linear relationship	015	1	903
Number of allowed observations	762		

[a]For 0 (0.0%) cells, expected values less than 5 are used. Minimum expected cell count is 47.30

heteroscedasticity (Durbin-Watson = 1,691). The model is highly significant ($p < 0{,}001$), indicating the reliability of the results (Tables 8 and 9).

Table 8. Summary for the model[f].

Model	R	R-square	Adjusted R-square	Standard estimation error	Change statistics					Durbin-Watson
					Changing the R square	F-change	df 1	df 2	F-Change Values	
5	297[e]	088	082	5185	007	6,136	1	756	013	**1,691**

The perception of high corruption risks influences the likelihood of accepting the opinion that professional development within a specific field is sufficient for career success. Respondents' age also impacts this opinion: young people are more likely to share this view. Students' appreciation of the qualifications of teachers, the practical orientation of learning, and the variety of extracurricular activities at modern universities reinforces their belief in the importance of higher education, regardless of external factors (Table 10).

Table 9. ANOVA[a].

Model		Sum of squares	df	Average square	F	Value
5	Regression	19,597	5	3,919	14,579	000[f]
	Remains	203,244	756	269		
	Total	222,841	761			

Table 10. Coefficients[a].

Model		Non-standardized coefficients		Standardized coefficients	T	Value	Collinearity statistics	
		B	The standard error	Beta			Allowance	VIF
5	(Constant)	1,854	077		24,118	**000**		
	What do you think, a Russian university in the PRESENT is... [Qualified teaching staff]	**070**	024	118	2,970	**003**	759	**1,318**
	Do you agree that a Russian university in the PRESENT is ... [Practice-oriented learning]	**061**	021	116	2,894	**004**	752	**1,330**
	Do you agree that a Russian university in the PRESENT is ... [Corruption threats]	**−049**	016	−108	−3,058	**002**	**975**	**1,026**
	Do you agree that the Russian university in THE PRESENT is ... [A variety of forms of extracurricular activities, leisure activities for students]	**059**	021	108	2,749	**006**	784	**1,275**
	newage2	**−058**	023	−088	−2,477	**013**	965	1,036

[a]Dependent variable: Do you consider higher education necessary and sufficient in the age of digitalization?

4 Conclusion

Based on the empirical research conducted among residents of Moscow, the capital of Russia, the following was revealed:

1. Most respondents share the opinion that higher education is necessary, but insufficient, for successful adaptation to the modern world, which is amplified by digitalization. Therefore, higher education, as in the USSR, should provide specialists not only with professional knowledge but also with universal skills in time management and information processing.
2. Education level and gender do not have a statistically significant impact on respondents' position regarding the necessity and sufficiency of higher education in the digital age. Age is statistically significant: older people are more likely to believe that higher education is sufficient, while younger generations are more inclined to believe that higher education is unnecessary for life in a digital space.
3. Respondents' perceptions of higher education are significantly influenced by their perception of the university. Respondents who perceive the university as a source of corruption risks are more inclined to believe that higher education is unnecessary.

References

Aflalo, E., Vaknin, M., Harband, Y., et al.: Using digital technologies to support active and self-directed learning. J. Comput. High. Educ. (2024). https://doi.org/10.1007/s12528-024-09428-y

Bali, T., Mann, B., Minosky, S.A.: Student perceptions of the use and effectiveness of engagement strategies in the online learning environment. Act. Learn. High. Educ. (2024). https://doi.org/10.1177/14697874241303154

Burner, K.J., Dennen, V.P., Jian, S.: Prior online learning experiences as incidental professional development: a cross-sectional survey of online instructor beliefs and strategies. Internet High. Educ. **64**, 100977 (2025). https://doi.org/10.1016/j.iheduc.2024.100977

Воротилина, Т.В.: Высшее юридическое образование как средство формирования профессионального правосознания в условиях цифровизации. Теория государства и права. **3**(19), 40–49 (2020)

Clayton, T., Rosenberg, S.L., Wyatt, J., VanOrsdale, B., Bisaillon, D.: Higher education and real-world training (HEaRT): a new ecosystem for learning. J. Contin. High. Educ. **72**(3), 392–404 (2024). https://doi.org/10.1080/07377363.2023.2291592

Emikh, N.A., Fomina, M.N.: Specifics of the new paradigm of higher education in the context of its digitalization. Sci. Educ. Today. **13**(4), 100–121 (2023). https://doi.org/10.15293/2658-6762.2304.05

Frølich, N., Bleiklie, I., Stensaker, B.: Leadership communication during organizational transformation. Eur. J. High. Educ., 1–16 (2025). https://doi.org/10.1080/21568235.2025.2456871

Karpov, E., Karpova, E.: Digitalization of higher education in the context of globalization. In: XXIV International Conference "culture, personality, society in the conditions of digitalization: methodology and experience of empirical research" (Ekaterinburg, Russia, 19–21 March, 2021), pp. 9–19. Knowledge E, Dubai (2022). https://doi.org/10.18502/kss.v7i3.10270

Díaz-García, V., Montero-Navarro, A., Rodríguez-Sánchez, J.-L., Rocío, G.-L.R.: Digitalization and digital transformation in higher education: a bibliometric analysis. Front. Psychol. **13** (2022). https://doi.org/10.3389/fpsyg.2022.1081595

Nešković, S.: Digitalization of higher education in the context of economic development of the region of south East Europe, with reference to Montenegro and Bulgaria. Int. Sci. J. Monte (ISJM). **9** (2024). https://doi.org/10.33807/monte.20243113

Nieminen, J.H., Yan, Z., Boud, D.: Self-assessment design in a digital world: centring student agency. Assess. Eval. High. Educ. **1–15** (2025). https://doi.org/10.1080/02602938.2025.2467647

Prakasha, G.S., Muttungal, P., Lapina, M., Taran, V.: Flipped classroom strategy in online teaching: challenges faced by higher education teachers. In: International Conference on Innovative Approaches to the Application of Digital Technologies in Education and Research, pp. 241–253. Springer, Cham (2025a)

Prakasha, G.S., Thamaraikannan, S., Lapina, M., Sanskriti, R.: Pre-service and in-service teachers' perceptions of using virtual reality tools in teaching. In: International Conference on Innovative Approaches to the Application of Digital Technologies in Education and Research, pp. 340–355. Springer, Cham (2025b)

Tam, A.C.F.: Interacting with ChatGPT for internal feedback and factors affecting feedback quality. Assess. Eval. High. Educ. **50**(2), 219–235 (2024). https://doi.org/10.1080/02602938.2024.2374485

Zhang, L., Xu, J.: The paradox of self-efficacy and technological dependence: unraveling generative AI's impact on university students' task completion. Internet High. Educ. **65**, 100978 (2025)

Alexandrova, L., Mozhaeva, G., Karabasheva, M., Lapina, M., Rugelj, J.: Use of IT for student-centered learning approach to education at modern university. In: Paper Presented at the CEUR Workshop Proceedings SLET-2020: International Scientific Conference Innovative Approaches to the Application of Digital Technologies in Education, vol. 2861, pp. 32–40 (2020)

Lapina, M., Shiriaev, E., Babenko, M., Istamov, I.: High-speed convolution core architecture for privacy-preserving neural networks. Program. Comput. Softw. **50**, 417–424 (2024). https://doi.org/10.1134/S0361768824700282

Rusanov, M., Babenko, M., Lapina, M., Sajid, M.: Identification of exploited unreliable account passwords in the information infrastructure using machine learning methods. Big Data Cognit. Comput. **8**, 159 (2024). https://doi.org/10.3390/bdcc8110159

Pedagogically Organized Socialization by Means of Digital Technology

Marina V. Katrenko[1(✉)], Nina S. Gorbaneva[1], Leonid V. Yukhtenko[1], Yulia I. Zhuravleva[2], and Rafael Riza oglu Zeynalli[3]

[1] North-Caucasian Federal University, Stavropol, Russia
mkatrenko69@bk.ru
[2] North Caucasus Institute—Branch of RANEPA, Pyatigorsk, Russia
[3] Volgograd State Medical University, Volgograd, Russia

Abstract. The successful self-realization of a person in life, first of all, depends on his health. The restoration measures to maintain health is motor activity. In terms of work on the designated topics, the article proposes a combination of health forms, comfortable environmental conditions, information technologies and pedagogically organized socialization. The proposed means of research showed a positive trend in use. The analysis of literary sources made it possible to identify the potential of pedagogically organized socialization among the city residents, in order to unite in interest groups, and the development of mentoring. Carrying out a sociological survey of residents of the metropolis, revealed the preference of citizens in choosing a recreational route for a healing run. The use of technologies of the geoinformation system in the work of the metropolis routes in terms of a healing run. The inclusion of AR-technology, Web-site, specialized sports electronic applications, revealed an increase in interest in the study of digital technologies and increasing the motivational component for sports involved in recreational places. The final results made it possible to conduct a comparative analysis of the studied routes. In addition, to formulate the conclusions that the visual consideration and use of the maps of routes, the use of the proposed funds, had a positive impact on the participants in the study and contributed to the active inclusion of people in recreational health activities.

Keywords: Health · Recreation · Socialization · Sociological Survey · GIS-Technologies · IT technologies · Involved · Map-scheme · Landscaping of Territories

1 Introduction

World socio-economic changes lead to fundamental transformations in the life of each individual person who cares about successful professional self-realization, well-being, which primarily depends on health. Consequently, measures to maintain the best physical form, do not lose their exceptional priority in a scale, both individual-personal development and socially organized groups of the population. In this regard, the relevance of the consideration of issues of restoration measures in the field of health maintenance is not in doubt.

M. Lapina et al. (Eds.): DET 2024, LNNS 1616, pp. 215–225, 2025.
https://doi.org/10.1007/978-3-032-04365-8_22

"The implementation of measures to preserve the life and health of citizens in the process of their training and labor activity in accordance with the legislation of the Russian Federation" [1] is controlled at the state level and is important in the process of transferring the experience of the social values of society.

2 Socialization as a Means of Organizing Recreation

Synergy of funds and forms of organization of leisure and functioning of recreation is possible within the framework of pedagogically organized socialization.

"Socialization is one of the processes that is necessary for the normal functioning and development of society" [2], which involves the search for fundamentally new solutions to pedagogical problems.

B.G. Ananyev characterizes the manifestation of socialization through "the formation of knowledge about people, society and about itself <...> The development of skills of norms, roles, positions of practical activity, attitudes, value orientations" [3].

A.S. Leontiev defines "the essence of socialization as follows: socialization is the acquisition of experience in society and the application of this experience in practice" [4].

T.E. Manger assists an important role in the process of socialization in solving the problems of "the formation of a personality of a new format" [5].

In the works of a number of authors, the processes of socialization (Kon I.S., Mudrik A.V., Parsonson T., Parygin B.D., etc.) and the pedagogical impact on the personality of a person (Ananiev B.G., Andreeva G.M., Kolominsky Y.L., Rean A.A. Shatsky S.T., etc.), intersect and continue all their lives.

Purposefully organized pedagogical socialization contributes to the unification of people into society that pursue the same values, solving the tasks of the social formation of a person, social relations in the systems of institutions and collective organizations.

The main type of recreational, restoration activity is an active rest that includes wellness, running and structured physical activity. Also, the presence of unique natural resources (park zones, installed terrenckers laid in the forest, treadmills, etc.) and infrastructure on the ground, relate to instruments to satisfy the recreational needs of those involved.

A.S. Belenko believes that "with the help of physical exercises, a person receives emotional unloading, negativity leaves, and he has positive thinking" [6], in addition, "an important motivational prerequisite for physical recreation in men and especially in women is the possibility of group communication in the conditions of motor interaction" [Cit. by: [7]].

According to L.M. Demyanova, G.A. Manukyan, "regular classes of recreational improvement running contribute to the synthesis of endomorphins in the body, improve their emotional state, enhance performance, reducing the number of negative behavioral reactions, expand functionality, increase the body's stability to adverse working conditions" [8].

Joint of leisure, participation in general events (in competitions on cross -country run, Scandinavian walking, sports orientation, etc.), allows you to form objective conditions for expanding the knowledge transmitted by participants to each other, gaining practical experience by the younger generation from the older and applying this experience in practice. Socializing, members of society, interact, and the communication component is an important aspect in the implementation of mutual activity. Tuition is based on "attitude of cooperation, mentoring, dialogue ..." [9].

Definitely, pedagogically organized socialization is the direction of a person's personality in a sociocultural environment, where processes such as "observation, imitation, communication, training, education, education, education, professional and creative activity" [5].

The above prompted the authors to develop the project "Recreational routes of the megalopolis" in order to study the preferences of people in choosing recreational routes of the metropolis and ways to motivate to increase motor activity.

3 Content of the Project "Recreational Routes of the Megapolis"

3.1 Sociological Survey

The project included:

- conducting a sociological survey among a healthy run about preferences in the choice of recreational routes of a metropolis;
- building the recreational routes of the metropolis to increase the motivational component for motor activity using AR technology, Web-site, specialized sports electronic applications.

The task of the study was in the development and conduct of a sociological survey of residents of the metropolis. Using a random sample, in the form of electronic questionnaires (Google-form), 48 people with various socio-demographic characteristics (different sexes, age, kind of activity) were interviewed.

The convenience of using Google-forms is that "the service is easy to use, does not require downloading, has a convenient, understandable and accessible time for users, the interface that allows you to send the resulting form by e-mail and store free in cloud storage (this is a free service for storing photos, videos, music and other files) along with statistical results. In addition, which is especially attractive, Google-forms are adapted to mobile devices, which allows you to use a lightweight mobile version of the service with full functionality for developing, editing, viewing forms and results, sharing forms between mobile devices, for example, a phone and a tablet" [10].

Intense interaction made it possible to collect and analyze information about the preferences of citizens in the choice of the most famous, publicly accessible, running routes of the metropolis and, in general, about their sports.

A conditionally survey had several main semantic blocks [11]:

1. general orientation issues;
2. questions with geographical binding (i.e., such matters proposed options for choosing the route that is most suitable for each respondent with a demonstration of this route on the map);
3. issues aimed at determining the personal preferences of the respondent.

3.2 Geoinformation Technologies in the Visualization of the Geographical Location of the Recreational Routes of the Metropolis

The solution to the next task of the study was aimed at drawing up running routes along the territory of the metropolis. The use of GIS-technologies in the work of, allowed geographic information modeling of routes and geographical analysis of the area in terms of cross-country training.

The integration of spatial data with physical culture and sports represents GIS-technologies, in the words of A.V. Orekhova, as a "multi-purpose tool for use" [12], because "both sports and geography are focused on how people move and interact in space ..." [13]. "Geoinformation modeling in the study is represented by the developed cartographic material that reflects the configuration, length and complexity of routes in different areas" [11]. Marshrut cards are presented in Figs. 1, 2 and 3.

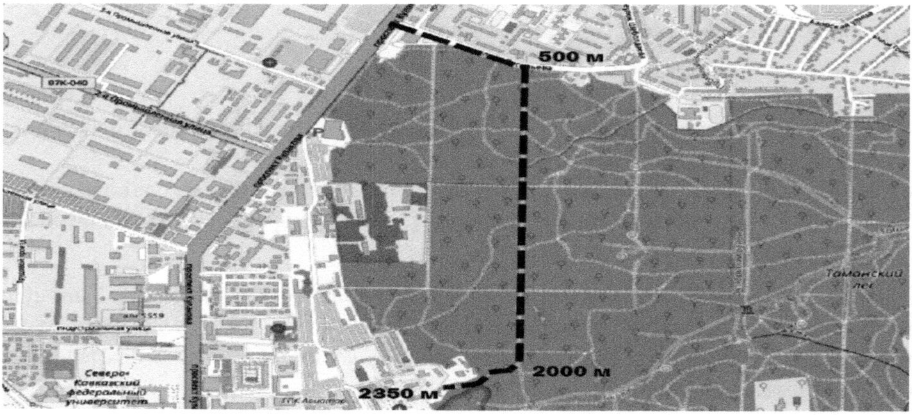

Fig. 1. Map of the route "Health Trail" (Taman Forest). *Source: compiled by the authors [11].

When developing routes, QGIS 3.34, Adobeillustrator2023.

Software products were used, in recent years, there has been a significant increase in the use of geoinformation systems, technologies and appropriate software that allows you to process large volumes of information that have temporary, spatial and geographically coordinated characteristics, since "... most of the information that a person uses for different purposes is connected with space ..." [12]. The introduction of GIS-technologies "makes it possible to more accurately determine the location of objects, conduct a detailed analysis of topographic, geological and climatic conditions on the ground" [14].

"GIS-system uses multi-level data, where each layer represents one of the forms of spatial data, such as a layer of streets, a layer of forest, a layer of vegetation and much more. All these layers are then combined into one, forming an integrated layer with all the data" [15]. The instrumentation of the geoinformation system allows us to "determine the coordinates of any point on the map with a sufficiently high degree of accuracy" [16, 17]. Thus, the user is exempted from calculations and spatial constructions. In other words,

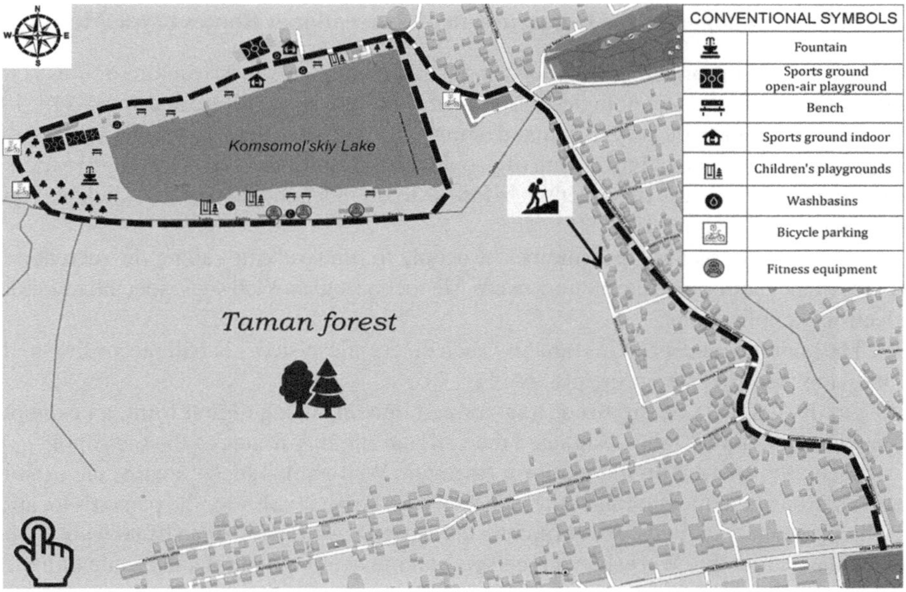

Fig. 2. Map of the route "Komsomolskoye Lake". *Source: compiled by the authors [11].

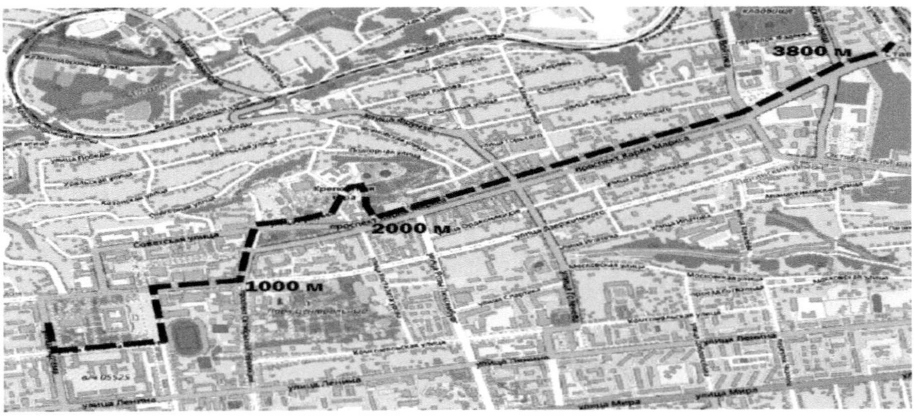

Fig. 3. Map of the route "Academic Theatre - Railway Station". *Source: compiled by the authors [11].

GIS-technologies "allow you to visualize data, analyze information, conduct research, also contribute to the expansion of knowledge in creating your own cartographic product" [14].

3.3 IT Technologies in the Construction of Recreational Routes of the Metropolis

One of the most promising areas for the development of wellness running routes is the integration of digital technologies into the sphere of physical culture and sports [18, 19].

IT technologies are very in demand among all age categories as a whole; It is a "most important tool for social integration of people with disabilities" [20, 21]; they are used in the educational process of students, "allow us to do independent work of students more affordable and mobile" [22, 23].

In order to attract a larger number of people to run (walking) along the recreational routes of the metropolis, the authors used: AR-technologies, Web sites, specialized sports electronic applications.

The general concept of introducing such electronic systems is built according to the following step -by -step principle (see Fig. 2).

Firstly, it should be digitized (transformed into an analog digital form, an example: "Yandex. Panorama") the designated recreational running routes of the metropolis.

Secondly, create a digital platform (thematic Web-site), which contains the existing information about the running routes themselves, the basic rules for the person's location in the forest, next to roads, about running classes, etc. A function of a virtual walk should be connected on this Web-site where users can move along the selected route online.

Thirdly, by analogy with existing electronic sports applications (Pumatrac, Simple Run, Sports Tracker for All Sports), several options were used to help everyone to try out the developed running routes. It is enough for the user to choose one of the three route options (levels: initial, medium and advanced) and the application will pave the way, direct and answer the questions of interest.

Fourth, several electronic links and QR codes for downloading special sports applications are placed on the Web Site itself, which will combine the functions of mobile consultants, shagomer applications (for users using SMART hours).

In the process of use, an electronic sports application is able to automatically monitor, correct the direction of movement and the general motor activity of the user with an accessible visualization of the dynamics of the results of the work.

The dissemination of information about the symbiosis "IT-gis-physical culture" occurs through the media, social networks, messengers, etc.

4 Results

The final stage of the study was mathematical processing of results and summing up.

The quantitative and qualitative analysis of the data showed the following patterns and features in the answers of people who took part in the survey (Figs. 4, 5 and 6).

For example, to the question "Are you playing sports?", 64.9 % of respondents answered positively, and these are people who play sports periodically, at an amateur level and in their free time; The remaining 35.1 % - gave a negative answer: "We are not engaged." In general, on the one hand, this situation shows that, most likely, the reason why more than a third of the surveyed people are not playing sports is a lot of workload and employment at work. However, the fact that more than half of the respondents find free time for sports creates a positive basis for future development in this direction. To

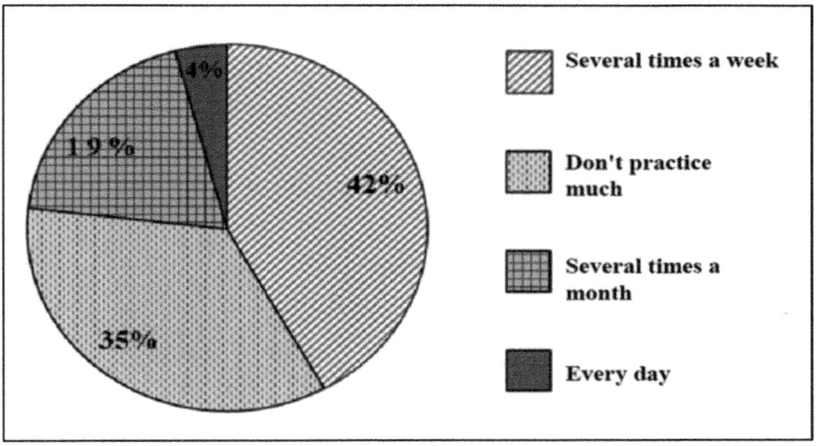

Fig. 4. Volume of motor activity. *Source: compiled by the authors.

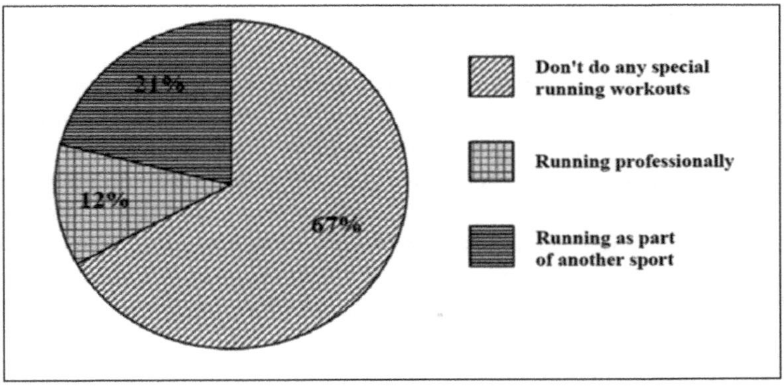

Fig. 5. Orientation of motor activity. *Source: compiled by the authors.

the question "How often do you go in for sports?", 42 % of the questionnaires replied that they are engaged in several times a week, but those who are engaged every day, unfortunately, are only 4 % (see Fig. 4).

Therefore, it can be assumed that a rather large percentage of the population is somehow engaged in motor activity, regardless of whether it is a run in the park or visiting the gym.

Further, the question was about the identification of what motor activity the respondents are engaged in. To the question "Are you doing special running training?", The most popular was the negative answer (67.3 %), since this part of the respondents is engaged only in amateur running; professionally, only 11.6 % of respondents are running; 21.1 % - do this within a different sport. From this it follows that the role of healing is very underestimated and has no great popularity among the population of the city (see Fig. 5). It is worth noting that the open question about the goals of running training,

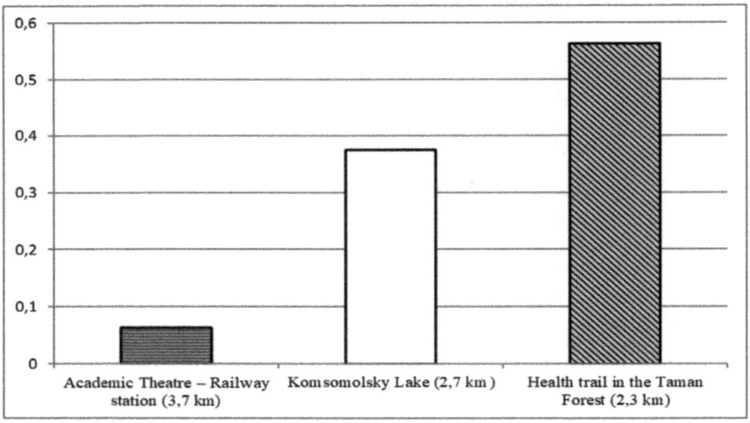

Fig. 6. Comparative analysis of the choice of running routes in Stavropol. *Source: compiled by the authors.

most often, could find answers about "maintaining the body in good condition", "to strengthen health", "training of general endurance". This means that people who really engage in running exercises consider wellness running, first of all, as a means or a useful and simple tool, which allows to relieve excess tension and maintain health in normal.

Special attention should be paid to the block of questions indicating the cross - country routes of the metropolis. A detailed consideration by the fulfillment running of the catra-scheme of routes (see Figs. 1, 2 and 3) showed the following results: more than half of all respondents (56.2 %), voted for the route along the "Health path" in the (Tamansky Forest); Another 37.5 % chose the "Komsomolskoye Lake" and only 6.3 % counted the city center, the route "Academic Theater - Railway Station", a suitable place for running training (see Fig. 6).

Along with this, the use of routes recommended by the authors in training, their consideration, as if from a height, contributed to confidence among those involved in overcoming the distance. Citizens involved in running visually imagined descents, ascents, turns, the name of the streets, the lanes, etc., and were mentally ready to pass the route.

The use of an application with augmented reality technology (-AR) contributed to an increase in the intellectual component of outdoor enthusiasts. However, the results of a sociological survey showed that for the organization of independent training activities, many people do not use such applications (28.6 %). They also did not express the desire to use any other digital technologies for additional physical activity and sports.

In a full-time demonstration of the proposed means (maps of routes, a software product AR-technology, Web-site, specialized sports electronic applications), many vacationers (67.2 %), showed a desire to have such cards and an applied technology with augmented reality (-AR); engage in healing walking or running in the near future and precisely along these routes (45.3 %); Learn to build a trajectory of your own route (21.8 %).

5 Conclusions

Thus, a sufficient amount of information provided on the subject, showed that a large role in the recovery measures of the psychosomatic sphere, in the prevention of various diseases, in counteracting hypodynamia, etc., is played by wellness physical culture, in particular, recreational classes using health -improving running, including in the conditions of the urban environment.

The targeted, pedagogically organized socialization of a healthy run, through the use of GIS technologies and proposed IT technologies, aroused great interest among lovers of wellness running and simply vacationers in recreational zones. The rapprochement of people with the same preferences in the choice of recreational routes and interested in the joint use of digital technologies and physical activity have contributed to the expansion of the general educational interests of athletes; Improving the intellectual potential of everyone separately and athletes, united in groups of runners, lovers of Scandinavian walking and any motor activity.

The study on the implementation of the project "Recreational routes of the metropolis" made it possible to draw the following conclusions:

1. the use of various means and forms of motor activity, motivates a person to strengthen health, maintain himself in good physical form;
2. the use of GIS technologies in the development of running routes has a positive effect on the psycho-emotional sphere of motor activity;
3. the use of GIS-technologies contributes to the development of interest in those who have certain knowledge in the field of informatization, in creating their card schemes and their use in synergy with motor activity;
4. pedagogically organized socialization, using IT technologies acting as pedagogical tools, contributes to increasing the digital literacy of vacationers and building educational space in the design of personal card schemes, strengthening motivation to study new software, uniting groups of like-minded people of all people who want to engage in physical culture and sports;
5. regularly conducting recovery measures in such conditions (including without financial costs), increases the recreational effect of health and walking.

However, despite the fact that the participants in the sociological survey did not give examples of little comfortable or poorly secured recreational places of the metropolis, according to the authors, the development and implementation of the road map by diversification of infrastructure, can contribute to the transformations of not only the routes presented for the study, but also be active participants in the national projects of the Russian Federation, healthcare and education.

References

1. Federal Law: On the basics of public health protection in the Russian Federation. No. 323-FZ. Accessed 21 Nov 2011
2. Mirdianova, B.: Socialization as a pedagogical phenomenon. In: Fundamental Research. Section: Pedagogical Sciences, vol. 3, pp. 87–88 (2005)

3. Andreeva, T., Ideas, B.: Ananyev and empirical study of the personality path. Bull. St. Petersburg Univ. Sociol. **1** (2008)
4. Leontyeva, A.: The essence of socialization as a pedagogical category. Young Scientist. **30**(320), 274–275 (2020)
5. Manger, T.: The pedagogical foundations of the socialization of personality as a subject of a sociocultural society. Bull. Tambov Univ. Ser. Humanities. **10**(138), 59–66 (2014)
6. Belenko, A., Andryushchenko, Y.: The influence of sports on the human psyche and consciousness. Current issues in the humanities and social sciences: from theory to practice. In: Materials of the All-Russian Scientific and Practical Conference with International Participation, pp. 223–225. Chief Editor J.V. Murzina, Cheboksary (2023)
7. Barabash, O., Gudkov, Y., Mazitova, N., Munirov, E., Pankrashin, D.: Features of recreational activities using recreational running equipment for participants of the running club of the first mature age. Sci. Notes Lesgaft Univ. **11**(189), 41–46 (2020). https://doi.org/10.34835/issn.2308-1961.2020.11
8. Demyanova, L., Manukyan, G.: The effect of recreational running on the human body. In: Modern education: current issues, achievements and innovations: a collection of articles by the winners of the VIII International Scientific and Practical Conference, pp. 116–118. Nauka i prosveshchenie, Penza (2017)
9. Katrenko, M., Zhuravleva, Y., Ageev, A., Kukhtareva, O., Desenko, O., Ivanova, I.: Model of open information education of university students in the field of physical education. In: Proceedings of the III International Conference on Advances in Science, Engineering and Digital Education. ASEDU-III 2022. Krasnoyarsk, p. 040018 (2024)
10. Katrenko, M., Kulikova, T., Poddubnaya, N., Zhuravleva, Y.: Digital technologies in the educational process in physical culture. Bull. North Caucasus Fed. Univ. **2**(95), 222–232 (2023). https://doi.org/10.37493/2307-907X.2023.2.26
11. Gorbaneva, N., Yukhtenko, L., Katrenko, M., Zhuravleva, Y.: Physical culture and sport: integration of science and practice. In: Proceedings of the XXI scientific and practical conference with international participation (November 22 2024), pp. 214–217. NCFU Publishing House, Stavropol (2024)
12. Orekhova, A.: Formation of geoinformation competence of students in the process of studying geography based on the use of GIS technologies. Bull. IYa. Yakovlev CHSPU. **1**(81), 150–157 (2014)
13. Suprunchuk, I.: Geography of sports: theoretical and methodological approaches to the formation of a scientific direction. Sci. Innov. Technol. **1**, 43–65 (2022). https://doi.org/10.37493/2308-4758.2022.1.3
14. Pupysheva, S., Zhadovskaya, D.: Application of geoinformation technologies in geography lessons. In: Collection of materials of the I Interregional scientific and practical conference dedicated to the Year of the teacher and mentor "Modern educational practices in student research", vol. 1, pp. 96–99. Publishing house of ANO DPO "Interregional Center for Innovative Technologies in Education", Kirov (2023). https://doi.org/10.24412/cl-37132-2023-1-96-99
15. Suyargulov, I.: The use of geoinformation technologies for mapping tourist routes. Sustain. Innov. Dev. Des. Manag. **17**, 111–128 (2021)
16. Gubeeva, S., Ulengov, R., Urazmetov, I.: Geography information systems and teaching and learning geography in the republic of tatarstan. In: В сборнике: 3rd International Multidisciplinary Scientific Conference on Social Sciences and Arts GEM 2016 Sofia, pp. 779–786 (2016). https://doi.org/10.5593/SGEMSOCIAL2016/B11/S03.099
17. Akhrenov, V., Akhrenova, N., Belous, E.: Distance training - 2020 and problems of cyber-socialization of participants in the educational process. Bull. Moscow. State Reg. Univ. Ser. Pedagogy. **3**, 6–13 (2020). https://doi.org/10.18384/2310-7219-2020-3-6-14

18. Katrenko, M., Zhuravleva, Y., Evmenenko, E., Savin, D., Tkacheva, N., Ivanova, S., Yaroshenko, E.: Information and communication technologies in the implementation of physical education in the University. In: Proceedings II International Scientific Conference on Advances in Science, Engineering and Digital Education (ASEDU-II-2021). Conference Proceedings. Krasnoyarsk, p. 50013 (2022). https://doi.org/10.1063/5.0104271
19. Katrenko, M., Nebytova, L., Shchekin, A., Zhuravleva, Y.: The role of the technology of augmented reality in the organization of independent work on physical culture at the university. Pedagogy Educ. **2**, 61–65 (2019). https://doi.org/10.7256/2454-0676.2019.2.28943
20. Nebytova, L., Katrenko, M., Savin, D., Zhuravleva, Y.: Augmented reality in the training process of children with hearing disorders. In: CEUR Workshop Proceedings. 4. Cep. DLT 2019 - Selected Papers of the 4th All-Russian Scientific and Practical Conference with International Participation "Distance Learning Technologies", pp. 330–339 (2021)
21. Nebytova, L., Katrenko, M., Savin, D., et al.: The use of augmented reality technology as a means of increasing the efficiency of the training process in athletics for children with hearing impairments. Bull. North Caucasus Fed. Univ. **5**(74), 169–173 (2019). https://doi.org/10.37493/2307-907x-2019-74-5-169-173
22. Kulikova, T., Poddubnaya, N.: Instrumental means of implementing distance educational technologies. Standards Monit. Educ. **7**(1), 32–35 (2019). https://doi.org/10.12737/article_5c540d3049a748.60754182
23. Kulikova, T., Poddubnaya, N., Ardeev, A., Alekseeva, P.: Formation of digital literacy of students by means of virtual and augmented reality technologies. In: Innovative Approaches to the Application of Digital Technologies in Education 2020, Stavropol, Russia, November 12-13, 2020, CEUR Workshop Proceeding, vol. 2861, (2020) http://ceur-ws.org/Vol-2861/paper_36.pdf

The Use of Artificial Intelligence in Education and University Student Scientific Research

Anatoliy Kazak[1(✉)], Victor Selivanov[1], Dmitry Nekhaychuk[2], Igor Bukreev[3], and Daria Syedneva[2]

[1] V.I. Vernadsky Crimean Federal University, Simferopol 295007, Russia
kazak_a@mail.ru
[2] Plekhanov Russian University of Economics, 115093 Moscow, Russia
[3] Kherson State Pedagogical University, 273003 Kherson, Russia

Abstract. The paper discusses the issues of activating comprehensive research of modern information technologies, in particular the basics of artificial intelligence, in the education system and scientific research of students of higher education in order to form a favorable, effective, qualitative and ethical process of obtaining knowledge in the context of the formation of an information society. The scientific study presents a detailed retrospective analysis of the conceptual and categorical apparatus of "artificial intelligence" from the perspective of generally recognized scientific theories and leading ideas of domestic and foreign scientists for further theoretical and methodological consolidation. The relevance of scientific research lies in the fact that university students strive to learn high-level technologies. These technologies help them to solve the problems of self-control and self-organization on their own. They help to shape the student's personality, variably thinking, creative personality with a new type and level of thinking. The purpose of the research is also to reveal and systematize the features of scientific discourse regarding the potential, trends and directions of adequate and ethical use of artificial intelligence technologies in the educational process and scientific research of students of higher education as an instrument of educational reform, as a means of increasing the effectiveness of their educational achievements, as a means of intensification and individualization of the formation of professional knowledge and skills of future specialists in the era of public informatization.

Keywords: information technology · artificial intelligence · scientific activity · educational process · student · higher education

1 Introduction

Considering the modern formatted component of the development of education and science among students, it is legitimate to come to the conclusion that, thanks to artificial intelligence technologies, computer systems can offer students personalized, adaptive or integrated learning, and automated assessment of achievements in the educational process, support and feedback during training [5]. Virtual reality as a newly organized social space leads to planned, consequential and systematized changes in the educational

M. Lapina et al. (Eds.): DET 2024, LNNS 1616, pp. 226–233, 2025.
https://doi.org/10.1007/978-3-032-04365-8_23

process and research work of students of higher educational institutions, improvement of existing and formation of hitherto unused organizational forms, types and methods of teaching, and also changes the interaction between the teacher and the student and the educational environment as a whole. It should be taken into account that the use of artificial intelligence as an innovative mechanism in modern pedagogy is important from the point of view of conceptuality and methodology. He has an active influence on the formation of a scientific and educational cluster [2].

2 Methods

In the context of the global transition of society and all spheres of human life to an innovative development paradigm, special attention is paid to rethinking the value-semantic boundaries of the existing national education system, its modernization, transformation and informatization through the actualization of new trends in the development of the labor market, as well as modern aspirations and values of society. The process is accompanied by significant changes in the practice of educational and upbringing activities and scientific metaphors of pedagogy [11]. Understanding the modern foundations of scientific work sets new goals and objectives for higher education institutions (HEIs) to meet the needs of individuals for knowledge that could be used depending on the new conditions of social transformation, emphasizing the priority of the individual characteristics of the student during training.

The modern labor market dictates the need to revise the depth of traditional approaches in the higher education system with the aim of introducing innovative technologies, as well as their perfect combination; the use of new progressive methods for organizing the educational process in higher education institutions; computerization of education and scientific research of higher education students; integration of all stages of the educational process based on the processing of the best achievements of science, which, in turn, will contribute to increasing the qualification level of graduates of higher education institutions. It contributes to the improvement of their professionalism and the qualities of a modern personality, which is in demand by the modern society of high technologies. It is of no small importance for the student organization of the educational process, especially in the context of an interdisciplinary approach and a practice-oriented direction in education [1–5].

Nowadays, the involvement of teachers in the educational process of tools generated by artificial intelligence (AI) is no longer science fiction. Since AI can perceive much more information than a person, its direct implementation in the field of education and scientific research of students of higher educational institutions thanks to interactive assistants and adaptive learning programs provides additional opportunities (Fig. 1) [9]:

3 Results

3.1 Main Advantages of Using AI in Education

Modern changes in the development of postmodern society have a somewhat abrupt nature due to the creation and implementation of artificial intelligence technologies. AI is a product of human activity. He has the opportunity to replace a person in the field of

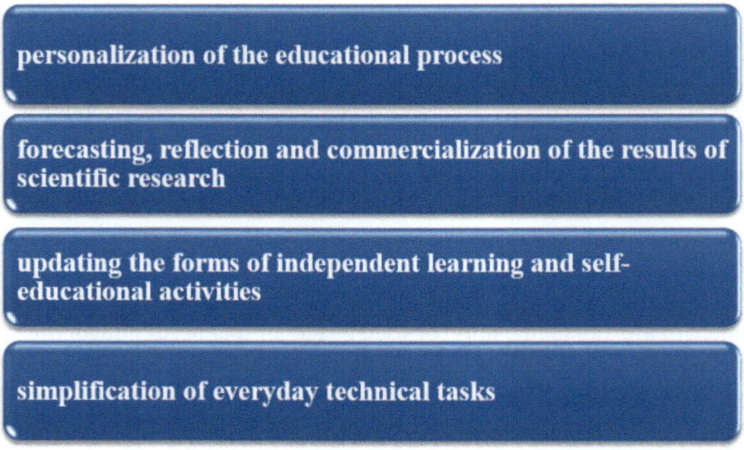

Fig. 1. Additional opportunities provided by AI in the field of education.

performing various tasks. Artificial intelligence is also actively working in the educational field and is integrated into the system of social interaction. It is with the help of AI that society expresses its readiness to move to a new qualitative level of its existence, actually opening a new era. This is an era of high information level, which actively affects the economy and the quality of life. It is able to change developing technological processes, approaches to cognition and the formation of a modern worldview on a new qualitative level, and creates new forms of communication. Along with these changes, the educational landscape is also changing, or more precisely, the process of acquiring knowledge, within this framework, questions of mobility, efficiency, time, space are raised [6]. If we compare information technologies at the initial stage of their formation and with the current state, we can notice the predominant role of technologies that determine the place of man in the social structure of society, and the state in the economic system. The development of high-tech has become a motive for the creation of AI and its full-scale implementation in everyday human life. Accordingly, now the sphere of higher education must meet the current needs of its users in the process of acquiring knowledge, transferring all the material into the virtual world created by AI, with the aim of developing an adaptive, inclusive, flexible, personalized and effective learning environment and a deep understanding of the behavior, reactions, and emotions of students [4]. Institutions that provide educational services should adapt their own development strategy and focus on promoting responsible, high-quality, ethical and transparent use of AI tools by their employees and students in accordance with applicable law, with subsequent mitigation of the negative consequences of its impact on society, culture and the economy.

In general, the evolution of the development of learning using information technology dates back to the early 1980s and the publication in 1989 of the first International Journal of Artificial Intelligence in Education, while the International Society for Artificial Intelligence in Education was founded only in 1993 [10]. Unfortunately, scientists have not yet formed a single definition regarding the essence of artificial intelligence.

In fact, it can be stated that the topic of artificial intelligence in the field of education is associated with its enormous global scientific potential, which is growing daily, and the disclosure of its essence awaits a practical solution. In accordance with this, today a significant array of scientific publications is being actively published. The functions related to understanding text messages are being improved; there are special training programs that are personalized and adapted. These programs have the opportunity to take corrective actions in relation to some of the educational programs. The scientific analysis of the level of education of university students helps in this. Scientific analytical forecasts are being formed, integration processes are underway, and assistance in providing information is being personalized are gaining relevance; virtual student assistants are used - chatbots (ChatGPT, Duolingo, Thinkster, Querium, Altaby Knewton, etc.) [7–9].

In recent years, many tools have emerged that change the lives of both teachers and students, providing convenience in acquiring knowledge. For example, Google Translate allows you to translate text into more than 100 languages; Siri and Google Assistant allow people to ask questions and get quick answers; Chat GPT provides assistance in scientific research, summarizing sources or directing to useful publications, as well as instant answers to any questions, explanations, giving examples, writing poems or stories. Stable Diffusion and Imagen allow you to enter text to create realistic images; Caktus provides personalized recommendations to students to monitor their learning; Gradescope helps teachers assess students faster and more effectively while reducing bias; Altaby Knewton allows you to adapt to the unique learning style of each higher education student; Education copilot offers to save teachers time and energy by quickly generating customized templates, and also provides the ability to collaborate between teachers, etc. [11–15].

AI in education is a revolutionary way to help students in their studies and research, which changes the process of acquiring knowledge, communicating with information and educational resource bases makes it possible to rationalize the use of the time budget and rapid spatial movement; it makes it possible to receive education in accordance with one's own plans, which does not allow one to misunderstand the large number of unknown problematic issues. Accordingly, it is worth focusing on the identification and detailed analysis of the key advantages and disadvantages of using AI in higher education and scientific activities in the Student Scientific Society. The main advantages of using artificial intelligence technologies include:

1. Artificial intelligence as a system education has the opportunity to selectively meet the needs of mastering the knowledge of specific students, to take into account their strengths and weaknesses. This makes it possible to form a variable system of individual education.
2. The technological features of artificial intelligence make it possible to analyze and control the features of the educational process of each individual student, take into account their abilities, and create supportive individual learning content.
3. Artificial intelligence as an evolving system makes it possible to evaluate not only closed-ended test answers, but also descriptive results.
4. The use of artificial intelligence in the educational process saves students from worrying about the possibility of making individual mistakes in the implementation of

academic disciplines. As part of the so-called feedback system, they can correct these errors in a timely manner.

5. Thanks to the available capabilities of artificial intelligence, students receive the kind of education they need. In particular, visually impaired students can receive the studied content using voice communication. There are special facilities for both deaf and hard of hearing students.
6. Artificial intelligence technologies can support the digitalization of content, thereby facilitating the extraction and use of necessary information.
7. Delegating routine tasks to AI (creating draft lectures, assignments, systems for automating the assessment of work, express answers to frequently asked questions, general answers to questions on writing, for example, a course project) allows you to speed up and relieve teachers of their workload.
8. The use of artificial intelligence by teachers at higher education institutions helps to expand the audience for obtaining knowledge without geographical restrictions and to create a new image of the teacher.

3.2 Experience of Using AI in the Educational Process in V.I. Vernadsky Crimean Federal University

Using the advantages of AI, V.I. Vernadsky Crimean Federal University implemented two ideas that illustrate the use of AI technology in educational assessment. One concerned the development of assignments; the second concerned the verification of open responses.

The automatic development of assignments in Russian language and literature for schoolchildren was implemented. It turned out that two types of AI could be used:

- Simple algorithms. They help to generate tasks to check the understanding of the rules of the Russian language. For example, choosing a letter that represents a sonorous sound from several suggested options. This method of creating test tasks is suitable for situations where there is a clear list of rules of the Russian language that can be described for AI.
- Advanced neural networks. They are able to generate tasks that test language literacy and text comprehension. For example, you can arrange the confused sentences of the text in order or choose the appropriate phrase to fill in the gap in the sentence. Russian Russian was based on the RUGPT-3 language model, which was pre-trained on the Russian language corpus and was non-specific for a certain age or topic, and powerful machine learning was used on a smaller amount of data so that the model could create assignments in the Russian language and literature for schoolchildren.

Checking assignments with open answers or essays. Here we have identified the great potential of AI. It's no secret that when an expert check open response, there is a problem of subjectivity. Psychometry has been working with this problem for many years. For example, there are recommendations on which processing methods can be used to remove individual distortions introduced by the examiner. Our experiment has shown that psychometric principles that work with people are also applicable to training neural networks, and it is very likely that they will eventually give the best result.

Along with these advantages, there are a number of serious problems that contradict the high-quality and sustainable use of technologies: issues of data privacy and security;

decline in cognitive abilities among both students and teachers; the risk of excessive everyday use of technologies (AI should be considered only as a supplement to educational materials developed by the teacher); there is no correlation with the possible sources of information received; one-sided interpretation of algorithms and analytical data; risk of plagiarism, violation of copyrights; reduced skills of independent search for scientific information by students, which will lead to a decrease in the level of education. All this may be an unsafe challenge for the scientific and learning community. This leads to the legitimate conclusion that AI as a system must adapt very quickly to reality [3].

4 Discussion

Taking into account the above advantages and problems of the process of implementing artificial intelligence technologies in the educational component and research activities of higher education students, we believe that such a transformation is an important component of the transition to a new type of educational process with the integration of various branches of science, reforming the organization of science and the cognitive process in it, which presupposes the possession of practical skills in processing information using AI technologies, the use of electronic libraries and scientometric databases, means of analyzing research data, the use of experience in effectively working with electronic documents of foreign language origin, etc. The practical application of artificial intelligence technologies helps students save time, which is essential for collecting important information, which stimulates their creative productivity, heuristic search, step-by-step and overall labor productivity [8].

Adequate, ethical and reasonable use of artificial intelligence technologies by students in the educational process and student research contributes to the formation of another world with a new educational landscape, algorithms for the development of civilization, as well as new individual, social and science-intensive technologies. The modern reality of a student is no longer possible without the daily use of AI, since the proposed technologies no longer play a secondary role, but carry important conceptual and methodological significance.

5 Conclusions

The technological future has arrived today. It is not without reason that it is noted that the 21st century is the century of education, characterized by the unfolding of the information revolution, where knowledge, information and intelligence are the driving vector of development. Artificial intelligence, which has massively entered into the daily life of a person and acts as a revolutionary useful assistant for optimizing his tasks and a phenomenon of creating and implementing the impossible possible in virtual space, has great potential in education. Today, the integration of two components - open science and information technology - is understood as a guarantee of the future intellectual potential of society, and the use of AI's technological capabilities to provide higher education significantly expands the possibilities of obtaining it, expands the individual possibilities of the knowledge acquisition process, and optimizes the methods and strategy of organizing education. It is impossible not to pay attention to the fact that artificial intelligence technologies do not lead to the substitution of traditional education. However, it can enrich the possibilities of improving the educational process, which can ultimately improve the professional level of university graduates.

Artificial intelligence is a modern pedagogical mechanism in the process of teaching and scientific research of students, thanks to the adequate and ethical application of which the ability of independent learning acquired under modern conditions turns into a vital necessity for everyone.

References

1. Martínez-Maldonado, R., et al.: Artificial intelligence in education. In: Encyclopedia of Education and Information Technologies (2023). https://doi.org/10.1007/978-1-4419-1428-6_3197
2. Kamalov, F., et al.: New era of artificial intelligence in education: towards a sustainable multifaceted revolution. Sustainability. (2023)
3. Kazak, A.N., Dorofeeva, A.A., Nyurenberger, L.B.: Problems and prospects of digitalization of education for industry 4.0 in Russian Federation. In: CEUR Workshop Proceedings, vol. 2522, pp. 17–25 (2019)
4. Shashkova, N., Chetyrbok, P., Lukyanova, Y., Kazak, A., Sidin, M.: Artificial intelligence in education and training. In: Lapina, M., Prakasha, G.S., Grigoriev, S. (eds.) International Conference on Innovative Approaches to the Application of Digital Technologies in Education and Research. SLET 2022. Lecture Notes in Networks and Systems, vol. 1222. Springer, Cham (2025). https://doi.org/10.1007/978-3-031-78776-8_36
5. Prakasha, G.S., Thamaraikannan, S., Lapina, M., Sanskriti, R.: Pre-service and in-service teachers' perceptions of using virtual reality tools in teaching. In: Lapina, M., Prakasha, G.S., Grigoriev, S. (eds.) International Conference on Innovative Approaches to the Application of Digital Technologies in Education and Research. SLET 2022. Lecture Notes in Networks and Systems, vol. 1222. Springer, Cham (2025). https://doi.org/10.1007/978-3-031-78776-8_34
6. Bhatt, C., et al.: Artificial intelligence in current education: roles, applications & challenges. In: 2023 3rd International Conference on Pervasive Computing and Social Networking (ICPCSN), pp. 241–244 (2023). https://doi.org/10.1109/ICPCSN58827.2023.00045
7. Chen, L., et al.: Artificial intelligence in education: a review. IEEE Access. **8**, 75264–75278 (2020). https://doi.org/10.1109/ACCESS.2020.2988510
8. Baidoo-Anu, D., Ansah, L.O.: Education in the era of generative artificial intelligence (AI): understanding the potential benefits of ChatGPT in promoting teaching and learning. SSRN Electron. J. (2023). https://doi.org/10.2139/ssrn.4337484
9. Grassini, S.: Shaping the future of education: exploring the potential and consequences of AI and ChatGPT in educational settings. Educ. Sci. (2023). https://doi.org/10.3390/educsci13070692
10. Farrelly, T., Baker, N.: Generative artificial intelligence: implications and considerations for higher education practice. Educ. Sci. (2023). https://doi.org/10.3390/educsci13111109
11. Pan, W.H., et al.: Assessing AI Detectors in Identifying AI-Generated Code: Implications for Education. ArXiv abs/2401.03676 (2024). https://doi.org/10.48550/arXiv.2401.03676
12. Dempere, J., et al.: The impact of ChatGPT on higher education. Front. Educ. (2023). https://doi.org/10.3389/feduc.2023.1206936
13. Aithal, P.S., Aithal, S.: Application of ChatGPT in higher education and research – a futuristic analysis. Int. J. Appl. Eng. Manag. Lett. (2023). https://doi.org/10.47992/ijaeml.2581.7000.0193

14. Bilad, M.R., et al.: Recent progress in the use of artificial intelligence tools in education. Jurnal Penelitian dan Pengkajian Ilmu Pendidikan: e-Saintika. (2023). https://doi.org/10.36312/esaintika.v7i3.1377
15. Yu, H.: The application and challenges of ChatGPT in educational transformation: new demands for teachers' roles. Heliyon. **10** (2024). https://doi.org/10.1016/j.heliyon.2024.e24289

Psychological Concepts and Pedagogical Approaches for Effective Online Courses Development

Oksana Konik(✉)

V. I. Vernadsky Crimean Federal University, 4, Acad. Vernadsky ave, Simferopol 295007, Russian Federation
oksana-yalta207@mail.ru

Abstract. The article considers one of the most relevant issues of modern education in general and university professional training in particular. The subject matter of the paper is the description of fundamental psychological notions and theories that can be implemented to foster online course development efficiency. The concepts described are cognitive theory and behaviorism. The authors justify a set of pedagogical approaches that are to be incorporated while creating online courses. The approaches discussed are the constructivist, the neurobiological and the postmodern ones. The article provides the deliberation and results of the conducted experiment to prove the theoretical assumptions on above mentioned psychological and pedagogical issues.

Keywords: Online Learning · Distance Learning · Psychological Concept · Pedagogical Approach · Behaviorism · Cognitivism · Constructivism · Neurobiology · Postmodern Education · Instructional Design

1 Introduction

Throughout history, there have been discussions about how people think and learn. As a result, different points of view have been developed and continue to develop influenced by philosophy and psychology. Successful online course developers are distinguished by having a clear understanding of the most important perspectives and how important they are for implementing distance learning instructions.

Scientists classify the process of thinking into the most complex categories, it is used to denote different concepts and in different contexts; in this regard, depending on the situation, determining what thinking means can be a difficult task.

Knowledge of the basics of the nature of the thought process, the formation of knowledge, and the methods by which people extract, process, and store information is of key importance to online course compilers. This knowledge helps them to better understand how people learn information, because learning and thought processes are closely interrelated.

M. Lapina et al. (Eds.): DET 2024, LNNS 1616, pp. 234–245, 2025.
https://doi.org/10.1007/978-3-032-04365-8_24

2 Purpose and Objectives of the Study

The purpose of the article is to describe the major psychological concepts and pedagogical approaches adding to the efficiency of the online courses.

The primary objectives included a review of psychological and pedagogical literature pertinent to the article's focus, devising strategies for incorporating those insights into the creation of digital learning experiences, and then evaluating those approaches through practical application.

3 Psychological Concepts

Before discussing thinking in terms of instructional design, it is important to consider what cognition is. The way people carry out the process of thinking is called cognition [1]. This refers to the mental process of acquiring knowledge, which includes perception, judgment, awareness, and reflection.

Essentially, cognition encompasses all the ways our minds handle incoming information. This broad process involves fundamental activities – like calculating costs or conversing while at the store – as well as complex choices, such as selecting a career path or crafting a research paper. It's worth mentioning that in the creation of online learning materials, "thinking" and "cognitive abilities" are frequently used interchangeably. Nevertheless, a wide range of specialized vocabulary surrounds this concept.

Numerous terms describe specific aspects of cognition, and many share overlapping definitions. The challenge in understanding these elements stems from their varied interpretations depending on the context [2]. While some convey significant scientific or theoretical distinctions, others are simply specialized jargon particular to a specific field or profession.

Let's look at the key aspects related to cognition. Memory is not just a repository of knowledge that operates in a passive mode. It is a complex of active processes encoding information. This also implies the organization of data storage. An important part of this process is the constant restructuring of stored information to integrate new knowledge with existing knowledge. In addition, it allows you to find and extract the necessary information as needed.

Distinct mental aptitudes – these encompass the total body of an individual's learned understanding and proficiencies. This covers areas like literacy, composition, understanding, physical coordination, and the capacity to perceive and navigate space.

Leadership capabilities involve advanced reasoning, encompassing the power to foresee upcoming requirements and devise suitable strategies, the capacity to establish importance, and the aptitude for self-assessment and behavioral adjustment. Essentially, these are the talents that empower someone to apply their intellectual resources and specific cognitive strengths to fulfill personal, vocational, and emotional desires [23].

Metacognition is the ability to control own cognitive processes. Using metacognition, a person looks introspectively at the thinking process. This "allows to critically consider the path to certain ideas, concepts and thoughts. Metacognition is a reflexive process, which helps to improve a person's control over" [23] their own learning process.

4 Historical Specific Features of Development of Theory of Thinking

Let's dwell on some historical specific features of development of theory of thinking. Initially, this idea resided within the realm of philosophical thought. Some of the most foundational explanations of mental processes originated with ancient Greek thinkers like Empedocles, Democritus, and Epicurus. Their perspective suggested that the mind interprets representations generated by external things. These sensed representations are subsequently recorded as sensory data and retained within memory.

A contrasting perspective, championed by Plato and revisited by I. Kant, is commonly referred to as idealism. Its core assertion suggests that the intellect doesn't grasp reality outright, nor does it simply mirror it. From this point of view, the mind is considered as a fundamental construct. The main statement of this point of view is that all data coming from sensory systems and organs is interpreted by the mind in accordance with innate inclinations. "This influence has led to so called scholasticism – a philosophical perspective based on a mixture of Aristotelian and Ecclesiastical writings and ideas" [23].

Scholastic philosophy was built around a highly organized system of truths that differed from the doctrines of the Church, but did not contradict them. Philosophers scholastics such as Thomas Aquinas believed that thinking was directly related to what God gave people through their inspiration (or "enlightenment").

A new way of thinking emerged at the beginning of the Renaissance and persisted through the 17th and 18th centuries. This movement shifted focus away from the idea that truth was solely a divine inspiration, suggesting instead that individuals could identify reality and truth through their own perceptions. Thinkers such as Descartes and Locke posited that personal observations and feelings were valid indicators of what is true. Individuals started to grasp the ways in which thinking and learning occur.

There was a prevailing belief that human intellect possesses the ability to differentiate between what is true and what is false. Consequently, specific approaches to uncovering the truth, as well as evaluations of evidence, began to be regarded as dependable and adequate.

"While philosophy continues to play a role in conceptualization thinking process, the birth of psychology (especially pedagogical psychology) in the middle – late 19th century continues to have the most significant impact on how pedagogical designers view thinking" [23]. Pedagogical psychology allowed focus on thinking and learning from a more scientific point of view based on research. Two main points of view, from several variations, arose from psychology and prevailed in how learning materials developers see thinking and learning processes. In this respect there are two major approaches – behavioral and cognitive.

Although, practice shows, that most online course developers tend to feel more comfortable with one point of view compared to another, it is important to use the principles and methods of both parties in the development of online academic materials. It must also be remembered that these two points of view are just two of many existing ones. They are described here because they are the two most influential points of view relevant to the process of developing online course at the university level.

5 Behavioral and Cognitive Views on Design of Online Courses

Behavioral and cognitive views on design of online courses are based on positivist and interpretative philosophy. Let's consider them.

Behaviorism dominated psychology in the first half of the 20th century.

"According to behaviorists, mental processes are invisible and therefore cannot be studied scientifically. External behavior can be observed" [23]; therefore, instead of thinking about internal reasons of current events, you should focus on how organisms respond to various stimuli [19].

Behaviorism proposes that the human mind is adaptable, capable of being molded to generate specific responses and actions when particular factors and situations are carefully managed. The development and validation of behaviorism stemmed from the groundbreaking work of key thinkers, including I. Pavlov (1800–1950), E. Thorndike (1874–1949), J. Watson (1878–1958), and E. Guthrie (1886–1959). Their investigations largely involved animal subjects.

Building upon the foundations laid by these earlier researchers, B. F. Skinner (1904–1990) became the most recognized proponent of behaviorism, formulating the operant conditioning model. Through careful study of pigeons and rats, Skinner noted that actions are more prone to repetition when swiftly followed by positive reinforcement – essentially, a reward. These ideas offered crucial perspectives, significantly contributing to our comprehension of how the mind operates ([23], p. 28).

After World War II, there was a shift in the field of psychology as researchers began to distance themselves from behaviorism and embrace a new understanding of the human mind and its cognitive functions. This alternative point of view became known as cognitivism. This approach emphasized the importance of internal mental processes, arguing that they could be identified and studied. Cognitive scientists view the human mind as a complex system, often using a computer as a metaphor to illustrate how it works. Like a computer, the brain processes information through various interconnected processes that function in concert. Several influential figures have made significant contributions to the development of the cognitive perspective.

6 Pedagogical Approaches for Effective Online Courses Development

Although behavioral and cognitive views have dominated way of perceiving thinking, later work influenced the formation of instructional design as a science of developing online courses. The fundamental approaches of this process are constructivist, neurobiological and postmodernist ones. Constructivism, being a kind of cognitivism, based on the principle that a person constructs their own understanding of the world in which they live, reflecting on their experiences.

R. Brandt and D. Perkins write: "Pedagogically, constructivism claims that the human mind not only perceives the world, but actively creates it".

J. Brooks develops the basic principle of constructivism: "We all build our understanding of the world by integrating new experiences with existing knowledge. Often, we encounter an object or concept, a belief, or an event that lacks immediate clarity.

Faced with this initially ambiguous information or representation, we either adjust our interpretation to match our current system of explanation, simplifying our approach, or we develop a new system that more effectively explains what we observe. Despite this, our concepts and guidelines are constantly evolving, shaping our understanding".

The mind is active in scientific research and artistic development, constructivists believe that constructive process is important even in routine mental operations such as perception and memorization.

Within the constructivist approach, a key element influencing how individuals develop understanding and acquire knowledge is collaborative engagement. Hence, the constructivists draw an extremely important conclusion for pedagogical design: – "through social interaction learners will have a deeper understanding of the material".

Hence, one must consider this while developing any online course. It's worth incorporating various opportunities for communication during the course, because when students study online, they do lack peer socializing. Also, such interactions should be implemented for synchronous participation (Zoom meetings adding to MOOCs etc.)

Considering the impact of neurobiology on the structure of online learning experiences, it's clear that neuroscientists are primarily dedicated to deciphering the brain's mechanisms for managing all its processes. The Society for Neuroscience states that brain scientists are driven by a desire to comprehend human actions, such as "how the brain's electrical pathways enable reading and communication, and the reasons behind the formation of interpersonal connections" Furthermore, researchers in this field examine neurological and physical ailments – their origins and potential for prevention or treatment. While neuroscience itself has a long history, the deliberate application of its principles to education is a more recent development.

According to M. Beauchamp Bouchamp, "modern neurobiological research has provided knowledge about the brain and its functioning, which could potentially affect how teachers build their interaction with students" ([23], p. 29). These results can also influence how educational designers create teaching materials. Thus, before working on any online course, one should conduct a thorough research of last developments in the field of neuroscience ideas devoted to effective learning process organization.

Emerging in the final decades of the 20th century and continuing into the 21st, postmodernism offered a novel lens through which to view human thought. Largely, this movement arose as a challenge to the previously held certainty surrounding scientific or objective efforts to define reality [13].

This shift stemmed from the understanding that reality resists straightforward capture within human comprehension, instead forming through the mind's engagement with its own subjective experience. Within a postmodern framework, interpretation reigns supreme; existence itself is born from individual understandings of the world's significance. Favoring lived experience over generalized theories, postmodernism acknowledges that any personal encounter will inevitably be flawed and contextual, rather than absolute and universal. As noted earlier, postmodern philosophy critically examines – and frequently dismisses – the established notions of truth and knowledge passed down from earlier philosophical schools. Ultimately, postmodern thought is a fluid, constantly evolving process shaped by an individual's unique perception of their surrounding environment.

It is worth noting that a person's view of thinking largely depends on his education, theoretical background, philosophical beliefs and profession. A learning experience architect's understanding of thought processes, and consequently, learning itself, is significantly shaped by their foundational theories and the demands of their role. Consequently, these designers often approach thinking practically, posing the question: what essential knowledge of thinking, and what relevant research, is required to craft impactful online learning experiences? So, it seems beneficial to investigate the targeted audience of the online course and to research their academic level, personal values, beliefs and aspirations together with their expectations of the course, which is called the needs analysis stage of the course development.

Unsurprisingly, most developers of educational materials are considered eclectic – they borrow from different points of view and use what works in given situations to obtain the desired results. When it comes to considering thinking (and learning), educational designers tend to take a systems theory approach, exploring it from several different points, rather than focusing on just one aspect of what is thinking.

7 The Analysis of Learning Theory

Let's move on to the analysis of learning theory. But what exactly constitutes learning? Similar to how we understand thinking, learning lends itself to varied interpretations. Typically, it's understood as a fairly stable shift in behavior, mental frameworks, or connections, stemming from exposure to something [17].

These alterations can be readily apparent, like a youngster mastering the skill of tying their shoes, or they might be more nuanced, such as developing a deeper fondness for classical compositions [17]. Furthermore, these modifications can be flawed or accurate, accidental or deliberate, and both aware and unaware. It's crucial to acknowledge that these alterations arise through the accumulation of experience – events, whether structured or spontaneous, encountered throughout a student's journey. Consequently, transformation happens as individuals engage with the world around them.

At the same time, we consider it important to emphasize that not all the changes that occur are internalized, and therefore they cannot be attributed to the learning process. Such changes are usually caused by physical development or temporary conditions of the body; they are not based on experience and, therefore, are not considered acquired through training.

When analyzing our definition of the learning process, it is assumed that two completely different types of changes can occur in this context.

One key aspect involves modifying actions, while another relates to cognitive frameworks or connections. Recognizing this difference is crucial, as it highlights contrasting perspectives within two foundational psychological schools of thought –behaviorism and cognitivism – explored previously. These approaches, and their various subfields, still significantly shape instructional design practices today.

Hence, from a behavioral point of view, "learning is defined as behavioral change due to experiences that can be measured" [23].

A. Wolffolk (2004) states that "Behaviorism seeks to clarify the process of learning by emphasizing external factors as the source of observable changes". This theory suggests that intricate behaviors emerge from simpler components or initial versions of the desired behavior, with reinforcement applied progressively to shape the behavior towards the goal. In everyday life, learning occurs frequently as the surrounding environment influences behavior in this manner. Both tangible benefits like food and subtle cues indicating potential rewards influence behavior by forming learned connections. Existing learning models give an explanation of how learning occurs in terms of behaviorism.

Skinner differentiated between two distinct methods of instruction: classical and operant. Operant conditioning involves modifying a behavior through the application of positive reinforcement. In this case, the answer is the result of an arbitrary choice. The individual (or other living being) who initiates the reaction completely controls its occurrence.

The term indicates that the individual voluntarily affects the environment and, therefore, may have an effect on their impact. Despite differences of opinion between behaviorists.

According to J. Ormrod, behaviorists in the past held a number of common beliefs:

- first, the principles of learning should be universal and applicable to all types of behavior and categories of people;
- secondly, the study of learning processes becomes most objective when research focuses on stimuli and reactions;
- thirdly, internal processes, as a rule, are not taken into account in scientific research.
- Fourth, learning involves a change in behavior;
- finally, learning is largely a consequence of environmental factors.

Considering this, we may infer that any online course should be aimed at developing some certain patterns of students' behavior, it should include strategies for stimulating students thinking and actions, and the educational environment where the course is going to be taken should be thoughtfully constructed.

Skinner's perspective on learning highlights the learner's actions, positing that students don't simply take in information; they need to participate. Consequently, learning occurs through active engagement. As J. Burton and others [19] explain, "the focus rests on the learner's responsive actions – learning and validating training both require behavioral involvement".

Let's move on to the analysis of learning theory from the point of view of cognitive scientists.

Here, learning is defined as a shift in cognitive frameworks and connections, resulting from lived experiences.

Contemporary researchers believe that cognitivists perceive learning as the process of reshaping the significant understanding we already possess. This perspective aligns with the conventional cognitive approach to learning, which posits that mental processes are real and can be examined through scientific methods.

R. Brandt and D. Perkins noted that early studies of cognitivism "were primarily focused on processing information, especially on pattern recognition, memory and solving problems. The mind was considered a computing device controlled by rules.

The scientist's objective centered on pinpointing the core mechanisms through which awareness manipulates symbols to achieve defined outcomes. This perspective aligns with a prevalent analogy embraced by numerous cognitive researchers – the notion that the mind operates similarly to a computer; thus, knowledge acquisition happens via programmed routines built into the framework.

Brandt and Perkins also point out that the scope of cognitive science has steadily broadened, now encompassing a remarkably diverse spectrum of human actions. Consequently, when designing an online course, it appears sensible to incorporate exercises focused on judgment creation, effective choices, sparking ingenuity, analytical reasoning, and cultivating emotional awareness.

Summarizing all of the above, we note that behavioral and cognitive points of view differ in their assumptions regarding what is being studied. From a cognitive point of view, knowledge is assimilated. It is the increase in knowledge that makes change possible reflecting in the behavior. According to the behavioral point of view, the new behavior itself is learned.

In addition to the two main teaching methods that we discussed earlier, another approach has emerged that relates to the constructivist paradigm. This approach has appeared relatively recently and has had a significant impact on the theory of learning. Constructivism is based on both psychological and philosophical concepts. According to R. Marzano, constructivism implies a general approach in which students, based on existing knowledge, form their own, meaningful idea of the new topic being studied. By receiving data from the outside world, the student processes it and uses it to build his own individual picture of the world [19].

M. Driscoll adds: that constructivism "is based on the assumption that knowledge is constructed by students when they try reflect on their experience. Thus, students are not empty vessels, awaiting fulfillment, but rather active organisms seeking meaning" [23].

Summarizing the above, we note that the most important conditions necessary for creating online courses are:

- to ensure effective interaction between students;
- to carry out integrated learning in realistic and relevant conditions;
- to accommodate varied perspectives and diverse methods of conveying information;
- the development of fresh insights requires cultivating self-reflection;
- inspiring learners and fostering their active involvement in education is crucial (see Fig. 1).

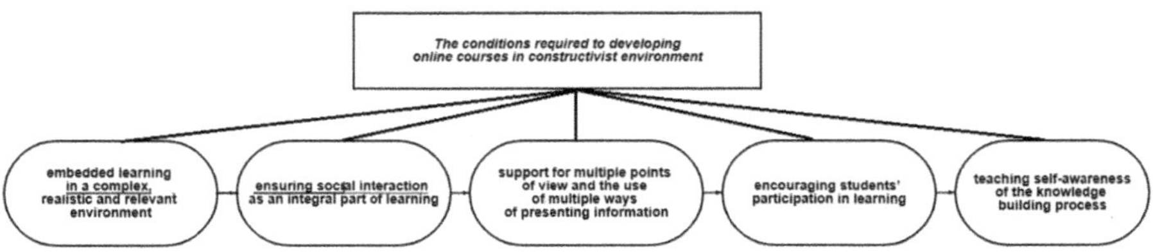

Fig. 1. The conditions required to developing online courses in constructivist environment.

8 Methodology

The methodological foundations of the concepts discussed in the article are formed by the general scientific approaches relating to using online tools in academic purposes, which follows from the theory of digitalization; considers the theoretical foundations and practical applications of the studied issues within the frames of environmental, constructivist and systemic approaches in professional pedagogy. The main methods are literature review, analysis, synthesis, hypothesizing and piloting experiment.

9 Results

In order to test efficiency of the discussed above approaches and concepts in the process of developing online courses, an experiment was carried out.

While developing the online courses for university Moodle platform and MOOCS (massive open online courses for the international Udemy platform), the authors incorporated the assumptions of cognitivism and behaviorism, constructivist, positivist approaches and interpretism.

The asynchronous Moodle courses were accompanied by scheduled Zoom group sessions for the experiment group, the educational chats in messengers were also introduced. Also, students of this group had the opportunity to take part in constructing the curriculum by filling in pre-developed questionnaires. They enumerated the core areas (topics) of the mentioned courses and some extra-topics to choose from. The control group during the experiment studied according to the general program approved by the university.

The criteria described were the following – motivational (students' engagement in the course), productive (students' achievements and results) and reflexive (students' critical opinions on their own success during the courses). There were three levels grounded – high, medium and low.

As a result, an obvious advantage in the dynamics of changes in the level of course efficiency was recorded: the number of students with a high level in the EG increased by 23%, and in the CG – by 4%. The indices of the medium level in the EG increased by 5%, and in the CG – by 3% (the difference is 2%). The decrease in the number of students with a low level in the EG occurred by 28%, while in the CG – by 7% (the difference is 21%).

The general results of comparing the level of course efficiency before and after the experiment are presented in Table 1.

Table 1. Dynamics of changes in the level of course efficiency (in %).

Level	Experiment Group		Control Group	
	Before	After	Before	After
High	11	34	10	14

(continued)

Table 1. (*continued*)

Level	Experiment Group		Control Group	
	Before	After	Before	After
Medium	56	61	52	55
Low	33	5	38	31

The results of working out the experimental data testify to the effectiveness of implementing described concepts and approaches to the process of developing online courses.

10 Conclusion

In general, different approaches to learning theory can provide very useful directions in the development of online courses. It's usually unwise to adhere rigidly to a single method, as every learning environment presents a unique set of factors that demand attention. No single strategy can possibly encompass every influencing element.

To effectively design online courses, the instructional designer must skillfully integrate various development techniques, carefully weighing all relevant factors.

One of the most important components of any educational context is the type of learning process. The compiler of online courses should clearly define these types to develop effective and efficient training materials in the online course.

Today, there are 3 main characteristics or types of learning: affective, cognitive, and psychomotor. These characteristics are based on the idea that the educational process should be organized in accordance with the transition from a low level of knowledge to a higher one.

References

1. Levicky-Townley, C., Garabedian, S., Zhang, J., Weatherford, E.: Exploring the impact of universal design for learning supports in an online higher education course. J. Appl. Instr. Des. **10**(1) (2021). https://doi.org/10.51869/101cltmcsjzew
2. Black, E.L.: Instructional design for single information literacy sessions. Public Serv. Q. **16**(3), 161–171 (2020). https://doi.org/10.1080/15228959.2020.1775753
3. Golovanova, I.I., Alipichev, A.Y., Ayupov, T.A., Fazlyeva, F.A., Khasanova, E.N.: Digital educational environment and online learning format through the lens of students: pros and cons. Educ. Self Dev. **17**(3), 202–221 (2022)
4. Gerasimova, E.K., Zorin, S.L., Kobeleva, G.A., Mamaeva, E.A.: Designing a personalized educational model while working with digital technologies. Perspect. Sci. Educ. **47**(5), 398–412 (2020). https://doi.org/10.32744/pse.2020.5.28
5. Soboleva, E.V., Suvorova, T.N., Vekua, N.N., Vasilenko, A.V.: Using game platforms of class management in online-learning for improve the quality of educational results. Perspect. Sci. Educ. **46**(4), 409–424 (2020). https://doi.org/10.32744/pse.2020.4.29
6. Kulikova, T.A., Poddubnaya, N.A., Bagdasaryan, L.S., Ardeev, A.H.: The technique for future teachers' digital literacy development. J. Phys. Conf. Ser. **1691**, 012129 (2020)

7. Soboleva, E.V., Suvorova, T.N., Masharova, T.V., Zaslavskaya, O.: The use of mobile tools when implementing the master's degree program "adaptive physical rehabilitation to improve the quality of students' professional training". Eur. J. Contemp. **12**(3), 987–1001 (2023)
8. AlDahdouh, A.: Jumping from one resource to another: how do students navigate learning networks? Int. J. Educ. Technol. High. Educ. **15**(45) (2018). https://doi.org/10.1186/s41239-018-0126-x
9. Despujol, I., Castañeda, L., Marín, V., et al.: What do we want to know about MOOCs? Results from a machine learning approach to a systematic literature mapping review. Int. J. Educ. Technol. High. Educ. **19**(53) (2022). https://doi.org/10.1186/s41239-022-00359-1
10. Kwon, S., Kim, W., Bae, C., et al.: The identity changes in online learning and teaching: instructors, learners, and learning management systems. Int. J. Educ. Technol. High. Educ. **18**(67) (2021). https://doi.org/10.1186/s41239-021-00304-8
11. Hung, M., Chou, C., Chen, C., Own, Z.: Learner readiness for online learning: scale development and student perceptions. Comput. Educ. **55**(3), 1080–1090 (2010). https://doi.org/10.1016/j.compedu.2010.05.004
12. Kazak, A., Oleinikov, N., Gorobets, D., Ryndach, M., Shamaeva, N.: Educational service robotics in the Russian Federation. J. Phys. Conf. Ser. **1691**(1) (2020). https://doi.org/10.1088/1742-6596/1691/1/012164
13. Huang, R., Tlili, A., Chang, T.W., et al.: Disrupted classes, undisrupted learning during COVID-19 outbreak in China: application of open educational practices and resources. Smart Learn. Environ. **7**(19) (2020). https://doi.org/10.1186/s40561-020-00125-8
14. Karagiannis, S., Magkos, E.: Adapting CTF challenges into virtual cybersecurity learning environments. Inf. Comput. Secur. (2020). https://doi.org/10.1108/ICS-04-2019-0050
15. Sergeyeva, A.I.: Activation of the cognitive processes of students with the use of neuropsychological exercises in online learning format. TSPU Bull. **2**(214), 112–118 (2021). https://doi.org/10.23951/1609-624X-2021-2-112-118
16. Zakharova, U.S., Vilkova, K.A., Egorov, G.V.: This cannot be taught online: applied specialties in a pandemic. Educ. Issues. **1**, 115–137 (2021). https://doi.org/10.17323/1814-9545-2021-1-115-137
17. Adedoyin, O.B., Soykan, E.: Covid-19 pandemic and online learning: the challenges and opportunities. Interact. Learn. Environ. (2020). https://doi.org/10.1080/10494820.2020.1813180
18. Solovyova, O.A., Alipichev, A.Y., Sergeeva, N.A., Zakharova, A.N.: Promoting online interactions of foreign language learners in a computer-mediated learning environment. AIP Conf. Proc. **2647**, 040053 (2022)
19. Van Wart, M., Ni, A., Medina, P., et al.: Integrating students' perspectives about online learning: a hierarchy of factors. Int. J. Educ. Technol. High. Educ. **17**, 53 (2020). https://doi.org/10.1186/s41239-020-00229-8
20. Nazarova, L., Kubrushko, P., Alipichev, A., Gryazneva, S.: Development trends in practical training of college students in the context of digital transformation of education. In: E3S Web of Conferences: 14th International Scientific and Practical Conference on State and Prospects for the Development of Agribusiness, INTERAGROMASH 2021, Rostov-on-Don, vol. 273, p. 12059. EDP Sciences, Rostovon-Don (2021). https://doi.org/10.1051/e3sconf/202127312059
21. Alipichev, A.Y., Takanova, O.V., Novikova, B.: Forming professionally important qualities of students of non-linguistic universities by means of the "foreign language" discipline. J. High. Educ. Theory Pract. **22**(7), 121–129 (2022). https://doi.org/10.33423/jhetp.v22i7.5277

22. Alexandrova, L., Mozhaeva, G., Karabasheva, M., Lapina, M., Rugelj, J.: Use of IT for student-centered learning approach to education at Modern University. In: Innovative Approaches to the Application of Digital Technologies in Education 2020 (SLET 2020), CEUR Workshop Proceedings, vol. 2861, pp. 32–40 (2020) https://ceur-ws.org/Vol-2861/paper_2.pdf
23. Brown, A.H., Green, T.D.: The Essentials of Instructional Design, 3rd edn, p. 220. Routledge, New York, NY (2016)

Educational Technologies in the Master's Course "Intellectualization of Information Processing"

Margarita Kozlova(✉), Vladimir Lukianenko, and Denis Savchuk

V. I. Vernadsky Crimean Federal University, Simferopol 295007, Russia
art-inf@mail.ru

Abstract. When studying in the field of training "01.04.02 Applied Mathematics and Computer Science" (focus "Models and Methods of Artificial Intelligence") at the Physics and Technology Institute of V. I. Vernadsky Crimean Federal University, master's students are traditionally taught the course "Intellectualization of Information Processing" (IIP). The name of the course was born based on the All-Russian scientific conference of the same name, which is traditionally held by the Federal Research Center "Informatics and Control" of the Russian Academy of Sciences. During practical classes in the IOI course, project teams are formed, work is organized in groups of three people on each of the projects. In recent years, students have quite successfully taken on solving problems in the field of intellectual analysis of big data, social network analysis, image analysis (memes), discrete optimization problems (search for optimal routes), etc. To solve such problems, neural network technologies, heuristic algorithms, and statistical data processing methods are used. It should be noted that some of these projects are continued in scientific papers and a number of works have been published on certain topics. One of the interesting areas was the development of augmented reality applications. This article presents the results of the project team's work on creating 3D models of interior augmented reality based on 2D schemes.

Keywords: design activities · augmented reality · object recognition · interior design · 3D technologies

1 Introduction

When studying in the field of training "01.04.02 Applied Mathematics and Computer Science" (focus "Models and Methods of Artificial Intelligence") at the Physics and Technology Institute of V. I. Vernadsky Crimean Federal University, master's students are traditionally taught the course "Intellectualization of Information Processing" (IIP). The name of this course was born based on the All-Russian scientific conference of the same name, which is traditionally held by the Federal Research Center "Informatics and Management" of the Russian Academy of Sciences.

Project teams are formed during practical classes in the IIP course, and work is organized in groups of three people for each project. In recent years, students have successfully taken on problems in the field of intelligent analysis of big data, social

M. Lapina et al. (Eds.): DET 2024, LNNS 1616, pp. 246–255, 2025.
https://doi.org/10.1007/978-3-032-04365-8_25

network analysis, image analysis (memes), discrete optimization problems (searching for optimal routes), etc. To solve such problems, neural network technologies, heuristic algorithms, and statistical data processing methods are used. It should be noted that some of these projects continue in scientific papers and a number of works have been published on individual topics [1–3].

The distribution of roles in the project group in the work on the IIP course depends on the competencies and desire of the student to master new technologies. Experience has shown that a leader is always identified in the group, who skillfully distributes the responsibilities of the other participants. With the help of a mentor and consultants, the project group in most cases achieves its goals in the allotted time: in 4 months it is necessary to understand the essence of the project, identify existing approaches to solving similar problems, find or create training data, process them, adapt existing or develop their own algorithms to solve the problem. When working in groups, projects not only reach their logical conclusion in the form of ready-made applications, but also are submitted to competitions, winning prizes.

One of the interesting areas was the development of augmented reality applications. This article presents the results of the project team's work on creating 3D models of interior augmented reality based on 2D schemes.

Modern augmented reality (AR) technologies are increasingly used in various fields, including architecture, interior design and space modeling. They allow visualization and exploration of spaces with a high degree of interactivity, which makes the design process visual and convenient [4]. However, it should be noted that the automatic construction of 3D models of premises based on two-dimensional schemes remains a technically complex task that requires the use of modern computer vision methods and optimal approaches to data processing.

The project presented in this paper develops an Android application with augmented reality (AR) support, which converts 2D room layouts into 3D models. The application uses the YOLOv11 model to detect key room elements (walls, windows, doors) and server architecture for image processing. The resulting 3D models are visualized in AR, providing the user with the opportunity to interactively explore the layout of the premises. The tests conducted showed high accuracy and stable performance of the application. The project team considered the prospects for further development, including optimization of 3D orientation of objects and the possibility of local data processing on mobile devices.

The aim of the group's work on the research project is to develop a system that combines the accuracy of modern machine learning models and the interactivity of augmented reality. The work examines methods for optimizing data processing, adapting the model to the task, and integrating 3D visualization into a mobile application. In addition, the main problems associated with the reconstruction of premises are highlighted and approaches to solving them are proposed.

The development of such a tool opens up prospects for the use of AR in professional and everyday tasks, providing a new level of convenience and accuracy in the design and study of premises.

2 Models, Methods and Technologies of 3D Visualization of Two-Dimensional Scenes

2.1 Analysis of Existing Approaches

Transformation of 2D diagrams into 3D models is a relevant task that can be solved by both traditional methods of computer geometry and modern approaches based on machine learning. The key point of these methods is the correct recognition of structural elements (walls, doors, windows) and their subsequent reconstruction in three-dimensional space. In this case, such parameters as identification accuracy and processing speed are of particular importance.

The article [5] shows that after pixel classification, the walls of the rooms are transformed into polygons using OpenCV. Morphological operations are used to remove small defects and thinning, as well as the Ramer-Douglas-Peucker algorithm to simplify the contours. Bounding boxes and the Hough transform are used to detect doors, windows and wall lines, and the Tesseract recognition system is used to extract room labels. The combined data is projected onto a 3D model using the Trimesh library. The authors claim that this technology shows high accuracy in reconstructing rooms: walls, doors and windows.

The article [6] presents a method for converting 2D floor plans into 3D models using Faster-RCNN and FPN architectures with ResNet. The process described by the author includes segmentation, vectorization (outline extraction, transformation into primitives), wall processing (orientation determination, bounding box construction, shape optimization using a compression algorithm) and intersection elimination. It is shown that the result can be exported in IFC format for use in computer-aided design (CAD) and building information modeling (BIM) systems.

In [7], an algorithm for converting 2D images into 3D models and an architecture for processing natural scenes are presented. The authors introduced the concept of a horizontal dividing line and based on it, an image depth map is created, which is used to create 3D images in real time. To reduce costs, the authors developed hardware consisting of a time interval management system. It is shown that only one frame of memory is required to create 3D images in real time using a depth map. The depth map is calculated based on the location of the dividing lines. This, in turn, allows for a reduction in storage size and I/O bandwidth.

2.2 Model Development, Training Data Preparation

2.2.1 Selection and Rationale

During the development of an approach to automatically convert 2D drawings into 3D models, the project team faced the need to select the most suitable model for the task of segmentation and detection of objects. Several modern architectures were considered: U-Net, Fast-RCNN and YOLOv11, which at that time was the latest version of a popular detection algorithm [9].

The dataset originally used for the instance segmentation task was chosen as the initial data. This made YOLOv11 the most suitable, since it was necessary to accurately distinguish objects such as walls, doors, and windows in the original images (drawings).

The main advantage of YOLOv11 is its ability to localize and classify objects in a single process, which ensures high speed and accuracy. Such characteristics are especially important for working with drawings, where every detail matters.

The alternative models, although showing good results in their areas, were rejected for the following reasons:

- U-Net excels at segmentation tasks thanks to pixel-wise classification, but the approach it uses makes it of little use for cases where accurate detection of individual objects is required;
- Fast-RCNN demonstrated excellent detection capabilities, but to fully implement the approach adopted in the project, an additional segmentation step would be required, which complicates the process and increases processing time.

Taking these factors into account, the choice was for YOLOv11, which provided a balance between accuracy, speed and ease of integration, becoming the optimal solution for the task at hand.

To train the YOLOv11 model, a set of images containing 2D floor plans with key elements: walls, windows, and doors is required. A search for publicly available datasets was conducted on platforms such as Kaggle, HugginFace, Roboflow. One user dataset was the most suitable for the task under study, which the project team members had to further process. The selected dataset (Korea) contains about one thousand photographs of architectural plans, with marked windows, doors, and walls. However, this dataset turned out to be extremely poorly marked, so the project team members had to look through each photograph and make sure the markup was correct.

Roboflow was used as a platform for working with the dataset, where you can quickly and collaboratively work on dataset photos, generate datasets with various filtering and augmentation parameters, and also easily and quickly download generated versions of the dataset via API.

Due to the limited volume of the initial dataset, augmentation methods were used, which allowed increasing the training sample and improving the generalization ability of the model. The choice of augmentation methods is due to the specifics of the task of recognizing structural elements on 2D diagrams:

1. rotations and reflections: rotation of images at arbitrary angles from $-45°$ to $45°$, also mirror reflection allowed to simulate different orientations of room plans;
2. scaling: changing the size of objects to simulate plans with different complexity and image quality;
3. adding noise: imitation of artifacts (noise, spots) to increase the model's resistance to low-quality and damaged images;
4. adding blur: blur is used to imitate a quick photo taken on a phone. It is not always possible to find the perfect focus, and the phone camera is not always perfectly clean, which makes blur in augmentation a great choice.

The use of these augmentations allowed the YOLOv11 model to effectively handle a variety of room layouts with different quality and design styles (Fig. 1). Rotations and scaling contributed to better adaptation of the model to new data, and the addition of noise and blur ensured resistance to real-world distortions.

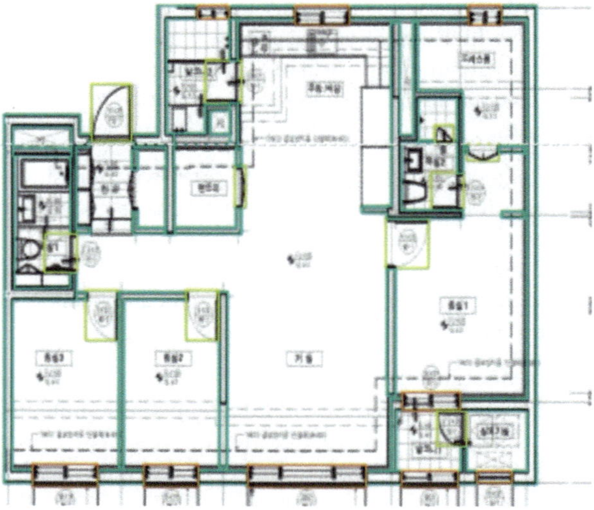

Fig. 1. Dataset image example.

2.3 Converting a Model into 3D Objects

During the process, many assumptions were considered on how to transform masks into a 3D model:

1. CityJSON format: used to describe 3D city models, but it is also suitable for this project task;
2. triangulation: break masks into triangles so they can be manipulated as polygons in a 3D model;
3. rectification: breaking down masks into rectangles, where the rectangles are pre-prepared models with textures.

Ultimately, the rectification option was chosen, since this method provided the simplest way to handle the textures of objects. To do this, an algorithm was used to search for rectangles in the image, then filter the found rectangles so that they did not overlap each other. Next, after obtaining the coordinates of these rectangles, 3D models of the corresponding classes were placed in their places by stretching them to the size of the rectangle. The result was a 3D model corresponding to the original layout.

The Python library Trimesh was used to generate 3D models, which allowed us to avoid using third-party technologies such as Blender and Unity.

3 Android Application Implementation. Results and Testing

3.1 Application Architecture and Development

The Android Studio platform was used to develop the Android application, which provides tools for working with augmented reality (AR) through the ARCore and Sceneform libraries. The project team members decided to choose a client-server architecture for the highest speed, as well as to minimize the size of the application. If everything were done locally, this would mean the need to store all models of walls, windows, doors locally. In addition, although YOLOv11 provides an excellent balance between its complexity and performance, the accuracy of processing is extremely important for the project task, which is quite difficult to achieve with the resources of a mobile device.

The application has the following functionality: it is possible to both upload photos as files from a local storage and take a photo of something, thereby sending a photo to the server. After the server receives the photo, it is processed and converted into a 3D model, after which the model is sent to the client as a response to its request. The user is notified of the processing results, and he is given the opportunity to click on the surface that the ARCore library has identified from the image from his camera. After the user has clicked, a 3D model of the layout begins to be displayed, which the user can then examine in detail.

3.1.1 Detection Model Quality

The efficiency of the YOLOv11 model was assessed by two parameters: the mAP50–95 metric, which was 0.43, and our own tests. The model copes well with the task. With a maximum number of detections of 300 (which is what it was trained on), there are difficulties with very large and complex layouts — the model simply does not have the ability to process everything found. At the same time, this number of detections is enough for small and medium layouts. When the number of detections increases, the implementation encounters problems on small and medium layouts — an excess of details, which spoils the quality of the layout description. Therefore, it was decided not to change the number of maximum detections in order to ensure optimal operation on most layouts.

3.1.2 Application Performance

The performance was assessed based on the processing time and system response. The client-server approach ensured stable operation of the application even on devices with limited resources. The average time for image processing and 3D model generation was 5–10 s when processed on the processor. As a result, it was possible to maintain the optimal image processing time, while retaining the possibility for potential acceleration of this issue.

3.2 Problems with the Chosen Solution

During testing of the trained model, very good results were obtained, but a number of problems arose.

The first problem was that wall masks sometimes absorbed window masks. What is the reason for this? Visually, in the drawing, a window differs from a wall by adding a couple of lines, and in general, we can say that a window consists of three wall figures. Such closeness of the geometric figures of windows and walls often gives them similar confidence parameters during automatic detection. But since windows are usually located between walls, their mask is somehow absorbed and preference is given to the wall.

How to deal with this? The thing is that, as described earlier, in areas with a window, the model must detect both the window and the wall. But in areas with walls, the model only looks for the wall. Therefore, it was decided to give priority to the window mask over the wall mask. This simple solution has shown impeccable efficiency.

Second: the coordinates of the rectangles should be consistent with each other. What does this mean? In the usual implementation, the ends of the rectangles overlap, which looks strange when viewing 3D models. In addition, sometimes the gaps between objects should be removed (Fig. 2).

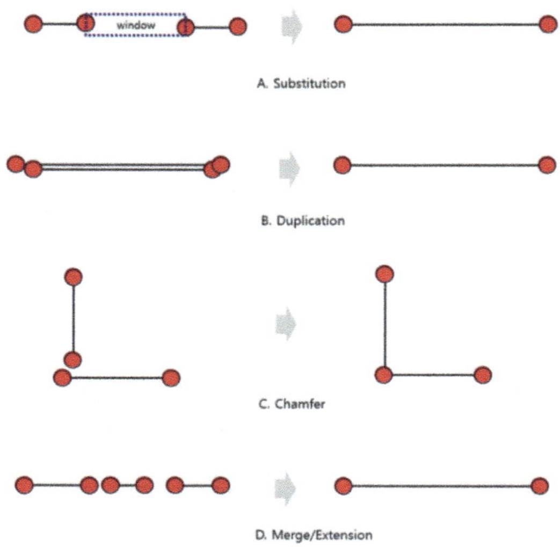

Fig. 2. Example of coordinate adjustment [10].

To solve this problem, all pairs of rectangles that are neighbors were compared, then they were checked for overlaps or gaps between each other, and their coordinates were corrected if problems were found.

Third: it is necessary to correctly determine the orientation of objects in the layout. The fact is that, for example, a window model has sides that a person should see, and sides that should "look" at the walls, that is, be invisible to the human eye. The same applies to doors — the door must be attached to the wall and they must be perpendicular to each other.

It was decided to use heuristic methods: for a window, this is finding its neighboring walls, which helps to understand how exactly the window should stand. For a door, we focused on the neighboring walls, and determined the direction in which the door moved away from the plane formed by the walls.

This approach allowed us to calculate the required normal vector and correctly represent the 3D model.

3.3 Examples of Final 3D Models

Testing showed that the application successfully generates 3D models that match the original schematics. In simple cases, such as rectangular rooms, the reconstruction was performed without visible errors. In complex scenarios involving overlapping elements, minor discrepancies were observed, such as incorrect positioning of windows or doors. These errors were eliminated by further training the model and adjusting the hyperparameters.

As an experiment, an attempt was made to implement adding an interior to the layout so that 3D models would be more interesting. To do this, the areas in which interior objects should be were designated with rectangles on the images and marked with color. During the generation of models, a search for such rectangles was performed and, if they were found, the required object was substituted in their place (Fig. 3).

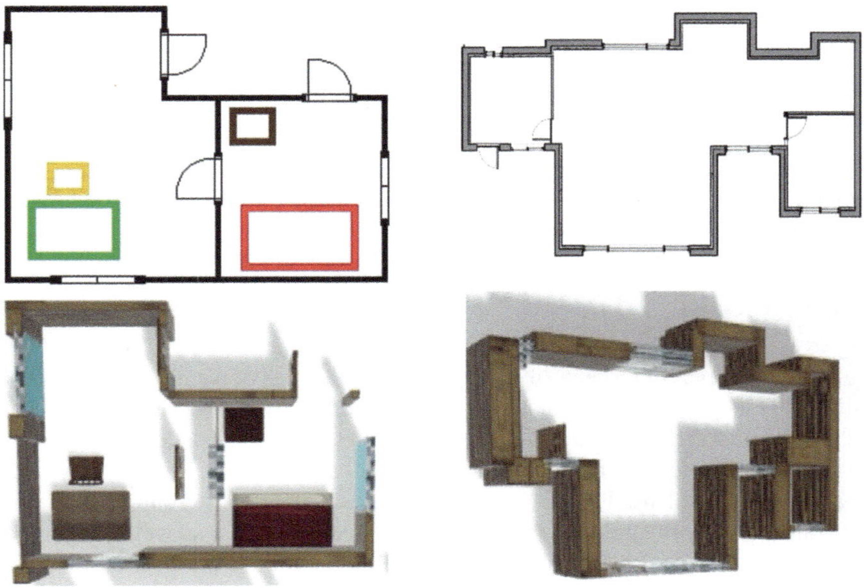

Fig. 3. Results of the application.

3.4 Main Conclusions of the Work on the 3D Visualization Project. Achieved Results

As part of this study, a client-server application with augmented reality (AR) support was developed, allowing 3D models to be automatically created and visualized in AR based on 2D room layouts. The application combines modern computer vision technologies and interactive augmented reality capabilities, providing users with an intuitive and convenient tool for exploring reconstructed spaces. A key stage of the work was the implementation of the YOLOv11 model for detecting objects on 2D layouts. The server architecture of the solution allowed achieving high performance in data processing, which is critical for mobile applications [11].

The following results were obtained:

1. an application was created that successfully solves the problem of converting 2D diagrams into a 3D model;
2. the YOLOv11 model demonstrated high detection accuracy, providing mAP50–95 at the level of 0.43;
3. the model output (mask) undergoes a rectification and filtering stage to ensure the best result;
4. when generating a 3D model, the normal vectors of windows and doors are determined for their correct display;
5. visualization of 3D models in AR allows the user to study reconstructed rooms from different viewing angles;
6. the application has proven its effectiveness on devices with ARCore support, providing stable performance and an intuitive interface.

The project has potential for further development in the following areas:

1. using other detection models: including alternative models (e.g. EfficientDet or Mask R-CNN) to improve recognition of complex structures;
2. improving filtering and rectangle correction for more accurate 3D model generation.

The developed application opens up new possibilities for using AR in architectural and design modeling. Its development will allow the introduction of innovative approaches to creating three-dimensional spaces based on two-dimensional data.

4 Conclusion

The proposed topics of projects in educational technologies of the master's course of the Institute of Geographical Informatics allow to significantly increase the level of knowledge, skills and abilities in the analysis and processing of big data. The applied educational technologies and developed software packages allow to significantly increase the competencies of masters. Essential in the work of project groups is the requirement to develop software implementation of project solutions, as well as to present a report (defend the project), prepare articles within the framework of the annually held All-Russian scientific and practical conference "Mathematics, Informatics, Computer Science, Modeling, Education" (micme.ru).

Acknowledgments. This work was financially supported by the Ministry of Science and Higher Education of the Russian Federation, project N 075-02-2025-1543.

References

1. Kozlova, M.G., Lukianenko, V.A., Germanchuk, M.S.: Development of the toolkit to process the internet memes meant for the modeling, analysis, monitoring and management of social processes. In: Recognition and Perception of Images: Fundamentals and Applications, pp. 189–219 (2021). https://doi.org/10.1002/9781119751991.ch6

2. Kozlova, M.G., Lukianenko, V.A., Gurzhiy, P.A., Makarov, O.O.: Intelligent analysis system for monitoring the spread and influence of political Internet memes. In: Proceedings of Voronezh State University. Series: Systems Analysis and Information Technologies, vol. 2, pp. 145–156 (2022). https://doi.org/10.17308/sait/1995-5499/2022/2/145-156
3. Kozlova, M.G., Lemtyuzhnikova, D.V., Lukyanenko, V.A., Makarov, O.O.: Models and algorithms for multiagent hierarchical routing with time windows. J. Comput. Syst. Sci. Int. **62**, 862–883 (2023). https://doi.org/10.1134/S106423072305009X
4. Sünger, İ., Çankaya, S.: Augmented reality: historical development and area of usage. J. Educ. Technol. Online Learn. **2**(3), 118–133 (2019). https://doi.org/10.31681/jetol.615499
5. Barrerio, A. C., Hilsmann, A., Trzeciakiewicz, M., Eisert, P.: Automatic reconstruction of semantic 3D models from 2D floor plans. 18th International Conference on Machine Vision and Applications (MVA), Hamamatsu, Japan, 2023, 1–5. https://doi.org/10.23919/MVA57639.2023.10215746. (2023)
6. Kippers, R.G.: Methodology for reconstruction of 3D Building Models using 3D-CityJSON and Floor Plans (2021). https://purl.utwente.nl/essays/87724
7. Hsia, S.C., Wang, S.H., Tsai, H.C.: Real-time 2D to 3D image conversion algorithm and VLSI architecture for natural scene. Circuits Syst. Signal Process. **41**, 4455–4478 (2022). https://doi.org/10.1007/s00034-022-01983-y
8. Zhang, X.: Technologies of creation and visualization of 3D animation. World Sci. Ser. Sociol. Philol. Cult. Stud. **14**(1), 27KLSK123 (2023) https://sfk-mn.ru/PDF/27KLSK123.pdf. (In Russ.)
9. Kashlev, D.: Efficient 3D Building Model Generation from 2D Floor Plans (2008). https://dspace.mit.edu/handle/1721.1/45649
10. 3DPlanNet: Generating 3D Models from 2D Floor Plan Images Using Ensemble Methods. https://www.mdpi.com/2079–9292/10/22/2729
11. Efficient 3D building model generation from 2D floor plans. MIT Libraries. https://dspace.mit.edu/handle/1721.1/45649

The Formation of Preservice Teacher's Competence in Using Distance Technologies for Teaching Children with Disabilities

Tatyana A. Kulikova[1(✉)], Olga V. Fedina[1], and Ksenia A. Kirichek[2]

[1] North-Caucasian Federal University, Pushkin Street, 1, Stavropol 355029, Russian Federation
t_a_kulikova@mail.ru

[2] Stavropol State Pedagogical Institute, Lenin Street, 417a, Stavropol 355029, Russian Federation

Abstract. The article gives proof of the relevance and needs for preservice teachers to develop the competence in using distance technologies to teach children with disabilities. To achieve this goal, the description of competence modules, normative and pedagogical literature, both school and university teaching experience have been analyzed, the development of competence has been proposed, and the corresponding experimental work has been designed and implemented. On this basis, the possibilities of distance technologies have been revealed, demonstrating their effectiveness as a means of organizing education for children with disabilities. Technological, psychological, pedagogical, and methodological modules for the formation of preservice teachers' competence in using distance technologies for teaching children with health limitations are shown to use distance technologies for teaching children with disabilities have been determined. Specifics of teacher training for the use of distance technologies for the organization of individual work of teacher-student, independent work of teacher-consultant, the inclusion of children in a lesson in real-time, inclusion of children in collective activity are disclosed. The developed competence will allow preservice teachers to use computer equipment and software for education and socialization of children with disabilities; it also helps build a system of work on psychological and pedagogical support for the training; to form an attitude of acceptance and tolerance among classmates; to develop methodological material, special teaching methods concerning special needs and capabilities of children with health limitations, as well as to organize, monitor and adjust their education.

Keywords: FBachelor of Education · Preservice Teacher · Professional Competence · General Professional Competence · Distance Technology · Children with Disabilities · Inclusive Education · Networking

1 Introduction

While global living standards have risen and advances in medicine and technology have proliferated in the 21st century, the overall health of populations, including children, is facing concerning trends. Specifically, there's a notable rise in the prevalence of chronic

M. Lapina et al. (Eds.): DET 2024, LNNS 1616, pp. 256–265, 2025.
https://doi.org/10.1007/978-3-032-04365-8_26

conditions and disabilities among children [1]. This phenomenon is partly linked to a decrease in infant mortality, meaning more children with pre-existing health issues are surviving [2]. Such children require a special approach when organizing the educational process [3]. The issues extend beyond the classroom, encompassing the socialization of children with disabilities, their preparation for both general academic pursuits and future vocational opportunities. Furthermore, the adaptation of curriculum and instructional materials to accommodate learners with special educational needs is becoming increasingly important. Likewise, preparing educators to create and sustain inclusive learning environments, and to work in partnership with the parents and families of students with disabilities, is gaining global urgency [4–8]. These efforts are particularly vital in a world where societal structures may not always fully accommodate the diverse requirements of every child [9]. For example, in some regions, schools are experimenting with assistive technologies like speech-to-text software for students with dyslexia, or interactive simulations for those with autism to practice social scenarios. These initiatives aim to make learning accessible and engaging for all students, fostering their independence and academic achievement [10].

The Law of the Russian Federation provides the right of children with disabilities to receive education in ordinary schools. This includes the active utilization of distance learning technologies (DLT). The Russian Education Law allows for instruction to be delivered partially or fully through DLT. These policies highlight the critical need to equip prospective teachers with the practical skills required to leverage DLT effectively in the education of children with disabilities. The Federal state educational standard for the training of future teachers indicates the need to form general professional competence, namely the provision of the opportunity to organize joint and individual educational measures for students, including children with special educational needs.

The digital transformation of society has triggered a corresponding digital transformation in education, including through the use of Information and Communication Technologies (ICT) to support the learning of children with disabilities [11]. The works of the authors [12] consider the possibility of using mobile devices in the preparation of children for school. Of particular interest is the introduction of supplemented and virtual reality technologies in inclusive learning [13]. J. M. Sishchuk, M. A. Fedorova, M. V. Tsyguleva, T. A. Vinnikova analyzed distance platforms used for studying foreign languages as related to children with special educational needs [14]. T. Kramarenko, K. Bondar, O. Shestopalova have identified opportunities for applying project-based learning and utilizing Information and Communication Technologies (ICT) in teaching plane geometry and stochastics to children with disabilities at the upper secondary education level [15]. The authors of Savelova, L., Konopko, E., Pankratova, O. the need to use distance technologies in the preparation of the future teacher is considered [16]. In the works of R. Bazhenova, Yu. Stepa, K. Kirichek, O. Bogdanova, I. Ledovsky, the existing problems of teaching students of pedagogical educational institutions were considered and the ways to eliminate them were proposed. The model developed by the authors allows you to instill in future teachers the ability to use modern methods and technologies for learning [17].

A model for developing competence in diagnosing (monitoring and evaluating) the results of students mastering the educational program among future teachers has also been compiled and tested [18]. A. Amirova, K. Buzaubakova, Z. Kashkynbayeva, M. Yelubayeva, Z. Kumisbekova, U. Elmira, Z. Genc assert the need to purposefully develop the creative abilities of preservice teachers, the desire for professional excellence and professional growth [19]. O.Budnyk, R. Zozuliak-Sluchyk, S. Nedilskyi, I. Chervinska, T. Malaniuk, N. Prevysokova, U. Ketsyk-Zinchenko offered digital instruments and applications for distance learning, as well as outlined the ways of teachers' training for the effective use of digital technologies of training [20]. The authors [21] studied the psychological aspect of the readiness of future teachers to teach children with disabilities. The model allows not only to prepare teachers for work with children with special educational needs, but also the diagnosis of their readiness is proposed by I. Demchenko, O. Lavrikova, N. Kuzmenko [22].

The analysis of literature and practical studies showed that the teachers are concerned, because their readiness to use distance technologies for the integration of children with disabilities into the educational process of the school requires advanced training. Students of pedagogical educational organizations understand that the possession of digital tools will contribute to the creation of an educational environment that excludes the barriers of the interaction of schoolchildren and children with special educational needs. Therefore, questions about the training of teachers not only owning distance technologies, but also capable of applying these technologies to integrate children with disabilities into classrooms remain open and require further study.

2 Preparing the Future Teacher for the Use of Distance Technologies for Teaching Children with Disabilities

Drawing upon the insights gleaned from this exhaustive evaluation, a competence-building paradigm was formulated, and a related empirical inquiry was structured and conducted at the North Caucasus Federal University and Stavropol State Pedagogical Institute (Stavropol).

The competence development framework follows a modular structure (see Fig. 1). The research indicated that the education of children with disabilities should be addressed in all relevant disciplines studied by future teachers. Only under this condition can the competence to effectively utilize distance learning technologies for the education of children with disabilities be successfully developed.

During the implementation of each module, pre-service teachers should develop readiness to organize the following types of activities:

- individualized instruction («teacher-student»);
- independent learning («teacher-consultant»);
- real-time lesson engagement; collaborative (group) activities.

The readiness of the future teacher to organize an individual work «teacher-student» takes place in the process of implementing the technological module. Students get acquainted with digital educational platforms (gttrbrains, Moodle, Teach.ru, Yandex.Chobnik, Core-App, Bigbluebutton, AIS Network City Education, Jacques, Spherem, Russian Electronic School, Foxford, FGS "My School"). During the methodological module, it's crucial not only to teach students how to organize the educational process but also to

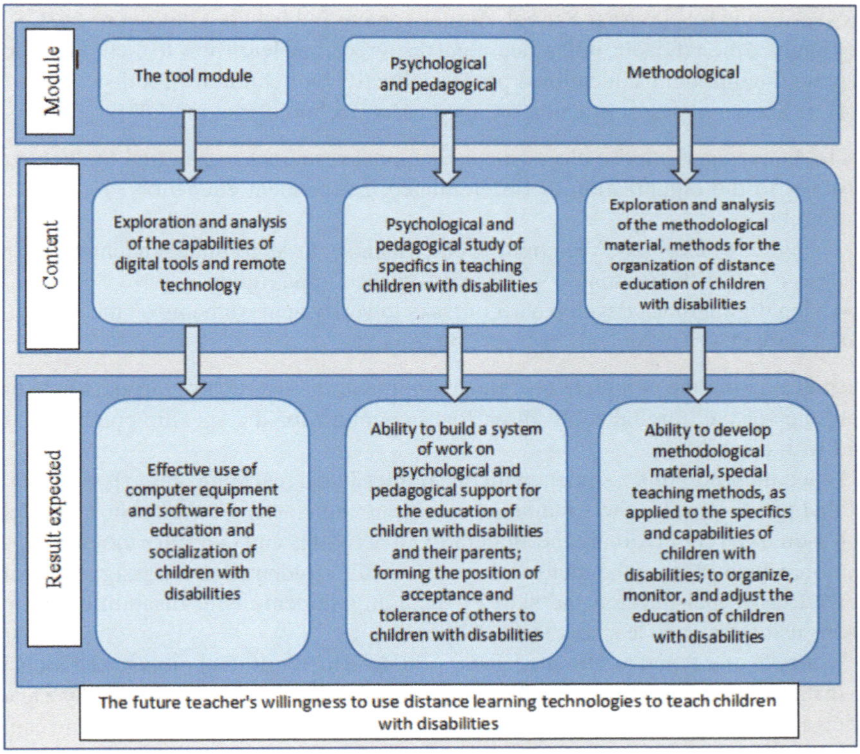

Fig. 1. The modular structure of the formation of students' readiness for the use of distance technologies in teaching children with disabilities.

incorporate students with disabilities into this process using distance learning technologies. The pre-service teacher must be able to provide timely feedback to a child with special educational needs. If the child is struggling or tired, the educator should quickly react [23] and adjust the educational process accordingly.

To this end, students are given practice-oriented assignments, for example (methodological module): develop an individual lesson plan for the subject «_» for a student in «_» grade with musculoskeletal disorders. When completing the assignment, the student needs to: choose the lesson topic, select appropriate software tools and digital educational resources, develop a lesson plan, and provide methodological support. When evaluating the assignment, the instructor emphasizes the student's ability to select software tools that are effective for organizing feedback at all stages of the individualized lesson and considers the depth of the student's reflection on their work.

Readiness to organize the activity type – independent learning («teacher-consultant») – is developed during the implementation of the technological module. Students explore electronic educational resources, such as: Russian Electronic School, Moscow Electronic School, Virtual School of Cyril and Methodius, «Eidos» Distance Education Center, «New Technologies», TeachPro.ru Educational Website,

«Prosveschenie.ru» Internet School, digital content publishers «Physicon», «1C», etc. They analyze their didactic potential, select resources, and learn how to methodologically integrate them into the educational process effectivelyof children with disabilities.

Here are examples of practice-oriented tasks (technological module):

- «To analyze and choose electronic educational resources and digital tools that contribute to the organization of differentiated independent education of a child with disabilities».
- «Using the video, AR-, VR-applications, visualize tasks for independent work on the subject «_» of the student «_» class with speech impairment».
- «Using digital tools, develop a set of tasks to verify achievements on the subject «_» of a student «_» class with hearing impairment».

When evaluating the completion of the assignment, the instructor considers how effectively the selected digital tools allow for accounting for the specific condition of the child with disabilities.

When incorporating a student with disabilities into a real-time online lesson, it's crucial that pre-service teachers maintain consistent engagement and attention throughout the lesson, actively involving the student in all activities and carefully monitoring their emotional state. Within the methodological module, students can be assigned practice-oriented tasks that develop the skill of including students with disabilities in online lessons using distance learning technologies.

Example of a Practice-Oriented Task: «Propose methods, techniques, and tools that, in a distance learning setting, would replace the actions typically performed by a teacher during a traditional in-person lesson: organizing students for the lesson, creating engagement with the new topic, checking individual student work, providing individualized assistance, regulating behavior, and focusing attention».

One activity for pre-service teachers is to role-play a scenario where one student acts as the teacher leading the lesson, another connects to the lesson remotely and plays the role of a student with disabilities, and the remaining students act as typical classmates. The analysis of these sessions is conducted collaboratively. Evaluation considers the pre-service teacher's command of the situation and their level of attentiveness, not only to all the students but particularly to the student with disabilities.

Readiness to organize collaborative (group) activities in pre-service teachers is developed during the implementation of the technological and methodological modules. Within the technological module, students are introduced to various collaborative tools and platforms (e.g., VK Board, iDroo, Yandex Concept, Unidraw, AWW board, Esboard, Draw Chat, etc.), as well as tools integrated into unified digital collaboration suites (Google Docs, Evernote, Miro, etc.). Within the methodological module, students learn how to use distance learning technologies to organize collaborative research projects; learn to design (or adapt) research tasks that require simple equipment and accessible materials [24]; and acquire the skills to coordinate collaborative activities that include a student with special educational needs, and to facilitate communication between participants in the process using digital technologies.

Every teacher should maintain a methodological toolkit that contains resources to support their preparation for various types of lessons. Practical, problem-solving tasks can help populate this toolkit. For example, considering the specific needs of a student with hearing or visual impairments, develop a collaborative (group) task and select software that can be used in a remote learning format by a group of students, including the student with disabilities.

Every teacher should maintain a methodological toolkit that contains resources to support their preparation for various types of lessons. Practical, problem-solving tasks can help populate this toolkit. Here are examples of practice-oriented tasksconsidering:

- «The specific needs of a student with hearing or visual impairments, develop a collaborative (group) task and select software that can be used in a remote learning format by a group of students, including the student with disabilities»;
- «Taking into account the specific needs of a student with hearing or visual impairments, develop a game task that promotes the development of thinking, and select software that can be used to organize joint play between a child with disabilities and his classmates».

In the student's completed work, the instructor evaluates the pre-service teacher's consideration of the selected software's capabilities for including a student with disabilities in collaborative (group) activities, as well as the appropriateness of the task's complexity to the specific needs of the student with health deviations.

A schematic representation of the development of competencies in pre-service teachers for using distance learning technologies to educate children with disabilities is presented (see Fig. 2).

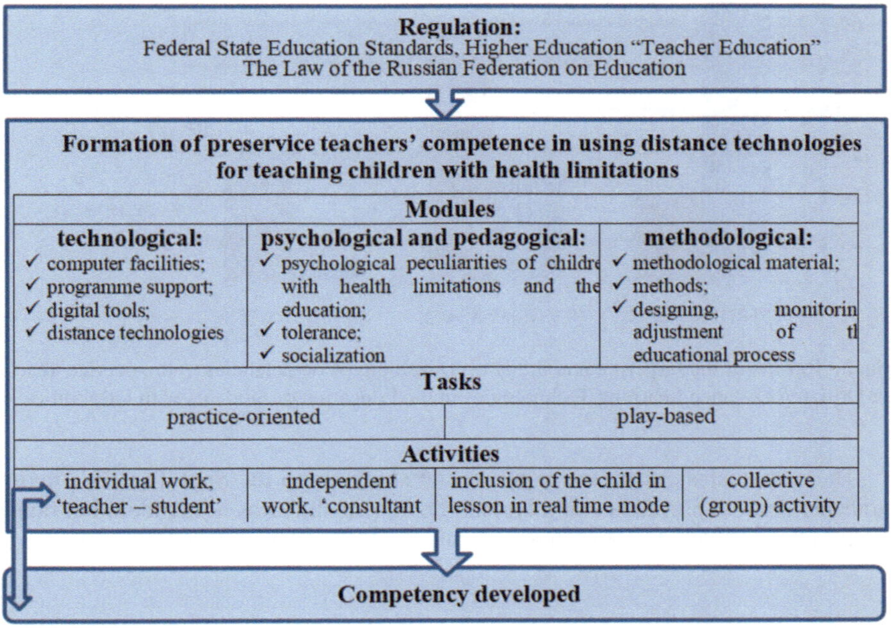

Fig. 2. Developing Competencies in Pre-service Teachers for Utilizing Distance Learning Technologies in the Education of Students with Disabilities.

In the process of forming competence, students realize that distance technologies are an effective means of organizing teaching children with disabilities. Since these technologies allow:

- teaching children in the usual home-made households specially equipped for them;
- include parents in the educational process;
- build work on an individual plan developed taking into account the characteristics and needs of children with disabilities;
- quickly monitor the result and timely adjust the learning process;
- choose the time of training, adjusting to well-being, procedures, treatment in medical institutions;
- to include a child in a lesson in real time, organize both independent, individual, and group work.

The diagnostic assessment of the competence development process was conducted through observation of student activities during class sessions and their performance on competence-based tasks. The study involved 132 undergraduate students in teacher education programs and 18 instructors. The results of the pedagogical experiment are presented (see Fig. 3).

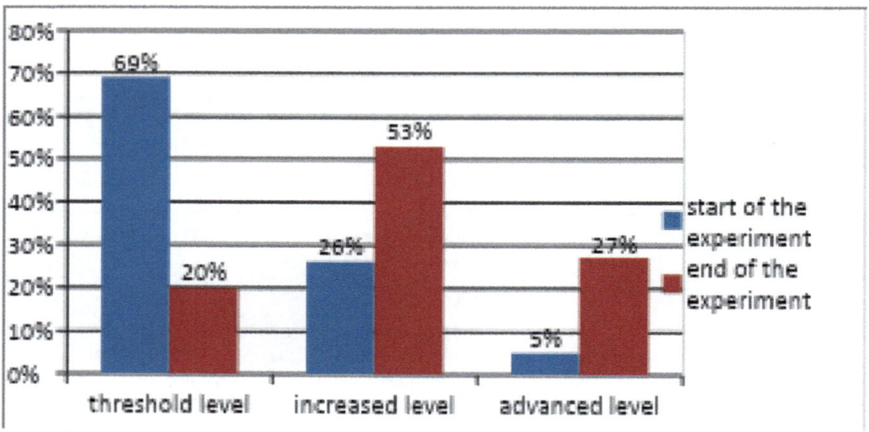

Fig. 3. Results of the Experimental Study on Developing Competencies in Pre-service Teachers for Utilizing Distance Learning Technologies in the Education of Students with Disabilities.

The data analysis from the experimental study, using the Sign Test (G-criterion), indicated that the prevalence of a typical directional shift was not random but rather a result of the teacher education students' participation in the proposed training program.

3 Conclusions

The experimental study conducted at the North Caucasus Federal University and the Stavropol State Pedagogical Institute demonstrated that developing competencies in pre-service teachers for utilizing distance learning technologies in the education of students with disabilities will be most effective if:

- continuous training on the topic of educating students with disabilities is implemented for students, integrating topics that explore the specific challenges and strategies for working with these students into the relevant disciplines of the curriculum;
- student training is structured in a modular format: technological, psychological-pedagogical, and methodological modules;
- the psychological-pedagogical module incorporates a moral dimension: teachers should not only be patient and attentive themselves when working with students with disabilities but also cultivate a tolerant attitude within the student body.

An educator who possesses competence in utilizing digital tools is capable of organizing group work for students, including students with disabilities, in lessons through distance learning technologies. This helps children with special educational needs, to gain social experience so necessary for the inclusion of a person in society.

Acknowledgements. We would like to express our gratitude to reviewers for reviewing the paper.

References

1. Pasichnyk, V., Khimenes, K., Pityn, M., Bas, O., Hlukhov, I., Hnatchuk, Y., Drobot, K.: Physical condition of preschool children with disabilities in psychological and physical development. J. Phys. Educ. Sport (JPES). **21**(1), 352–359 (2021). https://doi.org/10.7752/jpes.2021.01033
2. Kamiya, Y.: Current situation of children with disabilities in low-and middle-income countries. Pediatr. Int. (2021). https://doi.org/10.1111/ped.14904
3. Hutzler, Y., Meier, S., Reuker, S., Zitomer, M.: Attitudes and self-efficacy of physical education teachers toward inclusion of children with disabilities: a narrative review of international literature. Phys. Educ. Sport Pedagog. **24**(3), 249–266 (2019). https://doi.org/10.1080/17408989.2019.1571183
4. Mantey, E.E.: Discrimination against children with disabilities in mainstream schools in Southern Ghana: challenges and perspectives from stakeholders. Int. J. Educ. Dev. **54**, 18–25 (2017). https://doi.org/10.1016/j.ijedudev.2017.02.001
5. Cockburn, L., Hashemi, G., Noumi, C., Ritchie, A., Skead, K.: Realizing the educational rights of children with disabilities: an overview of inclusive education in Cameroon. J. Educ. Pract. **8**(6), 1–19 (2017)
6. Alshamri, K.H.: Challenges and experiences of high school teachers with students having intellectual disabilities in inclusive classrooms in Saudi Arabia. Univ. J. Educ. Res. **8**(5), 2191–2196 (2020). https://doi.org/10.13189/ujer.2020.080559
7. Al Hazmi, A.N., Ahmad, A.C.: Universal design for learning to support access to the general education curriculum for students with intellectual disabilities. World J. Educ. **8**(2), 66–72 (2018)

8. Stepanova, G.A., Tashcheva, A.I., Stepanova, O.P., Menshikov, P.V., Kassymova, G.K., Arpentieva, M.R., Tokar, O.V.: The problem of management and implementation of innovative models of network interaction in inclusive education of persons with disabilities. Int. J. Educ. Inf. Technol. **12**, 156–162 (2018)
9. Lazurenkoa, S.B., Solovyovaa, T.A., Terletskayab, R.N., Konovab, S.R.: Problems of health protection of students with health limitations in Educational Institutions of the Russian Federation. Int. Educ. **25**(1), 127–143 (2021). https://doi.org/10.15507/1991-9468.102.025.202101.127-143
10. Nebytova, L.A., Katrenko, M.V., Savin, D.I., Zhuravleva, Y.I.: Augmented reality in the training process of children with hearing disorders. In: CEUR Workshop Proceedings, vol. 2834, pp. 330–339 (2021)
11. Shevchenko, G.I., Kalashnikova, A.I., Rybakova, A.A., Brynza, S.Y.: Tasks, functions and basic activities of an online tutor. In: Proceedings of the International Scientific Conference on Innovative Approaches to the Application of Digital Technologies in Education (SLET 2020) Stavropol, Russia, November 12–13, pp. 285–291 (2020)
12. Chmiliar, L.: Improving learning outcomes: the iPad and preschool children with disabilities. Front. Psychol. **8**, 660 (2017)
13. Tesolin, A., Tsinakos, A.: Opening real doors: Strategies for using mobile augmented reality to create inclusive distance education for learners with different-abilities. In: Mobile and Ubiquitous Learning, pp. 59–80. Springer, Singapore (2017). https://doi.org/10.1007/978-981-10-6144-8_4
14. Fedorova, M.A., Tsyguleva, M.V., Vinnikova, T.A., Sishchuk, J.M.: Distance education opportunities in teaching a foreign language to people with limited health possibilities. Astra Salvensis. **6**, 631–637 (2018)
15. Kramarenko, T., Bondar, K., Shestopalova, O.: The ICT usage in teaching mathematics to students with special educational needs. J. Phys. Conf. Ser. **1840**(1), 012009 (2021). https://doi.org/10.1088/1742-6596/1840/1/012009
16. Savelova, L., Konopko, E., Pankratova, O.: Digital technologies and effective practices in teacher. In: Training CEUR Workshop Proceedings. "SLET 2020 - Proceedings of the International Scientific Conference on Innovative Approaches to the Application of Digital Techno. logies in Education", pp. 172–178 (2020)
17. Kirichek, K., Bazhenov, R., Bogdanova, O., Shtepa, Y., Ledovskikh, I.: Formation of competence of using modern methods and technologies of education in bachelors of education training. In: Inted 2018: Conference Proceedings, pp. 2403–2410. IATED Academy, Valencia (2018). https://doi.org/10.21125/inted.2018.0457
18. Kirichek, K., Bazhenov, R., Bogdanova, O., Ledovskikh, I., Pronin, A., Samorukov, A.: Formation of competency for implementation of diagnostics of schoolchildren outcomes by bachelors of education. In: INTED2019 Proceeding, pp. 4029–4037. IATED Academy, Valencia (2019). https://doi.org/10.21125/inted.2019.1012
19. Amirova, A., Buzaubakova, K., Kashkynbayeva, Z., Yelubayeva, M., Kumisbekova, Z., Elmira, U., Genc, Z.: Training the creative competence of future teachers. J. Educ. Teachers Trainers. **9**(2), 118–125 (2018)
20. Budnyk, O., Zozuliak-Sluchyk, R., Nedilskyi, S., Chervinska, I., Malaniuk, T., Prevysokova, N., Ketsyk-Zinchenko, U.: Modern digital distance learning technologies: challenges of future teacher training. Rev. Inclusion. **8**(1), 41–53 (2021)
21. Pilipchuk, L.S., Kovaleva, A.S., Karnauhova, Y.B.: Study of teachers' preparedness for professional activities in an inclusive educational environment. In: SHS Web of Conferences, vol. 113. EDP Sciences (2021). https://doi.org/10.1051/shsconf/202111300075
22. Lavrykova, O., Kuzmenko, N., Demchenko, I., Kovalchuk, O., Slipchuk, V., Pilevych, O.: Model of formation of future teachers' readiness to work in inclusive education. Int. J. Sci. Technol. Res. **9**(02), 410–414 (2020)

23. Kulikova, T.A., Poddubnaya, N.A., Bagdasaryan, L.S., Ardeev, A.H.: The technique for future teachers' digital literacy development. J. Phys. Conf. Ser. **1691**(1), 012129 (2020). https://doi.org/10.1088/1742-6596/1691/1/012129
24. Fedina, O.V., Zakinyan, A.R., Agibova, I.M.: Design of science laboratory sessions with magnetic fluids. Int. J. Mech. Eng. Educ. **45**(4), 349–359 (2017). https://doi.org/10.1177/0306419017708644

Research of Prospects and Challenges in Artificial Intelligence Technology Teaching and Learning

Maria Lapina[1], E. A. Mary Anita[2], Alina Bagautdinova[1](✉), and Marina Gordova[3]

[1] North-Caucasus Federal University, Stavropol 3550017, Russia
bagaauttdinova@mail.ru
[2] Christ University, Bangalore, India
[3] Financial University Under the Government of the Russian Federation, Moscow 125993, Russia

Abstract. Recently, knowledge in the field of artificial intelligence in order to modernize various aspects of human activity has played a significant role. The exploration of the opportunities and difficulties associated with the development of artificial intelligence technologies is becoming an important area of research, as it profoundly affects our perception of work, education, medicine and other spheres of existence. New methods of machine learning, deep learning and reinforcement learning are being developed. These technologies are changing our understanding of how machines can learn and adapt to the world around them. The application of artificial intelligence covers many areas, including healthcare, finance, education and industry. In medicine, for example, AI can improve diagnostic accuracy and develop customized treatments. In education, it is possible to create personalized learning plans for each student. While in industry, artificial intelligence technologies are able to optimize production processes and increase business efficiency. However, despite the potential benefits associated with learning artificial intelligence technologies, there are serious challenges that require careful analysis. These challenges include ethical dilemmas, such as issues of algorithm transparency and responsibility for making principled decisions. Data security and privacy are also among the key aspects that require innovative approaches to AI technology training.

The main purpose of the research is to deeply analyze the prospects and challenges in the field of artificial intelligence technology training, provide a comprehensive understanding of the current state of this field, identify key areas of development and propose practical strategies for effectively overcoming challenges. Taking into account both positive and negative aspects, it is necessary to have a meaningful look at the future of artificial intelligence technology education, taking into account social, ethical and technical aspects.

Keywords: Artificial intelligence technologies · learning · machine learning · deep learning · reinforcement learning · perspectives · challenges · AI ethics · data security · algorithm transparency · medicine · finance · education · industry · socio-cultural aspects · efficiency · innovation

M. Lapina et al. (Eds.): DET 2024, LNNS 1616, pp. 266–276, 2025.
https://doi.org/10.1007/978-3-032-04365-8_27

1 Prospects in Teaching Artificial Intelligence Technologies

One of the key aspects of the revolution in medical education is the introduction of virtual patients and clinical simulators created using artificial intelligence technologies. These virtual scenarios allow medical students and doctors to train in a variety of clinical cases, from classic to extreme. This learning experience provides an opportunity to develop diagnostic and therapeutic skills in a safe and controlled environment. Artificial intelligence technologies allow you to create personalized educational programs for medical students. Data analysis algorithms take into account individual needs, the rate of assimilation of information and the level of training of the student. This leads to the optimization of training and the formation of competent and individually trained medical professionals [1]. The development of AI-supported simulation technologies provides medical students with the opportunity to gain hands-on experience without having to use real patients. This is especially valuable when teaching complex procedures, surgery and manipulation, where precision and safety play a key role [2].

Artificial intelligence methods and machine learning algorithms are currently used both in manufacturing, to improve automation efficiency, and in medicine, medicine, and financial structures for big data analysis. With the growing popularity of language models and the launch of open source GPT, artificial intelligence methods have entered even more areas of human activity [26, 27].

The training of medical specialists in complex medical procedures and operations becomes highly effective in conditions where accuracy and safety of medical actions play an important role. Immersion in virtual clinical episodes, through the integration of developments based on artificial intelligence, significantly enriches the repertoire of practical skills. Medical students practice in a wide range of situations - from rare diseases to the most difficult surgical interventions [3]. In this way, students gain experience in meeting a variety of clinical challenges. The introduction of artificial intelligence into the medical training process significantly enhances the skills of making clinical diagnoses, providing diagnostic conclusions with high accuracy based on the analysis of extensive medical databases, thereby enriching the training of future doctors.

The process of learning using AI in production requires not only the improvement of technical skills, but also the development of project management skills [4]. Artificial intelligence tools facilitate project planning and management by automating routine operations and providing analytical support in making informed decisions (Fig. 1).

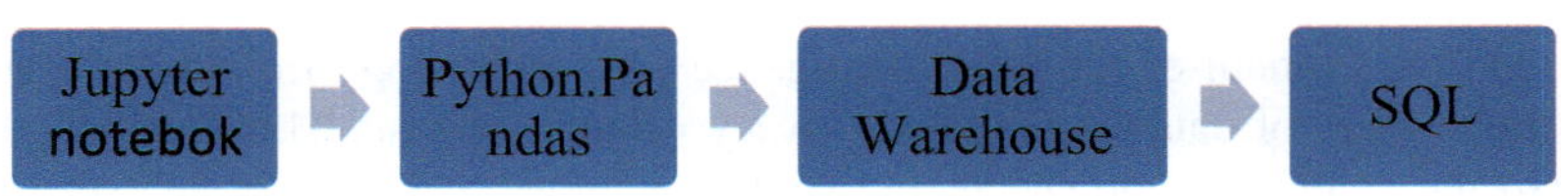

Fig. 1. A scheme for learning the basics of data analytics.

2 Ethical Issues in Artificial Intelligence Training

In the era of the introduction of artificial intelligence into educational processes, modern society faces complex ethical tasks that require detailed study and the creation of a framework for control.The fundamental ethical problems associated with the use of AI in pedagogical methods, the study of influences, difficulties, and ways to neutralize potential risks are analyzed [5]. The introduction of intelligent systems into the educational environment generates a number of ethical tasks that require careful analysis and the formation of standards aimed at the honest and effective use of these technologies. The transparency and clarity of the algorithms embedded in the learning platforms are the key points of concern to those involved in the educational process. When adapting educational material to the needs of an individual student, complex machine learning models can become an obstacle to understanding the underlying processes, leading to problems with clarity in educational decision-making and their implications for the curriculum [6].

The issue of equality in the use of AI systems in education also occupies one of the leading places in the discussion. Biased and unbalanced data sets can deepen existing social inequalities reflected in the educational opportunities of various groups of students. It is important to focus efforts on creating algorithms that are fair and take into account the diversity and individual needs of each student, as well as on implementing a system for monitoring and correcting algorithmic results when signs of inequality are detected [7]. The development of ethical standards that contribute to the organization of a more equitable educational space is becoming a critical element in solving these problems.

With the integration of intelligent modules into the education system, the need to combat inequality and bias is increasing, since algorithms trained on distorted data can contribute to uneven student access to educational resources [8]. In order to ensure equal educational opportunities for every individual, an active policy against discrimination and equity-oriented methods of using AI is necessary [24, 25].

Protecting the personal information and privacy of students' data when processing them to improve pedagogical processes also raises ethical issues. Transparent data processing methods, high standards of information protection and a strict privacy policy are needed as a basis for managing this ethical aspect [9]. It is important for students to be given the opportunity to control their personal information (Fig. 2).

3 Data Security and Privacy in AI Technology Education

With the development of artificial intelligence (AI) technology training in education, significant issues of data security and privacy protection arise. This section examines in detail the challenges, prospects and strategies aimed at ensuring data security and protecting personal information in educational contexts where AI technologies are used [10]. Modern learning technologies using artificial intelligence generate a large amount of personal data of students. This includes information about class attendance, learning progress, preferences, and even biometric data when using technologies such as facial recognition or fingerprint scanning. A significant amount of personal data makes educational institutions and systems vulnerable to cyber attacks and misuse. Educational institutions using artificial intelligence technologies are exposed to a variety of

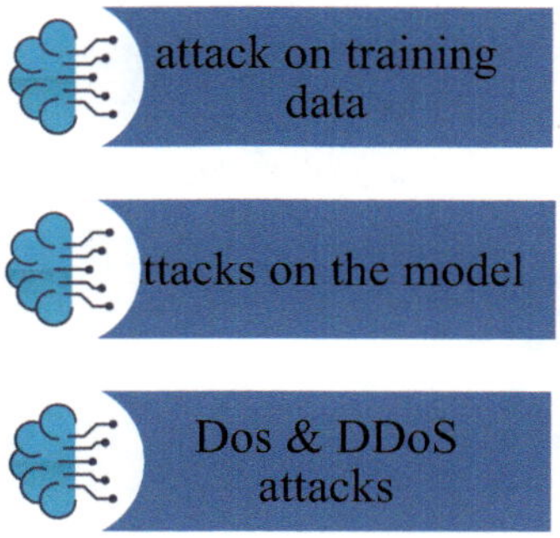

Fig. 2. AI vulnerabilities and malicious attacks.

cybersecurity threats. Attacks can be aimed at stolen personal data, interference in the educational process, or even refusal to provide training services [11]. Cyber attacks can disrupt the functioning of educational systems, affect the reliability of data and violate the privacy of students and teachers. The introduction of strict security standards and data encryption is a mandatory step to ensure security in educational systems using artificial intelligence technologies. Protecting data from unauthorized access, as well as ensuring their integrity and confidentiality, are key points in preventing information leaks and complying with legal privacy requirements. Regular security audits and monitoring are an integral part of the data security strategy in education [12]. This includes a systematic analysis of the system for vulnerabilities, as well as tracking unusual activity on the network. Effective audit measures help to identify and respond to security threats before they lead to serious consequences [13]. Cooperation with the information security industry provides additional resources and expertise for data protection in educational systems. Cybersecurity experts can provide recommendations to improve system security, as well as actively respond to threats and incidents [14]. Training staff and users about data security plays a key role in reducing risks. Staff at educational institutions should be trained in the basics of cybersecurity, including password management, phishing attack detection, and the safe use of technology (Fig. 3).

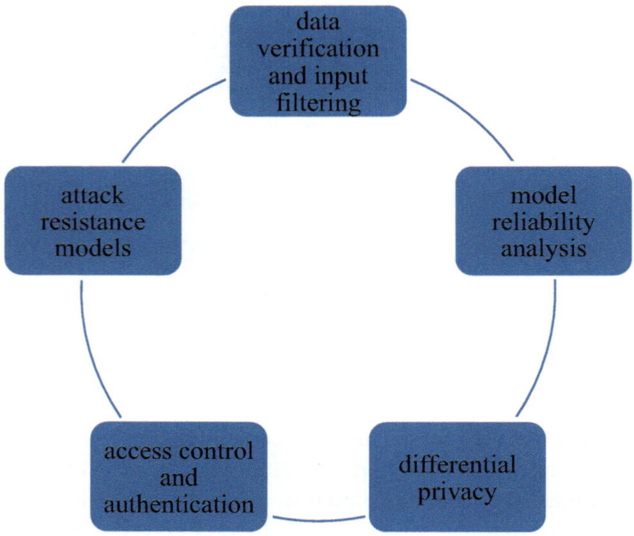

Fig. 3. Methods of protecting AI from malicious attacks.

4 Prospects for the Industrial Application of Artificial Intelligence Technologies

The use of artificial intelligence technologies in industry promises a sharp jump in production dynamics. The ability of machine-learning algorithms to analyze extensive data, to continuously streamline work procedures and to accurately predict potential equipment failures opens the door to minimizing costs, improving the quality of goods and enhancing overall productivity at enterprises [15].

Among other prospects for the application of AI in industry, the concept of intellectualization of production flows stands out. Intelligent AI systems are able to control and synchronize the operations of various mechanisms and automated complexes, applying optimization taking into account a variety of parameters, including the operability of equipment, the availability of raw materials and energy efficiency.

Machine learning and data analysis technologies play a key role in predicting and preventing production incidents, reducing the likelihood of downtime [16]. Thanks to real-time monitoring, potential malfunctions can be identified and eliminated in a timely manner, reducing the risk of downtime and increasing the overall reliability of production systems.

As AI technologies develop, the prospect of creating entirely independent production complexes appears. Control systems equipped with artificial intelligence are able to independently make decisions in accordance with current changes in the production arena, which is especially important for the operation of robotic devices, machines with autonomous abilities and systems based on feedback and data received from sensors (Fig. 4).

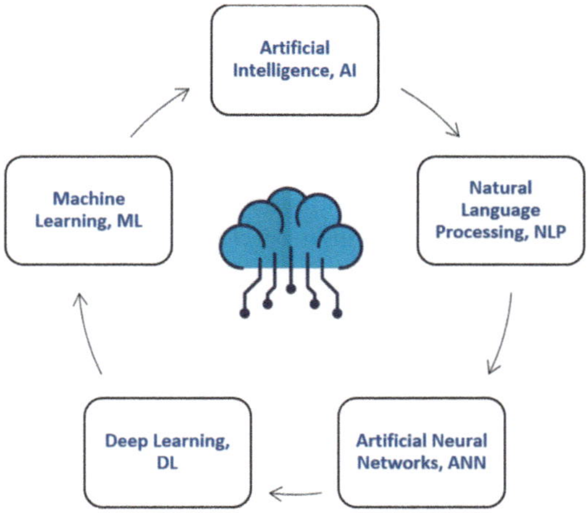

Fig. 4. The concept of studying data analysis.

5 Challenges in Teaching Artificial Intelligence Technologies

Classes on the practice and theory of artificial intelligence are considered one of the most relevant and promising sectors in the field of modern education [17]. However, the integration of artificial intelligence techniques into educational processes often faces obstacles that can complicate this process. In this chapter, we address the problems that educational institutions have to face when integrating these technologies, as well as consider possible ways to solve them.

The lack of teachers and experts with appropriate qualifications in the field of artificial intelligence is one of the main difficulties in the educational process [18]. In order to form and maintain educational programs that focus on creating a professional base for teachers and specialists in this field, educational institutions must make significant efforts. Without qualified and experienced specialists, the effectiveness of using the most modern methods and tools may be at risk.

Also, inequality in access to advanced educational infrastructures and computing resources can cause heterogeneity in the level of education [19]. Educational institutions, limited in financial and technical resources, face problems related to setting up and maintaining the necessary equipment, which affects the quality of the introduction of artificial intelligence into the educational process.

Bringing AI into education also activates a number of ethical dilemmas. Both the data collection process and its further use should be carefully monitored in order to avoid discrimination and ensure the protection of personal information [20]. The absence of certain standards leads to the risk of violations in the field of confidentiality and misuse of information databases.

Existing curricula and teaching methods may require revision and adaptation to integrate artificial intelligence. This includes updating courses, introducing new disciplines and revising teaching methods. Such a transformation of traditional educational approaches often requires significant energy and resources on the part of educational institutions.

Concerns about the introduction of artificial intelligence technologies are felt by both students and teachers. Some fear that automation will lead to job losses, while others express concern about the possible reduction of control over the educational process. To counter such anxieties, it is important to conduct educational campaigns and demonstrate the benefits and opportunities offered by integrating artificial intelligence into an educational context.

6 Trends in the Development of Artificial Intelligence Technologies in Education

The evolution of artificial intelligence in the pedagogical field. This segment is devoted to the analysis of modern development trajectories and promising innovations in integrating AI into educational processes. By studying these trends, one can imagine how innovations in the field of artificial intelligence will determine the education of the future. Thanks to the progress in AI, smart mentors have appeared who can customize educational content to the specific needs of each student, making personalized education a central innovation, which increases the effectiveness of studying the material taking into account the knowledge, style and pace of learning of the student [21].

Specialized AI algorithms are actively used to reduce the time for evaluating and analyzing student success, processing large-scale data in an expeditious manner and providing more neutral conclusions. This leads to a more insightful understanding of student performance and helps to timely notice obstacles in their learning.

One of the key areas is the use of virtual and augmented reality [22]. Supported by AI, these technologies create engaging learning environments where students can interact with virtual objects and situations, which improves the perception of the material.

Thanks to the analysis of educational data through AI, it is possible to create smart analytical tools capable of tracking educational trajectories, evaluating the effectiveness of educational programs and offering tips for optimizing them.

The field of AI is also revolutionizing the process of adapting learning materials by offering individually tailored materials based on students' interests, knowledge and skills, which makes learning more comprehensive and exciting (Fig. 5).

7 Conclusions

The process of mastering techniques related to artificial intelligence radically changes the foundations of learning, revealing unexplored expanses of possibilities to educational institutions, teachers and students [23]. The development of AI in pedagogy is characterized by a deep diversity and a wide range of applications. AI-rich pedagogical courses embody an individualized approach to learning, which contributes to the creation of customized learning trajectories that meet the unique needs of each student. Automation of assessment processes and performance analytics provides a new perspective on the achievements and difficulties of individual students, clarifying the picture of the educational process [24].

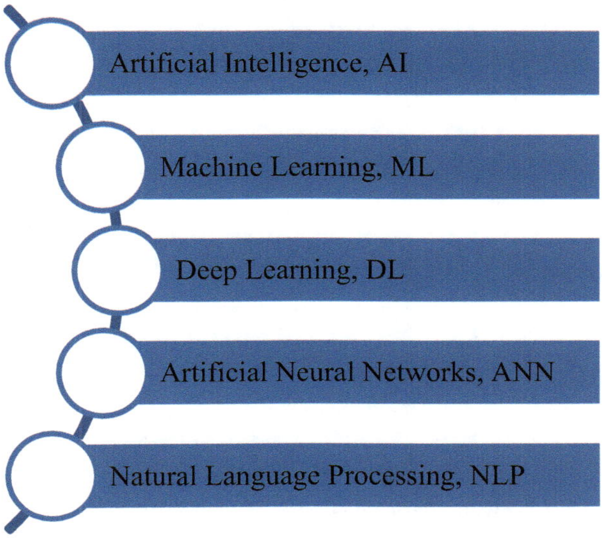

Fig. 5. Key aspects of AI in education the necessary skills to work with AI.

Virtual and augmented reality technologies enhanced by artificial intelligence are transforming the educational environment, opening up previously unattainable opportunities. Immersion in interactive learning worlds and work with visually rich materials enrich the learning process, making it attractive and effective. Work in the field of data mining helps to identify educational trends and improve courses based on accumulated experience and information.

For advanced AI training, it is necessary to have the following skills in programming and mathematics, respectively (Figs. 6 and 7).

But when faced with huge potential, it is important to notice the obstacles on the way. The shortage of AI specialists highlights the task of training professional teachers and experts. The unequal distribution of access to computing resources requires increased efforts to create an infrastructure accessible to all.

The ethical aspects of the use of AI in the educational field will require careful calibration and standardization to ensure the safety of personal data and eliminate discrimination and unauthorized actions with information. The revision of curricula to the new realities of AI is key, but it meets with resistance from established educational systems [24]. Upgrading professional skills and updating curricula is a time-consuming and investment-intensive process.

For the harmonious implementation of artificial intelligence technologies in education, a number of measures can be proposed: continuation of research within the framework of creating advanced training courses for teachers, taking into account the dynamic growth of technology; contributing to the establishment of standards and ethical principles for the application of AI in the educational field, ensuring the integrity of data; development of educational programs aimed at training personnel focused on AI, while focusing on interdisciplinary knowledge and practical skills; support for scientific

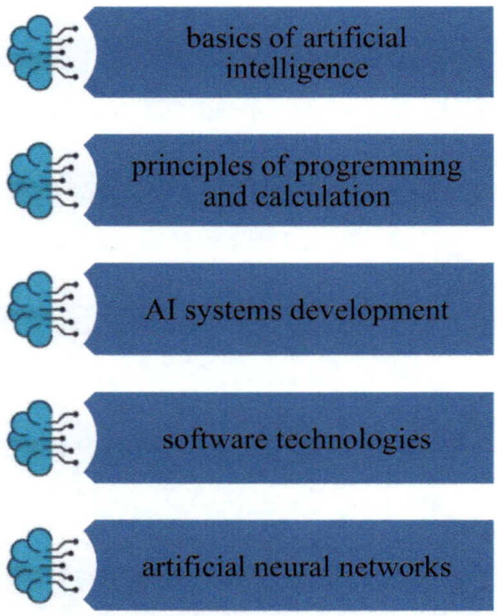

Fig. 6. Basic knowledge of AI programming.

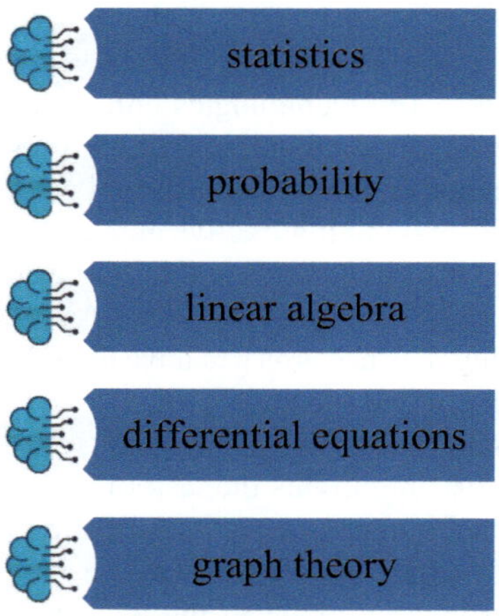

Fig. 7. Basic knowledge of AI in mathematics.

developments to create an open infrastructure for the study of AI, which will give equal chances to every student.

Li emphasizes that the future of AI depends not only on technological advancements but also on training a skilled workforce capable of using these technologies responsibly [25].

References

1. Baker, R.S.J.D., Inventado, P.S.: Educational data mining and learning analytics: applications to constructionist research. Technol. Knowl. Learn. **19**(1-2), 205–220 (2014)

2. Hinton, G., Deng, L., Yu, D., Dahl, G.E., Mohamed, A.R., Jaitly, N., et al.: Deep neural networks for acoustic modeling in speech recognition: the shared views of four research groups. IEEE Signal Process. Mag. **29**(6), 82–97 (2012)
3. Holmes, W., Aggarwal, P.: Artificial intelligence in education: what is the future? Science. **360**(6385), 968–970 (2018)
4. Blikstein, P., Worsley, M.: Multimodal learning analytics and education data mining: using computational technologies to measure complex learning tasks. J. Learn. Anal. **3**(2), 220–238 (2016)
5. Proshkin, N., Basan, E., Lapina, M.: Radio frequency method for emulating multiple UAVs. In: 17th International Conference on Intelligent Environments, IE 2021 – Proceedings, p. 9486599 (2021)
6. Rugelj, J., Lapina, M.: Game design based learning of programming: CEUR workshop proceedings SLET- 2019. In: Proceedings of the International Scientific Conference Innovative Approaches to the Application of Digital Technologies in Education and Research, vol. 2494, pp. 29–42 (2019) http://ceur-ws.org/Vol-2494/
7. Parfenov, D., Zaporozhko, V., Lapina, M., Sora, D.: Development and Research of Algorithms for the formation the Individual Educational Trajectories of Students in the Digital Educational Platform: CEUR Workshop Proceedings SLET-2019. In: Proceedings of the International Scientific Conference Innovative Approaches to the Application of Digital Technologies in Education and Research, vol. 2494, pp. 258–265 (2019) http://ceur-ws.org/Vol-2494/
8. Zhang, C., Lu, Y.: Study on artificial intelligence: the state of the art and future prospects. J. Ind. Inf. Integr. **23**, 100224 (2021)
9. Mukhamediev, R.I., Popova, Y., Kuchin, Y., Zaitseva, E., Kalimoldayev, A., Symagulov, A., et al.: Review of artificial intelligence and machine learning technologies: classification, restrictions, opportunities and challenges. Mathematics. **10**(15), 2552 (2022)
10. Pedro, F., Subosa, M., Rivas, A., Valverde, P.: Artificial Intelligence in Education: Challenges and Opportunities for Sustainable Development (2019)
11. Hussain, A.: Use of artificial intelligence in the library services: prospects and challenges. Libr. Hi Tech News. **40**(2), 15–17 (2023)
12. Chen, L., Chen, P., Lin, Z.: Artificial intelligence in education: a review. IEEE Access. **8**, 75264–75278 (2020)
13. Chen, X., Zou, D., Xie, H., Cheng, G., Liu, C.: Two decades of artificial intelligence in education. Educ. Technol. Soc. **25**(1), 28–47 (2022)
14. Schuur, F., Rezazade Mehrizi, M.H., Ranschaert, E.: Training opportunities of artificial intelligence (AI) in radiology: a systematic review. Eur. Radiol. **31**, 6021–6029 (2021)
15. Fiok, K., Farahani, F.V., Karwowski, W., Ahram, T.: Explainable artificial intelligence for education and training. J. Defense Model. Simul. **19**(2), 133–144 (2022)
16. Maity, S.: Identifying opportunities for artificial intelligence in the evolution of training and development practices. J. Manag. Dev. **38**(8), 651–663 (2019)
17. Morreale, F., Bahmanteymouri, E., Burmester, B., Chen, A., Thorp, M.: The unwitting labourer: extracting humanness in AI training. AI Soc. **39**(5), 2389–2399 (2024)
18. Basan, E., Abramov, E., Gladkov, N., Lapina, M., Lapin, V., Anita, E.A.M., Kumar, S.: UAV security analysis framework. In: Kumar, S.K., Kim, J.H., Bansal, J.C. (eds.) Fourth Congress on Intelligent Systems. CIS 2023. Lecture Notes in Networks and Systems, vol. 865, pp. 379–391. Springer, Singapore (2024) Tkachenko, V., Kuzior, A., & Kwilinski, A. (2019). Introduction of artificial intelligence tools into the training methods of entrepreneurship activities. Journal of Entrepreneurship Education, 22(6), 1-10
19. Dubarry, M., Beck, D.: Big data training data for artificial intelligence-based Li-ion diagnosis and prognosis. J. Power Sources. **479**, 228806 (2020)

20. Zhai, X., Chu, X., Chai, C.S., Jong, M.S.Y., Istenic, A., Spector, M., et al.: A review of artificial intelligence (AI) in education from 2010 to 2020. Complexity. **2021**(1), 8812542 (2021)
21. Holmes, W., Tuomi, I.: State of the art and practice in AI in education. Eur. J. Educ. **57**(4), 542–570 (2022)
22. Spielberg, S.: AI: Artificial Intelligence (2001)
23. Grigoryan, K., Olefirenko, E., Basan, E., Lapina, M., Mecella, M.: Analysis of Security Problems in Groups of Intelligent Sensors. In: Lecture Notes in Computer Science (including subseries Lecture Notes in Artificial Intelligence and Lecture Notes in Bioinformatics) 12689 LNCS, pp. 26–37 (2021)
24. Kormakova, V.N., Klepikova, A.G., Lapina, M.A., Rugelj, J.: ICT competence of a teacher in the context of digital transformation of education. In: CEUR Workshop Proceedings, vol. 2914, pp. 138–150 (2021)
25. Kormakova, V., Klepikova, A., Musaelian, E., Baybikowa, G., Lapina, M.: Formation of ICT competencies of postgraduate students of Teacher Education based on interactive techniques: CEUR Workshop Proceedings SLET-2019. In: Proceedings of the International Scientific Conference Innovative Approaches to the Application of Digital Technologies in Education and Research, vol. 2494, pp. 11–21 (2019)
26. Lapina, M., Shiriaev, E., Babenko, M., Istamov, I.: High-speed convolution Core architecture for privacy-preserving neural networks. Program. Comput. Softw. **50**, 417–424 (2024). https://doi.org/10.1134/S0361768824700282
27. Rusanov, M., Babenko, M., Lapina, M., Sajid, M.: Identification of exploited unreliable account passwords in the information infrastructure using machine learning methods. Big Data Cognit. Comput. **8**, 159 (2024). https://doi.org/10.3390/bdcc8110159

Research and Comparison of Analytical Platforms in the Field of Education

Maria Lapina[1(✉)], Dmitry Lukyanov[1], Marina Tokmakova[1], and Mohammed Saleh Ali Muthanna[2]

[1] North Caucasus Federal University, Stavropol 355017, Russia
mlapina@ncfu.ru

[2] Tashkent State University of Economics, Tashkent 100066, Uzbekistan

Abstract. The article discusses educational platforms, examines their role in the modern educational process through a comparative analysis. The concept of a data analysis platform, types and advantages of use in the educational process are described. The purpose of the study is to identify students' opinions on the use of analytical platforms in educational activities. During the study, a survey of students working with data analysis platforms was conducted. Empirical methods were implemented in the process of conducting an online survey of students within the framework of the discipline "Methods and Ways of Data Analysis" - students were given the task of analyzing the data provided by the teacher using the following platforms: Python, Tableau and Anaconda. During the work, students were tasked with conducting a comprehensive assessment of the data of specialized data analysis platforms, including an analysis of their functional characteristics, ease of use, effectiveness in solving the assigned tasks and accessibility. This task was set for second-year students of two groups: the specialty "Information Security of Automated Systems" of the North Caucasus Federal University, and the specialty "Computer Science" of Prince Mohammad Bin Fahd University. It was found that students currently believe that the best application for working with machine learning is Anaconda, which provides an interface and supports a huge number of languages.

Keywords: machine learning · artificial intelligence · data analysis · analytical platforms · quality of education · educational process · questionnaire survey

1 Introduction

The relevance of this study is because today in many areas a person has to deal with a huge amount of data, the analysis of which takes a huge amount of time. Analytical platforms can speed up the information processing process in order to reduce the time required by the user for a complete and detailed analysis. The use of these solutions in education will allow students and teachers to receive and analyze information faster and more efficiently: data on academic performance, attendance, etc.

Data analytics platforms are a comprehensive software solution designed to collect, store, process, visualize and interpret data in order to extract useful information and support decision making.

M. Lapina et al. (Eds.): DET 2024, LNNS 1616, pp. 277–288, 2025.
https://doi.org/10.1007/978-3-032-04365-8_28

In the modern world, machine learning is widely used in education and science. For example, the Duolingo application, designed for learning foreign languages, uses machine learning to optimize the difficulty level of tasks and tests for users. Large companies such as Google and Microsoft are developing their own platforms for implementing machine learning in the educational process. People working and actively using machine learning highlight a number of the best platforms and programs for using machines to analyze and predict data. These are the Python programming language [1, 3], software for visualizing data in tables and charts Tableau [4, 6], Anaconda [8]. Comparison and identification of the best platform is the main objective of the study.

Over the past decade, online content developers have offered students various educational projects, and in this regard, publications have begun to appear examining the compliance of platforms with modern educational standards [9–11].

2 Related Works

The use and impact of information technology (IT) on education is one of the key issues at present. Many studies on the impact of IT on education show an increase and active use of modern technologies in the pedagogical sphere.

In their research [12] they investigated the impact of modern teaching methods on education and pedagogy. The results of the work showed that modern teaching methods are aimed at developing personal qualities in students, thanks to which they will be able to gain their own experience and analyze professional activities that suit them.

The report [13] collected research on the use of digital technologies in modeling and managing complex systems in various fields, including education, as well as their protection and security. This report contains the results of many years of scientific research demonstrating the impact of information technology on various areas of life.

The study [14] examines the impact of information technology on education from the perspective of students at Russian universities. This work includes a survey of thousands of students at higher education institutions, which shows the best solutions for creating a comfortable and effective digital environment.

3 Analytical Platforms

Python is a high-level programming language. Because this language is a general-purpose language, it is widely used to work with machine learning. A large number of libraries allow you to analyze matrices, tables, files, use different machine learning models, and visualize them in diagrams, tables, graphs. The main libraries are NumPy (for working with mathematical calculations), Pandas (for data analysis and processing), Scikit-learn (for using machine learning models), Matplotlib (for creating and displaying various types of graphs). This set of libraries is enough to work with machine learning in Python. Next, you need to learn how to get started with machine learning using Python.

Working with machine learning in Python opens up a wide range of possibilities thanks to the availability of libraries and frameworks (Figs. 1, 2, 3 and 4).

```
import numpy as np

# Создание данных
X = np.array([[1], [2], [3], [4], [5]])
y = np.array([1\.5, 3.5, 5.5, 7.5, 9.5])
```

Fig. 1. Creating data for analysis.

The first step is to create data for analysis. The information that will be fed into the machine learning model will then be processed and analyzed so that the artificial intelligence can eventually find a pattern in it and predict the data.

Next, you need to create a machine learning model. Using different libraries, you can choose a huge number of machine learning models, all of them are suitable for different tasks: classification, clustering, etc.

```
from sklearn.linear_model import LinearRegression

# Создание модели
model = LinearRegression()
model.fit(X, y)
```

Fig. 2. Model creation and training.

Then comes data forecasting. The model uses the analyzed data and finds certain patterns in it, which allows it to forecast data with high accuracy.

```
# Прогнозирование
new_X = np.array([[6]])
pred = model.predict(new_X)
print(pred)  # [11\.5]
```

Fig. 3. Function for data forecasting.

The final step is to evaluate the model for accuracy. The model is evaluated by comparing the data, comparing the data that the user fed to the machine and the data that the model predicted itself. The ratio of the data obtained yields the accuracy percentage of the machine learning model.

Python is considered one of the easiest programming languages, which allows anyone interested to easily underset and how to work with machine learning using a programming language. Due to its simplicity, this language can be used to analyze and predict data in educational institutions. For example, to monitor the progress of students without using manual calculations, predicting student grades by changing any conditions of their learning.

Tableau is a software for data visualization and business analytics [4, 6]. This platform provides a wide choice of how to visualize data: graphs, charts, distribution by regions of

```
from sklearn.metrics import mean_squared_error

# Прогнозирование на обучающих данных
y_pred = model.predict(X)

# Оценка модели
mse = mean_squared_error(y, y_pred)
print(f"Среднеквадратичная ошибка: {mse}")
```

Fig. 4. Evaluation of the model accuracy.

the world, pie charts. Due to the built-in machine learning models, this program can solve small and simple tasks related to forecasting and data analysis. Based on the information that the user uploads, the platform is capable of creating dashboards and presentations [5].

First, you need to upload the data to Tableau, after uploading the file; the platform will display it on the desktop (Figs. 5, 6 and 7).

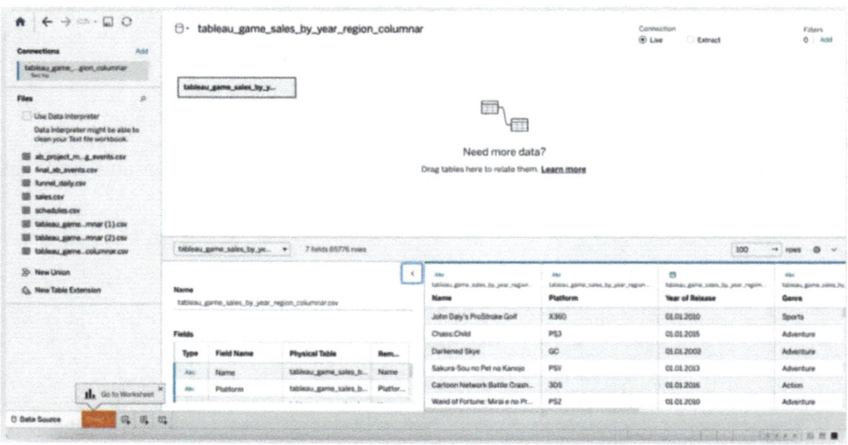

Fig. 5. File on the Tableau desktop.

To visualize data, you need to open a new sheet and select the data you want to visualize. Change the type of charts and graphs as needed.

Tableau is a platform for visualizing ready-made data and is able to use machine learning to predict simple and small data. Its main drawback is that this program is paid.

Anaconda is a distribution of Python and R programming languages that combines libraries related to data science and machine learning. This platform helps save a huge amount of time, as it is completely created for machine learning. All the necessary libraries for working with artificial intelligence are already installed [7, 8].

We launch the application, select the language in which the machine learning model will be created and the obtained data will be visualized.

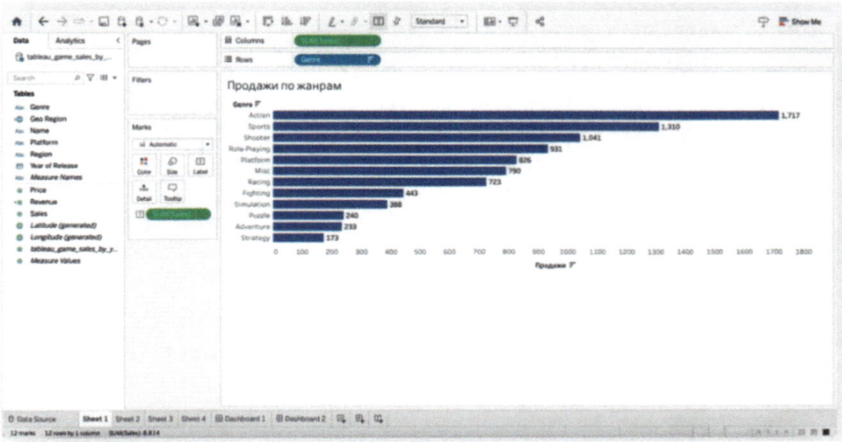

Fig. 6. Data visualization.

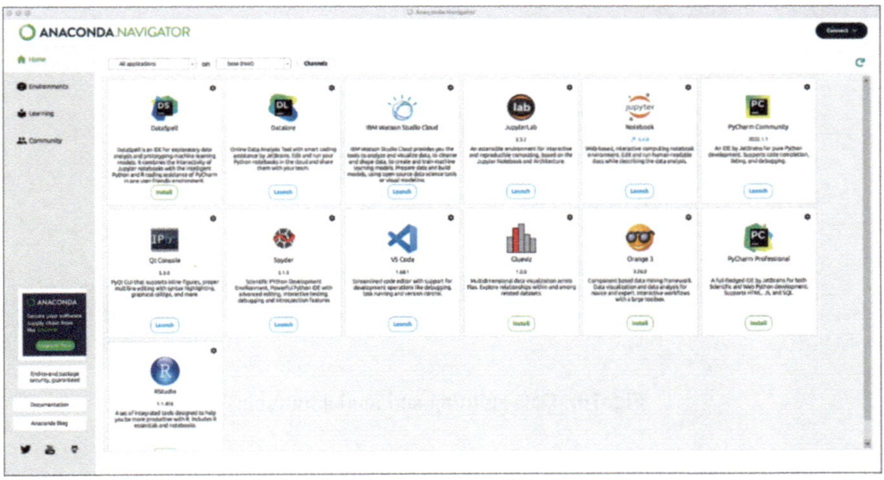

Fig. 7. Language selection menu.

Next, you need to create a development environment. Creating a development environment is not mandatory, but it is highly desirable to avoid conflicts between different versions of the programming language and libraries (Fig. 8).

In Python, we load libraries and data. The choice of libraries depends on the goal the user wants to achieve and the type of visualization of the predicted data (Fig. 9).

We divide the data into training and testing, create a machine learning model and load the data into it. Dividing the data is necessary so that the model learns and predicts data of the same type, to avoid conflicts and errors (Fig. 10).

```
conda create -n ml_env python=3.11
conda activate ml_env
conda install numpy pandas scikit-learn matplotlib seaborn jupyter
```

Fig. 8. Creating a virtual environment.

```
# Импорт библиотек
from sklearn.datasets import load_iris
from sklearn.model_selection import train_test_split
from sklearn.ensemble import RandomForestClassifier
from sklearn.metrics import accuracy_score
import pandas as pd

# Загрузка датасета
iris = load_iris()
X = iris.data
y = iris.target
```

Fig. 9. Loading libraries and data.

```
# Разделение на обучающую и тестовую выборки
X_train, X_test, y_train, y_test = train_test_split(X, y, test_size=0.3, random_state=42)

# Создание и обучение модели
model = RandomForestClassifier(n_estimators=100, random_state=42)
model.fit(X_train, y_train)
```

Fig. 10. Data splitting and model training.

As a final step, we evaluate the accuracy of the model. We perform the accuracy of the model in order to understand whether further and more detailed adjustment of our model is necessary or not (Fig. 11).

```
# Предсказание и оценка качества
y_pred = model.predict(X_test)
accuracy = accuracy_score(y_test, y_pred)

print(f"Точность модели: {accuracy:.2f}")
```

Fig. 11. Evaluation and output of model accuracy.

Based on the research conducted, it is necessary to compare platforms for using machine learning and identify the best one (Table 1).

Table 1. Comparative characteristics of analytical platforms.

Characteristic	Python	Anaconda	Tableau
Interface	–	+	+
Availability	For free	For free	Paid
Flexibility	One language	A large number of languages	Data visualization
Libraries	Imported	Built-in	No
Audience	Schoolchildren, students studying the language	Students studying the language at a more advanced level	Schoolchildren, students, business analysts, large companies
Feedback	–	+	+
Platform type	Local	Local	Hybrid
AutoML	Yes (via libraries)	Yes (via libraries)	Yes, but limited

Based on this table, we can conclude that the best application for working with machine learning is Anaconda, which provides an interface and supports a huge number of languages.

4 Methodological Basis of the Research

When working on the article, theoretical methods were used, which consisted of analyzing and systematizing the provisions of scientific literature on the use of analytical platforms in the educational process of modern universities.

Empirical methods included a questionnaire method, as well as pedagogical observation and analysis of the results of university students' activities.

During the study, a survey of students working with data analysis platforms was conducted. Students in the discipline "Methods and methods of data analysis" were given the task of analyzing the data provided by the teacher using the platforms presented above: Python, Tableau and Anaconda. During the work, students were tasked with conducting a comprehensive assessment of specialized data analysis platforms, including an analysis of their functional characteristics, ease of use, effectiveness in solving the tasks and availability.

This task was set for 2nd year students of two groups: the specialty "Information Security of Automated Systems" of the North Caucasus Federal University (group No. 1), and the specialty "Computer Science" of Prince Mohammad Bin Fahd University (group No. 2).

A total of 115 people took part in the study (group No. 1–50 people, group No. 2–65 people).

In response to the questionnaire question "Indicate your gender," the majority of respondents indicated that they belong to the male gender – 74% (group No. 1) and 88% (group 2), and to the female gender – only 26% (group No. 1) and 12% (group 2) (Figs. 12 and 13).

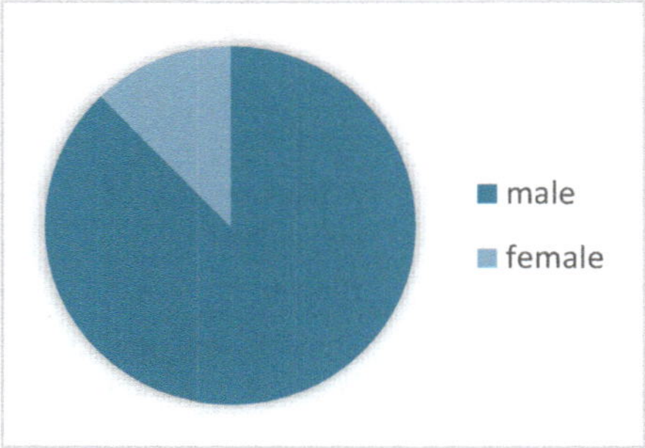

Fig. 12. Gender of respondents in group No. 1.

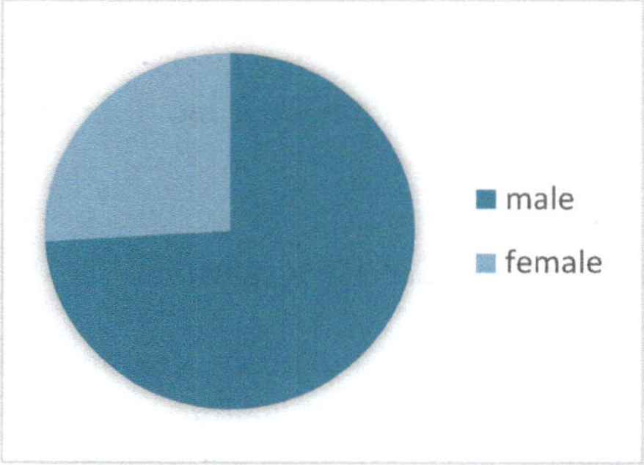

Fig. 13. Gender of respondents in group No. 2.

The work included the following stages: 1. selecting a data set for analysis; 2. installing software; 3. setting up platforms for data analysis; 4. completing the task set by the teacher.

During the work, the students studied and analyzed the principles of operation of the provided analytical platforms; upon completion of the work, a survey of students was conducted, which displays the number of students who chose one or another platform to complete the assigned task. The survey results are shown in the diagrams (Figs. 14 and 15).

The survey results show that the vast majority of students choose Anaconda as their analytical platform: 60% (group 1) and 62% (group 2).

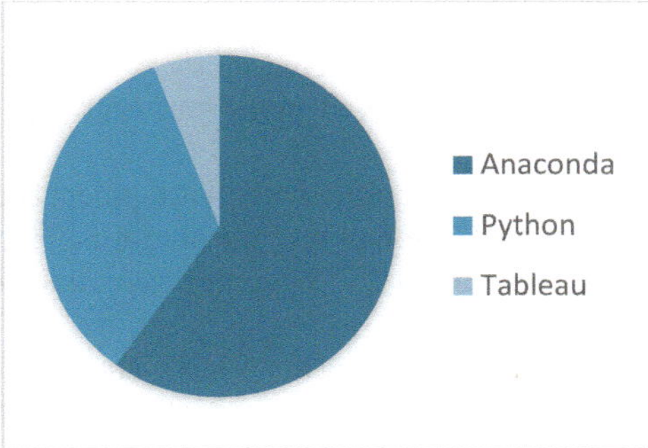

Fig. 14. Choice of platform by respondents of group No. 1.

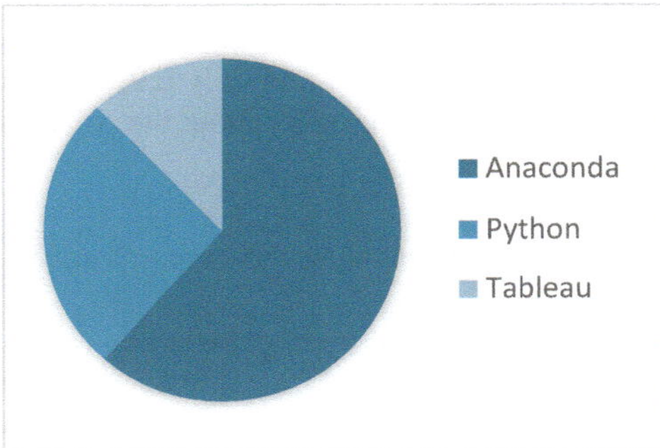

Fig. 15. Choice of platform by respondents of group No. 2.

5 Research Results

In order to identify students' opinions about the Anaconda analytical platform used to complete their assignment, a series of questions were compiled, to which all students who chose Anaconda as their main platform responded.

The very first question concerned the platform functionality. To the question "how often did you use the platform's capabilities to complete the task?" 62.61% (72 people) quite often used the capabilities offered by the platform to complete the task. 13.04% (15 people) also answered positively "more yes than no". A small portion of respondents, namely 8.7% (10 people), answered "more no than yes". And a categorical "no" was answered by 15.65% (18 people) of students.

The next question concerned the ease of use of the interface. The majority of respondents 49.57% (57 people) answered "yes". 12.17% (14 people) tend to give a positive assessment "Rather yes". However, 26.09% (30 people) experienced difficulties "Rather no", and 12.17% (14 people) were categorically dissatisfied "no".

To the question "Is the platform functional enough?" 86.96% (100 people) confidently answered "yes". 8.7% (10 people) chose the option "Rather yes". Only 4.34% (5 people) expressed doubt "Rather no".

On the question "Did the platform work stably?" 52.17% (60 people) did not notice any problems. 9.57% (11 people) encountered minor problems. But 13.04% (15 people) and 25.22% (29 people) encountered platform instability "rather no" and "no" respectively.

The next question touched on difficulties with development and revealed that 26.09% (30 people) had difficulties with development of the platform. 43.48% (50 people) admitted "rather yes". The remaining 30.43% (35 people) were divided between "rather no" and "no".

The question "Are there enough teaching materials?" showed that 69.57% (80 people) indicated their lack "no" and only 17.39% (20 people) and 13.04% (15 people) satisfied the need for teaching materials "yes" and "rather yes".

When asked "Have you used other platforms?" it was found that 78.26% (90 people) answered "yes", while 21.74% (25 people) did not use any alternatives.

The next question shows whether the educators would recommend this platform to their friends. 67.82% (78 people) answered "yes", 18.26% (21 people) "rather yes", 6.96% (8 people) were unsure "rather no", and only 6.96% (8 people) refused to give recommendations "no". Respondents' opinions about the Anaconda analytical platform is shown in Table 2.

Table 2. Respondents' opinions about the Anaconda analytical platform.

Questions/Statements	Yes		More yes than no		More no than yes		No	
	N/P	%	N/P	%	N/P	%	N/P	%
How often did you use the platform's capabilities to complete the task?	72	62.61	15	13.04	10	8.7	18	15.65
Was the app's interface intuitive?	57	49.57	14	12.17	30	26.09	14	12.17
Is the platform functional enough to handle your task?	100	86.96	10	8.7	5	4.34	0	0
Was the platform stable enough during operation?	60	52.17	11	9.57	15	13.04	29	25.22
Were there any difficulties in mastering the platform?	30	26.09	50	43.48	17	14.78	18	15.65
Are there enough training materials?	20	17.39	15	13.04	0	0	80	69.57
Have you used other platforms to complete the task?	90	78.26	0	0	0	0	25	21.74
Can you recommend this platform to your classmates?	78	67.82	21	18.26	8	6.96	8	6.96

*N/P – number of people

Based on the results in Table 2, the platform received a high efficiency rating, but requires improvement in the following:

– Stability of work (28% of negative responses)

- Usefulness of the interface (40.5% of criticism)
- Training materials (62.5% of dissatisfied).

6 Conclusion

In conclusion, it is worth saying that the use of modern visualization methods in classes develops the individual pace of educational and cognitive activity of each student. Organizing a course using an educational platform will allow students to distribute time for studying the material and completing assignments, to understand the information studied more deeply and to additionally study material not included in the main program.

Therefore, in order to ensure complete acquisition of knowledge, skills and abilities, develop the students' learning and intellectual abilities, make lessons livelier and more interesting, it is advisable to use distance learning platforms. They, in turn, allow: to take modern education to a completely new level; to raise the status of the teacher; to expand the possibilities of descriptive support for learning; to create conditions for the implementation of various forms of training, types of activities and effective organization of control over the knowledge, skills and abilities of students; to promote and improve the development of creative activity.

The widespread use of online educational services is due to the intensive informatization of education. The existing experience of researchers and teachers in working with educational platforms, the analysis of their capabilities allow us to conclude that the resources are of interest to the educational system, and also develop and improve their content for students every year. The use of online services should not replace traditional education. However, the resources in question are a good tool for supplementing the educational process, contributing to the formation of both subject results and meta-subject, in particular, regulatory universal educational actions.

When choosing a platform, it is important that each teacher relies on their own experience and knowledge, as well as the opinion of researchers and other teachers, and takes into account the age characteristics of students. It is important to understand that the resource should be available not only on a computer, but also on tablets and phones. For the convenience of students, it is desirable that this provision covers all levels of training to ensure continuity.

References

1. Official Python Website: https://www.python.org
2. Machine Learning Training in Python: https://sky.pro/wiki/python/obuchenie-mashinnogo-obucheniya-na-python/
3. Python: https://ru.wikipedia.org/wiki/Python
4. Official Tableau Website: https://tableau.com
5. What is the Tableau system and why do analysts need it. https://practicum.yandex.ru/blog/funkcii-i-vozmozhnosti-tableau/
6. Tableau: https://ru.wikipedia.org/wiki/Tableau
7. Official Anaconda Website: https://www.anaconda.com
8. Anaconda: https://ru.wikipedia.org/wiki/Anaconda_(Python_distribution)

9. Boriseyko, V.A.: pos$sibilities of an open digital educational environment in the context of organizing training in a comprehensive school. Soc. Sociol. Psychol. Pedagog. **2** (2020)
10. Goryunova, M.A., Lebedeva, M.B.: Mobile learning in the context of the implementation of the Federal State Educational Standard. Man Educ. **4**(49) (2016)
11. Karimov, F.R., Kaimova, M.B.: Important aspects of using interactive educational platform. Universum: Tech. Sci. **5**(110), 41–43 (2023)
12. Lapina, M., Tokmakova, M., Movzalevskaya, V., Karabasheva, M., Balakrishnan, D.: Analysis of modern methods and approaches using it in education and pedagogy. In: Lapina, M., Prakasha, G., S., Grigoriev, S. (eds.) SLET 2022: International Conference on Innovative Approaches to the Application of Digital Technologies in Education and Research. SLET 2022. Lecture Notes in Networks and Systems, pp. 40–53. Springer, Cham (2024)
13. Prakasha, G.S., Lapina, M., Balakrishnan, D., Sajid, M. (eds.): Digital Technologies in Modeling and Management: Insights into Education and Industry. IGI Global (2024)
14. Golovanova, I.I., Alipichev, A.Y., Ayupov, T.A., Fazlyeva, F.A., Khasanova, E.N.: Digital educational environment and online learning format through the lens of students: pros and cons. Educ. Self Dev. **17**(3), 202–221 (2022)

Distance Learning Technologies in Supporting and Developing Gifted Children

Elena P. Linnik[1,2(✉)], Ivan I. Linnik[2], Tatiana V. Shushara[1], Vladimir V. Aleshov[1], and Marina E. Tokmakova[3]

[1] Kherson State Pedagogical University, Kherson 273003, Russia
eplinnik@yandex.ru
[2] V.I. Vernadsky Crimean Federal University, Simferopol 295007, Russia
[3] North Caucasus Federal University, 355017 Stavropol, Russia

Abstract. The article explores the potential of utilizing distance learning technologies to support and encourage the development of gifted children. Analyzed are the advantages of distance learning technologies as exemplified by the Junior Academy of Sciences, such as individualization of learning, access to unique educational resources and flexibility in organizing the educational process and development of independent work skills. Surveying the students of the Junior Academy of Sciences have shown a high degree of satisfaction with the new format, which provides flexibility and accessibility. The research analyzes the advantages of blended learning formats allowing for effectively combining the traditional teaching methods with the modern distance technologies. Particular attention is paid to the adaptation of curricula, the use of interactive platforms, as well as support for students during the transition to the new learning formats. The article considers the methods for improving the process of teaching gifted children mathematics, computer science and physics, as well as the possibilities of distance learning technologies in achieving this goal. Developed and tested is a website for the distance support of students in mathematics, computer science and physics. Recommendations have been compiled for the creation and use of teachers' personal websites in order to improve the process of teaching mathematics, computer science and physics.

Keywords: Gifted Children · Distance Learning Technologies · Digitalization of Education · Individualization of Learning · Potential Development · Educational Platforms

1 Introduction

In the process of modern society's development, special attention is paid to the problems of organizing the educational process aimed at gifted children [1–3]. The teaching and learning process at educational organizations should focus on flexibility, dynamicity and personalization [4–6]. The education system faces the task of developing self-study skills (independent acquisition of knowledge) in students, which is essential under the conditions of constant and rapid information updating [7].

M. Lapina et al. (Eds.): DET 2024, LNNS 1616, pp. 289–300, 2025.
https://doi.org/10.1007/978-3-032-04365-8_29

The distance learning technology is one of the most effective and promising systems for developing a person's productive thinking, intellectual potential and creative design. Teachers need to create a learning environment motivating students to independently search for information, process and exchange it. One of the ways to improve the educational process is to use distance learning technologies in the teaching practice [8].

The distance learning technology allows for individualization and differentiation of educational activities, and provides the opportunity to organize the teacher - student interaction in a new way [9]. In this educational system, the student is the subject of learning and an active and equal participant in the educational process.

In distance learning, the traditional communication links are replaced by the telecommunication means providing interactive communication between the participants of the learning process and provide access to study information [10]. Thanks to the intensive development of distance learning, the use of technical means in the educational process has increased significantly.

2 Theoretical Framework for Enhancing Mathematics Education by Means of Distance Learning Technologies

The theoretical foundation for enhancing the teaching of mathematics through the use of distance learning technologies lies in leveraging learning resources and platforms to create more engaging, flexible, and personalized learning experiences.

By integrating these technologies, educators can provide students with access to interactive resources, real-time feedback, and collaborative opportunities, which can deepen their understanding of mathematical concepts. This approach aligns with modern educational theories that emphasize active learning, self-paced progress, and the development of critical thinking and problem-solving skills. Distance learning technologies also enable teachers to tailor instruction to individual student needs, making mathematics education more inclusive and effective.

In today's educational landscape, a key requirement for modern educators is the ability to utilize information technologies to ensure a high standard of training and to support the individual development and professional growth of students [11]. Among these technologies, distance learning stands out, rooted in the application of information and communication tools. Its primary advantage lies in its independence from geographical boundaries and the physical distance between educators and learners. This has given rise to the concept of "distance learning," which emphasizes education delivered remotely [12].

In the field of distance learning technologies, a unified conceptual framework has not yet been fully developed. Academic literature often uses a variety of terms, such as distance learning, distance education, online learning, e-learning, and distance learning technologies. Despite the differing terminology, all these concepts share a common foundation: the use of modern information technologies to facilitate learning at a distance [13].

The concept of "distance learning" can be examined in detail, as researchers often assign varying meanings to its definition. Based on these interpretations, the following approaches can be distinguished:

- Distance Learning as a Method. In this approach, participants in the educational process are physically separated and rely on audio, video, internet, and satellite communication channels to facilitate interaction.
- Distance Learning as a Process. This perspective defines it as a purposeful and organized process of interactive communication between the educator (subject) and the learner (object), utilizing specific learning tools regardless of their location in space or time.

When learning is conducted through telecommunications, participants engage remotely, leading to internal changes in their understanding and skills. Distance learning primarily relies on internet technologies and resources to enable this interaction.

Tracing the history and evolution of distance learning globally, it can be defined as a form of educational activity conducted by institutions using media technologies and diverse educational resources. This approach transcends the traditional boundaries of time and space associated with conventional educational settings [14].

Distance learning places the main emphasis on the targeted two-way activity of participants in the educational process, who also interact with the training resources (see Fig. 1).

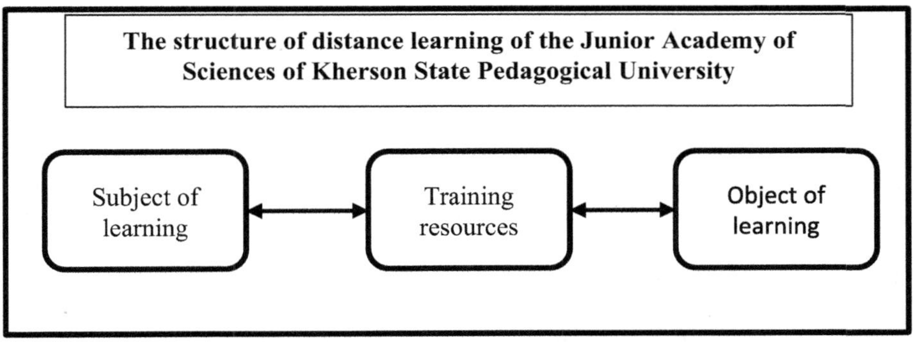

Fig. 1. The structure of distance learning of the Junior Academy of Sciences of Kherson State Pedagogical University.

Proceeding from this diagram, it can be noted that training resources form the link between the interaction of the subject and the object of learning. Apart from presenting the main material, the teacher provides the student with a list of additional literature oriented to his/her needs. The use of the necessary training resources places certain demands on teachers: they need be able to navigate the Internet freely, select accessible electronic texts and use audio and video material in the teaching process [15]. To effectively organize the process of teaching mathematics to gifted children using distance learning technologies, it is essential to outline the goals of their implementation:

1. Meeting Educational Needs. Addressing the demand for accessible and flexible education tailored to the unique needs of gifted learners.

2. Enhancing the Education System. Leveraging modern technologies to facilitate interaction between teachers and students, independent of geographical location or time constraints.
3. Providing Unrestricted Access. Ensuring free and open access to information and educational resources at any time.
4. Personalizing Learning. Individualizing the educational process, emphasizing independent student work, and improving learning outcomes through the integration of innovative technologies.
5. Promoting Knowledge and Literacy. Encouraging the dissemination of knowledge across the population and elevating overall educational and cultural standards.
6. Ensuring Quality Control. Establishing the necessary conditions for implementing systems to monitor and maintain the quality of education [16].

Based on the above is the following table revealing the essence of distance learning (Table 1). Distance learning is considered from the point of view of information and productive approaches.

Table 1. The essence of distance learning of the Junior Academy of Sciences of Kherson State Pedagogical University.

Comparison criterion	Information approach	Productive approach
The essence of distance learning	Distributing high-quality information using information technologies	Developing original products through the use of information technologies
Characteristics of knowledge (as a result of activity)	Knowledge is presented as transmitted information	Knowledge emerges as the outcome of productive activities
The role of the learner	The student acts as a receiver of information	The student becomes a creator of new information
Features of educational materials	Relies on textbooks and traditional face-to-face teaching methods	The teacher serves as a collaborator and facilitator, organizing the learning environment and process
The essence of distance learning	Distributing high-quality information using information technologies	Developing original products through the use of information technologies

From the perspective of the information approach, the primary objective of education is to create an information model within the student's mind, equipped with a specific vocabulary, and to cultivate their ability to perceive and extract information from both natural and artificial sources [17]. In contrast, productive learning is centered on achieving tangible outcomes from the student's efforts. These outcomes can be internal, such as qualitative improvements in the student's cognitive processes, or external, such as the creation of a material or informational product. The characteristics of productive activity can serve as valuable tools for assessing the development of a child's skills, cognitive abilities, and their capacity to plan and execute tasks effectively [18].

3 Distance Learning Methods at the Junior Academy of Sciences of Kherson State Pedagogical University

A key aspect of categorizing distance learning is the system of methods and techniques employed in the educational process. These methods can be classified based on the mode of communication between the teacher and the student [3].

According to this classification, distance learning methods (or techniques) can be grouped into the following categories:

1. Self-Study Methods. These methods involve students primarily interacting with educational resources independently. These resources include printed materials, audio and video content, as well as various learning tools accessible through telecommunication networks, such as computer-based training systems.
2. One-on-One Interaction Methods. In this approach, the student engages directly with a single teacher or another student. This interaction is facilitated through tools like smartphones, email, instant messaging platforms, and similar communication technologies.
3. Teacher-Centered Methods. These methods emphasize the active role of teachers in guiding and facilitating the learning process, often through structured instruction and support. They present the educational materials, while students are supposed to comprehend. These methods are most common in the traditional educational system, but with the development of modern information technologies they have received a new path to development. For instance, audio and video lectures can be supplemented by their electronic versions distributed over computer networks using bulletin board systems. An electronic lecture can present a selection of scientific articles, as well as a list of educational materials for further discussion.
4. The methods involving active interaction of all the participants in the educational process. With the development of telecommunication technologies, the importance of these teaching methods has increased significantly, since time and space boundaries no longer matter. Within this group of methods, the research and problem-based teaching methods are widely used, and the technology of personality-oriented teaching is implemented more effectively. Depending on their individual abilities and inclinations, students are given the opportunity to achieve certain results in various fields of knowledge, as well as to comprehend the knowledge they receive for further application in life.
5. Project-Based Method. These methods involve a comprehensive learning process in which students showcase their abilities in planning, organizing, and self-monitoring their educational activities. The outcome is typically the creation of a tangible product or the exploration of a specific phenomenon.
6. Problem-Based Learning Method. This approach centers on engaging students with complex cognitive tasks that stimulate their critical thinking and problem-solving skills. Teachers act as facilitators, providing guidance and directing students' reflections to ensure productive outcomes.

7. Research Method. This method is characterized by clearly defined and meaningful objectives for all participants, a diverse range of research techniques, and a meticulously planned and logically structured framework [19].

We shall proceed to examine in detail the main areas for which distance learning is relevant.

- Tutors specializing in a certain academic discipline. Recently, online tutoring using the Skype program has become increasingly popular, which facilitates educational materials' explanation, as well as knowledge control and testing. There is also an offline learning option, in which students study pre-prepared video lessons. The playback of video lessons is accompanied by teachers' explanations and note-taking on an interactive whiteboard [14].
- Preparatory courses for schoolchildren. As a rule, such courses are developed by teachers of higher educational institutions in order to prepare students for admission. The learning process begins with the familiarization with the theoretical material, watching a video lesson, then it is necessary to complete a test assignment in writing and then send it by e-mail to the teacher. After checking the assignment's correctness, assigning points and writing a review with an analysis of the tasks, the teacher sends a response letter to the student. Some platforms allow communication with the teacher during the learning process via chat, which significantly increases activity during classes [14].
- Distance higher education. Nowadays, there are numerous higher education institutions (both foreign and domestic ones) implementing distance higher education programs. For the most part, training takes place by means of sending educational materials to students, while students take the final exam in person. This direction is relevant for students already working in their chosen specialty, so they can combine both activities without detriment to their main activity.
- Educational projects for advanced training of already established specialists. This direction is accompanied by high motivation of students working in their fields of specialization. The advantage of such training is the absence of the need to leave the workplace, spend money on employees' travel and accommodation. There is also a significant difference in the cost of distance courses and face-to-face classes. Now many higher education institutions implement distance course programs for schoolchildren, students and professionals in various activity areas.

4 Forms of Distance Learning at the Junior Academy of Sciences of Kherson State Pedagogical University

Along with the teaching methods, the forms of distance learning should be considered (Fig. 2).

Similar to traditional education, the most commonly used formats include lectures, seminars, laboratory sessions, tests, and similar activities.

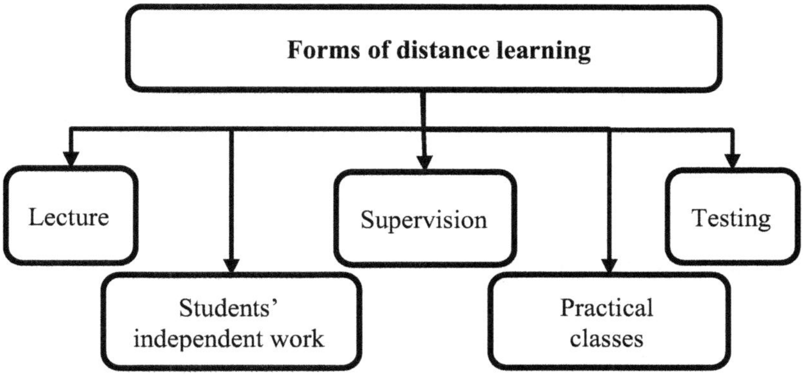

Fig. 2. The forms of distance learning at the Junior Academy of Sciences of Kherson State Pedagogical University.

5 Distance Learning Training Resources at the Junior Academy of Sciences of Kherson State Pedagogical University

They determine the content of training - the composition, structure, material of educational information, as well as a set of practical tasks, assignments and exercises. In traditional training, technical means of training are widely used; in relation to distance learning, we should talk about means of new information technologies.

According to [14], the following distance training resources can be singled out (see Fig. 3).

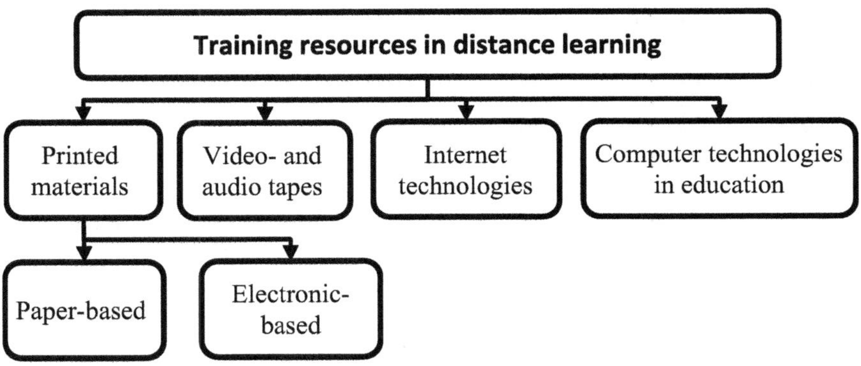

Fig. 3. The classification of training resources.

Printed materials can be presented in a variety of formats to accommodate different learning environments and technological access (See Table 2). These formats include:

Computer training technologies. Multimedia computer aids are based on hypertext, which gives the educational material a clear structure and also provides access to any

Table 2. The formats of printed materials.

Format	Description
Paper Version (Printed)	Traditional hard copies of textbooks, handouts, and other educational resources that can be physically distributed to students. This format is accessible without the need for electronic devices
Electronic Version on External Media	Digital copies of learning materials stored on external devices such as USB drives, CDs, or DVDs. This allows students to access the materials on their computers or other devices without requiring an internet connection
Electronic Version Sent by Email	Digital textbooks and resources that are distributed directly to students via email. This method ensures that students receive the materials quickly and can store them on their personal devices for offline access
Freely Accessed Electronic Version on the Internet	Online resources and e-textbooks that are available for free on educational websites or platforms. This format provides easy access to a wide range of materials and can be updated regularly to ensure the most current information is available

part of it. This is the main advantage of computer technologies over printed materials, video and audio tapes [11].

Internet technologies. This group of training resources can also be divided into the following subgroups:

- *World Wide Web* is a system of placing information on the Internet using hypertext. This allows placing textbooks, testing programs, in particular, those featuring multimedia elements in the public domain on the Internet.
- *File Transfer Protocol (FTP)* Thanks to this network service, students can download any files (textbooks, computer tests, etc.) from the computer of the educational institution to their computers.
- *E-mail is the electronic mail.* In distance learning, e-mail is used for communication between students and teachers, as well as for sending any files.
- *Teleconferences* are used for conducting supervising students online at a certain time.
- *Videoconferences* are used for conducting joint educational activities but require high-quality technical equipment and high-speed data transfer.

Based on the analysis of various sources of information, the following distance learning training resources can be singled out [5]:

- Textbooks (available in both printed and digital formats);
- Online textbooks hosted on the internet;

- Computer-based training systems;
- Audio-based learning materials;
- Video-based learning materials;
- Remote laboratory training;
- Simulators accessible online;
- Databases with remote access;
- Electronic libraries accessible remotely;
- Learning resources utilizing expert systems;
- Learning resources incorporating geographic information systems (GIS);
- Learning resources leveraging virtual reality (VR) technology.

Based on the presented classifications of distance learning resources, we can single out the resources used in the process of teaching mathematics, computer science and physics:

1. the electronic mail (e-mail);
2. chatting (real time correspondence);
3. videoconferencing;
4. interactive communication on the Internet;
5. web-conferences, bulletin boards, registration forms, tests, etc.;
6. FTP-servers, file archives, etc.

Mobile applications can also be one of the means of distance learning. Since most students have mobile devices, they can be used for transmitting and receiving the necessary information. In educational practice, particularly for teaching mathematics, computer science, and physics, various applications and tools are widely utilized. These include GeoGebra, MathHelper, MalMath, Engineering Calculator + Graphs, Trigonometric Circle, Euclidean, Pythagoras, Mathematical Tricks, and software packages from Google.

It is worth mentioning that for the effective use of distance learning resources in the educational process, it is essential to consider the technical component, as well as the advantages and disadvantages of distance learning support in mathematics, computer science and physics.

6 Conclusions

The results of working with gifted children are the students' victories at various conferences, festivals, competitions and olympiads. As a result, students successfully completed and presented their final scientific projects in the fields of Mathematics, Computer Science, and Physics at the scientific and practical conference titled "World of Discoveries: Information Technologies, Mathematics, Physics, Engineering, and Inventions" organized by the Junior Academy of Sciences.

In order to facilitate the effective functioning of the Junior Academy of Sciences, certain information technologies are used allowing for a full cycle of the educational process. The following resources are used within the framework of educational activities: the Telegram messenger, the Yandex.Telemost web application and the Sferum web platform.

As part of this study, surveys were conducted to gauge students' perceptions of the blended learning format. The responses to the following questions were as follows (see Fig. 4).

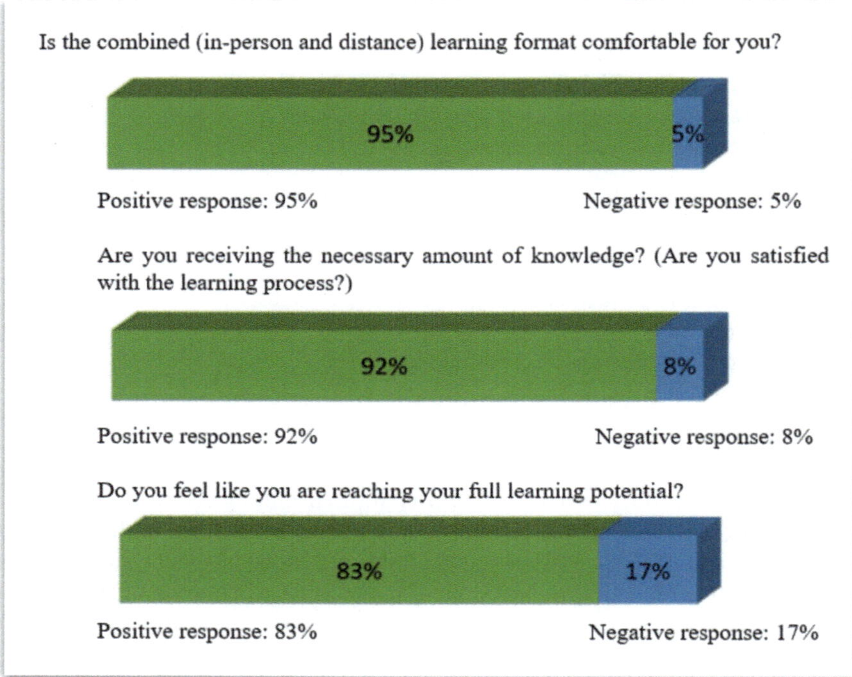

Fig. 4. Students' perceptions of the blended learning format.

The survey's results lead to the conclusion that the combined (in-person and distance) learning format satisfies the majority of students at the Junior Academy of Sciences. Consequently, the use of distance learning technologies in the educational process with gifted children allows for solving the following didactic problems: obtaining the necessary solid knowledge by students using information technologies, developing the cognitive abilities, forming independence in searching for information to deepen and expand knowledge in a certain field, and then apply it in practice. However, when creating an electronic educational course, it is necessary to foresee all possible options in advance in order to minimize the number of shortcomings and malfunctions in the training system.

Acknowledgements. This study was conducted under the directive of the Ministry of Education of the Russian Federation for 2025, as part of a state-funded research project titled "The Development of a Strategy for Building Workforce Capacity by Working with Gifted Children in Russia's New Regions, in the Context of Integration into the Socio-Economic Space of the Russian Federation."

References

1. Konopko, E.A., Pankratova, O.P., Abdullaev, D.A., Ediev, A.M., Taran, V.N.: Digital education toolkit and an overview of distance learning resources. In: Selected Papers of the V all-Russian Scientific and Practical Conference with International Participation “Distance Learning Technologies” (DLT 2020), Yalta, Crimea, September 22-25, pp. 374–382 (2020)
2. Serole, S.R.: The art of promoting blended learning through blended pedagogies in higher education classrooms. E. J. Human. Arts Soc. Sci. **6**(3), 274–287 (2025). https://doi.org/10.38159/ehass.2025633
3. Yadav, D.N.: The impact of digital learning on education. Int. J. Multidiscip. Res. Arts Sci. Technol. **2**(1), 24–34 (2024). https://doi.org/10.61778/ijmrast.v2i1.34
4. Tabassum, B., Moin, M., Abbas, Q., Kumbhar, M., Khan, M.: The impact of blended learning on student performance. J. Educ. Soc. Stud. **5**, 360–371 (2024). https://doi.org/10.52223/jess.2024.5217
5. Khenner, E.K.: Professional knowledge and professional competencies in higher education. Educ. Sci. J. **2**(20), 9–31 (2018). https://doi.org/10.17853/1994-5639-2018-2-9-31
6. Arinushkina, A.A., Morozov, A.V., Robert, I.V.: Contemporary Challenges in Education: Digitalization, Methodology, and Management, p. 392. IGI Global (2023)
7. Men, Q., Gimbert, B., Cristol, D.: The effect of self-regulated learning in online professional training. Int. J. Mob. Blend. Learn. **15**, 1–17 (2023). https://doi.org/10.4018/IJMBL.318225
8. Koppel, M., Akiva, N., Dagan, I.: A Corpus-Independent Feature Set for StyleBased Text Categorization. In: Proceedings of IJCAI’03 Workshop on Computational Approaches to Style Analysis and Synthesis, Acapulco, Mexico (2003)
9. Koppel, M., Argamon, S., Shimoni, A.: Automatically categorizing written texts by author gender. Literary Linguistic Comput. **17**(4), 401–412 (2002)
10. Stamatatos, E.: Ensemble-based author identification using character n-grams. In: Proceedings of the 3rd International Workshop on Text-Based Information Retrieval, pp. 41–46 (2006)
11. Schler, J., Koppel, M., Argamon, S., Pennebaker, J.: Effects of age and gender on blogging. In: Procceedings of AAAI Spring Symposium on Computational Approaches for Analyzing Weblogs (2006)
12. Sebastiani, F.: Machine learning in automated text categorization. ACM Comput. Surv. **34**(1), 1–47 (2002)
13. Ibrahim, M.M., Nat, M.: Blended learning motivation model for instructors in higher education institutions. International journal of educational technology in higher. Int. J. Educ. Technol. Higher Educ. **16**(1), 1–21 (2019). https://doi.org/10.1186/s41239-019-0145-2
14. Lutsenko, L., Testova, I., Chirkov, M., Taran, V., Mezentseva, O.: Methodical specifics of the subject distance camp organization “young cryptograph”. In: Lapina, M., Prakasha, G.S., Grigoriev, S. (eds.) International Conference on Innovative Approaches to the Application of Digital Technologies in Education and Research. SLET 2022. Lecture Notes in Networks and Systems, vol. 1222. Springer, Cham (2025). https://doi.org/10.1007/978-3-031-78776-8_21
15. Taran, V.N., Usmanov, M.I.: Learning Technologies through Augmented Reality Projects in the VI International Scientific and Practical Conference “Distance Learning Technologies” (DLT 2021).Yalta, Crimea, September 20-22, vol. 3057, pp. 253–259 (2021) ceur-ws.org/Vol-3057/paper30.pdf
16. Mantulenko, V.V.: Perspectives on the use of the digital footprint in higher education. In: Teacher XXI Century, vol. 3-1, pp. 32–42 (2020). https://doi.org/10.31862/2073-9613-2020-3-32-42
17. Shevchenko, G., Kalashnikova, A., Rybakova, A., Brynza, S.: Tasks, functions, and basic activities of an online tutor. In: CEUR Workshop Proceedings, “SLET 2020 - Proceedings

of the International Scientific Conference on Innovative Approaches to the Application of Digital Technologies in Education", pp. 285–291 (2020)
18. Arancibia Muñoz, M.L., Halal Orfali, C.: The role of learning management systems (LMS) in educational environments: didactic and technical use. In: ICERI2018 Proceedings, pp. 10473–10479 (2018). https://doi.org/10.21125/iceri.2018.0976
19. Kulikova, T.A., Poddubnaya, N.A.: Training of the future teacher in applying digital tools for the organization of project activities. In: Selected Papers of the V all-Russian Scientific and Practical Conference with International Participation "Distance Learning Technologies" (DLT 2020), Yalta, Crimea, September 22-25, pp. 391–397 (2020)

Theoretical Aspects of Using Augmented Reality-Based Technologies in the Process of Teaching Mathematics

Kostyantyn A. Malyshenko, Dmitriy V. Anashkin, and Marina V. Malyshenko(✉)

V.I. Vernadsky Crimean Federal University, Simferopol, Russia
iriska_3640@mail.ru

Abstract. Scientific and technological progress significantly influences education system development today. This paper explores integrating augmented reality technology into mathematical education. Our research proceeded in two phases: first, developing software that generates model meshes for augmented reality applications; second, demonstrating this application's implementation and discussing its advantages in mathematics instruction. The paper examines and illustrates the use of ArUco marker-based augmented reality technology. We provide theoretical background on developing AR-based mobile applications for mathematics education at both school and university levels. The research presents a Python-based mobile application that utilizes device cameras to implement augmented reality when studying "Surfaces of the second-order," enhancing visualization and comprehension of complex mathematical concepts.

Keywords: Augmented reality · Mathematics · Spatial Thinking · Marker · Visualization · Python · ArUco

1 Introduction

Modern societal development places special demands on educational processes and quality, creating new challenges for governments to form effective education policies. Technological advancement is fundamentally transforming education systems, prompting exploration of new approaches to traditional teaching functions and raising expectations for educator competencies. Traditional educational forms are increasingly superseded by more progressive, timely methods, as continued reliance on outdated approaches significantly diminishes educational outcomes.

Students face evolving expectations as well. Growing information volumes necessitate better structuring and new motivational frameworks to enhance knowledge retention. Integrating emerging technologies into education addresses many related challenges by improving process organization.

The effectiveness of new educational technologies is validated through digital learning environment creation—interconnected elements, teaching methodologies, and

M. Lapina et al. (Eds.): DET 2024, LNNS 1616, pp. 301–313, 2025.
https://doi.org/10.1007/978-3-032-04365-8_30

approaches built on contemporary developments. Global education trend analysis confirms this direction. Mobile technology integration represents the most prevalent and proven effective development vector in modern education.

Artificial intelligence systems and advanced robotics have become essential components, simplifying both daily activities and student knowledge acquisition. Virtual and augmented reality technologies are increasingly prevalent in educational settings, enabling deeper immersion and facilitating comprehension of extensive materials.

This study aims to examine and systematize modern approaches to mathematics instruction using contemporary technologies, particularly augmented reality. Visual supplementation of foundational materials represents a hybridization of classical mathematical teaching approaches adapted to information society requirements.

Additionally, the article explores approaches to developing augmented reality mobile applications using open-source Python libraries, with the objective of creating a client interface that implements augmented reality functionality through webcam imagery.

2 Building of Augmented Reality Application

2.1 Subject Area Overview

Scientific and educational advancement depends on processing massive datasets beyond human capacity. The ever-growing volume of information makes its storage and processing increasingly challenging. Humans lack the ability to efficiently and accurately structure this data. Furthermore, beyond simply perceiving and memorizing information, effective analysis is crucial. To solve the problem of preserving and accumulating various kinds of knowledge, therefore, the use of special tools is required. One of these is visualization, which is a tool aimed at converting large amounts of information into images. Such a transformation facilitates its perception by humans and compresses the original information, since instead of voluminous text data, diagrams, algorithms, drawings, and graphs are formed. Thus, visualization becomes one of the most important tools for delivering final messages to the general public, and is used in various spheres of life. It becomes an element of transition from the verbal component to the visual one.

Educational visualization techniques enable modeling of studied phenomena and processes, making their essence, structure, and interrelationships more comprehensible and improving material retention. Furthermore, these approaches enhance assessment methods for knowledge, skills, and abilities evaluation. Significantly, visualization establishes meaningful connections between theoretical foundations and practical applications, bridging abstract concepts with real-world implementation.

AR technologies are new mobile technologies that can be actively used in education. At its core, it is a computer technology that allows you to overlay virtual objects on the real world, thus recreating their presence in the space around the user.

Virtual reality technologies act as a separate object, despite the fact that in the literature they are often replaced by the concept of "augmented reality". However, these concepts are not identical. Virtual reality completely immerses users in artificial environments, blocking out the real world entirely. Augmented reality, conversely, combines digital imaging and multimedia with artificial intelligence to enhance real-world perception.

AR applications feature two main display types based on immersion depth:

1. See-through displays allow users to directly observe the real world, providing maximum immersion. Virtual objects are optically superimposed on the user's natural perception of reality, creating straightforward AR imagery. This technology is commonly implemented in aviation systems through specialized panels or helmets.
2. Monitor-based displays overlay computer graphics onto live or recorded video using digital or analog methods. This approach doesn't require specialized display hardware but offers minimal presence. The AR image is generated indirectly. This display type is prevalent in mobile applications that utilize device cameras to capture real-world imagery.

Second, technology is more applicable for the implementation of augmented reality in mathematics lessons, since it uses the digital environment not as a whole, but in the direction of each specific student, which gives the required result at the level of mastering the curriculum. In [11], such type of presentation is marked as most convenient for a wide range of users and can be convenient to create educational applications such as "personal assistant".

In Russia, augmented reality is classified as a core digital technology essential for digital economy development, as established in the "Digital Economy of the Russian Federation" Government Program of July 27, 2017.

When applied to mathematics education, augmented reality enables visualization of three-dimensional geometric shapes. This technology allows various interactive manipulations of these shapes, including movement, rotation, scaling, 3D model overlays, multi-angle viewing, connection and disconnection of virtual objects, and result analysis and interpretation.

Currently, four augmented reality types exist:

1. Marker-based AR, which uses specific visual markers.
2. Markerless AR, which functions without predefined markers.
3. Projection-based AR, utilizing projected light forms.
4. Superimposition-based AR, which replaces real objects with augmented views.

A computer AR-marker is understood as an object in physical space that specialized software detects and analyzes to render virtual objects in the correct position and orientation within the augmented environment. In theory, any figure (object) can be a marker. But in practice, we are limited by the resolution of the webcam (phone), the features of color rendering, lighting, and the computing power of the equipment (after all, everything happens in real-time, and therefore must be done quickly), and therefore a black and white marker of a simple form is usually chosen. As a rule, it is a rectangle or square with an identifier-image inscribed inside (Fig. 1).

A fiducial marker is an object placed in the visual field of the rendering system that appears in the resulting image for use as a reference point or measure. It can be either something placed in/or on the subject of the image or a mark or a set of marks in the grid of an optical device.

In robotics and machine vision, a special class of fiducial visual markers called ArUco has become widespread.

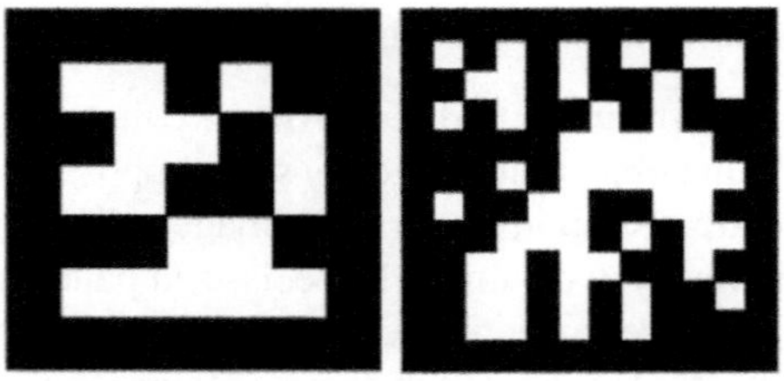

Fig. 1. Examples of different-size ArUco markers.

The first message about ArUco markers appeared in the work of S. Garrido-Jurado et al. "Automatic generation and detection of highly reliable fiducial markers under occlusion" [10]. ArUco stands for the Augmented Reality University of Cordoba.

An ArUco marker is a black square with distinct borders, containing a unique white pattern inside. The black border enhances visibility, making it easier to detect. These markers can be generated in various sizes, chosen based on the object's or scene's dimensions to ensure reliable recognition. If smaller markers are difficult to detect, increasing their size improves visibility and detection accuracy. Markers are built on a simple concept: there is the n-size binary grid that is filled with "0" and "1" which means that "1" is a filled cell and 0 is an empty cell. Each cell is called a bit. A dictionary of ArUco markers consists of some such markers. To detect a marker in a photo or video stream, one must first find the corners of the marker, correct perspective distortion to obtain an image that looks as if the marker were being seen from above, divide the resulting image into a grid, and compare this grid of white and black cells with the given dictionary to find out if there is a match.

ArUco markers were originally developed to solve the problem of evaluating camera orientation when used in a variety of applications, including augmented reality. But later, these markers became the main link of machine vision technologies and are now used everywhere for programming and automating complex robotic systems. This makes them indispensable in high-tech applications and also allows them to become part of the machine vision client libraries in widely-used programming languages such as Python [10].

It should be noted that the search for such markers on the image consists of several, in their way, important stages, at each of which the image from the camera (disk) is passed through a class of color filters to detect binary markers on it. A detailed scheme of the processing of the original image is shown in Fig. 2. Image capture is proceeding after finding of marker to impose a prebuilt model and make it rotatable.

Notably, markers function without requiring additional data uploads for new objects and can typically be applied to real-world objects with ease. However, a drawback of this approach is the necessity of placing a marker with relevant information on each real object intended for use in the AR system.

After the marker is found on the image, the next stage of building augmented reality begins. First, the perspective of the found marker is converted into some specific instructions that will be used in the next stages of work. After that, a search is made for key points on the marker, which will be further used for tracking an augmented reality

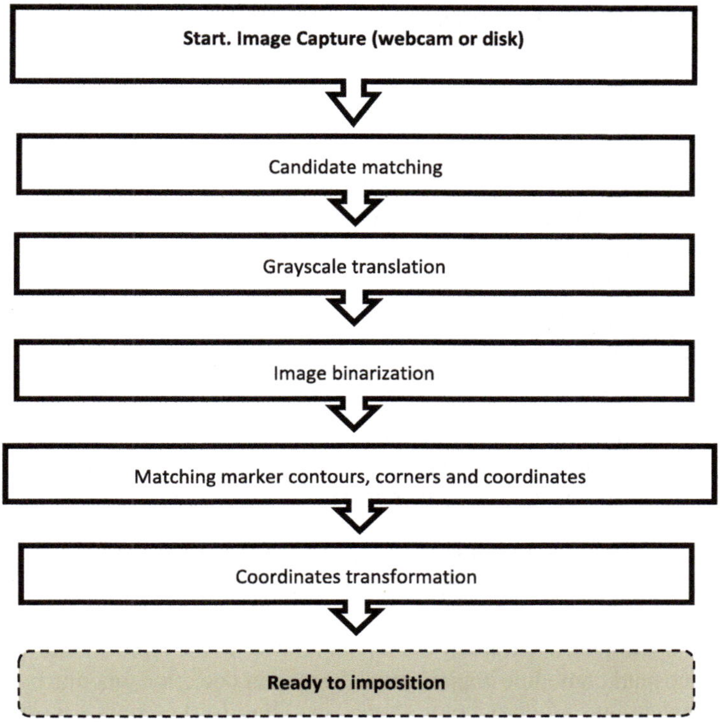

Fig. 2. Loop algorithm of marker processing.

object. The final stage is the construction of an augmented reality scene based on the perspective, position, and homography of the fiducial marker in space.

After the presentation of the theoretical basis, we can directly proceed with the implementation of the project.

2.2 Project Demonstration

Integrating augmented reality (AR) into education is technologically straightforward. For instance, mobile device cameras can trigger the display of text explanations, photos, videos, or combinations thereof when pointed at specific targets. We propose a structure for educational AR applications based on the following: using distinct markers; detecting these markers with a computer or mobile device; and overlaying additional information on the device's screen. AR implementation via mobile devices requires the following:

- AR mode should be able to be disabled, with the ability to transfer normal operating mode where the surrounding space is replaced with any "empty" background or static 3D scene, and the content is positioned to make it easier to view from the device.
- The 3D scene is initially prepared by the performer for correct display in AR mode and can have its controls depending on the scenario. Displayed content should be

organic "Framed" with backgrounds and other background objects to look natural in 360° viewing conditions.

When implementing the demo application, the authors followed the aforementioned requirements.

The work on the application began with the installation of the necessary libraries in the Python development environment. In the process of work, the OpenCV computer vision library was used as the main tool. With the help of this library, the broadcast of the image from the smartphone camera was realized, as well as the preparation of the layout of the application user interface.

Functionally, the project was divided into several interconnected modules: camera calibration module, three-dimensional object module, ArUco label module, constant value module. Each of them was subsequently integrated into the main script by an internal "import" command.

The camera calibration module included methods of the OpenCV library for reading the resolution parameters of a webcam of a mobile device and assessing the suitability of its quality characteristics for further work. Classical calibration methods involve assessing the position of the device in space, calculating the focal length and other optical characteristics required for the project. The obtained values were saved into a module containing the constants used in the project.

The module of a three-dimensional object consisted of parameters that set the size, position, and the imposed texture of the model relative to the ArUco marker.

The ArUco marker module implemented the marker detection function by translating the camera image into a binary color space, searching for key points (corners, closed areas), as well as evaluating the marker position relative to the observer by enumerating all possible rotations.

This was followed by the stage of preparing a 3D model of augmented reality for imposition on the image from the webcam. The second-order volumetric model of the surface was created in the computer simulation system "Blender". Special filters and texturing tools were applied to the modeled surface for correct display on the screen. The most important stage of the work was preparing the area for applying the fiducial marker and placing the readable image at a convenient angle without access to excess sunlight, which can make it difficult to recognize during color filtering (Figs. 3 and 4).

Figure 4 demonstrated the successful operation of the augmented reality algorithm, complementing the topic "Surfaces of the second-order". The object is projected onto the prepared area above the tracked ArUco marker. As you can see in the image, the algorithm ignores the extraneous marker shown for the example in the same plane.

The result of executing the algorithm built into the application is a ready-made augmented reality scene with a ready-made interface. The convenience of the application lies in the simplicity of its revision and implementation. Also, the application in the presented form allows the user to independently load and visualize mathematical and other volumetric objects, which increases the convenience of their detailed study.

The developed application opens up wide possibilities for its applied application in real mathematics lessons. Here are some examples:

1. Integral calculus. The application can be extremely useful in the presentation of the topic "multiple integrals". Overlaying a coordinate grid on a figure and adding

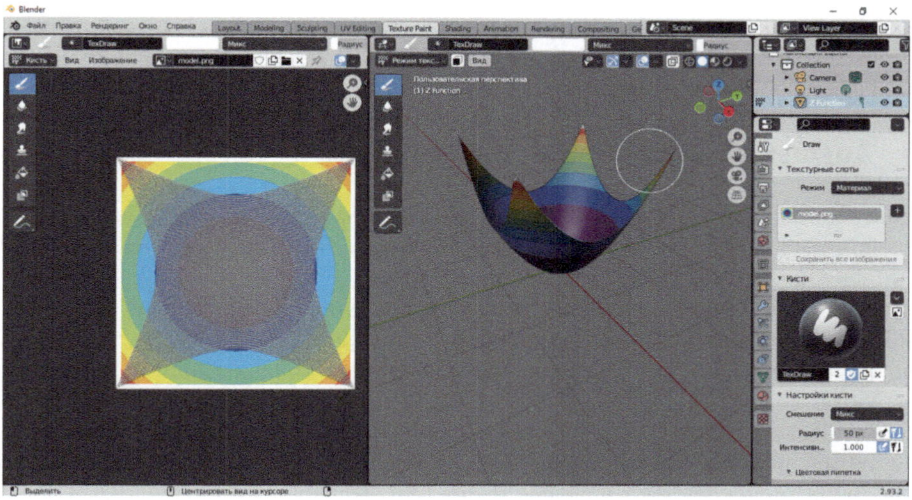

Fig. 3. Modeling and rendering of a superimposed three-dimensional surface in the program "Blender".

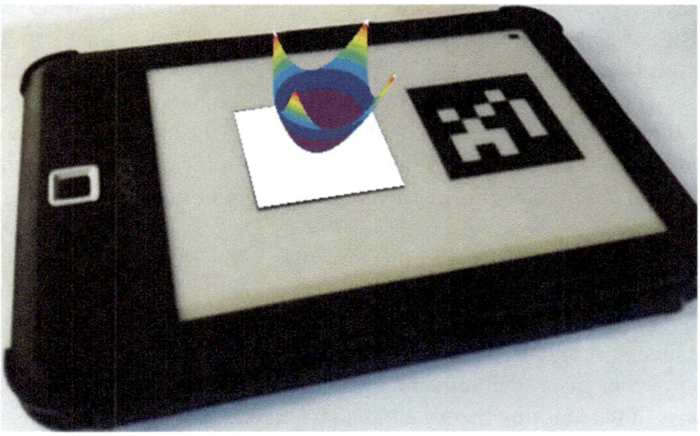

Fig. 4. Demonstration of the successful operation of the augmented reality algorithm.

a function to limit the area under consideration along the axes will allow you to visualize the integrated area for a better understanding of the theoretical material.

2. The application can be used to expand the visual component of the "surface of revolution" theme.
3. The application can supplement the presentation of the basic course of geometry, in particular, topics involving the consideration of three-dimensional spatial figures, for example: "sections of planes", "sections of volumetric surfaces", "section of geometric figures", can be visually supplemented.

4. Surfaces of the second order. Surfaces defined in a parametric view can be easily visualized and examined for maximum (minimum). It is also convenient to observe the breakpoints and boundaries of such functions.

3 Augmented Reality in Education Process

Education actively integrates advancements in science and technology, particularly modern information technologies, demonstrating their value in the learning process. One such cutting-edge technology is augmented reality (AR).

In the field of public relations, augmented reality and its applications can be considered primarily as end-to-end information technology [5]. So, by education, this technology can be considered as an auxiliary tool for the implementation of educational activities.

In education, various electronic resources and publications play a crucial role. One key advantage of these resources is their ability to integrate multimedia elements directly into the learning context. Augmented reality (AR) technology holds significant potential for enhancing education at both school and university levels. It diversifies content delivery methods, making lessons more engaging and information more accessible.

A primary goal of educators is to spark students' interest in their subjects. AR technology facilitates this by visualizing complex or abstract concepts through 3D models, making them easier to understand. In many professional fields, theoretical knowledge alone is insufficient for developing essential skills and competencies. AR helps bridge the gap between theory and practice, offering interactive and immersive learning experiences.

AR-based education can be classified as a form of virtual learning, enabling institutions to provide high-quality, personalized education regardless of location. Modern technologies further enhance this approach, ensuring an effective and student-centered learning experience.

The use of AR applications in education has been studied for a long time. One established methodology for integrating augmented reality into learning is the "MARE" (Mobile Augmented Reality Education) framework [13]. MARE is structured into three hierarchical layers: foundation, function, and results. It is designed to analyze different teaching approaches, from theoretical instruction to the development of practical skills.

Regardless of the learning theory adopted by an educational institution—such as the associative-reflex theory of learning or the problem-activity learning theory—the primary criterion for higher education is the practical application of acquired knowledge. The functional level of MARE is influenced by the student's approach, actions, and interactions with educational resources. "MARE" methodology schematic representation is shown in Fig. 5.

Since the average age of teachers in Russia is about 50 years old, for this reason, it is difficult for many teachers to accept modern IT technologies that are familiar to students. Application of the "MARE" method allows bridging the "digital gap", the so-called digital divide between generations.

Studies on AR in secondary education demonstrate its potential for enhancing learning. For example, one experiment showed that students using an AR physics application achieved a deeper understanding, higher performance, and increased interest and

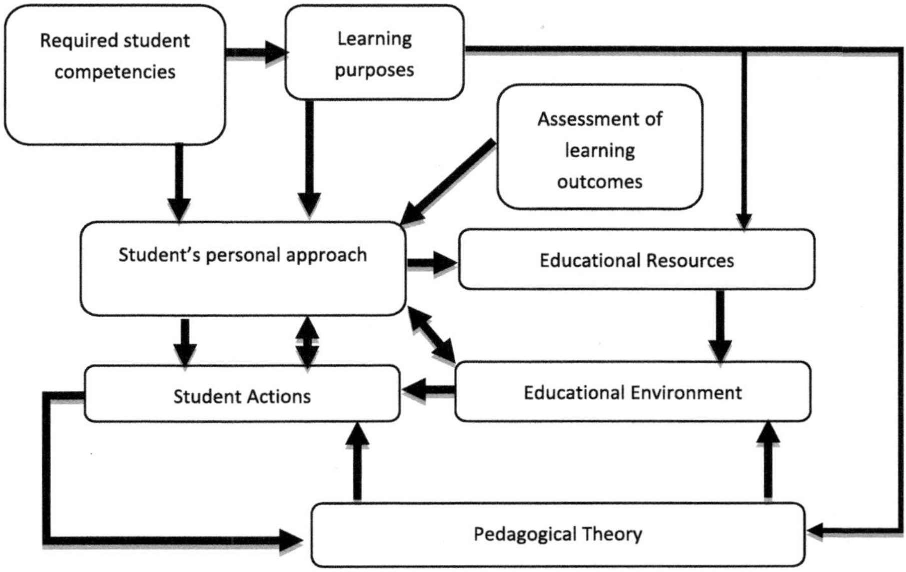

Fig. 5. "MARE" methodology schematic representation.

engagement compared to a control group [7]. Research on the motivational and cognitive impacts of AR [8] found it particularly effective for low-achieving students and beneficial for explaining abstract concepts.

Analysis of AR's use in higher education also reveals significant potential. Many researchers note increased student motivation and engagement with AR learning applications [9]. The ongoing development of mobile technologies contributes to the growing demand and popularity of mobile learning, which supports flexible, accessible, and personalized education, ultimately enhancing learning quality when implemented effectively.

Based on the reviewed literature [7–9], six key features of using AR applications in education emerge:

- Combination of the real and virtual environment. Augmented reality, unlike virtual reality, allows you to superimpose an interactive scene directly on the real environment, which allows the user not to lose sight of important events.
- Interactivity. Modern techniques allow you to create the effect of real presence and interaction with the virtual scene, which enhances the experience of using augmented reality applications.
- Visibility. The virtual scene, when directly superimposed, is easily distinguishable and does not impede the perception of any kind of information embedded in it.
- Complementarity. Augmented reality applications are difficult to develop, but some ready-made solutions allow you to modify them to fit your needs to get the maximum professional benefit from them.
- Three-Dimensional object (scene) imagination.

- Timeliness. Augmented reality technologies allow, if necessary, to quickly switch to another type of activity, without losing the overall process of the lesson.

The full inclusion of augmented reality in education sets itself some prospects that can be achieved only if all existing technical, financial, and other requirements are met [11], specifically:

1. At this stage, AR can only serve as a supplement to the educational process rather than a full replacement. The transition should be gradual, supported by relevant research, focus group studies, and careful analysis of outcomes. While AR has the potential to enhance certain aspects of education, excessive reliance on this technology could improve one area while unintentionally hindering another.
2. Working with such technologies has many factors that must be considered when forming a program. It is necessary to agree on all points of the curriculum so that they do not interfere with each other.
3. Insufficient number and variety of training programs. They are actively being worked on, but right now they are completely insufficient to be used as an alternative to classical teaching methods and tools.
4. Today, not only in Russia but also in the world, there are very few people who are well versed in augmented and mixed reality. And for the application of technology in the educational field, it is necessary to train teachers who will work with children, adolescents, and students. This also takes time.

4 Discussion

Augmented reality interfaces offer diverse and promising real-world applications, varying in their specific parameters. The choice of interface (input and output) depends on the task's requirements, available equipment, and technological feasibility. A well-designed, user-friendly interface is crucial for user satisfaction and commercial success. Current technological advancements enable the development of sophisticated and intuitive interfaces that offer immersive experiences.

From a psychological perspective, integrating AR technologies into education requires consideration of key factors: creating optimal conditions for information processing; structuring phased perception, beginning with concrete or materialized objects and progressing to internalization through stages of "socialized speech" and "private speech"; and systematically developing perceptual skills (observation, generalization, conscious awareness, etc.) throughout the learning process.

The problem of using augmented reality has become the subject of academic discussions not so long ago, referring to its short existence. However, augmented reality has already managed to show itself in several areas, from medicine to large-scale design. There are some contradictions on the way to introducing augmented reality into education that still needs to be eliminated.

In 2016, the Department of Informatics and Information Technologies of the Humanitarian and Pedagogical Academy of Crimean Federal University. V. I. Vernadsky in Yalta also addressed this topical issue. The department introduced the "Virtual and Augmented Reality Project Development Studio." One of the project's goals is developing augmented reality programs for Crimean secondary and higher education, along with methodologies for adapting software and hardware to the specific requirements of AR tools. The project is supposed to develop methods of introducing augmented reality technologies in education and application development applicable not only to the field of education but also to the sectoral needs of enterprises and organizations of Crimea.

The authors of the work plan to refine the existing solution for its subsequent possible implementation in educational programs in institutions of secondary general education, as well as during student internship in the discipline of "mathematics" and related to it.

5 Conclusions

One of the significant results of the work was the development of a mobile application that allows you to visualize pre-selected three-dimensional objects of mathematical nature as part of an augmented reality scene. Such an application can be used in real math lessons in secondary school for a deep specialized understanding of the subject. The main advantage of such an application is its free distribution and the ability to build, which will significantly reduce the additional costs of educational organizations to support this type of software. It is also important to note that this approach to classroom work significantly contributes to the development of interdisciplinary relationships and increases students' adaptability in the context of developing or using computer software. Thus, this may be a reason to revise the content of the computer science discipline in special and general educational institutions and add relevant thematic aspects and integrations to the existing curriculum.

The second important point of the work is the presentation of the theoretical and methodological foundations of the use of augmented reality in mathematics lessons. The existing approaches to the organization of educational activities using end-to-end learning technologies are considered and classified.

Research indicates that integrating augmented reality applications into mathematics education, particularly when teaching fundamental concepts, can enhance learning and knowledge acquisition. The introduction of this technology should be implemented under the guidance of a teacher, since it is necessary to follow the logic of mastering the elements of the course, as well as maintain a balance between theoretical material and its development using modern technologies.

While there are many positive aspects, at the same time, the use of technology has a number of disadvantages. Among them, it is worth highlighting the most significant:

1. Lack of understanding of the possibilities of using additional reality in the learning process and incorrect assessment of technology.
2. The need for appropriate technical capacities, namely high–speed data processing. The pace of development of society indicates that in the near future, the use of the proposed technologies will become an integral part of learning. Virtual reality and augmented reality technologies contribute to the development of spatial thinking and open up new teaching opportunities. They significantly increase the visibility of the materials and make it possible to understand their place in practice already at the stage of studying the theoretical foundations.
3. Problems with the psychological perception of technologies and the unavailability of adapting educational materials to them, as well as problems with the availability of methodological materials that facilitate the use of such technologies.

4. The need for strict control and stage-by-stage implementation and use of technology in teaching activities in order to avoid situations associated with uncertainty of perception and avoidance of reality as a result of an overabundance of technology.
5. Unavailability of technologies for acquisition by educational institutions due to their high cost.
6. The lack of the necessary information base for the implementation of augmented reality technologies – its collection requires significant time and labor costs.
7. However, these disadvantages will be eliminated as technology evolves. Virtual and augmented reality are effective teaching aids and their role in the educational process will be enhanced in the future. As they are actively disseminated, the learning processes will move to a new, more effective level.

The obtained results can be implemented into the educational process, ensuring increased motivation and interest of students, consolidation of acquired knowledge and their deepening.

The preservation and further development of this area in the educational process requires additional efforts on the part of higher authorities to introduce digital education, as well as the interest of teachers in an interesting and non-standard form of presentation of the material presented by them. Only through the continuous integration of existing solutions in the field of augmented and virtual reality can the educational system provide affordable, high-quality and visual education for every citizen.

References

1. Kravtsov, A.A., Loiko, V.I.: Features of the implementation of marker tracking on mass mobile devices. Polythem. Netw. Electron. Sci. J. Kuban State Agrarian Univ. **113**, 430–444 (2015)
2. Abylkassymova, A.E., Zhumagulova, Z.A.: On some aspects of the content of mathematical education in schools and pedagogical universities. All Russ. Sci. Pedagog. J. "Science and School". **1**, 157–160 (2016)
3. Alfayadh Rasha, G.H.: The use of augmented and virtual reality technologies in mathematical education. Sci. Leader. **3**(204), 104–110 (2025)
4. Derbush, M.V.: Innovative approaches to the use of information technologies in the process of teaching mathematics. Contin. Educ. XXI Cent. **2**(30) (2020)
5. Order of the Government of the Russian Federation of July 28, 2017. N 1632. http://static.government.ru/media/files/9gFM4FHj4PsB79I5v7yLVuPgu4bvR7M0.pdf
6. Dyulicheva, J.J.: On the application of augmented reality technology in the process of teaching mathematics and physics. Open Educ. **24**(3), 44–55 (2020)
7. Cai, S., Chiang, F.K., Sun, Y., Lin, C., Lee, J.J.: Applications of augmented reality-based natural interactive learning in magnetic field instruction. Interact. Learn. Environ. **25**(6), 778–791 (2017)
8. Salmi, H., Thuneberg, H., Vainikainen, M.P.: Making the invisible observable by augmented reality in informal science education context. Int. J. Sci. Educ., Part B. **7**(3), 253–268 (2017)
9. Merchant, Z., Goetz, E.T., Cifuentes, L., Keeney-Kennicutt, W., Davis, T.J.: Effectiveness of virtual reality-based instruction on students' learning outcomes in K-12 and higher education: a meta-analysis. Comput. Educ. **70**, 29–40 (2014)
10. Garrido-Jurado, S.: Automatic generation and detection of highly reliable fiducial markers under occlusion. Pattern Recogn. **47**(6), 2280–2292 (2014)

11. Ivanilova, I.V., Yurkevich, E.V., Kryukova, L.N.: Mechanisms of using augmented reality technologies in education. Mod. Teacher Educ. **1**, 84–89 (2020)
12. Sosnilo, A.V.: AR/VR technologies in management and education. Upravlenie. **9**(2), 114–124 (2021). https://doi.org/10.26425/2309-3633-2021-9-2-114-124
13. Onime, C., Uhomoibhi, J., Radicella, S.: MARE: Mobile augmented reality based experiments in science, technology, and engineering. Online Exp. Emerg. Technol. IoT Chap. **11**, 209–227 (2015)

Implementation of AI-Assisted Tools in Foreign Language Training

Tatiana Marchenko[1(✉)], Liana Vartanova[1], G. S. Prakasha[2], and Oksana Moshchenko[3]

[1] North-Caucasus Federal University, 1 Pushkin Street, Stavropol 355017, Russia
tatiana-marchenko-25@yandex.ru
[2] School of Education, Christ University, Bengaluru, India
[3] Financial University Under the Government of the Russian Federation, Moscow, Russia

Abstract. The article discusses didactic issues of implementing of AI-assisted tools into foreign language training of linguistics students. The purpose of the study evaluate the potential and analyze the case of integrating an AI-assisted tool into a foreign language course. Methodologically the study rests on two didactic approaches, namely a competence-based approach and a personal activity approach. The analysis of the functional scope of AI-assisted tools used for academic linguistic and practical linguistic purposes is carried out. A brief comparative analysis of the available options with an account of essential functional characteristics is given. Based on the needs-analysis the choice of a particular platforms for the training process is reasoned. A set of teaching principles was observed to master the target skills in the experiment. To estimate the efficiency of the training process and development of the target skills, a questionnaire was compiled and offered to students. The analysis of students' performance as regards the focus communicative interaction skills was carried out. Challenges in combining traditional and AI-assisted tools in the educational process are analyzed; the ways to overcome them are recommended. The study suggests that the use of AI-assisted tools in foreign language training of linguistics students can be based on a combined approach aimed at improving the quality of individual work and further consolidation of target skills in contact time.

Keywords: language training · AI-assisted language training tool · English as a foreign language · information technologies in linguistics · communication skills

1 Introduction

Rapid integration of AI tools in almost all areas of life is caused by the development of AI and augmented reality technologies, necessitated transition to online practices due to the extensive online communication and widely accepted regulations and strategies for AI technology development. The trend fosters the need to search for didactic means and methods of training foreign language acquisition in a combined way (AI-assisted) and independent mode. AI technologies have modified the educational practice in terms of assessment procedure, efficiency, customization of the forms of educational interaction

M. Lapina et al. (Eds.): DET 2024, LNNS 1616, pp. 314–322, 2025.
https://doi.org/10.1007/978-3-032-04365-8_31

and the proficiency level of training materials. They have promoted the concept of personalized learning trajectory and made it possible within the framework of contact and online courses.

Members of academic community and participants of the training process gain access to AI-powered tools that analyze big data to enhance the educational practices and tackle various training-related problems. According to [1, 2, 13], currently AI use in teaching foreign languages falls into several main areas. They are teaching based on oral and written dialogue speech with AI, writing and speaking performance evaluation with reference to automatic assessment, and development of intelligent learning systems.

Artificial intelligence methods and machine learning algorithms are currently used both in manufacturing, to improve automation efficiency, and in medicine, medicine, and financial structures for big data analysis. With the growing popularity of language models and the launch of open source GPT, artificial intelligence methods have entered even more areas of human activity [14, 15].

The key difference in choosing the proper tool and means in traditional (contact) and technology-driven (AI-assisted including) language training lies in the selection criterion. The former primarily rests on didactic properties and functions of the technologies involved, the latter is more functional and focuses on the areas of didactic activity. They include:

- organizing the language training process. This domain incorporates preliminary work on developing assignments, tests and media materials, intermediary monitoring as well as follow-up assessment and reflection;
- training process management. This area of application goes beyond the limits of the immediate training process and integrates the analysis of performance, outlining an individual training trajectory and the customization of the available ones, development of assistance with AI-powered tools.
- teaching and studying processes proper. A plethora of digital AI-based tools are on offer to assist the learning process. In many cases, it entails independent training of foreign language skills and abilities.

The complexity and interrelated character of languages skills stipulates the need for integrated tools focusing on combined development of linguistic competences. It predetermines the necessity to operate interactive technologies, and apparently results in gaining the experience of diversifying and contextualizing communication.

2 Purpose and Objectives of the Research

The study aims to analyze the case of integration of AI-assisted tools in foreign language training and its testing within the academic course in oral and writing skills development with linguistics students.

The goal of the study has determined the range of tasks to be completed. They include:

- the analysis of operational potential of AI-assisted tools for foreign language learning;
- pilot assessment of the tool and tuning it in line with the academic purposes of the course;

- customizing the content of the course based on available functional features of the selected tool;
- testing the tool in developing the focus skills in the training process;
- carrying out critical analysis of performance productivity in terms of the acquired skills.

3 Literature Review

AI technologies are used in teaching foreign languages in a number of ways. First, these are traditionally machine-driven dictionaries of different types and text corpora predominantly for academic purposes. The latter serve a rich source for tracking written and oral speech formation patterns, concordance models and frequency data. Along with tools of purely formal nature, there are pragmatics-oriented options. These are neuro-linguistic programming technologies for text analysis and speech strategy/tactics detection, speech synthesis and text generation. Another developed area applied in language training is machine translation, which is a supportive means in enhancing vocabulary and grammar patterns awareness, text analysis and translation skills [3].

Textological approach is actualized in text-related software offering options of classifying, transforming and customizing didactic material along with language levels, interactivity and other relevant requirements [4]. It is also employed for development and evaluation of the quality of didactic materials for language courses.

Graphic, speech and audio recognition technologies are notably used to form socio-cultural and social competences of students based on audio and visualization [5].

Apart from the above-mentioned applications of AI in language training, the didactic ones are used for control, feedback, correction and automatic assessment of writing and speaking assignments. The relevance of automated checking is due to the reduction of knowledge testing subjectivity.

Another option is a chatbot operating on AI algorithms on large language models (LLM), exponentially increasing their knowledge and conversational skills. These include natural language processing programs, speech recognition and synthesis, and imitation of communication in the form of oral or written dialogue [6]. Technological solutions such as ChatGPT and other LLMs can generate texts that are almost indistinguishable from human-produced texts [7]. A number of studies evaluate the use of chatbots in terms of producing and editing written works, focusing on saving time and effort spent by teachers and students, revealing the ability of the chatbot to generate ideas and improve the quality of texts [8]. While others reflect on the impact of GPT chat on cognitive activity and the possibility of extensive integration of smart technologies into the educational process [9].

As for the benefits, it is admitted that the use of chatbots in education provides a personalized and results-oriented online learning environment, enhances interactivity and motivation, provides feedback, and contributes to the development of language skills [10].

Nevertheless, there are limitations of the system and the inconsistency of chatbots as co-authors of training materials. In addition, some experts focus on the analysis of the impact and the result of students' learning [11]. Yet, further study of the GPT chat potential remains relevant, since the system continues to develop and offers alternative results during interaction. There remain methodological problems associated with a number of essential issues. They mostly concern the advisable level of foreign language proficiency allowing the introduction of chatbots for training. Other debatable topics are the efficiency of bots in developing interaction and mediation skills, methodology for correct formulation of didactic requests (prompts), general methodology for teaching students foreign language speech interaction with a didactic chatbot, instructions how to avoid communication breakdowns during a dialogue with a bot, etc.

Note should be taken about technological solutions for developing didactic materials based on methodological models. These digital tools have significant didactic potential since they are capable of prompt and efficient development of:

- semiotically variable tasks,
- materials oral and writing practical activities,
- materials varying in genre/level of proficiency/topic, etc.,
- a lesson planning in line with the provided input, etc.,
- other didactic purposes.

The strong point about designing and implementing didactic material using AI tools is that they can be tuned and customized for each student in line with proficiency in a foreign language, their individual needs and abilities. It is appears vital in terms of language level when teaching heterogeneous groups.

4 Methodology

The methodology to implement AI-based tools into the process of teaching a foreign language rests on two didactic approaches:

- a competence-based approach. Competence concept is admitted to be the leading one both in contact and technology-assisted second language acquisition. It involves the enhancement of target skills and abilities immediately through contextualized experience of real life communication or its simulation.
- an individual activity approach. It implies immediate personal involvement into training assignments. The approach is purely student-centered as it incorporates subject-oriented tasks varying in linguistic nature, pragmatic purpose and complexity. Since by their nature linguistic skills have integral character this approach enables the development of several competences rolled up in one activity, e.g. personal interaction, subject-related skills and general communicative skills.

The selected approaches stipulated a set of methodological and linguodidactic principles to be observed in the use of AI-assisted tools in language training. These principles presuppose the development of cognitive abilities that are responsible for mastering the mechanisms of speech (grammar, vocabulary, phonetics) and types of speech activity (reading, speaking, writing and listening). They are:

- contextuality principle focusing on contextualization of purely linguistic and culture-specific material. The principle fosters the recognition skills considered to be in charge of the ability to recognize sounds, letters, parts of words, morphemes, parts of speech, etc., to distinguish one linguistic phenomenon from another;

- interactivity principle promotes analytical skills expressed in the ability to perform phonetic, morphemic, word-formation, morphological, syntactic, and stylistic analysis. At the communication level the principle is actualized in the interaction of students with a teacher, within a group to solve group/team tasks, in operation with online resources and applications;
- modularity principle at a linguistic level promotes the ability to use words and grammatical constructions in particular communication situations. At a more general level, it offers the possibility to expand and specify individual components of the curriculum, facilitate the skills and communication tasks.

The analysis of AI-assisted language training tool and the best practices of integrating these tools into teaching a foreign language, highlights the possibility to form a number of skills in the learning process, namely:

- productive speech activities (both oral and written ones);
- listening comprehension, lexical and grammar skills;
- critical analysis ability to analytically perceive oral or written texts;
- self-management since training incorporates the combination of traditional training resources and applications;
- operational experience with the selected tool and proper methods.

A set of training-related settings and the above-mentioned principles ensure the enhancement of the training process and its final goal, e.g. professional competences.

5 Results

The choice of an AI-assisted tool for the development of communicative interaction skills required a preliminary analysis of its functionality. The focus was on the resources employing authentic materials. One such option is an online platform based on up-to-date news – Newsela [12]. The project was initially launched as a resource for graded reading and has evolved into a comprehensive AI-assisted tool for language training and critical thinking development.

The resource is based on relevant media discourse materials offered by authentic media sources covering various subject areas (science, history, community, nature, environment, art, music, lifestyle, etc.). The materials supplied are in written form, yet each comes with an option of graded reading starting with the lowest level equal to grade 3 up to the highest one equal to grade 12. The grading is based on text transparency concept, which is composed of lexical and grammar complexity, literacy features and explicit/implicit manner of message delivery. With each grade up the expression of indicators grows in sophistication. It is claimed that grading is not AI-assisted, whereas the assignments are driven with AI-generated prompts.

Each text comes with two introductory sections ("Main, idea, key details and summarization" and "What the text says"), which could be used for preliminary screening and selection of didactic material and as an introductory summary for students. Count of words, text level and standards indicators serve as extra selection features. Graphic organizers are used to customize the sections.

All texts on the platform open with the share/save options, paragraph count (optional enabling), listening option and the presentation view. The text size is also available, which makes it all-inclusive.

One of the key options offered by the source is the "Apply to your lesson" section based on AI-generated suggestions for lesson content and inspiration. The suggestions fall into the main idea, key terms listed to match the level assigned to the text, before reading activity section and discussion questions section. The "before reading" part includes the warm-up questions foreshadowing the content of the text.

The check for understanding part comes on completion of a segment of the text. When assigned it is available at unlimited tries until the comprehension is reached. The rest of the text stays hidden until the correct answer is given.

Lexis, which is a significant aspect of text comprehension, is customized according to the grade chosen, starting with CEFR A1-A2 levels and up to approximately C1 at the highest lexical density of the text. AI-supported subject-specific vocabulary is tagged in the text (marked in bold and underlined) with an interactive option of pronunciation and simple definition. The feature could be disabled on teacher's request.

The feature that makes the tool flexible for didactic material customization is the assignment builder. The assignment comes under the title, enables selecting the activities for the article under consideration and assigning it to students. The most common ones are checks for understanding, annotations and student polls. A section on English language arts comes in the write prompts part and the quiz part. The quiz questions contain topic-related vocabulary, relevant grammar patterns and balance between explicit/implicit information.

The work with Newsela platform was integrated into Practical course of the English language in accordance with the curriculum of third-year Linguistics students. The tool was chosen to develop perception, interpretation skills and critical analysis of authentic oral/written texts. Two groups of students followed a regular course of studies and two more were provided with an algorithm to master the material, combining contact work with a teacher and individual work with the resource:

- introduction to the topic (contact);
- work with an authentic oral/written text (lexical, grammar, analytical skills) (individual);
- consolidation and contextualization (individual and contact);
- a control practical task (contact/individual).

The results of a pilot course were analyzed on the completion of the experiment. A questionnaire on the reflection on the course was offered to students on its completion. The results are given in Table 1. The indicators correlate with an average score (in accordance with a five-grade scale) for individual assessment of experience as compared to the training process feedback in the previous term.

Table 1 shows that generally students reflect in a positive way on completion of the course with integration of AI-assisted tools. The highest advance is in lexical skills and text comprehension, which is predictable and is obviously reasoned with the core material text-based training. Combined training has left the listening and communication skills at the same level. These indicators suggest that subjective assessment of the level of communication skills might be driven with a focus on individualized work. It

Table 1. Students' reflection on completion of the course.

No	Skill or experience	Previous term	Current term
1	Listening comprehension	4,7	4,7
2	Lexical skills	4,3	4,8
3	Grammar skills	4,7	4,9
4	Text comprehension, critical analysis	4,0	4,8
5	Communicative skills	4,8	4,8
6	Productivity	4,0	4,6
7	Time-consumption	4,8	4,4
8	Overall contentment with the course	4,7	4,9

is noteworthy that, according to students' feedback, critical analysis skills have significantly grown. This assessment suggests that a traditionally accepted approach to critical analysis development is immediately connected with confident command of vocabulary since the latter provides a more fluent text comprehension. This skill is complex and is developed on the basis of the primary ones: listening, grammar and lexical abilities. Yet, despite mainly positive assessment, the group admits higher time-consumption that could be explained with a number of reasons. First, the introduction of any new training tool or mode requires more efforts and, consequently, entails increase in time-consumption. Secondly, the ratio of contact and individual work is to be tuned to enhance time input and the obtained outcomes.

The experiment allows us to conclude that the use of AI-assisted tools in language training adds to the overall awareness of students, but also enhances the development of independent learning skills and fosters personal improvement. Generally, students reveal motivation in the process of training the language material in a combined manner, which results in the level of skills formed, the efficiency and expediency of results.

6 Discussion

Predominantly positive assessments obtained from students during the experiment on combined development of language skills and competences are accompanied with certain shortcomings and weak points to be addressed, analyze and serve for appropriate adjustments and tuning.

First, there is certain need for time to be spent on mastering the flow of work on platform or application and working out the algorithm of training. It could be tackled through the enhancement of the general digital culture of students, development of detailed instructions and training materials for combined work, getting more experience with AI-assisted tools prior to and within the course.

Second, target skills could be formed unevenly. It might require a wider range of activities for complex development of skills and expansion of the tools used. It appears reasonable to provide a more frequent change of activities within contact and individual modes of work.

Third, another drawback is connected with higher degree of individualization and lack of cooperative coordination. It requires the expansion of the scope of training materials to develop communicative interaction skills within the course.

7 Conclusion

Currently, AI-assisted tools offer a wide range of applications in language training. Apart from purely linguistic skills, the use of such tools enhances soft skills, overall motivation and commitment. As any new resource or tool, their introduction into the training process requires extra time to master it appropriately. Otherwise, the increase in the time spent on the task expands. AI-assisted tools offer constructive aspects of instant verification and testing, an opportunity to get immediate result, maintain individual pace of work and flexible time with no constraints in access to the resource from anywhere. Special attention is to be paid to integration of traditional contact language training and the AI-assisted one since so far the latter cannot substitute the former.

Thus, the experience of using AI-assisted resources in training a foreign language has shown that it appears more efficient in combined way to enhance the quality of training through the integration of different didactic modes and pedagogical support for students.

References

1. Zhou, C., Hou, F.: How do EFL teachers utilize AI tools in their language teaching? Theory Pract. Lang. Stud. **15**, 403–413 (2025)
2. Rani, B.: Artificial intelligence tools in learning English language and teaching. How can be AI used for language learning. Edumania. **02**, 230–234 (2024)
3. Marchenko, T., Kolobkova, A.: Design and implementation of a CAT training course. In: SLET-2020: International Scientific Conference on Innovative Approaches to the Application of Digital Technologies in Education, pp. 324–331. CEUR Work (2020)
4. Marchenko, T., Bredikhin, S., Lapin, V.: Application of text analysis instruments in foreign language training. In: Lapina, M., Prakasha, G.S., Grigoriev, S. (eds.) International Conference on Innovative Approaches to the Application of Digital Technologies in Education and Research. SLET 2022. Lecture Notes in Networks and Systems, vol. 1222, pp. 323–332. Springer, Cham (2025)
5. Rugelj, J., Lapina, M.: Game design based learning of programming. In: Rugelj, J., Lapina, M. (eds.) International Scientific Conference Innovative Approaches to the Application of Digital Technologies in Education and Research. CEUR Work. Proceedings, vol. 2494, pp. 29–42 (2019)
6. Çakmak, F.: Chatbot-human interaction and its effects on EFL students' L2 speaking performance and speaking anxiety. Novitas-ROYAL. **2**(16), 113–131 (2022)
7. Taecharungroj, V.: What can ChatGPT do? Analyzing early reactions to the innovative AI chatbot on Twitter. Big Data Cognit. Comput. **1**(7), 35 (2023)
8. Belda-Medina, J., Calvo-Ferrer, J.: Using chatbots as AI conversational partners in language learning. Appl. Sci. **7**(12), 1–16 (2022)
9. Bione, T., Cardoso, W.: Synthetic voices in the foreign language context. Lang. Learn. Technol. **24**(1), 169–186 (2020)

10. Huang, Y.-M., Chiu, P.-S., Liu, T.-C., Chen, T.-S.: The design and implementation of a meaningful learning-based evaluation method for ubiquitous learning. Comput. Educ. **4**(57), 2291–2302 (2011)
11. Corti, K., Gillespie, A.: Co-constructing intersubjectivity with artificial conversational agents: people are more likely to initiate repairs of misunderstandings with agents represented as human. Comput. Hum. Behav. **58**, 431–442 (2016)
12. Newsela: https://newsela.com/. Accessed 20 Jan 2025
13. Kormakova, V.N., Klepikova, A.G., Lapina, M.A., Rugelj, J.: ICT competence of a teacher in the context of digital transformation of education. In: CEUR Workshop Proceedings, vol. 2914, pp. 138–150 (2021)
14. Lapina, M., Shiriaev, E., Babenko, M., Istamov, I.: High-speed convolution core architecture for privacy-preserving neural networks. Program. Comput. Softw. **50**, 417–424 (2024). https://doi.org/10.1134/S0361768824700282
15. Rusanov, M., Babenko, M., Lapina, M., Sajid, M.: Identification of exploited unreliable account passwords in the information infrastructure using machine learning methods. Big Data Cognit. Comput. **8**, 159 (2024). https://doi.org/10.3390/bdcc8110159

The Impact of Digitalization of Preschool Education on the Interaction of Natural Social and Artificial Intelligence

Natalya N. Novik(✉)

Federal State Autonomous Educational Institution of Higher Education "Kazan (Volga Region) Federal University, Institute of Psychology and Education", Kazan, Russia
Natalya.Novik@kpfu.ru

Abstract. Relevance. In the modern world, characterized by rapid digitalization of all spheres of life, preschool education also does not remain aloof from the influence of technological innovations. However, this process requires a balanced and critical approach that takes into account both the potential opportunities and risks associated with the use of digital technologies in preschool age. The problem of this study has as its source the contradictions arising between the conscious positive influence of the digital educational environment on educational results and significant shortcomings in methodological support. The purpose of the study is to analyze existing digital products that can be used in the educational process of kindergarten for the comprehensive development of senior preschoolers in five different educational areas. Method: a theoretical review of studies that have been published over the past five years and concern the impact and structuring of the digital educational environment in kindergarten, as well as the potential of digital tools for the comprehensive development of senior preschool children. Conclusions: digitalization of preschool education requires a balanced and conscious approach. Digital tools can be a useful addition to traditional forms of education, training and development of preschoolers, but should not replace them completely. At the same time, the risk increases that artificial intelligence can replace natural intelligence, since children will not learn to think independently. It is important to integrate digital resources into the educational process in a measured and appropriate manner, using them to diversify activities, increase children's motivation, individualize learning, but at the same time maintaining the priority of live movement, active games, direct interaction with the outside world and peers.

Keywords: digitalization · digital environment · preschool education · senior preschooler · kindergarten · artificial intelligence · natural intelligence · child development

1 Introduction

Modern preschool education is a reflection of complex and contradictory processes occurring in society, which also find their expression in the educational sphere. One of such processes is digitalization.

M. Lapina et al. (Eds.): DET 2024, LNNS 1616, pp. 323–333, 2025.
https://doi.org/10.1007/978-3-032-04365-8_32

Digitalization of preschool education is the process of using digital technologies in the educational process with preschool children. This process involves the use of various digital tools and resources. These can be:

- electronic educational platforms and applications that help children learn through interactive tasks and games;
- digital resources that imply the use of video lessons, animation, interactive books, books with augmented reality, etc.;
- the use of tablets, computers and other devices for teaching and developing skills in children;
- digital tools for tracking children's progress and providing feedback to parents and teachers.

The main goal of their implementation is to improve the effectiveness of learning and increase the child's interest in the materials being studied.

Artificial intelligence methods and machine learning algorithms are currently used both in manufacturing, to improve automation efficiency, and in medicine, medicine, and financial structures for big data analysis. With the growing popularity of language models and the launch of open source GPT, artificial intelligence methods have entered even more areas of human activity [34, 35].

The trend towards digitalization of the educational process is focused on the introduction of new technologies and the use of innovative methods in the education and upbringing of preschool children. These areas contribute to the formation of a digital educational environment that allows exploring familiar topics, but along unusual routes. New challenges of our time pose new tasks for preschool teachers in the field of development, training and education.

2 Background

Digital technologies and resources are used in the educational process of a preschool educational organization as an additional tool for the full development of a preschooler in various educational areas.

Most scientific papers are devoted to studying the impact of digitalization on the cognitive development of preschoolers. Thus, E.A. Gorobets et al considered the features of cognitive development in preschoolers using gadgets [13]. P.I. Taruntaev identified a connection between the use of digital devices by older preschoolers and the level of development of their mental abilities [29]. M.A. Chernenko and L.L. Lashkova studied the organization of children's cognitive and research activities in the digital educational environment of a kindergarten [6]. V.A. Buzni and S.Z. Abduraimova scientifically argued and proved that the use of computer technologies is necessary for intellectual development, which can also include spatial thinking [4].

At the same time, there are not many studies devoted to the development of social and communication skills in kindergarten. For example, S.I. Shuminskaya and A.M. Permyakova analyzed the results of an experiment on the development of children's social and communicative skills with the targeted and systematic use of digital educational tools in preschool educational institutions. The researchers claim that while working with digital resources, children develop skills in analyzing and solving problems, learn to collaborate in a group, formulate their thoughts and ideas, and navigate the information flow [7]. Yu.V. Batenova et al identified possible ways of constructive interaction between adults and children, and between children in the digital space, taking into account the inclusion of various types of joint activities for emotional impact [2].

I.D. Yemelyanova came to the conclusion that the effectiveness of the speech development process in preschoolers can be influenced by: a specially designed speech development program with the regular use of digital technologies in educational activities; enrichment of the subject-spatial educational environment of the kindergarten with digitalization tools; work with kindergarten teachers on the use of digital technologies [8]. The study by A.P. Galysheva and N.Sh. Syrtlanova shows that the use of digital technologies contributes to the successful acquaintance of preschool children with various literary genres [12]. O.V. Pustovoitova identified the features of the speech-thinking activity of older preschoolers in the digital era and came to the conclusion that children's frequent use of digital devices for immersion in a virtual environment can lead to the formation of a simplified model of speech interaction and the creation of a digital perception of the world in the future [24]. P.A. Kurbonmamedov states trends in the growth of interaction between children and artificial intelligence [18]. E.I. Trofimova determined that virtual communication begins to replace real contact between people, which leads to the loss of emotional connection. Children increasingly perceive speech as fragmentary, lacking clear grammatical frameworks and structural design. The importance of direct interaction with adults also changes and weakens, including aspects of emotional mutual understanding, individual communication and cognitive interaction in the process of joint activities. Children experience a noticeable decrease not only in the level of speech skills, but also in the need for speech communication itself [30, 32, 33].

E.E. Klopotova and E.B. Gazanchyan wrote about the peculiarities of preschool children's drawing using traditional and electronic means. The authors state that children get more pleasure and satisfaction from drawing on paper than from working on a tablet. Their actions when drawing on a tablet are more like experiments than real creative activity. Both quantitative analysis of the works and expert opinions confirmed that drawings made on paper are superior to works created on a tablet in such criteria as the presence of a plot, detail and compositional completeness [16]. Y. Chen & Z. Ding consider the possibilities of digitalization in preschool education on the creative and cognitive development of children [5]. According to N. Atış Akyol, R. Turanoğlu & N. Parpucu, teachers occupy a key position in assisting preschool children in the use of technology in the educational process. Therefore, there is a need to increase the volume of research focused on educators, and professional development programs focused on the use of digital technologies should be developed and evaluated [1].

M.N. Kalashnikova shares her experience in developing interactive digital educational content, which is implemented within the framework of the main educational program of the kindergarten, in order to ensure effective interaction with children and their parents [15]. Ze Li provided useful links for the practice of digitalization of preschool education. Digital education provides preschool children with opportunities to enrich their educational experience, promoting the development of their cognitive, linguistic and social skills. However, the integration of digital technologies into education faces certain challenges. This requires parents and teachers to collaborate to provide effective digital educational resources and create a favorable environment conducive to the progress of preschool education [19].

Today it is difficult to imagine the educational environment of a preschool educational organization without computers, interactive whiteboards, online platforms, etc. However, there are two contradictory opinions here. On the one hand, according to E.V. Filippova and V.S. Korenskova, preserving children's health remains important and a priority, on the other hand, the use of digital technologies contributes to a more productive and interesting process of preschoolers learning about the world around them [9]. In addition, we should not forget that modern preschoolers were born at a time when life without the use of digital technologies became impossible. N. Fedina et al defined the structure of the model for the formation of a digital educational environment for a kindergarten [11]. E.V. Korotaeva and Yu.E. Vodyakh found that the restrictive method, in which the use of devices in the educational process with children aged 5-6 in kindergartens is reduced to twenty minutes per week, creates an incorrect idea among educators of children's institutions about their role as supporters of the formation of the younger generation for life in society [17].

Currently, the use of digital technologies in preschool education is one of the priority and most researched areas. And the use of information and communication technologies and computer gaming tools is considered one of the most important conditions for increasing the productivity of the educational process in preschool educational institutions, believe Yu. M. Gorvits et al [14].

A. N. Veraksa et al note that digital technologies simplify the work of teachers in preschool educational organizations. They allow creating interesting and varied tasks and materials for classes, using new sources of information and resources. However, the introduction of digitalization in the educational process requires careful training of teachers and the ability to correctly use digital technologies in it [31].

The results of the study of the features of interaction in the game of senior preschool children with their parents, conducted by E.S. Belova and N.B. Shumakova, showed that it is necessary to develop methodological guidelines for parents. These guidelines should help in the formation of a family microenvironment favorable for development, taking into account issues related to its digitalization [3].

At the same time, we should not forget about the risks that digital resources may carry. V.V. Lupandin and M.S. Egorov, reflecting on the dangers of developing artificial intelligence systems, believe that, on the one hand, modern tools are capable of reducing the danger in the digital environment due to more effective detection of threats, but on the other hand, they can also increase the likelihood of their occurrence, especially if they fall into unscrupulous hands [20].

Within the framework of the federal initiative "Digital Educational Environment", an innovative digital environment is understood as an open set of information platforms developed to solve a variety of problems that arise in the learning process [10].

In their scientific work, N.N. Novik and A.A. Tvardovskaya presented an analysis of digital gaming resources in order to improve and optimize the digital educational environment in preschool organizations [22].

The results obtained in the study and determination of the level of readiness of educators to use digital tools in the educational process by T.A. Svatalova emphasize the importance of future research in the field of creating specialized advanced training programs focused on digital technologies [25].

The study by E. Yu. Sentemova of about thirty studies on the interaction of gadgets on the mental development of preschool children allows us to conclude that significant shifts are taking place in the cognitive sphere of modern children, manifested both in a positive and negative way [26].

L. B. Osipova identified key areas of application and future opportunities for the development of artificial intelligence in the educational environment. She emphasized the advantages and identified the probable dangers that arise when introducing the analyzed technology, noting that the limited practical experience of using artificial intelligence makes it difficult to fully assess them [23].

N.V. Sokolov and V.G. Vinogradsky note that the optimization of educational practice includes not only the improvement of the learning process, but also the improvement of educational approaches. The authors believe that the next stage in the development of the modern Russian educational environment using artificial intelligence should be the unification of individual educational spaces of different areas and regions into a single educational platform. It is important to take into account both the peculiarities of children's thinking and the specific features of educational organizations [28].

Digital technologies help not only in the development of individual characteristics of preschoolers, but also in the process of carrying out diagnostic measures. Thus, E.V. Slavutskaya and N.S. Vostretsova believe that the use of a neural network approach based on the principles of fuzzy logic allows identifying non-obvious relationships within the system, without taking into account the number of parameters in the original data set. This becomes especially valuable when analyzing the development of preschool children, since assessment methods for this age group are often limited. A significant advantage is that, unlike traditional statistical approaches, artificial neural networks make it possible to analyze the relationships of psychological characteristics not only in general for the entire data sample, but also individually for each subject [27].

An analysis of studies on the problem under study allows us to conclude that digitalization of preschool education is a dual process that carries both potential opportunities and risks. At the same time, the authors use the following concepts in their works: digital technologies, information and communication technologies (ICT), gamification technology, digital devices and resources, electronic means, gadgets, devices, artificial intelligence, neural networks, etc.

Thus, the use of digital technologies in preschool education is a relevant and promising area that can significantly improve the educational process, develop the necessary skills and prepare children for life in modern society. However, it is extremely important to approach this issue with caution and responsibility in order to extract the greatest benefit and minimize possible negative consequences.

3 Conclusion

An analysis of scientific papers on the research problem allows us to conclude that scientists have convincingly discovered a connection between the use of digital tools by preschoolers and an increase in their level of cognitive development. Whereas there is insufficient scientific research on the connection between interactions with gadgets and the development of social intelligence. In the era of rapid technological progress, it is extremely important not to lose sight of the social and spiritual-moral development of preschoolers. Modern children are increasingly immersed in the world of gadgets, which leads to a deficit of real interaction, and attempts to limit the use of devices cause irritation and even aggression in them. In this regard, there is an urgent need to develop in children not only the ability to establish contacts with peers and adults, but also the ability to empathize with others, to cultivate a sense of patriotism and respect for the history of their homeland.

Diagnostics of social intelligence was carried out in older preschool children at the ascertaining stage of the experiment. The study involved 300 senior preschoolers from several kindergartens in Kazan. The total number of children included two groups: the experimental group – 150 children and the control group – 150 preschoolers.

The psychodiagnostic tool known as the Guilford Four Factor Test of Social Intelligence is based on his theoretical model of the structure of intelligence [21]. This method is a set of standardized tests designed to measure various aspects of social intelligence, which is a unique mental phenomenon. Social intelligence is understood as the ability to interpret and predict the actions of others in various situations, as well as to recognize their intentions, feelings and emotional state, both through speech and through non-verbal signals. This technique includes four subtests that use non-verbal stimuli, and one - verbal. These subtests are aimed at assessing four different components of the structure of social intelligence: the ability to predict consequences; understanding emotions and expressing them in a particular situation; the ability to notice changes; understanding the meaning and consequences of behavior.

The following Table 1 gives the results of the ascertaining stage according to the Guilford method.

Table 1. Results of the ascertaining stage according to the Guilford method.

Level	Control group	Experimental group
High	36; 24%	36; 24%
Medium	90; 60%	96; 64%
Low	24; 16%%	18; 12%

In the control and experimental groups, the same number of children showed a high level of emotional development, it was assigned to 36 children in each group - this is 24% of the group. In the control group, 90 children were assigned an average level - this is 60% of the group, and in the experimental group the same level was assigned to 96 children - this is 64% of the group. A low level was assigned 42 times, in the control group 24 times - this is 16% of the group and in the experimental 18 times - this is 12% of the group.

The obtained results lead to the conclusion that the development of an additional program aimed at developing social intelligence and its testing in the experimental group will reveal its effectiveness.

The significance of this program is due to the fact that senior preschool age is a key stage for the formation of cognitive processes, including attention, memory and thinking. During this period, children reach a new level of development, and at the same time, social and emotional self-determination occurs, which plays a significant role in the child's future communication skills. The goal of the program is to ensure the child's social adaptation, establish effective interaction with others, and create favorable interpersonal relationships.

The approximate program for the development of social intelligence of senior preschoolers consists of several blocks: recognition and understanding of the emotional spectrum; development of empathy and humanity; improvement of interpersonal communication skills; development of social adaptation skills.

The first block, "Recognizing and Understanding the Emotional Spectrum," includes: developing emotion identification skills, which include the ability to see feelings, analyze facial expressions, gestures, and voice intonations (happiness, sadness, anger, fear, surprise, etc.) using visual cards depicting emotions and short play situations; understanding the causes of emotions, which includes identifying the connection between the situation and the feelings it causes ("Why is he sad? Because he can't find his toy?"); identifying and expressing emotions, which includes developing the ability to accurately identify and express in words both his own emotions and the emotions of other people, and helping to verbally describe the feelings experienced; assessing the intensity of emotions—the ability to distinguish between the levels of emotion expression (e.g., "a little sad" and "very sad"); regulating emotional state. Teaching how to cope with negative emotions (anger, fear, resentment) using socially acceptable methods (breathing exercises, counting slowly, talking to an adult).

The second block, "Development of empathy and humanity," included the following content: recognizing the emotional state of others: the ability to mentally put yourself in someone else's shoes, experiencing their emotions in specific circumstances; demonstrating empathy – an emotional reaction to someone else's suffering (expressing sympathy, providing emotional support); providing support and assistance, expressed in an independent offer of practical help to those in need; tolerance, manifested in the development of a respectful attitude towards the emotions and views of others, even if they differ from your own.

The third block "Improving interpersonal communication skills": communication skills: the ability to listen attentively and understand the interlocutor; clear and understandable expression of one's own thoughts and feelings; the ability to ask and answer questions; skills for establishing, maintaining and ending a conversation; using nonverbal means of communication (facial expressions, gestures, intonation); cooperation skills: the ability to work in a team; the ability to share resources and reach compromises; resolving conflicts peacefully; respectful attitude to the opinions of others; adaptation to different social situations: knowledge and observance of rules of conduct in public places (kindergarten, store, transport, etc.); adequate behavior in conflict situations; the ability to establish new contacts and communicate with strangers.

Contents of the fourth block "Development of adaptation skills in society": assessment of the situation in society: identification of the main actors, their aspirations, motivations and connections; forecasting the results of one's own actions in the form of awareness of the impact of the child's behavior on other people around; mastering social standards and laws, manifested in the development of common patterns of behavior in various circumstances; interpretation of the reasons for the actions of others, that is, the desire to understand why people behave in a certain way.

The following forms and methods of working with children were chosen:

- variable approaches to learning through play (story-oriented, role-playing, developmental and active games, theatrical performances);
- practical classes (exercises recreating typical situations, role-playing games, tasks aimed at increasing attention, improving memory);
- communication and discussion (consideration of life situations, analysis of cartoons and feature films, analysis of literary works, study of the characters' actions, study of the motives and feelings that guide them);
- mini-scenes: reproduction of the spectrum of emotional states and co-event situations;
- use of visual materials: cards with images of emotions, photographs, illustrations;
- project activities: implementation of projects that contribute to the development of effective communication skills.

Work on this program was carried out during one academic year.

Upon completion of the program, we again diagnosed preschoolers using the Guilford test. The following Table 2 gives the results of the control stage according to the Guilford method.

Table 2. Results of the control stage according to the Guilford method.

Level	Control group	Experimental group
High	72; 48%	96; 64%
Medium	66; 44%	54; 36%
Low	12; 8%	–

During the control diagnostics in the experimental group, positive results are observed. In the experimental group, 96 children (64%) of the total number of the group showed a high level of social intelligence, and 54 children (36%) showed an average level of social intelligence. Consequently, in the experimental group, after testing the program, no child was identified who showed a low level of social intelligence. The results of the control group at the control stage showed small positive results: in the group, 72 children (48%) showed a high level of social intelligence, 66 children (44%) showed an average level, and 12 children (8%) showed a low result.

Summing up the results of the control stage, we can conclude that testing the additional program for the development of social intelligence showed a high result of effectiveness in the experimental group. The program contributed to an increase in social intelligence, a respectful attitude towards peers and adults.

The conducted study provides an answer to the question of whether artificial intelligence (AI) can replace natural social intelligence in preschoolers. The results of the experiment prove this: social intelligence inherent in preschool children cannot be completely replaced by artificial intelligence, since the former covers a wide range of complex, subtle nuances conditioned by human experience, feelings and interpersonal relationships.

Social intelligence is based on the ability to understand and empathize with emotions, both one's own and others. In preschool age, children learn to recognize and express feelings through interaction with others, which is impossible for AI, which does not have an emotional base.

The development of social intelligence in children occurs through the accumulation of experience and the analysis of social interactions. AI can process data and draw conclusions, but it lacks the intuitive understanding that arises from personal experiences.

A key element of social intelligence is the ability to show empathy and help others. From early childhood, children learn this through interactions with family, friends, and adults. AI can imitate empathy, but it does not have genuine experiences and awareness of human feelings.

Children learn social norms and rules through observation and communication. AI can be trained to follow given rules, but it cannot adapt to changing social circumstances as dexterously as a person.

Preschoolers acquire conflict resolution skills through their interactions with each other and through past experiences. Although AI can offer solutions, it is not able to sense a conflict situation on an emotional level.

Thus, despite the probable benefits of AI as a tool for supporting the learning and development of a preschooler, it cannot replace social intelligence, which is formed naturally in communication and life circumstances. Supporting the development of social intelligence in preschool children remains an important task for parents, educators and society as a whole.

Acknowledgment. This paper has been supported by the Kazan Federal University Strategic Academic Leadership Program (PRIORITY-2030).

References

1. Atış Akyol, N., Turanoğlu, R., Parpucu, N.: Digital technology in preschool education: a systematic review. Int. J. Educ. Res. Rev. (2023). https://doi.org/10.24331/ijere.1260176
2. Batenova, Y.V., Filippova, O.G., Safonova, A.K.: Organization of constructive emotional interaction of preschoolers in the context of digitalization of education. Concept. **2**, 204–217 (2024) https://cyberleninka.ru/article/n/organizatsiya-konstruktivnogo-emotsionalnogo-vzaimodeystviya-doshkolnikov-v-usloviyah-tsifrovizatsii-obrazovaniya. Accessed 10 Feb 2025

3. Belova, E.S., Shumakova, N.B.: Playful interaction of senior preschoolers with parents as a resource of the developing microenvironment in the context of digitalization. Mod. Presch. Educ. **18**(4), 24–36 (2024). https://doi.org/10.24412/2782-4519-2024-4124-24-36
4. Buzni, V.A., Abduraimova, S.Z.: Using digital tools in teaching preschoolers. In: Problems of Modern Pedagogical Education, vol. 81, pp. 9–12 (2023)
5. Chen, Y., Ding, Z.: Effects of digitalization in preschool education on the creative and cognitive development of children. Educ. Inf. Technol. **29**, 21567–21591 (2024). https://doi.org/10.1007/s10639-024-12730-y
6. Chernenko, M.A., Lashkova, L.L.: Organization of cognitive and research activities of preschoolers in the conditions of the digital educational environment of kindergarten. Bull. Surgut State Pedagog. Univ. **4**(73), 70–77 (2021)
7. Shuminskaya, S.I., Permyakova, A.M.: Digital kids: digital educational tools as a tool for developing communication skills in kindergarten. In: Bulletin of Perm State Humanitarian Pedagogical University. Series No. 1. Psychological and Pedagogical Sciences, vol. 2, pp. 122–131 (2023)
8. Emelyanova, I.D.: Speech development of preschool children in the context of digital transformation. Human. Stud. Central Russia. **4**(29), 44–52 (2023)
9. Filippova, E.V., Korenskova, V.S.: Implementation of health-saving technologies in the digital environment of preschool education. In: Digitalization of Modern Preschool Education: Main Approaches, Problems and Development Prospects: Materials of the All-Russian Scientific and Practical Conference, pp. 67–69. LSPU Named After P.P. Semenov-Tyan-Shansky, Lipetsk (2022)
10. Federal project "Digital educational environment". https://edu.gov.ru/national-project/projects/cos/. Accessed 10 Feb 2025
11. Fedina, N., Lazareva, M., Zvezda, L.: The structure of the model for the formation of a digital educational environment for a preschool educational organization. In: 4th International Conference on Technologies for Enhanced Learning in Higher Education (TELE). Lipetsk, pp. 386–390 (2024). https://doi.org/10.1109/TELE62556.2024.10605646
12. Galysheva, A.P., Syrtlanova, N.S.: The use of digital technologies in introducing older preschoolers to literary genres. Bull. Bashkir State Pedagog. Univ. Named After. M. Akmully. **S1**(67), 115–117 (2023)
13. Gorobets, E.A., Gamirova, R.G., Skhirtladze, A.V., et al.: Features of cognitive development in children of early and preschool age using gadgets. Russ. Bull. Perinatol. Pediatr. **66**, 163–167 (2021). https://doi.org/10.21508/1027-4065-2021-66-5-163-167
14. Gorvits, Y.M., Chainova, L.D., Poddyakov, N.N., Zvorygina, E.V., et al.: New information technologies in preschool education. LINKA-PRESS, Moscow (1998)
15. Kalashnikova, M.N.: Application of the LEARNING platform and the PADLET board in the practical work of a preschool teacher. Perm. Pedagog. J. **14**, 74–76 (2023)
16. Klopotova, E.E., Gazanchyan, E.B.: Features of drawing of preschoolers with the help of traditional and electronic means. Mod. Presch. Educ. **8**(80), 24–31 (2017)
17. Korotaeva, E.V., Vodyakha, Y.E.: Digital devices for preschool age: an aspect approach. Pedagog. Educ. Russia. **1**, 20–26 (2022)
18. Kurbonmamadov, P.A.: On the issue of new communications: artificial intelligence and preschoolers. In: Academic Research in Educational Sciences. NUU Conference 1, pp. 96–99 (2021)
19. Li, Z.: Study of the use of digital technology in preschool education. Trans. Econ. Bus. Manag. Res. **5**, 270–274 (2024). https://doi.org/10.62051/sxmcmq74
20. Lupandin, V.V., Egorov, I.S.: Risks of development of artificial intelligence systems. Theory Pract. Mod. Sci. **1**(43), 346–349 (2019)
21. Mikhailova, E.S.: Methodology for studying social intelligence. In: Adaptation of the Test by J. Guilford and M. Sullivan: Guide to Use. Imaton, St. Petersburg (1996)

22. Novik, N.N., Tvardovskaya, A.A.: The potential of digital tools for the development of executive functions in preschoolers. In: Lapina, M., Prakasha, G.S., Grigoriev, S. (eds.) International Conference on Innovative Approaches to the Application of Digital Technologies in Education and Research. SLET 2022. Lecture Notes in Networks and Systems, vol. 1222. Springer, Cham (2025). https://doi.org/10.1007/978-3-031-78776-8_6
23. Osipova, L.B.: Artificial intelligence in education: real possibilities and prospects. Bull. PNIPU Soc. Econ. Sci., 1. (2024). https://cyberleninka.ru/article/n/iskusstvennyy-intellekt-v-obrazovanii-realnye-vozmozhnosti-i-perspektivy (дата обращения: 12 Feb 2025)
24. Pustovoitova, O.V.: Features of speech-thinking activity of children aged 5-7 years in the digital era. In: Philological Sciences. Theoretical and Practical Issues, vol. 4, pp. 1317–1323 (2024)
25. Svatalova, T.A.: Research of preschool teachers' readiness to use digital technologies in pedagogical activity. Scientific Support of the System of Advanced Training of Personnel 1 (46), 38–44 (2021).
26. Sentemova, E.Y.: Review of studies devoted to the study of the development of the cognitive sphere of modern preschoolers, in the context of the transformations taking place in the world. Psychol. Pedagog. Sci. **1**, 97–107 (2024)
27. Slavutskaya, E.V., Vostretsova, N.S.: Neural network as a tool for analyzing the intellectual and emotional sphere of preschoolers. Kazan Pedagog. J. **4**(135), 61–65 (2019)
28. Sokolov, N.V., Vinogradsky, V.G.: Artificial intelligence in education: analysis, prospects and risks in the Russian Federation. In: Problems of Modern Pedagogical Education, vol. 76, pp. 166–169 (2022)
29. Taruntaev, P.I.: Use of digital devices by senior preschoolers and the level of development of mental abilities. Mod. Presch. Educ. **5**(119), 56–67 (2023). https://doi.org/10.24412/2782-4519-2023-5119-56-67
30. Trofimova, E.I.: The role of an adult in the development of speech in children of early preschool age and the specifics of interaction with adults in the era of digitalization. In: Personality in a Changing World: Health, Adaptation, Development, vol. 3, pp. 263–274 (2022)
31. Veraksa, A.N., Veresov, N.N., Sukhikh, V.L., et al.: Play to foster children's executive function skills: exploring short-and long-term effects of digital and traditional types of play. IJEC. **56**, 687–709 (2024). https://doi.org/10.1007/s13158-023-00377-8
32. Sherbakov, S., Lapina, M., Lapin, V., Rugelj, J.: Methodological support of game modeling in the educational process. In: Paper presented at the SLET-2020: International Scientific Conference Innovative Approaches to the Application of Digital Technologies in Education, CEUR Workshop Proceedings, vol. 2861, pp. 351–361 (2020)
33. Kormakova, V.N., Klepikova, A.G., Lapina, M.A., Rugelj, J.: ICT competence of a teacher in the context of digital transformation of education. In: CEUR Workshop Proceedings, vol. 2914, pp. 138–150 (2021)
34. Lapina, M., Shiriaev, E., Babenko, M., Istamov, I.: High-speed convolution core architecture for privacy-preserving neural networks. Progr. Comput. Softw. **50**, 417–424 (2024). https://doi.org/10.1134/S0361768824700282
35. Rusanov, M., Babenko, M., Lapina, M., Sajid, M.: Identification of exploited unreliable account passwords in the information infrastructure using machine learning methods. Big Data Cognit. Comput. **8**, 159 (2024). https://doi.org/10.3390/bdcc8110159

The Impact of Neural Networks on the Educational Outcomes of Students of Pedagogical Specialties

Natalya Novik[1(✉)], Valerian Gabdulhakov[1], Maria Lapina[2], and B. Deepanraj[3]

[1] Kazan (Volga Region) Federal University, Institute of Psychology and Education, Kazan, Russia
Natalya.Novik@kpfu.ru
[2] North Caucasus Federal University, Stavropol, Russia
[3] Prince Mohammad bin Fahd University, 617, Al Jawharah, 34754 Dhahran, Saudi Arabia

Abstract. The relevance of the research is associated with the lack of objective analysis of the positive and negative aspects of the influence of neural networks on the educational outcomes of university students. Educational results are recorded through quantitative data (scores) which indicate how well students have assimilated knowledge and developed necessary skills, competencies, and personal qualities during their study of both cultural and professional blocks. The study of the influence of neural networks on educational outcomes in several universities in the Republic of Tatarstan showed that the use of neural network resources significantly affects cognitive activity and the ability to independently acquire new knowledge. The reliability of the results was assessed using Pearson's $\chi 2$ (chi-square). It was found that in higher education institutions, the negative effects of neural networks manifest as a decrease in student activity in scientific and technological events, as well as a decline in interest in creativity. This occurs despite the existence of numerous government programs aimed at developing students' creative potential. The conclusion was made that the effectiveness of using neural networks to enhance students' educational outcomes depends on the presence of methodological and technological approaches, which are currently being developed by many programmers and pedagogical researchers.

Keywords: neural network · artificial intelligence · educational outcomes · students · digital environment

1 Introduction

In the 21st century, digital technologies have begun to play a significant role in scientific progress. Modern technologies are actively penetrating all areas of life, including education and upbringing, stimulating rapid societal development. One of the significant trends in recent years has been the use of neural networks in educational activities. Educational organizations are implementing intelligent interactive devices, educational

M. Lapina et al. (Eds.): DET 2024, LNNS 1616, pp. 334–341, 2025.
https://doi.org/10.1007/978-3-032-04365-8_33

software, neural network capabilities, artificial intelligence and other tools that help students in acquiring knowledge and broadening their horizons. This method of learning not only stimulates interest in studies but also enhances learning efficiency and develops independent learning abilities. Neural network technologies, due to their interactivity, visual appeal, and accessibility, are becoming effective tools for improving educational outcomes.

2 Literature Review

Student achievements in education represent specific skills and successes that learners acquire during their studies. These achievements can include knowledge, skills, and competencies developed through completing an educational program or mastering a specific course. Generally, results are formulated as goals or expected achievements in educational activities.

When considering the issue of monitoring educational outcomes, researchers view results not only as a volume of knowledge, but also as a structure based on systematic-methodological professional thinking (Feshchenko, 2014; Rubtsova, 2023). Thus, the question arises about improving the evaluation procedure for students' educational outcomes (Kupriyanov & Semenov, 2018; Kim et al., 2022; Amrenova, 2023). Educational outcomes encompass not only knowledge, skills, and abilities but also personal characteristics of students as well as competencies developed during the educational process, which form the core of personal and professional identity (Yermolaeva et al., 2020; Gabdulchakov and Scishova 2024; Malygin, 2023).

Artificial intelligence methods and machine learning algorithms are currently used both in manufacturing, to improve automation efficiency, and in medicine, medicine, and financial structures for big data analysis (Lapina et al., 2024). With the growing popularity of language models and the launch of open source GPT, artificial intelligence methods have entered even more areas of human activity (Rusanov et al., 2024).

The issue of assessing educational outcomes has been concerning the pedagogical community for many centuries, while active discussions about the capabilities of neural networks and artificial intelligence have taken place over the past 2–3 years (Martynova & Pfetzer, 2022; Kurbanova and Ismailova 2023). Artificial intelligence technologies help students understand and visualize their personal learning outcomes, track academic debts, view schedules and can assist with any questions that arise (Khvarzhev & Tsebrenko, 2024). Additionally, students utilize neural network capabilities in higher education for generating texts and images (Samarina & Boyarinov, 2023; Novik et al., 2023; Gabdulkhakov & Zinnurova, 2024; Gorina & Ufimtseva, 2024).

The advantages and risks of using neural networks in higher education remain ambiguous (Chapkin, 2023; Voronina, 2024; Bermus, 2024). Students describe their experiences interacting with artificial intelligence in educational activities (Garayev and Novik 2023).

Analyzing research on this topic allows us to highlight both the risks and advantages of integrating neural networks into the educational process. Both aspects are contentious.

In addition to the controversial opportunities presented by the use of neural networks in education, attention can be drawn to the complex factors (risks) associated with student interaction with artificial intelligence:

- technological dependence and loss of critical thinking: students are losing the ability to solve educational tasks by relying entirely on neural networks, resulting in a decrease in critical evaluation of information and independent thinking;
- unresolved ethical issues related to copyright and plagiarism: the use of neural networks raises questions about personal data protection, and the application of neural network materials may lead to doubts regarding compliance with intellectual property rights;
- lack of preparedness among educators to use neural network tools: educators do not yet fully understand the principles upon which to build the educational process and how to effectively integrate neural networks into teaching.

The list of factors (risks) may also include low-quality information, lack of human interaction (personal involvement of the educator), and unequal access to neural network capabilities among users.

Addressing these risks requires the development of a strategy aimed at mitigation the negative consequences of using neural networks in education while enhancing their educational process.

Literature analysis indicates that the use of neural network technologies in education opens numerous technological paths for increasing the effectiveness of student learning:

- individual approach and adaptation of the learning process: neural networks can analyze data about each student, identifying their strengths and weaknesses, automatically adjusting content according to knowledge levels and pace of information absorption, as well as forming personalized educational programs and materials based on this data;
- analysis of large volumes of data and forecasting: neural networks can process vast amounts of information, identifying trends and patterns that help pinpoint problems and optimize educational processes, as well as predict student success and identify those who may need help;
- support for learning and motivation: neural network chatbots provide round-the-clock responses to student inquiries, increasing accessibility to help, while the use of game elements and rewards enhances the motivation of students;
- access to high-quality educational materials: neural networks assist in developing and disseminating educational resources from experienced educators and experts;
- optimization of administrative process: neural networks improve organizational tasks, including scheduling, resource management, and automatic grading of student work, saving time and making assessment more objective.

Thus, the implementation of neural networks in education, with proper methodological guidance, can significantly enhance its effectiveness and quality. However, it requires attention to issues of ethics, data security, and technology accessibility.

3 Methodology

The study of the impact of neural network technologies on students' academic achievements represents an important area for future research. Observations show that today's students often demonstrate higher skills in digital, legal, and cultural aspects compared to their educators, although their subject-specific successes may remain average.

The aim of this study is to conduct a detailed analysis of the impact of neural network technologies on students educational outcomes.

To assess the effects of these technologies, a comprehensive approach and the application of various methodologies are required.

Let's consider some of them: testing knowledge through standardized assessments before and after the implementation of neural networks; comparing students' performance before and after using adaptive learning systems; conducting surveys among students and instructors about neural networks and their impact on learning; analyzing student activity on platforms, including completed assignments; qualitative methods, such as interviews to understand interaction experiences; examining specific examples of the application of neural network technologies in education; long-term monitoring of results to assess the sustainability of changes; analyzing behavioral data to identify patterns in learning.

This set of methods will help gain a deeper understanding of the influence of neural networks on educational outcomes.

In this article, we evaluated the impact of neural networks on learning outcomes through several main steps and methods.

The first step was to define criteria for assessment: specific educational achievements were selected (for example, academic performance, engagement levels, student satisfactuon) and skills that are planned to be developed using neural networks (such as critical thinking, creativity, problem-solving, etc.).

The next stage involved benchmarking: research with controlled groups where one group works with neural networks while the other uses traditional teaching methods. In the end, the results were compared.

Data analysis included collecting quantative (scores, tests) and qualitative (surveys, interviews) data to obtain comprehensive information about the impact, using statistical methods to identify significant differences.

It was also important to identify factors influencing the results. For this purpose, surveys were conducted among the participants (to gather their opinions on neural networks in education) to understand how these technologies affect student development in the long term.

We used a package of digital cases for students (first, second and third-year students), which were employed as assessment tools for diagnosing competencies (general cultural competencies and professional competencies).

Results were evaluated on a hundred-point scale: 0–10 points – no level; 11–35 points – low level; 36–70 points – average level; 71–100 points – high level.

A total of 573 students and 34 university instructors were involved in the study.

4 Research Results

From 2023 to 2025, pedagogical experiments were conducted to access learning outcomes. Pedagogical faculties at universities focused on the most effective integration of neural network capabilities into the educational process, while students in the control group participated in traditional classes without the involvement of artificial intelligence. In the experimental groups, instructors offered students creative and game-based assignments, emphasizing the use of neural network technologies. In contrast, no artificial intelligence tools were applied in the control classes, allowing for a comparison of learning outcomes under different conditions. Thus, the study aimed to identify the impact of implementing neural networks on students' educational achievements and to evaluate the potential of creative assignments that utilize modern software compared to classical teaching approaches.

The educational results of students in the absence of interaction with neural network resources (the observational phase of the experiment) are presented in Table 1.

Table 1. Educational results of university students in the absence of interaction with neural network Resources (in %).

Courses	First year		Second year		Third year	
Levels	experimental	control	experimental	control	experimental	control
High	**30**	31	**28**	33	**31**	35
Average	**36**	25	**33**	27	**22**	29
Low	**28**	30	**25**	38	**24**	32
Zero	**6**	14	**14**	2	**33**	4

Let's examine the indicators of high educational results. Here (see Table 1), we observe a decrease in the level in the experimental groups from 30% in the first year to 28% in the second year, followed by an increase to 31% in the third year. At the same time, control groups show a slight increase in educational results from 31% in the first year to 33% in the second year and up to 35% in the third year.

The program for integrating neural networks to improve educational outcomes may consist of several key elements (Fig. 1).

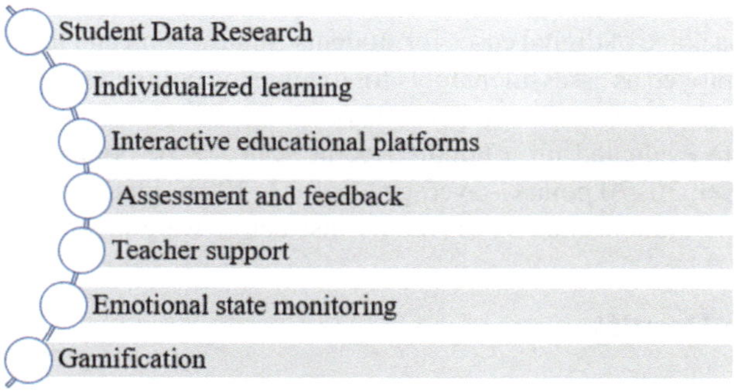

Fig. 1. Elements of the Program for integrating Neural Networks to improve educational outcomes.

The study of student data included: the collection and analysis of information on grades, attendance, and student engagement; the application of machine algorithms to identify patterns and predict successes or difficulties.

Individualized learning involved: the creation of adaptive learning programs tailored to the unique needs and knowledge levels of each student; the preparation of recommendations for learning materials based on the analysis of previous results.

Interactive educational platforms included: the development of platforms for generating interactive assignments and tests using neural networks; the implementation of chatbots to answer students' questions and provide necessary information.

Support for teachers was related to: tools for assessing the quality of education and identifying areas for improvement; advice on teaching technologies based on data about the student group.

Monitoring emotional states involved: the use of technologies to analyze student engagement and emotional well-being; recommendations for improving the educational atmosphere.

Gamification involved: the use of game elements in the learning process to enhance motivation and interest among students; the adaptation of game assignments to knowledge levels using neural networks.

Assessment and feedback, research, and improvements were aimed at: automated checking of student works using neural networks for text analysis; providing constructive feedback based on error and achievement analysis; regular analysis of program results and adjustments based on collected data; conducting research to assess the effectiveness of neural networks in education.

These elements can be adapted depending on the set goals and specifics of the educational program.

The educational outcomes of students in conditions involving interaction with neural network resources (the control phase of the experiment) are presented in Table 2.

Table 2. Educational outcomes of university students in conditions involving interaction with neural network resources (in %).

Courses	First year		Second year		Third year	
Levels	experimental	control	experimental	control	experimental	control
High	**36**	30	**43**	32	**48**	34
Average	**30**	22	**34**	25	**36**	27
Low	**31**	33	**20**	30	**14**	34
Zero	**3**	15	**3**	13	**2**	5

Table 2 shows hoe neural networks influenced the level of educational outcomes for students. When analyzing high-level results (see Table 2), a consistent growth is evident: in experimental groups, it increased from 36% in the first year to 43% in the second year and 48% in the third year; in control groups, the increase was less significant but still present - from 30% in the first year to 32% in the second year and 34% in the third year.

Correlation analysis showed that neural networks have a significant impact on improving students' educational outcomes. As a result, the educational outcome level for students in the experimental group increased by 20% ($\chi2 = 11{,}641$; $p < 0{,}01$), while the educational outcome level for students in the control group increased by only 4% ($\chi2 = 0{,}771$; $p > 0{,}05$). The obtained data indicate that the conducted work gave positive results.

5 Conclusions

The pedagogical experiments conducted in higher educational institutions in Tatarstan allowed for determining how artificial intelligence affects students' educational outcomes.

Modern digital didactics requires a clear definition of methodological principles and requirements for constructing content and assignments for effective use of neural networks in education.

Given the vast capabilities of neural networks for generating various products (projects, presentations, reports), conducting research activities, including simulating human creative activity, it is necessary to develop programs that ensure the identification of generated products (similar to "Anti-Plagiarism" programs).

Students who have access to neural network tools tend to stop thinking creatively in a digital environment, losing their skills in independent research and working with literature, etc. There is a need for new didactic tools that allow the neural network to function as a "compliant textbook" rather than as a "covert prompter".

The modern digital environment should be explored not only from the perspective of developing didactics but also from the standpoint of developing anthropology - specifically, the role and place of the educator within this environment, capable of influencing students through their personality and altering or controlling the trajectory of the student's interaction with neural network resources.

References

Amrenova, M.M.: Problems of improving the procedure for assessing the educational results of university students. Probl. Mod. Pedagog. Educ. **80–2**, 14–16 (2023)

Gabdulchakov, V.F., Scishova, E.O.: The anthropology of teacher training in interdisciplinary and digital education. Chapter 8. In: Progress in Education, vol. 79, pp. 181–194. Nova Science, New York, NY (2024) https://shelly.kpfu.ru/e-ksu/docs/F1588758500/The.Anthropology.of.Teacher.Training.pdf?p_random=66936

Gabdulkhakov, V.F., Zinnurova, A.F.: Artificial intelligence in the process of forming subject and functional literacy of students. International forum 9–11.09.2024 KAZAN DIGITAL WEEK – 2024: collection of materials / compiled by: r.Sh. Akhmadiyeva, R.N. Minnikhanov; under the general editorship of corresponding member of the Academy of Sciences of the republic of Tatarstan, doctor of technical sciences, professor R.N. Minnikhanov. Kazan: OSP "NTsBZhD RT AS". Part 1. 994–999 (2024)

Garayev, T.K., Novik, N.S.: Artificial intelligence in higher education through the eyes of students. The world of education - education in the world. Sci. Methodol. J. **3**(91), 221–229 (2023)

Gorina, E.V., Ufimtseva, S.M.: Features of the use of neural network texts in media and education. Russ. Linguistic Bull. **1**(49), 1–4 (2024)

Ermolaeva, M.V., Isaev, E.I., Lubovsky, D.V.: Assessment of personal educational results of students in the process of studying in the Master's degree. Educ. Iss. **4**, 193–203 (2020)

Khvarzhev, V.M., Tsebrenko, K.N.: Development of the concept of a student's personal account using artificial intelligence technologies. Int. J. Human. Nat. Sci. **5–4**(92), 147–150 (2024)

Kim, T.K., Kuzmenko, G.A., Essebbar, K.M., Maryin, I.S.: Specification of the system of requirements for quality and performance results as a condition for improving the educational process of students of the Institute of Physical Culture. Sport Tourism Teacher XXI Cent. **3**, 188–199 (2022)

Kupriyanov, R.B., Semenov, A.L.: Analysis of the Dynamics of Educational Results of Students of a Large Pedagogical University. Bulletin of Moscow State Pedagogical University. Ser. Comput. Sci. Inf. Educ. **1**(43), 66–71 (2018)

Kurbanova, Z.S., Ismailova, N.P.: Neural networks in the context of digitalization of education and science. MNKO. **3**(100), 309–311 (2023)

Martynova, T.N., Pfetzer, A.A.: Psychological and pedagogical support for the professional and personal development of students in the digital educational environment of the university. Prof. Educ. Russ. Abroad. **1**(45), 145–152 (2022)

Malygin, A.A.: Certification of students in modern conditions: from increasing the effectiveness of assessment to trust in assessments. Higher Educ. Today. **6**, 8–16 (2023)

Novik, N.N., Evseeva, O.P., Zaitseva, K.D.: Use of artificial intelligence in preschool education. The world of education - education in the world. Sci, Methodol. J. **1**(89), 102–110 (2023)

Rubtsova, T.P.: Quality of education: monitoring and evaluation of educational results. Bull. Samara Univ. Hist. Pedagog. Philol. **4**, 66–70 (2023)

Samarina, A.E., Boyarinov, D.A.: Neural networks for image generation: pedagogical potential in higher education. Sci. Methodol. Electr. J. "Concept". **11**, 161–179 (2023). https://doi.org/10.24412/2304-120X-2023-11116

Feshchenko, T.S.: On the issue of monitoring the educational results of student teachers. New view. Int. Sci. Bull. **6** (2014) https://cyberleninka.ru/article/n/k-voprosu-o-monitoringe-obrazovatelnyh-rezultatov-obuchayuschihsya-pedagogov. Accessed 4 Feb 2025

Lapina, M., Tokmakova, M., Movzalevskaya, V., Karabasheva, M., Balakrishnan, D.: Analysis of modern methods and approaches using it in education and pedagogy. In: Lapina, M., Prakasha, G.S., Grigoriev, S. (eds.) SLET 2022: International Conference on Innovative Approaches to the Application of Digital Technologies in Education and Research. SLET 2022. Lecture Notes in Networks and Systems, pp. 40–53. Springer, Cham (2024a)

Lapina, M., Shiriaev, E., Babenko, M., Istamov, I.: High-speed convolution Core architecture for privacy-preserving neural networks. Program. Comput. Softw. **50**, 417–424 (2024b). https://doi.org/10.1134/S0361768824700282

Rusanov, M., Babenko, M., Lapina, M., Sajid, M.: Identification of exploited unreliable account passwords in the information infrastructure using machine learning methods. Big Data Cognit. Comput. **8**, 159 (2024). https://doi.org/10.3390/bdcc8110159

Artificial Intelligence Systems for Automated Scientific Research

Ulyana Osipova[1(✉)] and Vyacheslav Zolotarev[2]

[1] Siberian State University of Telecommunications and Information Science, Kirova Avenue, 86, 630102 Novosibirsk, Russia
uljana.pavlova2012@yandex.ru

[2] IT Security Department, Siberian State University of Science and Technology, Krasnoyarskiy rabochiy Avenue, 31, 660000 Krasnoyarsk, Russia

Abstract. In today's rapidly evolving landscape, where economic progress hinges on scientific and technological advancements, the integration of Artificial Intelligence (AI) into scientific research is gaining momentum. This paper explores the potential of AI systems to automate and enhance various stages of the scientific research process, addressing the increasing demands placed on young scientists and graduate students. It examines how AI tools can be leveraged to improve efficiency, objectivity, and innovation across key research activities, from literature review and hypothesis generation to experimental design and data analysis. The study analyzes the benefits of AI in facilitating tasks such as automated data collection, pattern identification, and the interpretation of complex datasets, highlighting the advantages of using platforms like Semantic Scholar for targeted literature searches. However, the paper also critically assesses the challenges and limitations associated with AI-driven research, including concerns about data quality, algorithmic bias, security vulnerabilities, and ethical considerations related to data privacy and transparency. The analysis emphasizes the need for robust data validation, expert human oversight, and the development of appropriate training and skill sets to ensure the responsible and effective implementation of AI in scientific research. Ultimately, the paper advocates for a balanced approach that combines the power of AI with the critical thinking and domain expertise of human researchers, paving the way for more impactful and groundbreaking discoveries. Further, it puts forth a series of core principles to help guide effective implementation of AI tools for research.

Keywords: Artificial Intelligence · Scientific Research · Education

1 Introduction

In today's world, where the pace of economic development is accelerating rapidly, employers place high demands on specialists, including their qualifications and the organization of corporate work processes. Young scientists and graduate students are challenged to initiate significant projects, conduct research and development (R&D), and solve specific practical problems using tools that align with modern scientific and

M. Lapina et al. (Eds.): DET 2024, LNNS 1616, pp. 342–350, 2025.
https://doi.org/10.1007/978-3-032-04365-8_34

technological advancements. The outcome of such research can include intellectual property and the potential for participants to earn academic degrees. Consequently, one of the most important tasks for those involved in education, from schoolchildren of all ages (especially high school students) to university students, is to develop their intellect, master scientific research methodologies, and acquire the ability to apply research findings in practical settings.

Artificial intelligence methods and machine learning algorithms are currently used both in manufacturing, to improve automation efficiency, and in medicine, medicine, and financial structures for big data analysis. With the growing popularity of language models and the launch of open source GPT, artificial intelligence methods have entered even more areas of human activity [9, 10].

The use of artificial intelligence (AI) in this context not only allows for the delegation of certain tasks, leading to more efficient time management, but also enables exploration of problems from diverse perspectives, potentially revealing conclusions that are not immediately apparent. Currently, it's also crucial to consider that continuous education models must automatically adapt to the evolving landscape of requirements for the application of various AI tools [1]. By applying specific concepts and best practices, this challenge can be addressed comprehensively. An example of a conceptual framework illustrating such an application is shown below (Fig. 1).

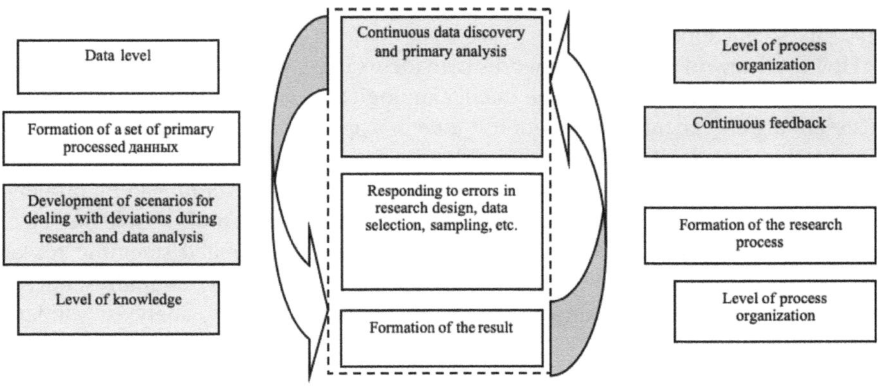

Fig. 1. Concept of AI application in scientific research considering deployment levels.

It is interesting that automation of processes in the field of education concerns both educational processes themselves and science, therefore, further in the text of the work, related tasks are also considered.

Obviously, the use of artificial intelligence tools, such as large language models, should not be episodic. If such tools are introduced as an element of process digitalization [2], then it is necessary to approach the implementation as an expansion of the automation area to a framework, a set of related tools that implements not only the task of automating scientific research, but also data collection or other related tasks. The article contains a number of recommendations in this area, but, of course, it needs further development.

2 Field Analysis

The use of artificial intelligence in education is a relatively nascent and evolving field. Consequently, the body of existing literature describing specific applications and experiences is rather limited. In works [3, 4], the authors provide examples of higher education institutions internationally where artificial intelligence is actively integrated into the educational process. For example, the Educational Testing Service (ETS) employs an electronic assessment system, also known as the Electronic Assessment Engine (EAE), which is an automated system used to evaluate graduate students' writing. Artificial intelligence within the system analyzes the characteristics of a student's essay writing, providing assessment and feedback.

The Buryat Institute of Infocommunications SibSUTIS utilizes its own AI-driven development to determine the most suitable IT specialization for individual students [3]. A combination of established neural networks was selected as the foundation for constructing a decision model.

Tomsk State University (TSU) has developed its own policy framework for the ethical integration of AI in the educational process [4]. In the near future, through the implementation of this policy, TSU intends to establish a mechanism for addressing ethical disputes arising from the use of artificial intelligence in educational activities. The AI ethics policy is directed at all stakeholders in the educational process: students, faculty, administrative and managerial personnel, and external parties involved in the university's educational activities.

However, few of the existing works provide examples of AI being used to support real scientific research tasks. Given that technological development is an indispensable and essential component of any economy, there is a pressing need to identify and describe tasks in which artificial intelligence could serve as an assistant to researchers.

To effectively meet the demands of scientific research, it would also be prudent to consider the following addition: Primarily, artificial intelligence tools should unlock new avenues for data access and processing. By leveraging automated scientific research frameworks, researchers gain the ability to work with data more rapidly and accurately, accessing a broader range of data sources for collection, analysis, validation, and verification.

From the perspective of performing scientific research, the development of digital twins of objects relevant to the researcher's work could also be explored. Throughout the research process, and even considering the continuous influx of new source data, these digital twins could dynamically evolve under the direction of, among other things, artificial intelligence tools.

In general, the effectiveness of scientific research can have a direct dependence on the quality and quantity of sources of source data [5], but the complexity of integrating various data sources for such solutions for automating scientific research can become unacceptable due to possible incomplete or inappropriate approaches and solutions.

Therefore, the concept should, at a basic level, address:

1. Data sources;
2. Data management policies for automated scientific research;
3. Automated processes and managed digital objects;

4. Knowledge management.

The concept of applying AI in scientific research taking into account processes is shown in the Fig. 2.

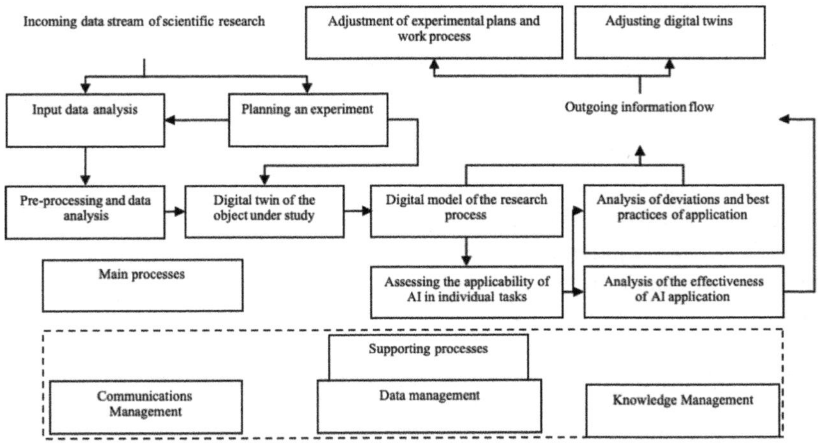

Fig. 2. The concept of applying AI in scientific research taking into account processes.

It is interesting that the automation of processes in the field of education, as well as scientific research, concerns both educational processes themselves and science, therefore, further in the text of the work, related tasks are also considered.

3 Methodology of Automated Scientific Research Systems

3.1 Selecting a Research Topic

Most often, a senior colleague or a teacher helps a postgraduate student choose a research topic. Having assessed the area of your own interests, you can try to independently determine the range of topics that would be interesting from the point of view of further research. However, quite often young scientists are not able to accurately formulate a suitable topic. In this matter, artificial intelligence is quite capable of helping. Having made the right requests, you can ask it to review current issues on the topic of interest, as well as generate specific formulations for the title of future scientific papers.

3.2 Analysis and Processing of Scientific Data

After defining the topic, identifying the problem, setting the goal and objectives, the stage of searching, analyzing and processing scientific data begins. Artificial intelligence systems have an undeniable advantage in this matter, as they are able to search for the requested data, generate algorithms for their processing, and also look for dependencies that may not be visible at first glance.

However, even if the most suitable data is obtained, it is necessary to carry out independent primary processing, as well as verify the reliability of the data obtained, since this is the criterion for evaluating the entire scientific work. Independent processing can reveal missing or not entirely correct results, or, conversely, confirm the correctness of the actions taken.

It is optimal to choose an integrated approach for processing, including several methods at the same time. Correct processing of results entails the formulation of specific conclusions on the tasks set.

3.3 Automatic Hypothesis Generation

The introduction of automated hypothesis generation entails a transition to more objective knowledge. AI-based tools are capable of moving away from standard ideas and, in some cases, biased judgments, and identifying patterns in input data arrays that an ordinary person would find insufficiently expressed. The ability to process huge data arrays, identifying patterns that may go unnoticed during manual analysis, will allow graduate students to formulate innovative hypotheses based on empirical data, rather than intuitive assumptions. The introduction of artificial intelligence into the research process will ensure the optimization of time resources, freeing up time for more in-depth analysis, experimentation, and critical evaluation of the results.

3.4 Automated Design of Experiments

The experiment design is a careful planning and organization of the process and includes stages from formulating hypotheses to analyzing the obtained results. The key goal of planning is to obtain the most informative results at minimal cost. By automating the design of scientific experiments using artificial intelligence, graduate students will be able to focus on the substantive part of the study, and entrust routine tasks to intelligent systems.

Beginning researchers can miss important stages of implementing a scientific experiment, while many popular language models based on artificial intelligence will suggest which stages can be added to the experiment, and describe the features of each of them. This process can be iterative, with repeated revision and adjustment of plans based on new data or emerging problems, but for researchers with little experience, it is important to create a foundation of key stages.

3.5 Automatic Search and Analysis of Scientific Literature

When searching and processing scientific texts, it is necessary to assess the relevance of the materials found. To improve the relevance of results, it is possible to use search engines developed on the basis of artificial intelligence specifically for scientific texts, for example, Semantic Scholar [5].

Search queries that return tens of thousands of results in popular search engines return several hundred in Semantic Scholar, all directly relevant, which helps limit the distraction to less relevant studies. At the moment, Semantic Scholar predominates in the biomedical literature, but the nature of the search may change as more links are added to this platform.

3.6 Automatic Interpretation of Experimental Results

Traditional approaches relying on manual statistical analysis and subjective researcher experience are often ineffective when working with complex multidimensional data arrays. In this regard, the introduction of artificial intelligence tools into the process of interpreting experimental results opens up new prospects for increasing the speed, accuracy, and objectivity of scientific research, and also poses the challenge for higher education to train specialists capable of effectively using these advanced technologies.

With the help of artificial intelligence, it is possible to identify anomalous data points that indicate errors in the process of collecting or obtaining data or the existence of previously unknown phenomena, and also to compare the obtained results with existing scientific literature and knowledge bases to determine whether they correspond to generally accepted concepts or represent a new discovery.

Despite numerous advantages, the use of AI in the interpretation of experimental data is associated with a number of difficulties. The effectiveness of artificial intelligence directly depends on the quality of the data used to train models, so incomplete, inaccurate, or biased data can lead to erroneous conclusions. The use of artificial intelligence in scientific research may also raise ethical issues related to data privacy, transparency of algorithms, and responsibility for hypotheses put forward.

3.7 Key Considerations Regarding the Concept

In brief, the plan for automated scientific research driven by the constant integration of artificial intelligence tools shows both considerable benefits and built-in limitations that might affect its overall success. While automation promises greater efficiency and scale, a complete evaluation of its downsides is essential for a responsible rollout. The following points deserve careful consideration:

- It is assumed that the data obtained by automated information collection is sufficient to perform scientific research. This does not fully correspond to the empirically obtained results. As a result of data collection, distorted, incomplete or falsified data may be obtained;
- It is assumed that the data obtained at the stage of automated collection will have a structure suitable for scientific research. At the same time, there is a problem of deduplication of data obtained in this way, and some other difficulties [7, 8];
- The problem of defining the boundaries of the area of data collection and analysis, as well as experimental research, is not solved only by drawing up the necessary instructions (programming, configuration) of artificial intelligence tools. For direct work with configuration and programming, it is necessary to involve a highly qualified specialist. In addition, in particular, when working with neural networks, one should also take into account the randomness factor that operates during neural network training, as well as training errors (for example, when forming a training sample);
- The problem of countering security threats, which should also be taken into account during the continuous integration of artificial intelligence tools. In particular, it is possible to deliberately distort the results by influencing the neural network training process. Thus, it can be assumed that the specified scheme may be ineffective in the event of attacks on the tool or the technology it uses;

– The issue of data visualization during automated scientific research also requires the need to generate synthetic data (additionally taking into account factors that allow you to control the problems of using artificial intelligence tools). The creation, use and visualization of such data is a non-trivial task.

The scheme of application of tools taking into account possible problems is shown in Fig. 3.

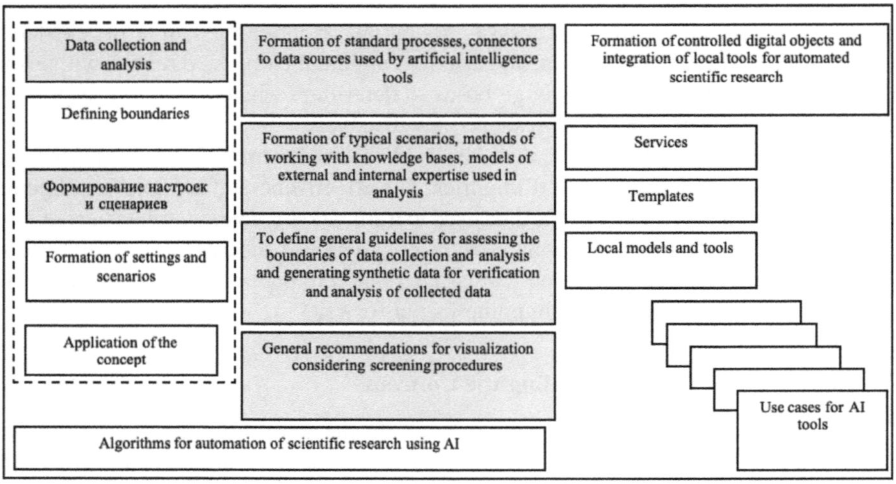

Fig. 3. Scheme of application of tools taking into account possible problems.

Considering the points above, we recommend these core principles to ensure successful AI tool implementation in automated scientific research:

– The data and knowledge used by artificial intelligence tools must be taken into account by the researcher. The procedures for working with them must meet the standard requirements of scientific research, such as reproducibility;
– When managing automated scientific research, the boundaries, control parameters, criteria, and verification procedures must be clearly defined by the researcher to minimize various errors. This part of the work cannot be completely (at the moment) performed automatically;
– At the same time, if digital objects are used that imitate research objects, it must be taken into account that the research is abstracted from the state of the real object. In this case, the data collected by artificial intelligence tools can have a higher level of abstraction than when using a real physical object;
– External requests for data and knowledge must be controlled and must not change the accumulated data and knowledge from the outside;
– Working with the context of requests and the formation of scenarios must not distort the boundaries of the study or the collected primary data, since a situation is possible when the collected primary data can be used to verify the results of the experiment.

In addition, there should not only be a repository of collected primary data, but also (within the framework of automated scientific research) there should be an automated way to access primary retro data, including the results of verification and analysis, to check the result of a specific query or action. It is also worth noting that the researcher must have sufficient qualifications to formulate accurate and well-written queries to artificial intelligence tools. Moreover, to ensure constant updating, it is necessary to regularly provide the AI with relevant theses, hypotheses, etc., obtained as a result of the research.

4 Conclusion

Using artificial intelligence to automate scientific experiments can significantly improve the efficiency and quality of research, it is important to remember that it is just a tool, and its successful use depends on your knowledge of the subject area and critical thinking, and it cannot replace the expert knowledge of a researcher and his critical thinking.

So, the main processes of automated scientific research using artificial intelligence tools should include (Fig. 4):

- data requirements management, including collection, storage, processing, verification and access to retrodata;
- research environment requirements management;
- development of scenarios for the automated use of AI tools (playbooks);
- development of training samples and datasets for local models and AI tools.

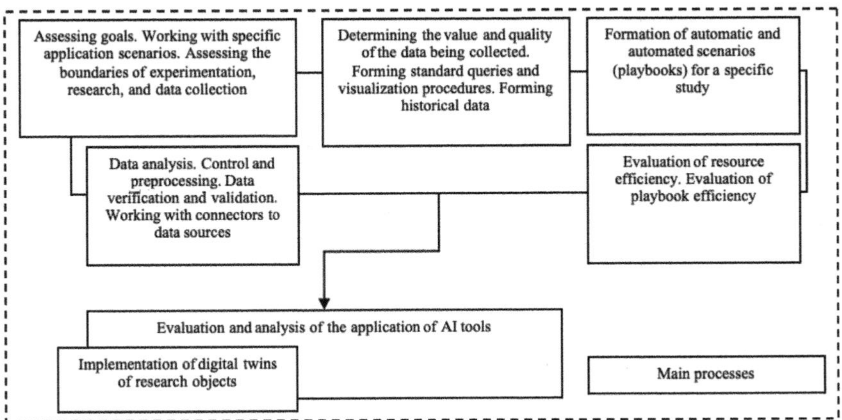

Fig. 4. General requirements for the implementation of the concept (processes).

References

1. Skatova, M.M.: Assessment of requirements of regulatory documents on the use of artificial intelligence in higher education. Technoeconomics. **2**(9), 22–33 (2024)

2. Tikhomirov, V.: Development of university's web-services smart education and smart e-learning. Smart Innov. Syst. Technol. **41**, 265–271 (2015)
3. Sirotyuk, O.L.: Use of artificial intelligence in Western education. Lang. Cult. Soc., 126–139 (2019) [in Russian]
4. Spichak, K.D.: World experience of using artificial intelligence in the field of education. In: Place and Role of Russia in the World Economy, pp. 175–180 (2020)
5. Jain, A., et al.: Overview and importance of data quality for machine learning tasks. In: Proceedings of the 26th ACM SIGKDD International Conference on Knowledge Discovery & Data Mining, pp. 3561–3562 (2020)
6. Fricke, S.: Semantic scholar. J. Med. Libr. Assoc. JMLA. **106**(1), 145 (2018)
7. Navon, R., Goldschmidt, E.: Monitoring labor inputs: automated-data-collection model and enabling technologies. Autom. Constr. **12**(2), 185–199 (2003)
8. Lapina, M., Prakasha, G.S., Grigoriev, S.: International Conference on Innovative Approaches to the Application of Digital Technologies in Education and Research: SLET 2022 - Conference Proceedings (2022)
9. Lapina, M., Shiriaev, E., Babenko, M., Istamov, I.: High-speed convolution Core architecture for privacy-preserving neural networks. Program. Comput. Softw. **50**, 417–424 (2024). https://doi.org/10.1134/S0361768824700282
10. Rusanov, M., Babenko, M., Lapina, M., Sajid, M.: Identification of exploited unreliable account passwords in the information infrastructure using machine learning methods. Big Data Cognit. Comput. **8**, 159 (2024). https://doi.org/10.3390/bdcc8110159

Modern Sources Application of Open Education by a Mathematics Teacher in Classrooms and Extracurricular Activities

Marina V. Ovchinnikova[1(✉)], Irina N. Zinenko[1], Olga V. Panisheva[2], Olga V. Kudelina[3], and Elena V. Korneyeva[4]

[1] V.I. Vernadsky Crimean Federal University, Simferopol 295007, Russia
mv_ovchinnikova@bk.ru
[2] Lugansk State Pedagogical University, Lugansk, Russia
[3] Crimean Engineering and Pedagogical University the Name Fevzi Yakubov, 295015 Simferopol, Russia
[4] Kerch State Maritime Technological University, Kerch 298309, Russia

Abstract. The article discusses the results of some studies of distance learning. Distance learning has developed rapidly due to the coronavirus pandemic. The forced use of this form of education has shown that teachers are not ready to work in such conditions. The main advantages and disadvantages of using distance learning, the main areas of using the experience of their application in the post-pandemic period are analyzed. The development of distance learning stimulates the development of open education as a modern phenomenon. The need and demand for affordable, free education is ensured by the use of electronic distance courses based on modern learning management systems with strict regulation of the content and process of learning. At the same time, open education provides students with the freedom to choose the time and place of study. The article provides an analysis of modern domestic open education resources available for use in teaching mathematics to schoolchildren, using the capabilities of developing digital technologies in lessons and extracurricular activities in mathematics. The Russian Electronic School, Sferum and others are considered as the main domestic platforms. The first of the named platforms was developed within the framework of the departmental target program of the Ministry of Education and Science of the Russian Federation "Russian Electronic School". The capabilities of the most common software tools in activating the educational and cognitive activity of students and increasing the level of subject knowledge are also shown.

Keywords: Open Education · Teaching Mathematics · Digital Technologies

1 Introduction

At the present stage of development of education, the final result of the educational process is the formation of a competent schoolchild. The goals and objectives of mathematical education, which are formulated in the Concept of Mathematical Education

M. Lapina et al. (Eds.): DET 2024, LNNS 1616, pp. 351–362, 2025.
https://doi.org/10.1007/978-3-032-04365-8_35

[1], are specified and expanded in connection with the constantly changing needs of society. The 2020 pandemic gave impetus to the development of distance learning, as well as the means for its implementation [2–4]. In addition, these conditions have most acutely determined the need for a teacher's ability to organize training in all formats: face-to-face, distance and blended.

The use of the obtained and new developments allows increasing the effectiveness of a modern mathematics lesson. The effectiveness of a mathematics lesson is characterized by the degree of achievement of the set goal of pedagogical activity, taking into account the optimality of the efforts, resources and time spent. The teacher is always learning, improving himself, mastering new ways to increase the effectiveness of lessons, new teaching technologies. In addition, the teacher needs to be able to respond to all sorts of changes in any area of life, since students react to everything instantly. This requirement has become more relevant than ever after 2020, when all educational institutions were forced to switch to a distance learning format. Learning using digital technologies required significant efforts from teachers and students. Many were not ready for such learning. The results shown by ninth and eleventh grade students in the main and unified state exams in mathematics were lower than in previous years.

Now, when returning to face-to-face forms of organizing education, a modern mathematics teacher has the opportunity to use the positive experience gained during the pandemic in the educational process in the classroom.

It should be noted that the preparation of a mathematics teacher for the use of digital technologies in the learning process cannot be considered separately; it is a subsystem of the integral system of his professional training in the conditions of a university and the system of advanced training. The pedagogical reality is changing at an exponential rate, so the principle of "Education throughout life" remains relevant.

2 Task Setting

The development of distance learning forms has prompted the development of digital platforms that allow these forms to be used most effectively. In turn, the development of such platforms has made it possible to more actively use their capabilities in the application of open education resources.

A modern mathematics teacher must be ready to use all the capabilities of existing and emerging digital tools in their professional activities. The formation of such readiness in working and future teachers, as experience shows, does not occur spontaneously. A special system of activities is needed that will allow the teacher to navigate the problem.

In connection with the inextricable link between the digitalization of education, the development of distance learning forms and open learning, we will analyze some studies on these issues.

We analyzed modern publications devoted to the professional training of mathematics teachers [2–8]. Thus, the study [4] displays the main aspects of the use of multimedia technologies for the formation of professionalism of a future mathematics teacher. The authors raise the question of the place, time and expediency of using multimedia tools in the holistic pedagogical process. The publication [5] describes a set of features of preparing a teacher to use HTML 5 in creating their own web applications and multimedia presentations for mathematics lessons.

The study [6] is devoted to the problem of postgraduate teacher training. The authors defined the requirements for the information and educational environment of the educational institution and the qualifications of the teacher for organizing work in it; the content of courses to ensure the required level of qualification of teaching staff; developed and described technologies for improving the qualifications of teachers in the information and educational environment. Cloud and distance technologies were used as basic ones. The author's content of the advanced training course, designed for teachers with different levels of training in the field of IT, is posted in the cloud data storage and distance learning system Moodle.

The use of cloud technologies in the professional training of a mathematics teacher was also discussed [7]. The results obtained are successfully used by the authors at the present stage.

As the analysis shows, the organization of distance learning in modern realities is devoted to the publications of many authors [8–12].

The positive and negative aspects of forced distance learning are analyzed. We agree with the opinion of A.V. Solovov and A.A. Menshikov, who, based on a system analysis, considered two main didactic approaches to distance learning. These are classes in a virtual classroom using traditional methods of the classroom-lesson system of teaching, and independent cognitive activity of students in learning management systems using specially prepared digital educational resources. The authors specified the requirements for the structure of an electronic distance course in modern learning management systems, which provide for the possibility of freely choosing the time and place of educational work with the necessary regulation of the content and process of training inherent in formal education institutions [13]. However, we cannot support the opinion that the main disadvantage of using distance learning is attempts to "implement the rudiments of the usual classroom-lesson system of teaching with the help of innovative electronic technologies" [13]. There are specialties for which distance learning is an exclusively forced measure, regardless of the quality of teaching, the skill of the teacher, the quality of the digital platform, technical equipment [14]. It is impossible to remotely train a doctor, engineer, artist, actor and many other professions. The conducted study shows, in our opinion, the main disadvantage of distance learning - the lack of universality.

The study [15] examines additional organizational conditions that must be met when organizing distance learning. These conditions remain relevant for the use of open learning tools, both in person and in other formats. These include: the ability to quickly communicate with the teacher (psychological comfort); duplication of information on various resources (taking into account the technical capabilities of each student); a common platform for various disciplines (allows you not to scatter attention to various interfaces); modern and effective technical support. A separate point is the need to encourage teachers who apply and adapt digital forms of organizing learning with high professionalism.

Adaptation to modern conditions and the use of open education sources both as information resources and as means of organizing learning are considered in [16].

Research into digital educational environments has gained momentum [17]. To build a conceptual model of a modern digital educational environment of an educational institution, the authors used a systems approach. The digital educational environment is presented as an organizational and technical system, the components of which are marketing subsystems, a set of technical means, various types of support and an organizational subsystem. Digital technology-based learning tools are becoming a focus of attention for higher education methodologists. Augmented reality is becoming a full-fledged learning tool in the educational process of a university [18], school [19], and in additional education [20, 22].

3 Results

In this article, we plan to analyze the current state of the use of digital open education platforms by mathematics teachers from the standpoint of using digital technologies in their maximum possible manifestations.

According to the curriculum for students studying the program 44.03.05 Pedagogical Education with two training profiles (Mathematics, Computer Science) and 44.04.01 "Pedagogical education", the Master's program "Mathematics in professional education" does not provide for the study of academic disciplines related to the digitalization of education, distance learning technologies, open education. Only individual topics within the framework of methodological disciplines are considered. Therefore, to form a holistic picture, the capabilities of all types of educational pedagogical practice in the real conditions of educational institutions are used. At the same time, students, practice supervisors and mentors simultaneously learn and teach each other, share experiences.

Various research tools helped us determine the state of use of open educational sources and modern digital technologies in the professional activities of mathematics teachers. Observations, conversations, questionnaires, analysis of attended lessons, lesson plans, electronic journals, subject chats on Telegram, VKontakte, Sferum platforms and other resources available to a modern mathematics teacher were used. To clarify the attitude of mathematics teachers to the use of open education tools, we conducted an online survey using https://forms.yandex.ru/.

Dear mathematics teachers, please take part in a survey dedicated to the use of modern, relevant open education resources in teaching mathematics in 2024 (Table 1).

Table 1. Application form.

No	Dear Colleagues! Please answer the questions posed
Question 1	Do you use modern open learning platforms in your professional activities? Yes No
Question 2	What modern services for real-time communication have you used to conduct online consultations? Sferum Vkontakte Telegram Other

(*continued*)

Table 1. *(continued)*

No	Dear Colleagues! Please answer the questions posed
Question 3	What was the duration of one online consultation (in real time)? 20 min 30 min 35 min No online consultations were conducted
Question 4	What modern services for communication in correspondence mode have you used? Sferum Vkontakte Telegram Other
Question 5	You spent time preparing for an online consultation... More than usual Less than usual The same amount You did not conduct online consultations
Question 6	What tools did you use to organize elements of open learning in a distance format? Test shells and platforms where you had to fill in (create) the content yourself Ready-made developments (online courses, already created lessons or assignments on various platforms, etc.) Videos from RuTube Materials from the Yaklass and Foxford platforms Materials from the Sferum and Vkontakte platforms Materials from the Reshedu platform Multimedia content was not used, we worked according to the textbook other
Question 7	What learning platforms have you used in your work with students to learn mathematics? Yaklass FIPI Sdamgia Foxford Other:
Question 8	Have you conducted tests and independent work in a blended learning environment (distance learning)? Yes No

(continued)

Table 1. *(continued)*

No	Dear Colleagues! Please answer the questions posed
Question 9	In what form did you offer students independent and test assignments? You sent photos of assignments for students, and the student in response sent photos of the completed work You created the work in a certain test shell, students were given time to complete it Students wrote these assignments during online lessons in real time + sent photos of the completed work immediately after completing the allotted time Used ready-made independent or test assignments in various shells Did not offer to write assignments or independent work remotely Other:
Question 10	Have you checked the students' assimilation of theoretical material in distance learning? Yes No
Question 11	Is there a lag in accordance with the program this year? Yes No In some classes no
Question 12	Do you have time to review the course material for the year? Yes No In some classes no
Question 13	What has changed in students' attitudes towards learning in the context of using open learning platforms? Responsibility Adaptation Activity Self-organization Other:
Question 14	What are the results of using open education sources for you as a teacher? Mastered new forms of work Increased the level of computer literacy Improved the atmosphere in the classroom Improved relationships with parents Other:
Question 15	In which classes is it more effective to use open education sources? 5–6 7–9 10–11

(continued)

Table 1. *(continued)*

No	Dear Colleagues! Please answer the questions posed
Question 16	Which subjects were more difficult for students to learn using open education sources? Algebra Geometry Probability theory and mathematical statistics No difference was noticed
Question 17	What difficulties do you experience when using open education sources in your professional activities? My answer
Question 18	Have you completed any advanced training courses in the 2023–2024 academic year on the use of digital technologies in your professional activities? Yes No
Question 19	What features of the use of digital platforms, in your opinion: contribute to increasing the effectiveness of teaching mathematics negatively affect the effectiveness of teaching mathematics My answer

Thank you for your help and invite you to scientific cooperation! Respectfully yours, the staff of the Department of Mathematics and Informatics and of the Department of Higher Mathematics and Methods of Teaching Mathematics.

The survey was conducted among mathematics teachers of schools and institutions of secondary vocational education in Greater Yalta, schools in other regions of the Republic of Crimea. As well as graduate students of universities in Crimea, specialties related to the pedagogy of mathematics. A total of 170 respondents took part in the survey. We did not single out students separately, since all of them already work in schools. Analysis of the survey results showed that 150 (88.2%) of them use modern open education platforms in their professional activities.

Almost all of the respondents switched to the Sferum platform in their communication with students. This is due to the fact that Sferum has become the main free digital platform for schools associated with the electronic journal. However, there is some duplication in communication (100 respondents (58.9%) left the opportunity to communicate with students in Telegram chats. The reason is that not all parents agree to involve their children in work in Sferum. Most respondents (148 (87%)) spent no more than 20 min on online consultations, which, in our opinion, is in line with the norms.

The rest of those who used them (2 (13%)) spent 30 min. 20 (19.8%) did not conduct online consultations, who do not use digital platforms in principle. Of those who conducted online consultations, almost all answered that it did not take them much time. Note that this indicator has changed compared to 2020 [11]. According to our data, previously, a teacher needed more time to prepare for an online lesson than to prepare for face-to-face interaction.

As a tool for distance learning, 123 (72.3%) respondents used ready-made content, and only 27 (27.7%) developed the content themselves. Note that 8 of them were teachers of mathematics and computer science, as well as working students. Accordingly, 82 (48.2%) respondents used videos on Rutube, 54 (31.8%) used materials from the Yaklass and Foxford platforms, 57 (33.5%) used materials from the Sferum and Vkontakte platforms, and 95 (55.9%) used materials from the Reshedu platform. Many respondents indicated the use of all of the listed platforms. As other sources, 72 (42.4%) respondents named materials from the Videouroki digital platform (https://videouroki.net). Directly in class (not for homework, not in extracurricular activities) for studying mathematics, 135 (79.4%) respondents used the Sdamgia platform (https://sdamgia.ru).

For control during the lesson (current and final), 100% of respondents preferred traditional forms. However, some respondents noted that they use Skysmart for promptly checking homework. Most of them were teachers (34 (20%)), whose workload was more than 18 h per week.

Theoretical material in a distance form (time-limited closed tests) was used by 114 (67%) respondents. In conversations, this was explained by the fact that the theory needs to be checked, but there is no way to listen to each definition of a concept, formulation of a theorem, formula, proof.

The respondents did not note any lag or advancement in the program, but they also did not note any changes in the students' attitude to learning. All age categories were chosen as the best age for use, with respondents noting that there are many more resources that can be used for preschoolers and primary school children. As a result of using open education sources for respondents as teachers, the majority (150 (88.2%)) chose to master new forms of work.

All respondents highlighted the same difficulty in studying subjects in the mathematical cycle, although later in the conversation they noted that developments in probability and statistics are in great demand, since educational and methodological kits for this subject have not yet been sufficiently tested.

Each of the teachers improves their qualifications in the use of digital technologies for professional activities once every 3 years.

As positive features of digital platforms in the work of a mathematics teacher, the following were highlighted: a possible reduction in the teacher's workload (with a fully equipped classroom with the necessary equipment), the ability to quickly select the necessary material for lessons, control and homework.

The negative aspects were: an abundance of mandatory digital resources on which students and teachers must be registered, forced mandatory use of certain platforms, constant presence in some resource, unavailability of full functionality on paid platforms, lack of skills in working with regular textbooks among students and much more.

At the same time, many teachers believe that the use of digital platforms harms the educational process more than it improves it.

4 Discussion

One of the ways to improve the quality of knowledge, as well as to develop students' interest in studying mathematics is the use of digital technologies in lessons and extracurricular activities in mathematics. Turning to open education sources increases students' interest in acquiring new knowledge through the resources of these platforms.

We have analyzed many resources that allow a math teacher to increase the intensity of teaching in the classroom, at home and in extracurricular activities. Modern educational platforms that can be used by a math teacher in their professional activities are Urait (https://urait.ru/), Yaklass (https://yaklass.ru/), Foxford (https://foxford.ru/), Skysmart (https://edu.skysmart.ru/). The experience of using the Urait platform is discussed in [21]. The listed platforms are useful for mathematics teachers, but unfortunately not all of their capabilities are openly available, free of charge, to anyone who wants them.

Thus, for full-fledged work with the Yurait platform, teachers and students must have an agreement with an educational organization (subscription of an educational organization). Higher education institutions conclude such agreements, which allows organizing the educational process on this platform. Teachers of V.I. Vernadsky Crimean Federal University, as part of a full subscription, have access to content search on the platform, the ability to design flexible courses, organize full-fledged work with a group of students in lectures, practical and seminar classes, organize all types of control (incoming, current and final assessment). Teachers also have the opportunity to post their own developments on the platform. Summer and Winter Teacher Schools are held twice a year for teachers and lecturers. The topic of the last winter school is "Smart Education: Digital Content, Services and Data". There are three types of participation in such schools: "Expert", "Qualified Participant", "Free Listener". Of these, only the last one is free. The portal works closely with the MTS ecosystem. For the first group "Expert" online webinars are held on "MTS Link", full access to all chats, full opportunity for personal communication with speakers and organizers in virtual rooms, all video recordings of the webinars are available. Individual trajectory is ensured by contacts with tutors and individual feedback. Control activities provided for obtaining a certificate of advanced training (108 h) include testing, completing a comprehensive task on working in the digital environment of the Yurait platform, as well as preparing an individual project. If the project meets the requirements, it is published in the School's collection with subsequent posting in the Russian Science Citation Index system. For the second group "Qualified Participant" online webinars are held on "MTS Link", all video recordings of the webinars are available. The form of control for obtaining a certificate of advanced training (72 h) includes testing, completing a comprehensive task on working in the digital environment of the Yurait platform. Participants of both groups receive full access to the Yurait platform for a year. For the "Free Listener" category, webinars are available as part of the VK broadcast. The final certification for obtaining a participant certificate involves working in the digital environment of the platform.

As the survey showed, none of the working teachers in Yalta used the services of the Yurait platform before getting to know it during our experiment. Students who were doing their internship regularly accessed this resource, since the faculty of the Department of Mathematics and Informatics and the Department of Higher Mathematics and Methods of Teaching Mathematics use the Yurait platform on a permanent basis.

The Foxford, Yaklass and Skysmart platforms also provide free access to incomplete functionality. Most of the capabilities of these platforms are not available without payment. Some materials that are available on these platforms for free are related to the electronic journal. They can be used even without registration. For example, a teacher can share with students a link to theoretical material posted on the Foxford platform. Foxford also organizes free subject Olympiads for students by age groups (preschoolers, grades 1–4, grades 5–8, grades 10–11).

On the Videouroki platform (https://videouroki.net), free unlimited location access (after registration) is opened to create classes, personal accounts of teachers and students, view statistics on student activities. You can distribute educational materials to your classes (created by other teachers and posted on the blog), demonstrate video content created by colleagues, use tests developed independently or by other teachers from the blog. You can use the electronic online notebook twice for free.

Teachers thank their colleagues who share their developments on the platform blog Unfortunately, free access to materials developed by platform professionals is available only with a paid subscription.

As the survey showed, many math teachers buy a monthly subscription because they can use electronic online notebooks, tests prepared by experienced professional methodologists, and professionally filmed video lessons in the classroom. It is also convenient to use a subscription to organize homework not only to receive online testing results. The teacher has the opportunity to receive information about the study of video content by each student. The platform has a control system that not only records all student actions when viewing each video lesson, but also formats statistical data on this viewing accordingly.

We analyzed the materials posted on free domestic digital platforms Russian Electronic School (Reshedu) (https://resh.edu.ru) and Sferum (https://sferum.ru). The materials presented on the first platform are available to both registered and unregistered users. Using the teacher's link or independently, the student can select a class, subject (mathematics, algebra, geometry, algebra and the beginning of mathematical analysis), and a lesson topic. Each lesson contains a theoretical part, a small but very informative video lecture, practical assignments in a test form with the possibility of verification and hints, control tasks in two versions, for the completion of which a limited time is allotted. The student finds out the result immediately.

Unfortunately, the platform does not have separate materials on probability and statistics, they are included in the materials on algebra. It turns out that the platform's functionality has not been updated in the last two years.

The platform is actively used to determine the level of functional literacy, prepare for the final state certification, and all-Russian tests. However, according to many teachers, the platform's functionality and interface are significantly inferior to platforms with paid content. The Sferum platform is still in development and is used mainly for safe, ad-free communication between students, teachers, and parents. Requires mandatory registration. Allows for video conferences, online consultations on subjects. There are not enough materials to organize mathematics teaching on the platform yet.

5 Conclusion

The use of open education resources in teaching mathematics has become firmly established in pedagogical reality. A modern teacher has the opportunity to turn to various digital platforms to improve the effectiveness of their lessons. At the same time, the issue of equalizing the quality of content provided on free and paid digital platforms is acute.

References

1. The Concept of Development of Mathematical Education in the Russian Federation (approved by the order of the Government of the Russian Federation of December 24, 2013, No. 2506-p). Moscow (2013)
2. Gorev, D., Gurevich-Leibman, I.: Experience of integrating various technological tools into the study and future teaching of mathematics to education students. Int. J. Math. Educ. Sci. Technol. **46**(5), 737–752 (2015). https://doi.org/10.1080/0020739X.2014.1002550
3. Smirnov, E., Tikhomirov, S., Dvoryatkina, S.: Self-organization technology of student's mathematical activities based on intelligent management. Perspect. Sci. Educ. **3**(45), 77–86 (2020). https://doi.org/10.32744/pse.2020.3.6
4. Gluzman, N., Sibgatullina, T., Galushkin, A., Sharonov, I.: Forming the basics of future mathematics teachers' professionalism by means of multimedia technologies. EURASIA J. Math. Sci. Technol. Educ. **14**(5), 1–13 (2018). https://doi.org/10.29333/ejmste/85034
5. Gluzman, N.A., Gorbunova, N.V.: Forming future teachers' competence in developing electronic educational resources using HTML5 as an alternative to adobe flash. In: CEUR Workshop Proceedingsthis link is disabled, vol. 2834, pp. 159–169 (2021) http://ceur-ws.org/Vol-2834/Paper14.pdf
6. Konopko, E., Pankratova, O., Nersesyan, E., Abdullaev, J.: Training of teachers for professional activity in the digital environment of the educational space. In: Proceedings of SLET-2019 - International Scientific Conference Innovative Approaches to the Application of Digital Technologies in Education and Research, Stavropol - Dombay, Russia, pp. 205–212 (2019) http://ceur-ws.org/Vol-2494/paper_18.pdf
7. Ovchinnikova, M., Shilova, L., Linnik, E.: Using means of Cloud Technologies of Training in the Study of Discipline "Discrete Mathematics". Professional-Pedagogical Training of Math Teachers. Information Systems and Technologies in Modeling and Control 2019: Selected Papers of the IV All-Russian scientific and practical conference with international participation "Information Systems and Technologies in Modeling and Control" (ISTMC'2019). (2019). http://ceur-ws.org/Vol-2522/paper15.pdf
8. Pankratova, O., Konopko, E., Konopko, P., Kormakova, V., Savelova, L.: Introduction and development of innovative methods and technologies of E-Learning at the University. In: CEUR Workshop Proceedings, vol. 2861, pp. 261–267 (2020) http://ceur-ws.org/Vol-2861/paper_30.pdf
9. Yadav, D.N.: The impact of digital learning on education. Int. J. Multidiscip. Res. Arts Sci. Technol. **2**(1), 24–34 (2024). https://doi.org/10.61778/ijmrast.v2i1.34
10. Tabassum, B., Moin, M., Abbas, Q., Kumbhar, M., Khan, M.: The impact of blended learning on student performance. J. Educ. Soc. Stud. **5**, 360–371 (2024). https://doi.org/10.52223/jess.2024.5217
11. Men, Q., Gimbert, B., Cristol, D.: The effect of self-regulated learning in online professional training. Int. J. Mob. Blend. Learn. **15**, 1–17 (2023). https://doi.org/10.4018/IJMBL.318225

12. Batrakova, I.S., Glubokova, E.N., Pisareva, S.A., Tryapitsyna, A.P.: Changes in University teacher's pedagogical activity in the context of digitalization of education. Higher Educ. Russia. **30**, 9–19 (2021). https://doi.org/10.31992/0869-3617-2021-30-8-9-9-19
13. Solovov, A.V., Menshikova, A.A.: Coronavirus zigzags of electronic distance learning. Higher Educ. Russia. **30**(6), 60–69 (2021). https://doi.org/10.31992/0869-3617-2021-30-6-60-69
14. Zakharova, U.S., Vilkova, K.A., Egorov, G.V.: This cannot be taught online: applied specialties in a pandemic. Educ. Issues. **1**, 115–137 (2021). https://doi.org/10.17323/1814-9545-2021-1-115-137
15. Blagoveshchenskaya, E., Garbaruk, V., Popova N.: Organizational and Pedagogical Problems in the Process of Distance Learning at a University. (2020). http://ceur-ws.org/Vol-2861/paper_23.pdf
16. Ovchinnikova, M.V., Linnik, E.P., Zinenko, I.N., Shilova, L.I.: The application of open education resources as a tool and object of study in the professional training of a mathematics teacher. In: CEUR Workshop Proceedings this Link Is Disabled, vol. 2914, pp. 224–233 (2021) http://ceur-ws.org/Vol-2914/paper19.pdf
17. Solovov, A.V., Menshikova, A.A.: Models of design and functioning of digital educational environments. Higher education in Russia. **30**(1), 144–155 (2021). https://doi.org/10.31992/0869-3617-2021-30-1-144-155
18. Taran, V., Savelova, L., Abdullaev, J.: Increasing learning motivation through the use of educational and technological tools. In: Lapina, M., Prakasha, G.S., Grigoriev, S. (eds.) International Conference on Innovative Approaches to the Application of Digital Technologies in Education and Research. SLET 2022. Lecture Notes in Networks and Systems, vol. 1222. Springer, Cham (2025). https://doi.org/10.1007/978-3-031-78776-8_7
19. Taran, V.N., M.I., U.: Learning technologies through augmented reality projects. In: VIth International Scientific and Practical Conference "Distance Learning Technologies" (DLT 2021).Yalta, Crimea, September 20–22, 2021, vol. 3057, pp. 253–259 (2021) ceur-ws.org/Vol-3057/paper30.pdf
20. Lapina, M., Tokmakova, M., Movzalevskaya, V., et al.: Analysis of modern methods and approaches using it in education and pedagogy. In: International Conference on Innovative Approaches to the Application of Digital Technologies in Education and Research, vol. 1222, pp. 40–52. Springer, Cham (2025). https://doi.org/10.1007/978-3-031-78776-8_5
21. Prakasha, G.S., Muttungal, P., Lapina, M., Taran, V.: Flipped classroom strategy in online teaching: challenges faced by higher education teachers. In: Lapina, M., Prakasha, G.S., Grigoriev, S. (eds.) International Conference on Innovative Approaches to the Application of Digital Technologies in Education and Research. SLET 2022. Lecture Notes in Networks and Systems, vol. 1222. Springer, Cham (2025). https://doi.org/10.1007/978-3-031-78776-8_24
22. Ovchinnikova, M.V., Zinenko, I.N., Linnik, E.P., Shilova, L.I., Panisheva, O.V.: Educational platform "Urait" as a resource of open education in the professional training of a mathematics teacher. In: CEUR Workshop Proceedings. DLT 2021 - Selected Papers of the 6th International Scientific and Practical Conference "Distance Learning Technologies", vol. 3057, pp. 190–196 (2021) https://ceur-ws.org/Vol-3057/paper21.pdf

Development of an Adaptive Mathematical Education System for Middle Grades Using Machine Learning

Tatyana Peleshenko[1], Artem Lidzhiev[1(✉)], Daniel Shkirya[1], and D. Vinodha[2]

[1] North Caucasus Federal University, Stavropol 355017, Russia
artemlidzhiev.4444@gmail.com

[2] Christ University, Bangalore, Karnataka 560074, India

Abstract. The aim of the study is to improve the study of mathematics topics for middle school children by developing a software implementation of an adaptive educational system using machine learning. During the research, the topic of quadratic equations was chosen as the basis for the research and development of an adaptive system. During the testing of the adaptive system, mistakes were specifically made to simulate the consolidation of knowledge during the educational process and make sure that it works and is able to adapt to the individual level of each student, increasing the level of knowledge gained and contributing to the consolidation of the material.

To achieve this goal, Python code was developed in the Jupyter Notebook development environment. Python libraries were also used, in particular the scikit-learn library for implementing machine learning. The presented approach and software implementation can be used both by teachers to check students and track their progress, and by students themselves to assimilate and consolidate the material and knowledge gained in the lessons.

The results obtained during the study demonstrate an increase in the effectiveness of adaptive learning methods using machine learning.

Keywords: Adaptive educational system · Adaptive learning methodology · Individual learning · Logical regression algorithm · Machine learning · Python · Quadratic equations · Scikit-learn library

1 Introduction

In the modern world, traditional teaching methods are ineffective, as each student requires an individual approach, and tests designed to test knowledge in the shortest possible time are not able to cover all the diversity of the topic being studied. An alternative approach was proposed by researchers Zhikharev A. G., Korsunov N.I., Mamatov R.A., Shcherbinina N.V. and Ponomarenko S.V. in their study on the development of an adaptive educational platform that uses machine learning technology. From this study, it is worth emphasizing the principle of generating academic content, focusing on the level of knowledge and needs of the student, where machine learning technology will be the tool to achieve this goal [1, 26].

M. Lapina et al. (Eds.): DET 2024, LNNS 1616, pp. 363–373, 2025.
https://doi.org/10.1007/978-3-032-04365-8_36

Similarly, in his work Kuzovlev D. D. pays attention to innovative approaches to mathematical modeling in the field of intellectual learning systems. The author emphasizes that the use of machine learning and deep neural networks in such systems contributes to a significant increase in the competitiveness of future specialists. This confirms the promising methodology for creating individual educational trajectories using adaptive educational systems [3].

Artificial intelligence methods and machine learning algorithms are currently used both in manufacturing, to improve automation efficiency, and in medicine, medicine, and financial structures for big data analysis. With the growing popularity of language models and the launch of open source GPT, artificial intelligence methods have entered even more areas of human activity [28, 29].

Besler A.I., Myachinov A.A., Seryakov A.V. and Korchevskaya O.V. in their article also emphasize the potential of using artificial intelligence (AI) and machine learning in the field of education, as they make it possible to effectively automate the processes of data collection and analysis, which in turn can contribute to the creation of individual educational programs for each student. The material of this scientific publication indicates the need to develop a system that, analyzing the student's interaction with the educational material, will make it possible to draw up an individual learning plan taking into account his knowledge and skills [4].

Continuing the discussion about the student's interaction with the material being studied, it is worth paying attention to the scientific article authored by N. A. Nazhova and A.V. Nazhov, which describes in detail the system of automatic verification of the program code of students of IT specialties, which also uses an adaptive approach. By giving students prompt feedback, the system frees teachers from a lot of work, allowing them to focus on other aspects of learning. This is very valuable during the educational process [2].

It is important to create a simple and intuitive interface for users, both students and teachers, as this is a tool for achieving the implementation of adaptive education methods. In this context, the work of Kanash A.V. and Mezina A. S. confirms the importance of data analysis and building machine learning models in the field of education. The authors of the work claim that these models can improve the effectiveness of education by providing an objective assessment of knowledge and the subsequent development of an individual curriculum [5, 27].

Mukhamadieva K.B. in her work "Machine learning in improving the educational environment" draws attention to the fact that machine learning strongly influences the pedagogical industry. The research focuses on the use of machine learning in teaching in order to improve the learning environment. But the main thing in this work is an individual approach to learning. This provides feedback to the teacher, and the tool in the form of real-time data processing facilitates this task and contributes to the implementation of this teaching methodology [6].

Jagwani A. in his work "A review of machine learning in education", he argues that the introduction of AI and machine learning directly affects education, improving its quality [7]. This point of view is confirmed by a study conducted by Deming Li, in which he studied an interactive assessment system using machine learning algorithms. The system demonstrated a classification accuracy of 92.1%, which confirms the effectiveness of using machine learning in education [12].

Lidia Sandra, Ford Lumbangaol and Tokuro Matsuo, in their paper "Machine Learning Algorithm to Predict Student's Performance: A Systematic Literature Review," explore predicting student success using data and machine learning. After analyzing the research articles, classification algorithms were identified that were most suitable for the task of predicting success. These include: ANN, Naive Bayes, Logistic Regression, and Decision Tree [8].

In the scientific paper "Application of Machine Learning in Higher Education to Predict Students' Performance, Learning Engagement and Self-Efficacy: A Systematic Literature Review", Juntao Chen, Xiaodeng Zhou, Jiahua Yao and Su-Kit Tang consider current ways of predicting students' interest and academic performance in online learning using machine learning. The study offers recommendations for improving the level of online education, where machine learning can also help to implement an individual approach [9].

Carlos R. Cunha, André Moreira, Sílvia Coelho, Vítor Mendonça and João Pedro Gomes in the article "Converging Extended Reality and Machine Learning to Improve the Lecturing of Geometry in Basic Education", considering the introduction of augmented reality and machine learning into the study of geometry, prove the effectiveness of machine learning in the study of exact sciences [10].

In their scientific work, Nga Pham and his team study the issues of assessment fairness in educational systems using machine learning. Precise setting of conditions in machine learning is able to implement automatic grading, which is included in the implementation of adaptive education methodology [11].

The work of Gerardo Ibarra-Vazquez and his colleagues analyzes how machine learning is used in education to predict the level of competence. Data on 326 students from 26 countries confirm that taking into account the perception of their knowledge can be used to predict competence. This highlights the importance of feedback from students when implementing adaptive education techniques using machine learning [13].

A study by Kingsley Okoye and his co-authors examines the automatic prediction of student retention and graduation. The results based on the use of machine learning show high accuracy, which confirms the need to use new technologies to build an individual learning plan to increase the level of knowledge of lagging students[14].

The article by Lu Zheng and his colleagues is devoted to methods of using evolutionary algorithms to create an educational platform. The authors propose a personalized course recommendation system, which makes it possible to adapt the educational process to the individual characteristics of students using big data [15].

The review conducted by Ekene Francis Okagbue and his colleagues demonstrates how the use of AI and machine learning in education is changing the teacher's interaction with students. A bibliometric analysis of 1,138 documents confirms the need to maintain this trend and the relevance of introducing innovative teaching methods [16].

The study by Li Zhu and his coauthors identifies the factors influencing student academic achievement in mathematics in the United States. Using machine learning in the course of their work, the authors identified key performance factors, which became an important step for applying the adaptive education methodology [17].

Nathan McJames and his colleagues studied the benefits of homework in mathematics and other sciences. Data on 4,118 eighth grade students in Ireland demonstrated that daily homework contributes to success in mathematics. The implementation of adaptive education can contribute to constant and enthusiastic problem solving [18].

The scientific work of Abdellatif Sellami and his co-authors examines how teachers relate to the integration of new technologies into the learning process. The study highlights the importance of supporting teachers when learning about innovations in the field of education for the implementation of adaptive teaching methods [19].

The work of Caroline Hall and Martin Lundin explores the impact of the "one computer per pupil" principle on the academic performance of elementary school students in Sweden. Using data from before the COVID-19 pandemic, the study found no significant effects from the introduction of technology on the results of national standardized tests in mathematics, Swedish and English, which indicates the primary importance of an adaptive approach rather than the technology itself, which is only a tool [20]. The scientific article by Natalia Zakharova and her colleagues is also devoted to the integration of distance learning into the standard educational process. The authors of the article propose an algorithm for developing educational programs for the implementation of distance learning, which is well in line with the implementation of adaptive learning [21].

In the work, Sung-Hee Jin and her colleagues are exploring ways to successfully apply SRL in an online environment. The result of the study is a list of recommendations for building AI applications in online learning for self-regulation during the educational process [22].

A study conducted by Victoria Mirata and her colleagues reveals the obstacles that arise in the integration of adaptive education in educational institutions. Despite the desire to apply the methodology of an individual approach to each student, it remains ineffective using traditional teaching methods, which emphasizes the need to introduce adaptive systems [23].

Similarly, the work of Alisa Lincke and her team examines the use of machine learning to predict student responses based on their interaction with the studied material. The study highlights the advantages of using machine learning, allowing not only to competently analyze the student's progress and level of knowledge, but also to develop an individual training plan for him [24].

In conclusion, Mustafa Yağcı in his research raises the issue of using data mining to predict students' academic performance. By analyzing the results of midterm exams using machine learning, it is possible not only to determine the level of academic achievement, but also to characterize each student for the subsequent construction of an individual learning trajectory [25].

Thus, an adaptive educational system in the field of mathematics using the principles of machine learning can improve the level of education and make the learning process more individual for each student, facilitating the work of teachers.

2 Methodology

To implement adaptive learning in a programmatic form, it is worth paying attention to the following key stages.

The first stage is the generation of unique examples of tasks for each student. To do this, you need to create a template in the program code that will replace variables with random numbers in a certain range, depending on the type and conditions of the task.

The second stage is the collection and analysis of student data by solving assignments generated by them on a given topic. To implement an adaptive approach, the system must be universal and able to interpret input data in the case of responses with non-integers.

The third stage is the generation of individual tasks that meet the needs and abilities of the student. By analyzing the correct and incorrect answers, the machine learning algorithm will be able to generate additional examples to consolidate the material based on the initial data of each of the solved tasks. Additional examples will be divided into relatively simple and relatively complex ones. Simple examples include all tasks that were generated based on tasks with correct answers. They are necessary to consolidate the material and provide an opportunity to make sure that the correct answers were not accidental and are an additional test of students' knowledge. Conditionally complex additional examples are tasks that were generated based on data from examples that were solved incorrectly. Along with conditionally complex examples, a detailed description of their solution should be displayed so that the student can familiarize himself with the information received and improve his problem solving skills.

The last stage of the adaptive learning approach will be the construction of an understandable system for all participants in the educational process. Using the program code to achieve this goal, a simple interface will be created that will be convenient for both students and teachers. The teacher will not only be able to easily track the progress of his students, but also easily analyze their gaps in learning the material using an algorithm. In turn, students will be able to interact with the program without any problems, which will have a positive effect on building their individual learning plan.

The topic of quadratic equations was chosen as an example for testing and studying the hypothetical possibilities of adaptive education methods in mathematics implemented as a software adaptive system. The code was written in Python in the Jupyter Notebook development environment. Libraries for mathematical operations (numpy) and symbolic calculations (sympy), as well as libraries directly related to data storage (pandas) and the implementation of machine learning algorithms (scikit-learn) were also used in the creation.

The function shown in Fig. 1 generates quadratic equations. It returns random coefficients a, b, c and a string of a quadratic equation in the format $ax^2 + bx + c = 0$:

```
def generate_quadratic_equation():
    a = np.random.randint(1, 10)
    b = np.random.randint(-10, 10)
    c = np.random.randint(-10, 10)
    return a, b, c, f"{a}x^2 + {b}x + {c} = 0"
```

Fig. 1. The quadratic equation generation function.

The function in Fig. 2 solves the equations. The input data for the function are the coefficient x^2(a), the coefficient x(b), and the free term (c) generated in the generate_quadratic_equation() function. After execution, the function returns two roots, one root, or complex roots, depending on the value of the discriminant.:

```
def solve_quadratic(a, b, c):
    x = sp.symbols('x')
    solutions = sp.solve(a*x**2 + b*x + c, x)
    return solutions
```

Fig. 2. The function of solving a quadratic equation.

The analyze_answers() function, whose code is shown in Fig. 3, compares the student's answers with the correct ones. The input data is taken in the form of a list of student's answers and a margin of error. At the output, the function outputs a list of logical answers, which indicates whether the example was solved correctly or incorrectly:

```
def analyze_answers(user_answers, correct_answers, tolerance=0.1):
    results = []
    for user_answer, correct_answer in zip(user_answers, correct_answers):
        if len(user_answer) != len(correct_answer):
            results.append(False)
            continue
        match = all(abs(u - c) < tolerance for u, c in zip(sorted(user_answer), sorted(correct_answer)))
        results.append(match)
    return results
```

Fig. 3. Response analysis function.

In this case, the machine learning model learns from mistakes made. After entering the answers, the system analyzes their accuracy. All responses are recorded in an array, the data from which is then used for a logistic regression model. This happens based on features (equation coefficients) and labels (correct or incorrect answers), which allows you to predict which equations will be difficult for the user and which ones will be simple.

Let's consider the operation of the algorithm from a mathematical point of view. Let a, b, and c be the coefficients of a quadratic equation. They are used to create an equation of the form:

$$ax^2 + bx + c = 0 \tag{1}$$

The error data set E is an array:

$$E = \{(a_i, b_i, c_i, y_i)\}, \tag{2}$$

where $y_i \in \{0,1\}$ is the mark of the correct answer (0 is false, 1 is true).

Based on the data obtained, a logistic regression is trained:

$$X = \begin{bmatrix} a_1 & b_1 & c_1 \\ a_2 & b_2 & c_2 \\ \vdots & \vdots & \vdots \\ a_n & b_n & c_n \end{bmatrix}, y = \begin{bmatrix} y_1 \\ y_2 \\ \vdots \\ y_n \end{bmatrix}, \quad (3)$$

where X is the feature matrix and y is the label vector.

Logistic regression minimizes the logarithmic loss function, which is used in training the model:

$$L(\beta) = -\frac{1}{N}\sum_{i=1}^{N}\left[y_i \log(p_i) + (1 - y_i)\log(1 - p_i)\right], \quad (4)$$

where p_i is the probability that $y_i = 1$, calculated using the formula:

$$p_i = \frac{1}{1 + e^{-(\beta_0 + \beta_{1ai} + \beta_{2bi} + \beta_{3ci})}} \quad (5)$$

The maximum likelihood method is used in the learning process:

$$\hat{\beta} = \arg\max_{\beta} L(\beta) \quad (6)$$

Next, the logistic function builds a prediction based on the input data:

$$\hat{y} = \begin{cases} 1, \textit{if } p \geq 0.5 \\ 0, \textit{else} \end{cases} \quad (7)$$

Figure 4 shows a simple and intuitive interface for the software implementation of an adaptive educational system. First, the generated equations are displayed in the compiler along with additional information about the rules and the input format. Even lower, the first task from the list is displayed and the user has the opportunity to enter the answer for each of the quadratic equations sequentially.

3 Results

In the process of entering answers, mistakes were intentionally made in cases where the discriminant was negative. Instead, the minus sign was deliberately ignored and all calculations were performed taking into account that the discriminant was taken modulo. This was done to test how the algorithm, extracting information from the student's answers, would issue assignments. Thus, it was possible to evaluate how the adaptive educational system processes the data received from the student and takes it into account when building an individual learning plan. The list of equations and the answers entered to them is shown in Table 1.

After analyzing the students' answers, all correct and incorrect answers are displayed (Fig. 5). Then, thanks to the analysis of the received data and through machine learning,

Fig. 4. Program interface.

Table 1. Quadratic equations and the answers entered in them.

Equation number	Quadratic equation	Entered response	Correctness
1	$3x^2 + 3x + 1 = 0$	0.56, −0.79	True
2	$5x^2 - 7x - 7 = 0$	0.8, −1.46	True
3	$3x^2 - 8x - 4 = 0$	−0.7, 1.42	True
4	$1x^2 - 8x + 6 = 0$	0.7, −2.37	True
5	$9x^2 - 2x - 6 = 0$	no real roots	False
6	$5x^2 - 10x - 2 = 0$	no real roots	False
7	$4x^2 - 5x - 1 = 0$	1.08, −2.33	True
8	$4x^2 + 9x - 1 = 0$	no real roots	False
9	$4x^2 - 5x - 3 = 0$	−0.76, 1.76	True
10	$9x^2 - 1x + 0 = 0$	−0.48, 2.08	True

the adaptive system displays two additional tasks for fixing, the number of which is determined and can be adjusted using the num_similitar_examples variable. It can be noted that the complex example was generated taking into account the difficulties of a particular student, since during the study we specifically made mistakes with negative values of the discriminant, suggesting that if the student did not know the rules for solving quadratic equations, then such a problem could arise. Below is also a simple example for correcting correctly completed tasks.

To summarize, an adaptive educational system developed using machine learning will make it possible to effectively study the topic of quadratic equations by adapting to the level of knowledge of a particular student and generating individual tasks for him to consolidate the material.

```
Equation 9x^2 + 2x + -4 = 0: Correct answer! Correct answers: [0.56, -0.79]
Equation 6x^2 + 4x + -7 = 0: Correct answer! Correct answers: [0.8, -1.46]
Equation 7x^2 + -5x + -7 = 0: Correct answer! Correct answers: [-0.7, 1.42]
Equation 3x^2 + 5x + -5 = 0: Correct answer! Correct answers: [0.7, -2.37]
Equation 8x^2 + 2x + 2 = 0: Incorrect answer! Correct answers: no real roots
Equation 7x^2 + 7x + 7 = 0: Incorrect answer! Correct answers: no real roots
Equation 4x^2 + 5x + -10 = 0: Correct answer! Correct answers: [1.08, -2.33]
Equation 6x^2 + -6x + 2 = 0: Incorrect answer! Correct answers: no real roots
Equation 3x^2 + -3x + -4 = 0: Correct answer! Correct answers: [-0.76, 1.76]
Equation 5x^2 + -8x + -5 = 0: Correct answer! Correct answers: [-0.48, 2.08]
Equation 9x^2 + -1x + 9 = 0: Incorrect answer! Correct answers: no real roots
Equation 7x^2 + -3x + -5 = 0: Correct answer! Correct answers: [-0.66, 1.09]
Equation 2x^2 + -7x + 6 = 0: Correct answer! Correct answers: [1.5, 2.0]
Equation 8x^2 + 8x + -10 = 0: Correct answer! Correct answers: [0.72, -1.72]
Equation 8x^2 + -9x + 8 = 0: Incorrect answer! Correct answers: no real roots

Similar examples for reinforcement:
Difficult example: 8x^2 + -4x + 6 = 0
Solution of the equation 8x^2 + -4x + 6 = 0:
Step 1: Calculate the discriminant D = -176
Step 2: The discriminant is negative, no real roots.
Step 3: No roots.
Example: 3x^2 + -10x + -4 = 0
```

Fig. 5. Solutions of quadratic functions and additional examples.

4 Conclusion

The study concluded that the adaptive learning approach can not only be integrated into the educational process, but can also improve understanding of topics in mathematics by creating an individual learning plan for each student. Getting quick feedback and an individual approach would not have been possible without the use of machine learning to build an adaptive educational system, which underscores the importance of innovative technologies when using adaptive learning techniques.

The approach to learning discussed in the article can be applied to many topics of studying mathematics, therefore, promoting the use of machine learning in the development of adaptive educational systems is a promising and relevant direction.

References

1. Zhikharev, A.G., Korsunov, N.I., Mamatov, R.A., Shcherbinina, N.V., Ponomarenko, S.V.: Development of an adaptive educational platform using machine learning technologies. Econ. Comput. Sci. **49**(4), 810–819 (2022)
2. Nazhova, N.A., Nazhimov, A.V.: An automated code verification system for training specialists in the field of information technology. Mod. High Tech Technol. **6**, 37–42 (2024)
3. Kuzovlev, D.D.: Innovative approaches to mathematical modeling in intellectual learning systems to increase competitiveness in the labor market. Innov. Invest. **5**, 490–493 (2024)
4. Besler, A.I., Myachinov, A.A., Seryakov, A.V., Korchevskaya, O.V.: Possibilities of using artificial intelligence and machine learning technologies in the field of education. In: In: Proceedings of the XXVII International Scientific and Practical Conference. Krasnoyarsk, pp. 125–124 (2023)
5. Kanash, A.V., Mezina, A.S.: Data mining for building machine learning models in education. Q: Digital transformation is a step into the future. In: Proceedings of the II International Scientific and Practical Conference of Young Scientists. Minsk, pp. 135–139 (2021)

6. Mukhamadieva, K.B.: Machine learning in improving the educational environment. Educ. Probl. Soc. Dev. **4**(13), 70–77 (2020)
7. Jagwani, A.: A review of machine learning in education. J. Emerg. Technol. Innov. Res. (JETIR). **6**(5) (2019)
8. Sandra, L., Lumbangaol, F., Matsuo, T.: Machine learning algorithm to predict student's performance: a systematic literature review. TEM J. (2021)
9. Chen, J., Zhou, X., Yao, J., Tang, S.K.: Application of machine learning in higher education to predict students' performance, learning engagement and self-efficacy: a systematic literature review. Asian Educ. Dev. Stud. (2025)
10. Cunha, C.R., Moreira, A., Coelho, S., Mendonça, V., Gomes, J.P.: Converging extended reality and machine learning to improve the lecturing of geometry in basic education. J. Eng. Res. (2024)
11. Pham, N., Ngoc, H.P., Nguyen-Duc, A.: Fairness for Machine Learning Software in Education: A Systematic Mapping Study. (2024)
12. Li, D.: An interactive teaching evaluation system for preschool education in universities based on machine learning algorithm. Comput. Hum. Behav. (2024)
13. Ibarra-Vazquez, G., Ramírez-Montoya, M.S., Buenestado-Fernández, M., Olague, G.: Predicting open education competency level: a machine learning approach. Heliyon. (2023)
14. Okoye, K., Nganji, J.T., Escamilla, J., Hosseini, S.: Machine learning model (RG-DMML) and ensemble algorithm for prediction of students' retention and graduation in education. Computers and education. Artif. Intell. (2024)
15. Zheng, L., Wang, C., Chen, X., Song, Y., Meng, Z., Zhang, R.: Evolutionary machine learning builds smart education big data platform: data-driven higher education. Appl. Soft Comput. (2023)
16. Okagbue, E.F., Ezeachikulo, U.P., Akintunde, T.Y., Tsakuwa, M.B., Ilokanulo, S.N., Obiasoanya, K.M., Ilodibe, C.E., Ouattara, C.A.T.: A comprehensive overview of artificial intelligence and machine learning in education pedagogy: 21 years (2000–2021) of research indexed in the Scopus Database. Soc. Sci. Human. Open. (2023)
17. Zhu, L., You, H., Hong, M., Fang, Z.: Predictive Insights into U.S. Students' Mathematics Performance on PISA 2022 Using Ensemble Tree-Based Machine Learning Models (2024)
18. McJames, N., Parnell, A., O'Shea, A.: Little and Often: Causal Inference Machine Learning Demonstrates the Benefits of Homework for Improving Achievement in Mathematics and Science (2024)
19. Sellami, A., Santhosh, M.E., Bhadra, J., Ahmad, Z.: High school teachers' perceptions of technology integration in instruction. Horizon. (2024)
20. Hall, C., Lundin, M.: Technology in the classroom: personal computers and learning outcomes in primary school. Econ. Educ. Rev. (2024)
21. Zakharova, N., Frumina, S., Lobuteva, L., Alwaely, S.: The specifics of integrating distance learning technologies with traditional classroom instruction: how to design educational curricula in modern education? Heliyon. (2024)
22. Jin, S.-H., Im, K., Yoo, M., Roll, I., Seo, K.: Supporting students' self-regulated learning in online learning using artificial intelligence applications. Int. J. Educ. Technol. High. Educ. **20**, 37 (2023)
23. Mirata, V., Hirt, F., Bergamin, P., van der Westhuizen, C.: Challenges and contexts in establishing adaptive learning in higher education: findings from a Delphi study. Int. J. Educ. Technol. High. Educ. **17**, 32 (2020)
24. Lincke, A., Jansen, M., Milrad, M., Berge, E.: The performance of some machine learning approaches and a rich context model in student answer prediction. RPTEL. **16**, 10 (2021)
25. Yağcı, M.: Educational data mining: prediction of students' academic performance using machine learning algorithms. Smart Learn. Environ. **9**, 11 (2022)

26. Anita, E.A.M., Jenefa, J., Vinodha, D., Lapina, M.: Advancements in Sybil attack detection: a comprehensive survey of machine learning-based approaches in wireless sensor networks. In: Lapina, M., Raza, Z., Tchernykh, A., Sajid, M., Zolotarev, V., Babenko, M. (eds.) AISMA-2024: International Workshop on Advanced Information Security Management and Applications. AISMA 2024. Lecture Notes in Networks and Systems, vol. 863, pp. 67–75. Springer, Cham (2024)
27. Kormakova, V., Klepikova, A., Musaelian, E., Baybikowa, G., Lapina, M.: Formation of ICT competencies of postgraduate students of teacher education based on interactive techniques. In: CEUR Workshop Proceedings SLET-2019 – Proceedings of the International Scientific Conference Innovative Approaches to the Application of Digital Technologies in Education and Research, vol. 2494, pp. 11–21 (2019)
28. Lapina, M., Shiriaev, E., Babenko, M., Istamov, I.: High-speed convolution core architecture for privacy-preserving neural networks. Program. Comput. Softw. **50**, 417–424 (2024). https://doi.org/10.1134/S0361768824700282
29. Rusanov, M., Babenko, M., Lapina, M., Sajid, M.: Identification of exploited unreliable account passwords in the information infrastructure using machine learning methods. Big Data Cognit. Comput. **8**, 159 (2024). https://doi.org/10.3390/bdcc8110159

Exploring Social Network Usage and Well-Being Among College Students

G. S. Prakasha[1], R. Sanskriti[1], Jing Lei[2], and Vitalii Lapin[3](✉)

[1] School of Education, Christ University, Bangalore, India
[2] Syracuse University, Syracuse, NY, USA
[3] North Caucasus Federal University, 355017 Stavropol, Russia
mlapina@ncfu.ru

Abstract. Social media plays a significant role in bringing the world together with just a click. However, social media addiction is rapidly increasing, especially among the youth, which affects their mental health and well-being. Instagram is one of the social media sites that is used mainly by students. Due to increasing exposure to Instagram, including the short videos, called Reels, today's students consume the content at an alarming rate, thus becoming addicted to it. College students tend to relentlessly use Instagram, without considering the time and energy wastage or addressing its impact on their mental health and well-being. The present study aimed to (1) explore the relationship between Instagram usage and mental well-being among college students in India, and (2) examine whether Instagram usage serves as a meaningful predictor of students' mental well-being. The findings reveal a significant negative correlation ($r = -0.332$, $p < .001$) between Instagram usage and students' well-being, indicating that higher Instagram use is associated with lower levels of mental well-being among college students. Further, a simple linear regression analysis confirmed that Instagram usage is a significant predictor of well-being ($F(1, 154) = 19.099$, $p < .001$), explaining a meaningful portion of the variance. The results highlight the need for educators, parents and other stakeholders to recognise the influence of social media, especially Instagram, on students' mental health and well-being and develop strategies to promote healthier usage patterns.

Keywords: Instagram addiction · college students · mental health · academic performance

1 Introduction

Instagram, a popular social media platform, reached its usage peak since the introduction of Reels by targeting most young people who like to watch content made by others (Menon, 2022). Today's youth and young adults, particularly college-going students who belong to Generation Z (GenZ), prefer to consume Reels more than traditional forms of entertainment such as television or even streaming platforms (Doloi, 2024). Moreover, Instagram falls under the category of up-and-coming technology that enables users to interact in different forms, be it through posts, reels, direct messaging,

M. Lapina et al. (Eds.): DET 2024, LNNS 1616, pp. 374–388, 2025.
https://doi.org/10.1007/978-3-032-04365-8_37

live streaming etc. For educational purposes, applications like Instagram can prove to be helpful. Universities integrated disruptive technologies like social media platforms to connect with students and even use them as teaching tools (Prakasha et al., n.d.). Although Instagram has a lot to offer for students, its over-usage can disrupt students' well-being. Due to an extreme consumption of content on Instagram, various studies have been conducted to explore its impact on students' mental health. Instagram negatively impacts students' well-being by presenting a pseudo world and opening platforms for cyberbullying (Moreton & Greenfield, 2022). Additionally, university students' mental well-being and academic performance are negatively affected by the increasing use of Instagram content, particularly Instagram Reels (Rasheed & Hussain, 2024). Various studies suggest that loneliness among youngsters is increasing, which is causing a particular addiction to social media platforms like Instagram. In addition, young people in sheer need of validation have started creating content to influence others (Jabeen et al., 2023). Moreover, there is a positive relationship between Instagram addiction among students and loneliness, which is negatively impacting their academic performance and results (Ponnusamy et al., 2020). For many people, especially students, Instagram has become an escape from reality (Kircaburun & Griffiths, 2019). But the downside of this escape is that their academic performance is deteriorating due to this unhealthy addiction (Behera & Gartia, 2023). Moreover, time spent on Instagram is directly proportional to fear of missing out (FOMO) and loneliness (Rajan et al., 2022). Stress and anxiety due to Instagram addiction may affect academic performance, which subsequently may result in depression (Malak et al., 2022). Artificial intelligence methods and machine learning algorithms are currently used both in manufacturing, to improve automation efficiency, and in medicine, medicine, and financial structures for big data analysis (Lapina et al., 2024). With the growing popularity of language models and the launch of open source GPT, artificial intelligence methods have entered even more areas of human activity (Rusanov et al., 2024).

2 Gender Perspective

Gender plays a significant role in determining patterns of usage of Instagram[1] by male and female students (Sharma et al., 2022). According to gender-specific research, females using Instagram have an increased amount of anxiety about their physical image, dissatisfaction with beauty, and self-esteem, and a leading result is depression (Sherlock & Wagstaff, 2019). Moreover, studies prove that the tinge of extreme liking about self is a resultant variable of Instagram use (Kircaburun & Griffiths, 2018). Moreover, Instagram usage impacts females' moods and negatively impacts their life satisfaction causing issues such as self-objectification (Garcia et al., 2022). On the other hand, male youngsters suffer from severe ideal muscular body image addiction due to the influence of Instagram posts (Graham, 2022). In addition, cases of narcissism have a positive relation with popularity on social media platforms like Instagram (Kurunç & Şahin, 2023). The physical image-perfecting tendency due to Instagram addiction is resulting in a destructive quality of life among both males and females (Simon et al., 2022).

3 Impact of Instagram Usage on Mental Health

The addiction level of students on social media is already under check in Malaysian Universities (Chee et al., 2022). Social media platforms like Facebook affect the mindset and induce students' loneliness. Moreover, they tend to develop inevitable depression (Casingcasing et al., 2023). The fear of missing out, or FOMO, is notably seen as measured by various internet addiction scales that conclude that increased academic performance is inversely proportional to social media addiction (Avci & Kula, 2022). Apart from this, children and adolescents' addiction to social media is negatively impacting their moral behaviour (Muhammad & Yusuf, 2022). 50% of students are using WhatsApp and are using it for hours the whole day, whilst 22.4% are reporting serious addiction towards social media. Social media creates panic, obsession and FOMO (Hattingh et al., 2022). Stress is a result of FOMO, which in turn increases social media addiction among students (Bernard, 2020). A study reveals that 73% of students use social media for non-academic purposes (Uddin et al., 2022). Thus, social media addiction results in fatigue and lowers learners' performance (Malik et al., 2020). Young adults upload pictures frequently, and the narcissistic frequency of self-image shows the addiction picture (Casale & Banchi, 2020). Social media usage is inducing unhygienic bingeing habits on foods that create obesity problems. Thus, students' mental and physical well-being diminishes (Huang et al., 2023).

Moreover, over-usage of Instagram can affect the minds of the users, especially students. Various content creators and travellers post on their feeds, which becomes a source of envy for the audience. This may negatively impact the users' attitudes (Asdecker, 2022). Moreover, Instagram impacts people's self-esteem as they start comparing themselves to others, giving way to negativity (Jiang & Ngien, 2020). Moreover, content like travel vlogs instigates risk-taking travel among youth, which may fatal. The guardians should regulate such actions (Mohanan & Kushe Shekhar, 2021). Self-promotion is ravaging rapidly due to increasing Instagram usage (Honda et al., 2021). Social media addiction is raising ethical concerns that should be taken into record by social media policymakers, and it is high time that action should be taken (Bhargava & Velasquez, 2021). It is the most dangerous for students as social media usage worsens academic performance (Cao & Tian, 2022). Cases of obsessive-compulsive disorder (OCD), stress, hyper-anxiety, and depression disorders are increasing due to the excessive exposure of students to social media (Şentürk et al., 2021). As a result, the mental health of users, especially students, is taking a toll (Mackson et al., 2019). Moreover, technological abuse affects face-to-face interactions and social behaviours (Ali Mohamed Ahmed et al., 2019). The passive users of Instagram suffer less from loneliness and depression but have increased body satisfaction (Zhao et al., 2023). Social media platforms are changing people's personalities (Ezung & Baksh, 2023). People with a stable timetable of sleep, food habits, and minimal usage of computers are in a good mood, and on the other hand, excessive addiction to social media platforms like Instagram is creating temper fluctuations (Gwiaździński, 2022). People use it to improve their mood, but initially, it leads to escapist shells and increased screen time (Ahmed & Dixon, 2023). 75% of loneliness and 40.15% dissatisfaction towards life result from Instagram usage (Rogowska & Libera, 2022). Moreover, a temporary gratification of mood elevation after

Instagram usage results in escapism and various disorders, thus revealing the dark facet of its addiction (Masrom et al., 2023).

4 The Good and Bad of Instagram

Instagram as a social media platform has both positive and negative sides. While some studies reveal that it can prove to be helpful for students, others argue that it leads to stress, anxiety and even bad mental health. Students actively using Instagram had positive self esteem and better well-being (Trifiro & Prena, 2021). Moreover, Instagram improves the oral communication of students (Çakmak, 2020). Due to the advent of artifical intelligence (AI), and large language models (LLMs), the integration of AI in social media applications has become common. Regarding Instagram, it has integrated MetaAI with Llama 3.2 to give a platform for the users for creating multimodal content. Moreover, there are certain filters on the app that support augmented reality (AR). AR platforms such as these can prove to be helpful for students as well as teachers in teaching and learning (Prakasha et al., 2025). Moreover, generative AI integrated in Instagram can help create games and other resources for students, which can be beneficial in teaching-learning (Prakasha et al., 2024). Although the extroverted nature of individuals is affected positively by the contents of Instagram, but as for life satisfaction, it is showing a negative impact (Stalimerou, 2023). In addition, an idealised showcased fashion and beauty is upgraded on Instagram, which results in parasocial relationships (Lozano-Blasco et al., 2023).

Moreover, a never-ending search for social acceptance and cyber-venting is happening due to the excessive use of Instagram (Moreton & Greenfield, 2022). Instagram usage affects the mental health of young people and there should be a check on the same to improve mental health issues among students (Schwarz et al., 2022).

5 Instagram Reels and Lifestyle

The lived experiences of individuals are somehow influenced by Instagram content, especially reels (Budenz et al., 2022). In the world of reels, people live in false self-presentations (van der Schyff et al., 2022). Moreover, the use of Instagram among college students during classes and mealtimes is generating a lot of stress (Adeel and Hayat 2022), leading to lifestyle disorders like stress, panic, and eating disorders. In addition to this, the academic results are negatively affected (Faelens et al., 2021). The increased use of Instagram is destroying students' self-esteem, which has a significant role in their performance (Pekpazar et al., 2021). The present study aims to explore the effects of Instagram addiction on the mental health and well-being of college students in India.

6 Theoretical Framework

The uses and gratification theory facilitates the present research. Instagram content satisfies the students' needs by focusing on user preferences, psychological needs and lifestyle choices (Katz, et al., 1973). In addition, social information processing theory guides the present study in demonstrating that students using social media are developing interpersonal relationships to a certain extent (Walther, 2008). However, in this century, students consider Instagram content to be the actual mirror of society by responding to social media, which may affect their thoughts. This is a notion of media equation theory (Griffin 2003).

7 Context of the Study

Instagram content, especially reels, is a significant source of addiction among college students. Moreover, the increasing usage of Instagram may result in unhealthy food habits and health problems (Huang et al., 2023). Additionally, it may lower one's self-esteem, indirectly affecting one's academic performance (Pekpazar et al. 2021). Researchers conclude that most Instagram-addicted students are developing stress regarding perfect body image and narcissism, which results in a low quality of life (Simon et al., 2022). Apart from this, many shy and lonely youngsters are using Instagram as an escape, which is resulting in depression (Kircaburun & Griffiths, 2019). Thus, it is essential to study the impact of Instagram addiction on the mental health of college students and understand if that is affecting their academic performance. In the 21st century, when everything is technology-oriented, and social media has undeniably impacted our lives, it is essential for the present generation to know its pitfalls and use them constructively rather than making them an addiction.

8 Research Questions

The present study seeks to answer whether Instagram usage among college students affects their mental well-being. Furthermore, the study also investigates whether Instagram usage among college students is a significant predictor of students' mental well-being.

9 Methodology

This study employed a descriptive correlational design and a quantitative descriptive survey research design. It adopted a convenient random sampling method to collect research data by administering two standardized survey tools. The tools chosen for the study are The Instagram Addiction Scale (TIAS) (Sholeh & Rusdi, 2019), comprising 20 items, and The Mental Health Continuum-Short Form (MHC-SF) (Keyes, 2009), comprising 14 items. Researchers' pilot tested the tools to establish reliability on a small sample of 50 college students studying science and subjects of other streams. Researchers sent the survey tools to college students across India for field data and obtained 156 completely filled responses. It included 111 female students and 45 male students. The survey included demographic information and an informed consent form. The study sought institutional review board clearance from the university to conduct the present research. The researchers recorded the data in MS Excel and protected it with a password available only to them. The study employed statistical analysis of the data by using SPSS software version 21. The result section below presents the tables and graphs obtained from the statistical analysis (Field, 2009).

10 Results and Discussion

10.1 Demographic Results

Out of 156 respondents, most of them were females (71.2%), while males were 28.8%. Figure 1 shows the distribution of participants based on gender (Fig. 1).

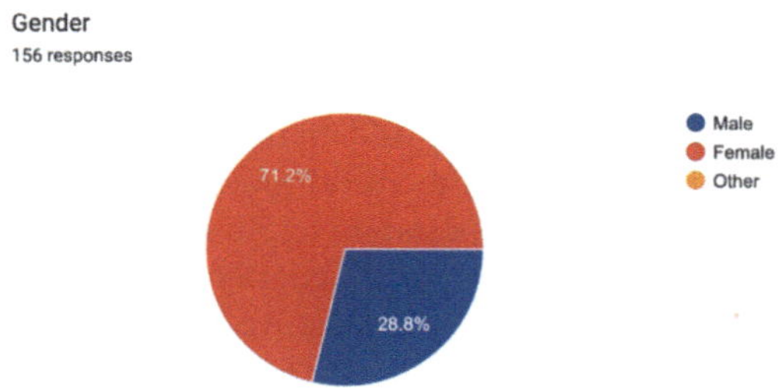

Fig. 1. Gender wise distribution.

Next, based on age, most of the respondents fell under the age group of 17–19 years (48.7%), followed by 32.7% respondents who belonged to the age group of 20–22 years and 18.6% of the respondents who belonged to the age group of 23–25 years. Figure 2 shows the distribution of participants based on age (Fig. 2).

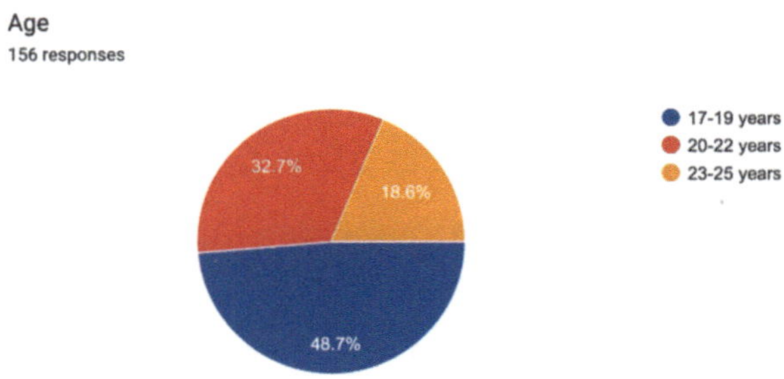

Fig. 2. Age wise distribution.

10.2 Relationship Between Instagram Addiction and Well-Being

To measure the relationship between Instagram addiction and mental well-being among the respondents, the researchers conducted a Pearson correlation test using SPSS software (version 29). The results revealed a moderate negative correlation between Instagram usage and students' well-being ($r = -0.332$, $p < .001$), indicating that higher Instagram use is significantly associated with lower levels of well-being. This correlation was found to be statistically significant at the 0.01 level (2-tailed). The sample size for the analysis was $N = 156$. Table 1 presents the results of the test.

Table 1. Pearson correlation between Instagram use and well-being.

Instagram usage and well-being	Pearson correlation	-.332[a]
	Sig. (2-tailed)	<.001
	N	156

[a]Correlation is significant at the 0.01 level (2-tailed).

The results in Table 1 highlight a significant inverse relationship between Instagram usage and well-being among college students in India. The moderate negative correlation ($r = -0.332$) suggests that students who spend more time on Instagram, particularly engaging with its features like creating posts and scrolling reels, may experience a decline in their mental well-being. This aligns with majority of previous research indicating that excessive social media use can lead to negative psychological outcomes, such as increased anxiety, stress, and reduced self-esteem.

Furthermore, to assess whether the regression model significantly predicts the dependent variable (well-being) based on the independent variable (Instagram usage), ANOVA test was conducted, the result of which is presented in Table 2.

Table 2. ANOVA Results for the Effect of Instagram Usage on Well-Being.

Model		Sum of Squares	df	Mean Square	F	Sig.
1	Regression	3243.222	1	3243.222	19.099	<.001[b]
	Residual	26151.137	154	169.813		
	Total	29394.359	155			

[a]Dependent Variable: Well-being.
[b]Predictors: (Constant), Instagram usage.

The results of ANOVA as shown in Table 2 show that the effect was statistically significant, $F(1, 154) = 19.099$, $p < .001$, indicating that Instagram usage accounts for a significant portion of the variance in well-being. This suggests that the amount of time spent on Instagram is a meaningful predictor of students' mental health and well-being status.

The significant ANOVA result confirms that Instagram usage has a notable impact on college students' well-being. Moreover, increased Instagram usage could be a contributing factor to lower mental health outcomes among students. These findings align with existing literature that highlights the negative psychological effects of excessive social media use, including feelings of anxiety, depression, and social isolation among young adults.

Further, the researchers conducted a regression statistical test to understand the correlation between Instagram usage and mental well-being. Table 3 presents the results of the regression in the form of model summary statistics.

Table 3. Model summary statistics of regression.

Model	R	R Square	Adjusted R2	Std. Error of the Estimate	R2 Change	F Change	df1	df2
1	.332[a]	.110	.105	13.031	.110	19.099	1	154

[a]Predictors: (Constant), Instagram usage.
[b]Dependent Variable: Well-being.

The results in Table 3 present the regression analysis explaining the total variance in the student well-being (dependent variable) as determined by Instagram usage (independent variable). The model summary statistics revealed that Instagram usage significantly predicts well-being ($R = 0.332$, $R^2 = 0.110$, adjusted $R^2 = 0.105$, $F(1,154) = 19.099$, $p < .001$), indicating that Instagram usage explains approximately 11% of the variance in well-being. Although other factors likely contribute to well-being, the findings suggest that higher Instagram usage is associated with lower mental well-being among college students in India.

Next, to find out the auto-correlation in the residuals from regression analysis, researchers ran a Durbin-Watson statistical test, the results of which are presented in Table 4.

Table 4. Pearson correlation between Instagram use and well-being.

Model	Sig. F Change	Durbin-Watson
1	<.001	2.257

[a]Predictors: (Constant), Instagram usage
[b]Dependent Variable: Well-being

The results highlighted in Table 4 yielded a Durbin-Watson statistic of 2.257, which is close to the ideal value of 2. It indicates that the residuals are independent and do not exhibit significant autocorrelation. This suggests that the regression model meets the assumption of independent errors, ensuring the reliability of the results.

The regression analysis histogram presented in Fig. 3 represents the distribution of standardized residuals data and explains the presence of skewed data and outliers. It explains whether the assumptions underlying regression analysis were met.

Figure 3 shows that the residuals of the regression line are approximately normally distributed. Therefore, the data is suitable for regression analysis and statistical assumptions are successfully met. Further, Fig. 4 represents the relationship between Instagram usage and well-being in the form of a scatterplot. It indicates the linearity or non-linearity of the relationship. Moreover, it indicates whether the variance of the residual in the regression model is constant.

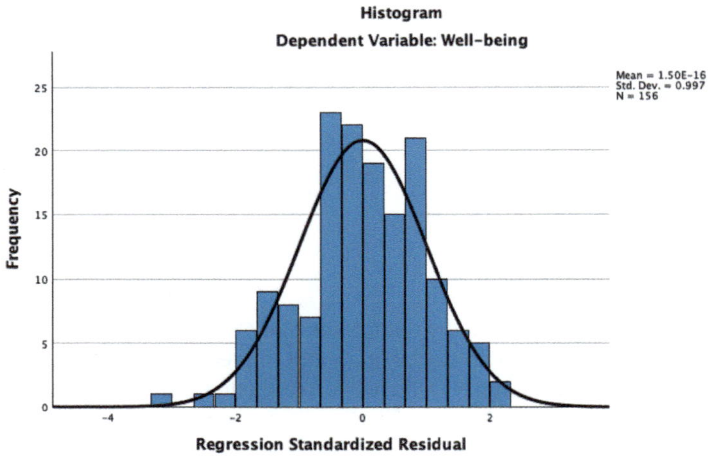

Fig. 3. Histogram of regression.

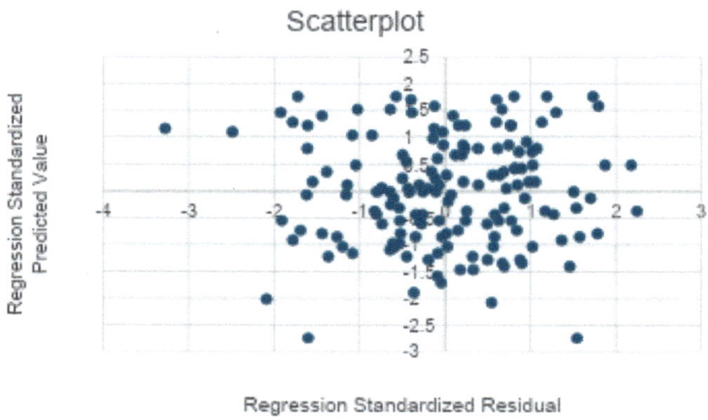

Fig. 4. Scatterplot of regression.

The data in Fig. 4 is homoscedastic, thus indicating a linear relationship between the variables Instagram usage and well-being. This further indicates equal variances which means that even if the data comes from a different population, the variance will be equal.

10.3 Discussion

The present study examined whether Instagram usage affects the mental well-being of college students. Additionally, the study explored whether Instagram usage is a significant predictor of students' mental well-being. The findings of the study contribute to research on social media, especially Instagram's impact on students' mental well-being, particularly among college students. The results revealed a moderate negative correlation between Instagram usage and well-being, suggesting that as time spent on

Instagram increases, students tend to experience a decline in mental well-being. Furthermore, the regression analysis confirmed that Instagram usage is a significant predictor of well-being, highlighting the potential psychological risks of excessive engagement with Instagram content. One plausible explanation for this relationship may be the culture of social comparison fostered by social media sites such as Instagram (Jabeen et al., 2023). The platform's curated nature and algorithm, where users showcase idealized or beautified aspects of their lives, may lead to unrealistic self-evaluations among college students, especially at such a crucial age (Sherlock & Wagstaff, 2019). Constant exposure to such content may amplify feelings of inadequacy, anxiety, and decreased self-esteem, aligning with previous research suggesting that frequent social media use can contribute to mental distress (Jiang & Ngien, 2020). Additionally, the introduction of Instagram reels and the culture of doom scrolling may reinforce addictive behaviors (Menon, 2022). The design of Instagram encourages prolonged usage, with algorithm-driven content recommendations keeping users engaged for extended periods. This excessive screen time may interfere with essential aspects of students' personal and academic lives, such as sleep quality, academic performance, and face-to-face social interactions, ultimately affecting their overall well-being (Walubita, 2017). The findings of the study also resonate with the theory of problematic social media use, which suggests that individuals who struggle with self-regulation may be more prone to excessive engagement with platforms like Instagram (Kircaburun & Griffiths, 2019). The ease of access and constant availability of content can lead to compulsive scrolling, escapism, and avoidance of real-life responsibilities, further amplifying stress and mental fatigue (Malik et al., 2020).

Given the above-mentioned insights, there is an urgent need to promote digital literacy and mindful Instagram consumption among college students. Educational institutions should incorporate awareness programs on responsible Instagram use, emphasizing the importance of self-regulation and critical engagement with online content. Moreover, mental health interventions addressing the psychological effects of Instagram could help students develop strategies to balance online engagement with their well-being. Moreover, stakeholders such as parents, educators, and policymakers must also recognize the risks associated with Instagram addiction and implement strategies to support students. Encouraging offline activities, fostering open and candid conversations about scrolling habits, screen time, cyberbullying, and promoting digital detox initiatives may help mitigate the negative impact of excessive Instagram use. Nevertheless, qualitative studies focusing on students' lived experiences with Instagram could offer a more in-depth understanding of the emotional and behavioral consequences of engagement with Instagram content. Future research may also examine the specific motives of content consumption (e.g., reels, posts, leisure, entertainment) and their differential impact on mental well-being.

11 Conclusion

The study examined the relationship between Instagram usage and mental well-being among Indian college students, revealing a significant negative correlation between the two. Regression analysis confirmed that Instagram usage is a meaningful predictor of well-being, suggesting that excessive engagement with the platform contributes to stress, anxiety, and reduced self-esteem. Features like doom-scrolling and social comparison may further exacerbate mental health concerns. These findings underscore the need for digital literacy initiatives and mental health awareness programs to promote healthier social media habits. Educators, parents, and policymakers must recognize the impact of excessive Instagram use and encourage a balanced approach. Future research should explore additional influencing factors, such as personality traits and coping mechanisms, to better understand the long-term effects of social media on well-being.

While Instagram fosters connection and creativity, unregulated use poses risks. Raising awareness and promoting mindful engagement are crucial for helping students maintain a healthy balance between social media and mental well-being.

References

Adeel, A.A., Hayat, D.: Undergraduate students' predisposition from instagram and snapchat interactive effect on stress. Webology. **403** (2022)

Ahlawat, K., Anand, D.D.: Misuse of persuasive technology in social media with subsequent solutions. Int. J. Multidiscip. Innov. Res. **2**(4), 11 (2022)

Ahmed, S., Dixon, M.J.: Instagram, Depression, and Dark Flow – Using Social Media as a Maladaptive Coping Mechanism (2023). https://doi.org/10.2139/ssrn.4391734

Ali Mohamed Ahmed, E., Hamed Farahat, N., Mohamed Sayed, S.: Effect of electronic media abuse on health among university students. Egypt. J. Health Care. **10**(1), 545–559 (2019). https://doi.org/10.21608/ejhc.2019.240862

Roy, A.: A study on instagram addiction & procrastination among college students. Int. J. Indian Psychol. (2016). https://doi.org/10.25215/1004.076

Asdecker, B.: Travel-related influencer content on instagram: how social media fuels wanderlust and how to mitigate the effect. Sustainability. **14**(2), 2 (2022). https://doi.org/10.3390/su1402 0855

Behera, D., Gartia, R.: Addictiveness of Social Networking Sites Among Students: in the Globalized World, pp. 198–206 (2023)

Bernard, C.: Lonely zs: examining the relationships among time spent on social networking sites, the fear of missing out (fomo), and loneliness among undergraduate college students. Doctoral Dissertations and Projects (2020). https://digitalcommons.liberty.edu/doctoral/2457

Bhargava, V.R., Velasquez, M.: Ethics of the attention economy: the problem of social media addiction. Bus. Ethics Q. **31**(3), 321–359 (2021). https://doi.org/10.1017/beq.2020.32

Bhatia, S., Haokip, G., Babu, N., Bhardwaj, G.: Perceived parenting and self concept of Indian College boys and girls. Int. J. Humanit. Soc. Sci. **12**, 85–110 (2022)

Budenz, A., Klassen, A., Purtle, J., Yom-Tov, E., Yudell, M., Massey, P.: If i was to post something, it would be too vulnerable: university students and mental health disclosures on instagram. J. Am. Coll. Health J. ACH. **70**(2), 615–624 (2022). https://doi.org/10.1080/07448481.2020.175 9608

Çakmak, S.: Investigation of the relationship between social media addiction and obesity and body perception in university students [MasterThesis, Sağlık Bilimleri Enstitüsü] (2020). https://aci kbilim.yok.gov.tr/handle/20.500.12812/339948

Cao, G., Tian, Q.: Social media use and its effect on university student's learning and academic performance in the UAE. J. Res. Technol. Educ. **54**(1), 18–33 (2022). https://doi.org/10.1080/ 15391523.2020.1801538

Casale, S., Banchi, V.: Narcissism and problematic social media use: a systematic literature review. Addict. Behav. Rep. **11**, 100252 (2020). https://doi.org/10.1016/j.abrep.2020.100252

Casingcasing, M.L.S., Nuyens, F.M., Griffiths, M.D., Park, M.S.: Does social comparison and facebook addiction lead to negative mental health? A pilot study of emerging adults using structural equation modelling. J. Technol. Behav. Sci. **8**(1), 69–78 (2023). https://doi.org/10.1007/s41347-022-00289-9

Chee, W.Y., Lee, W.K., Boo, N.Q.H., Tan, G.L., Yong, E.Y.: The Influence of Motives to Use Social Media and Social Media Addiction [Other, UTAR] (2022). http://eprints.utar.edu.my/4945/

Doloi, G.: The influence of Instagram reels on content consumption trends among gen Z. SSRN Electron. J. (2024). https://doi.org/10.2139/ssrn.5000132

Ezung, S., Baksh, D.S.A.: The influence of social media on youth behaviour in Kohima district. J. Pharmaceut. Neg. Results, 2178–2184 (2023). https://doi.org/10.47750/pnr.2023.14.02.268

Faelens, L., Hoorelbeke, K., Cambier, R., van Put, J., Van de Putte, E., De Raedt, R., Koster, E.H.W.: The relationship between Instagram use and indicators of mental health: a systematic review. Comput. Hum. Behav. Rep. **4**, 100121 (2021). https://doi.org/10.1016/j.chbr.2021.100121

Garcia, R.L., Bingham, S., Liu, S.: The effects of daily Instagram use on state self-objectification, Well-being, and mood for young women. Psychol. Popular Media. **11**(4), 423–434 (2022)

Graham, S.: Picture perfect? How appearance-focused images on instagram affect mood, self-esteem, and body satisfaction. Thesis, University of Otago (2022). https://ourarchive.otago.ac.nz/handle/10523/12829

Griffin, J.: A First Look at Communication Theory. McGraw-Hill, Boston, MA (2003)

Gunawan, F.: Instagram and Mental Health of the Students at Higher Education of Indonesia (n.d.). https://www.academia.edu/38092385/Instagram_and_Mental_Health_of_The_Students_at_Higher_Education_of_Indonesia. Accessed 4 Apr 2023

Gwiaździński, E.: Gender as a Moderator in Relation Between Self-Esteem and Social Media Activity Based on Instagram Platform (2022)

Hattingh, M., Dhir, A., Ractham, P., Ferraris, A., Yahiaoui, D.: Factors mediating social media-induced fear of missing out (FoMO) and social media fatigue: a comparative study among Instagram and snapchat users. Technol. Forecast. Soc. Chang. **185**, 122099 (2022). https://doi.org/10.1016/j.techfore.2022.122099

Honda, H., Yoshikawa, K., Yamashina, Y., Yamato, Y., Terada, S., Goto, M.: Impact of social networking service usage on self-learning activity and physical activity in Japanese undergraduate physical therapy students: a cross-sectional study. J. Allied Health Sci. **12**, 8–15 (2021). https://doi.org/10.15563/jalliedhealthsci.12.8

Huang, P.-C., Latner, J.D., O'Brien, K.S., Chang, Y.-L., Hung, C.-H., Chen, J.-S., Lee, K.-H., Lin, C.-Y.: Associations between social media addiction, psychological distress, and food addiction among Taiwanese university students. J. Eat. Disord. **11**(1), 43 (2023). https://doi.org/10.1186/s40337-023-00769-0

Ingram, G.P.D.: Adolescent Use of New Media and Internet Technologies: Debating Risks and Opportunities in the Digital Age. Taylor & Francis (2023)

Jabeen, F., Tandon, A., Azad, N., Islam, A.K.M.N., Pereira, V.: The dark side of social media platforms: a situation-organism-behaviour-consequence approach. Technol. Forecast. Soc. Chang. **186**, 122104 (2023). https://doi.org/10.1016/j.techfore.2022.122104

Jiang, S., Ngien, A.: The effects of instagram use, social comparison, and self-esteem on social anxiety: a survey study in Singapore. Soc. Media Soc. **6**(2), 2056305120912488 (2020). https://doi.org/10.1177/2056305120912488

Katz, E., Blumler, J.G., Gurevitch, M.: Uses and gratifications research. Public Opin. Q. **37**(4), 509–523 (1973) http://www.jstor.org/stable/2747854

Keyes, C.L.M.: The nature and importance of mental health in youth. In: Gilman, R., Furlong, M., Heubner, E.S. (eds.) Promoting Wellness in Children and Youth: A Handbook of Positive Psychology in the Schools, pp. 9–23. Routledge, New York (2009)

Kircaburun, K., Griffiths, M.: Problematic instagram use: the role of perceived feeling of presence and escapism. Int. J. Ment. Heal. Addict. **17**, 909–921 (2019). https://doi.org/10.1007/s11469-018-9895-7

Kircaburun, K., Griffiths, M.D.: Instagram addiction and the big five of personality: the mediating role of self-liking. J. Behav. Addict. **7**(1), 158–170 (2018). https://doi.org/10.1556/2006.7.2018.15

Kurunç, S., Şahin, M.: Üniversite öğrencilerinde sosyal medya bağımlılığı ve narsisizm İlişkisi. J. Soc. Human. Admin. Sci. **61**(61), 2267–2275 (2023). https://doi.org/10.29228/JOSHAS.67710

Lozano-Blasco, R., Mira-Aladrén, M., Gil-Lamata, M.: Social media influence on young people and children: analysis on instagram, twitter and YouTube. Comunicar. **31**, 117 (2023). https://doi.org/10.3916/C74-2023-10

Mackson, S.B., Brochu, P.M., Schneider, B.A.: Instagram: friend or foe? The application's association with psychological Well-being. New Media Soc. **21**(10), 2160–2182 (2019). https://doi.org/10.1177/1461444819840021

Malak, M.Z., Shuhaiber, A.H., Al-amer, R.M., Abuadas, M.H., Aburoomi, R.J.: Correlation between psychological factors, academic performance and social media addiction: model-based testing. Behav. Inform. Technol. **41**(8), 1583–1595 (2022). https://doi.org/10.1080/0144929X.2021.1891460

Malik, A., Dhir, A., Kaur, P., Johri, A.: Correlates of social media fatigue and academic performance decrement: a large cross-sectional study. Inf. Technol. People. **34**(2), 557–580 (2020). https://doi.org/10.1108/ITP-06-2019-0289

Masrom, M., Busalim, A., Griffiths, M.D., Asadi, S., Mohd Ali, R.: The impact of excessive Instagram use on students' academic study: a two-stage SEM and artificial neural network approach. Interact. Learn. Environ., 1–20 (2023). https://doi.org/10.1080/10494820.2023.2184393

Menon, D.: Factors influencing instagram reels usage behaviours: an examination of motives, contextual age and narcissism. Telemat. Inform. Rep. **5**, 100007 (2022). https://doi.org/10.1016/j.teler.2022.100007

Mohanan, M., Kushe Shekhar, S.: A study on the mediating effect of fomo on social media (instagram) induced travel addiction and risk taking travel behavioral intention in youth. J. Content Commun. Commun. **14**(8), 57–67 (2021). https://doi.org/10.31620/JCCC.12.21/06

Moreton, L., Greenfield, S.: University students' views on the impact of instagram on mental wellbeing: a qualitative study. BMC Psychol. **10**(1), 45 (2022). https://doi.org/10.1186/s40359-022-00743-6

Muhammad, S., Yusuf, A.: Social media addiction and moral behaviour development of students of Shehu Shagari College of Education, Sokoto. Zaria J. Educ. Stud. (ZAJES). **22**(2), 2 (2022)

Pekpazar, A., Kaya Aydın, G., Aydın, U., Beyhan, H., Arı, E.: Role of instagram addiction on academic performance among turkish university students: mediating effect of procrastination. Comput. Educ. Open. **2**, 100049 (2021). https://doi.org/10.1016/j.caeo.2021.100049

Ponnusamy, S., Iranmanesh, M., Foroughi, B., Hyun, S.S.: Drivers and outcomes of instagram addiction: psychological well-being as moderator. Comput. Hum. Behav. **107**, 106294 (2020)

Prakasha, G.S., Rawat, S., Basak, I., Mathai, S.: Gamification in teacher education: teacher educators and trainee teachers' perspectives in India. In: Enhancing Education with Intelligent Systems and Data-Driven Instruction, pp. 132–162. IGI Global (2024a)

Prakasha, G.S., Sanskriti, R., Ishani, B.: Examining the relationship between university academic staff's attitudes towards artificial intelligence and technology integration. J. Educ. Fac. (n.d.) https://dergipark.org.tr/en/pub/pauefd/issue/89123/1267071

Prakasha, G.S., Thamaraikannan, S., Lapina, M., Sanskriti, R.: Pre-service and in-service teachers' perceptions of using virtual reality tools in teaching. In: Lecture Notes in Networks and Systems, pp. 340–355. Springer Nature Switzerland (2025)

Rajan, B., Jose, L.G., Sundar, T., Rajan, B., Jose, L.G., Sundar, T.: Are you hooked to the 'gram'? Exploring the correlation between loneliness, the fear of missing out, and instagram usage among young Indians. Palabra Clave. **25**(2), 1–26 (2022). https://doi.org/10.5294/pacla.2022.25.2.5

Rasheed, A., Hussain, S.: Instagram reels and ethical muteness among university students. Jahan-E-Tahqeeq. **7**(2), 44–60 (2024)

Rogowska, A.M., Libera, P.: Life satisfaction and instagram addiction among university students during the covid-19 pandemic: the bidirectional mediating role of loneliness. Int. J. Environ. Res. Public Health. **19**(14), 14 (2022). https://doi.org/10.3390/ijerph19148414

Schwarz, D., Steinau, K., Kraus, L., In-Albon, T.: The effect of a 1-week abstinence from instagram on mental health in youth and young adults. Kindheit Und Entwicklung. **31**(4), 200–210 (2022). https://doi.org/10.1026/0942-5403/a000392

Şentürk, E., Geniş, B., Coşar, B.: Social media addiction in young adult patients with anxiety disorders and depression. Alpha Psychiatry. **22**(5), 257–262 (2021). https://doi.org/10.1530/alphapsychiatry.2021.21242

Sharma, A., Sanghvi, K., Churi, P.: The impact of instagram on young adult's social comparison, colourism and mental health: Indian perspective. Int. J. Inf. Manag. Data Insights. **2**(1), 100057 (2022). https://doi.org/10.1016/j.jjimei.2022.100057

Sherlock, M., Wagstaff, D.L.: Exploring the relationship between frequency of Instagram use, exposure to idealized images, and psychological well-being in women. Psychol. Pop. Media Cult. **8**, 482–490 (2019). https://doi.org/10.1037/ppm0000182

Sholeh, A., Rusdi, A.: A new measurement of instagram addiction: psychometric properties of the instagram addiction scale (tias). CISAK, 91–97 (2019)

Simon, P.D., Cu, S.M.O., De Jesus, K.E.M., Go, N.T.S., Lim, K.T.F., Say, C.L.C.: Worried about being imperfect? The mediating effect of physical appearance perfectionism between instagram addiction and body esteem. Personal. Individ. Differ. **186**, 111346 (2022). https://doi.org/10.1016/j.paid.2021.111346

Stalimerou, A.: The impact of instagram on happiness—evaluation of the impact of instagram on happiness, measured as self-esteem, the closeness of relationships and life satisfaction. MasterThesis (2023). https://run.unl.pt/handle/10362/150754

Trifiro, B.M., Prena, K.: Active instagram use and its association with self-esteem and well-being. Technol. Mind Behav. **2**, 1 (2021)

Uddin, M.K., Huda, M.N., Ali, M.J., Alam, M.S., Bhuiyan, A.B.: Impact of social media on students' academic performance: a case study of islamic university, Bangladesh. Am. Econ. Soc. Rev. **10**(1), 1–10 (2022). https://doi.org/10.46281/aesr.v10i1.1822

van der Schyff, K., Renaud, K., Townes, J., P.-, & Tshiqi, N.: Investigating the mediating effects of phubbing on self-presentation and fomo within the context of excessive instagram use. Cogent. Psychol. **9**(1), 2062879 (2022). https://doi.org/10.1080/23311908.2022.2062879

Walubita, G.: Social media use and its perceived impact on their social life: a case study of the University of Zambia. Int. J. Multidiscip. Res. (2017)

Zhao, Y., Cingel, D.P., Xie, L., Yu, M.: Exploring the relations of subjective and objective instagram use on young adults' mental health. Telematics Inform. **77**, 101921 (2023). https://doi.org/10.1016/j.tele.2022.101921

Walther, J.B.: Social information processing theory. In: Engaging Theories in Interpersonal Communication: Multiple Perspectives, p. 391 (2008)

Prakasha, G. S., Muttungal, P., Lapina, M., and Taran, V. (2024b). Flipped classroom strategy in online teaching: challenges faced by higher education teachers. In: Lapina, M, Prakasha, G. S., Grigoriev, S. (eds) SLET 2022: International Conference on Innovative Approaches to the Application of Digital Technologies in Education and Research. SLET 2022. Lecture Notes in Networks and Systems 241–254, Springer, Cham

Prakasha, G.S., Thamaraikannan, S., Lapina, M., Sanskriti, R.: Pre-service and In-service teachers' perceptions of using virtual reality tools in teaching. In: Lapina, M., Prakasha, G.S., Grigoriev, S. (eds.) SLET 2022: International Conference on Innovative Approaches to the Application of Digital Technologies in Education and Research. SLET 2022. Lecture Notes in Networks and Systems, pp. 340–356. Springer, Cham (2024c)

Lapina, M., Shiriaev, E., Babenko, M., Istamov, I.: High-speed convolution Core architecture for privacy-preserving neural networks. Program. Comput. Softw. **50**, 417–424 (2024). https://doi.org/10.1134/S0361768824700282

Rusanov, M., Babenko, M., Lapina, M., Sajid, M.: Identification of exploited unreliable account passwords in the information infrastructure using machine learning methods. Big Data Cognit. Comput. **8**, 159 (2024). https://doi.org/10.3390/bdcc8110159

Rising from Covid-19: Hybrid Teaching Experiences of University Teachers

G. S. Prakasha[1], Maria Carolina Ordenes Guzman[2], and Maria Lapina[3](✉)

[1] School of Education, Christ University, Bengaluru, Karnataka, India
[2] Pontificia Universidad Catolica de Chile, Santiago, Chile
[3] North Caucasus Federal University, 355017 Stavropol, Russia
mlapina@ncfu.ru

Abstract. COVID-19 pandemic has affected higher education all over the globe. The unprecedented situation has brought changes in teaching at universities. Universities across the world decided to teach online. By the time, teachers' hone online teaching Skills Universities resorted to hybrid teaching. Teachers are oblivious to hybrid teaching and had situational anxiety. In spite of all teachers started their hybrid classes. The present study assumes transtheoretical model for hybrid learning derived from ecosystem theory. Recent research is yet to capture the hybrid teaching experiences of university teachers amid COVID-19. The present study employed qualitative research method to capture in-depth understanding of hybrid teaching experiences. Study interviewed eight University teachers to understand the hybrid teaching experiences and applied interpretative phenomenological analysis (IPA) to interpret the interview data. The findings of the study emerged with themes and sub-themes describing hybrid-teaching experiences and listed major challenges faced by them. Few of the challenges are handling online and offline students' group simultaneously, technology integration to classes, technological barriers, and COVID-19 anxiety. Study recommends more research in the area for broader understanding of nuances of hybrid teaching. Nevertheless, to find solution to the challenges faced by the teachers while conducting hybrid classes.

Keywords: COVID-19 · Higher education · Hybrid teaching · New normal · University teachers

1 Introduction

Education in 2021 looking for reaching the new normal. COVID-19 is yet present as a second wave and third wave in countries across the world. Livelihood is affected. Industrial and agricultural activities have gradually rising back from the crisis. Schools, Colleges, and Universities have been able to manage academic year 2020–21 online. Higher education contributes significantly to the nation's productivity. Therefore, educational governing bodies are trying hard to bring normalcy in education system. However, returning to new normal is not an easy one-time transition (Daniel, 2020). Fewer higher

M. Lapina et al. (Eds.): DET 2024, LNNS 1616, pp. 389–405, 2025.
https://doi.org/10.1007/978-3-032-04365-8_38

education institutes partially began their campuses offline. Faculty members and students had an option to choose between online and offline mode to work or attend classes based on their health and other conditions. Higher education institutes resorted to hybrid teaching. Teaching staff have little knowledge about hybrid teaching and were new to online teaching as well. Hybrid teaching and learning is expensive when compared to only online or onsite classes (Thomas, 2020). Teachers who started hybrid classes began to face newer challenges and did not derive much job satisfaction. Significant number of faculty members are oblivious to hybrid classes. That called for the present study to understand this unique experience of transition from online to hybrid teaching and the experiences of hybrid classrooms among teachers. In order to account detailed description of hybrid teaching experiences researchers adopted qualitative research design to carry out the present study.

Artificial intelligence methods and machine learning algorithms are currently used both in manufacturing, to improve automation efficiency, and in medicine, medicine, and financial structures for big data analysis (Lapina et al., 2024). With the growing popularity of language models and the launch of open source GPT, artificial intelligence methods have entered even more areas of human activity (Rusanov et al., 2024).

Johnson et al. (2020) found that due to COVID-19 emergency teaching staff started to practice online teaching completely. These staff were either having online teaching experience or were completely new to this mode. Study revealed that the faculty made a major shift in their teaching styles. These changes included learning new teaching methodology, reducing the workload of students, changing the pattern and style of teaching as well as assessment. Faculty reported that they felt a need for assistance when it comes to lecture delivery, learning new tools, and managing student support. This study also found that the non-teaching staff shifted swiftly to online mode. Raes et al. (2020) conducted a study, which applied self-determination theory (SDT) to different learning settings to find the student relatedness, intrinsic motivation, and learning achievements. Study revealed that, student relatedness to their classmates and their intrinsic motivation is very low during hybrid teaching. Online quizzes motivates students. A systematic review conducted by Rasheed et al. (2020) reviewed 407 article to find the challenges faced in the hybrid teaching. The inclusion criteria here consisted of empirical studies with blended learning as a central theme. The main challenges identified were characterised as challenges faced by students, teachers, and educational institutions. The challenges faced by students included; self-regulation, technological literacy, complexity, and isolation. The challenges faced by teachers included; technological literacy, operation and competency, belief, and online video challenges. The challenges faced by educational institutions included training the teachers and providing the latest technology to the teachers. This study also summarised the solutions to the problems faced during hybrid teaching learning. Some of the solutions included using intelligent tutoring System, increasing group awareness and peer assistance, using discussion board, and using web based decision-seeking tool.

Adedoyin and Soykan (2020) explored the shift in teaching-learning experiences as a response to pandemic among university faculty members and students. Authors have discussed in detail about the opportunities and challenges this pandemic has brought into the education system leading to an effective hybrid-learning environment. German researchers, Skulmowski and Rey (2020) conducted a case study on the introduction of digital classes at Chemnitz University of Technology within a matter of a week of Covid-19 outbreak and lockdown announcement. Study revealed that this pandemic has paved the way for the digitization of teaching and learning at this university. Professors from California Quezada et al. (2020) attempted to explore the depth of transformation that happened in educational institutions, especially among the teachers of the graduate teacher education programme at Liberal Arts college. They identified five themes namely technology based instructional strategies, technology based support office consultation, alternative technology based course assessments, feedback for teaching and learning improvement, and social-emotional engagement in Courses. Authors highlighted the requirement for 'instructional digital technology tool box' among the teacher education faculty to have more value addition in terms of teaching and learning experience.

Challenges in online teaching are to motivate students and creating engaging and interactive learning materials. Lack of motivation leads to low achievements and questions teacher efficiency (Marshall, 2020). It is interesting to understand and analyse the measures taken by university teachers to respond to the demand of hybrid teaching. However, it is also important to think about the possibilities and opportunities post pandemic state of higher education. Murphy (2020) conducted a study to describe the consequences of emergency e learning as a measure of securitization and the transformation it has brought in teaching, learning, and assessment. Though security concerns lead to remote teaching, there are advantages like education made available to all from the best institutions across the globe, which is not otherwise possible. Huang et al. (2020) explored the effectiveness of open educational practices and resources to analyse the state of undisrupted learning in the disrupted classes during pandemic. Study identified frameworks for open educational practices and resources.

One important tool for engaging online students is Gamification, as it enhances the quality and depth of the online discussion. Huang et al. (2019) probed into the social network patterns of the students in a gamification environment. This study used learning analytics to study the impact of gamification on the online learning behaviour of the students. Study revealed that the students of gamification based online courses had a thicker network and gave better feedback to their peers. Gamification based online courses are more promising in terms of teaching and learning. Butz and Stupnisky (2017) applied SDT on the teaching learning environment involving technology and student relatedness. Study revealed that, students prefer to relate with peers during hybrid course. Student relatedness is positively associated with learning outcomes. Nevertheless, discussion threads increased student relatedness in a Hybrid class. Bower et al. (2015) conducted a cross-case analysis of a blended synchronous learning environment. The study found that hybrid teaching needs systematic plan and rehearsal. Presence of a student varied according to the technology and its drawbacks. Hybrid mode complicates teacher task. Ilgu (2015) conducted a study on faculty perspectives on benefits and challenges of hybrid learning. Benefits revealed by the study are free time for complex problem solving, flexibility, increased student engagement. Challenges revealed by the study are lack of peer interaction, technical issues, and initial investment of time. Hege (2011) conducted a study on the online theology classroom: strategies for engaging a community of distance learners in a hybrid model of online education. Study revealed that, the online faculty should not replicate the traditional brick and mortar model of teaching. The design of the course and the strategies used should be in accordance with the online learning need. Hybrid classes can use tools like blackboard, Academic Suite, blogs, social networking websites, discussion forums, and videos.

Khan et al. (2012) conducted a study of blended learning process in education context. Study proposed four characteristics of blended learning process; students can learn basic course through face-to-face method, make use lab for practical session, instructor can use e-learning platform to post learning materials and use simulated practical session. Study revealed blended learning advantages like flexible timing, easy access to materials, constant review, and feedback. Study found challenges on blended learning such as being adaptability to learn new technology, infrastructure and change management. A study conducted by McFarlin (2008) on hybrid lecture- online format increases student grades in an undergraduate exercise physiology course at a large urban university. Study finds that, hybrid format of teaching directly affects the grades of the students and ultimately results in marked improvements in the performance. Instructional technology should sync with content delivery and students learning style. Study revealed hybrid class challenges such as; monitor students work, class preparation time, and technical glitches. Reasons et al. (2005) conducted a study on questioning the hybrid model: Student outcomes in different course formats. Study found that, student participation, interaction, and grades are high in hybrid mode than traditional and internet-based modes. Dowling et al. (2003) conducted a study on do hybrid flexible delivery teaching methods improve accounting students' learning outcomes. Study achieved better outcome through hybrid teaching mode compared to traditional mode.

From the reviews, researchers found that there have been attempts to understand challenges of online teaching and blended teaching in the past two decades. However, studies related to hybrid teaching practices and challenges are available from recent years across countries especially in the year 2020 onwards. In order to capture comprehensive details about hybrid teaching world needs more research studies at present. Interestingly, there are very few studies on hybrid teaching in developing countries and middle-income countries. Hybrid teaching involves investment from higher education institutions on video-conferencing platform, internet, learning management system, staff training, etc., and now with COVID-19 protocol spacious classroom infrastructure, Wi-Fi points, less teacher-student ratio, high bandwidth internet facility and so on. It is difficult for low-income countries to manage infrastructure for hybrid mode. In developing countries infrastructure will not come from country alone, but from institutions and individuals. Teachers and students also need infrastructural investment like personal computer, internet, headphone, electricity, uninterrupted power supply etc. Unless there is a complete willingness and acceptance from all the stakeholders, it is difficult to implement hybrid teaching in low-income countries. Therefore, transtheoretical model largely derived from ecosystem theory guides the present study while recording the lived experiences of teachers in hybrid classes and the challenges they face. The hybrid teaching experiences of teachers do have the elements of Panarchy or adaptive cycle like teachers' potential for adaptive behavioural change, teachers' connectedness in the immediate context, and resilience to rise from crisis amidst educational setting (Gunderson & Holling, 2002). Thus, there is a need for qualitative research especially in developing countries to understand the awareness on hybrid teaching arrangements and the challenges faced by the higher education teachers while conducting hybrid classes amid COVID-19 pandemic.

2 Research Questions

The present study framed following research questions based on the research gap emerged from the review of literature and the contextual experiences of all researchers involved in the present study. Are University teachers aware of Hybrid teaching? The concept, theoretical framework, and the process of hybrid teaching. What are the lived experiences of University teachers while conducting hybrid classes amidst COVID-19? And What are the challenges faced by the University teachers while conducting hybrid classes?

3 Methods

The present study followed qualitative design and conducted interviews to obtain in depth information on hybrid teaching experiences of University teachers and the challenges faced by them. The following paragraphs present the details of methods and procedures followed in the present study.

3.1 Sample

The present study employed qualitative research design address the research questions raised in the study. Study applied purposive sampling technique to choose the participants of the study. Accordingly, eight University teachers taking HyFlex type hybrid classes formed the sample of the study (Guest et al., 2006). Among them, four are male teachers and remaining four are female teachers. Study included teachers from age 30 to 45 years and excluded the beginner teachers and teachers above 50 years old due to their vulnerability to COVID-19 infection. Study collected the data in the month of March 2021 when hybrid teaching began at Universities. Table 1 below presents the demographic details of the participants.

3.2 Ethical Considerations

Researchers obtained approval of institutional review board (IRB) of the University to conduct the present research. At the beginning of each interview researcher briefed the purpose of the study to each of the participants and clarified their doubts. Researcher obtained consent from each participant for participating in the study. Researcher assured participants of their anonymity and confidentiality of the data collected from them. Researcher gave participants an option to withdraw from the study at any point if they are not finding it comfortable. Researchers assigned pseudonyms P1, P2, P3, P4, P5, P6, P7, and P8.

Table 1. Showing Participants demographic data.

Participant	Gender	Age	Teaching	Department
P1	M	37	UG and PG	Commerce
P2	F	31	UG and PG	Commerce
P3	F	44	UG	School of Business & management
P4	M	45	UG and PG	School of Business & management
P5	M	30	UG	Psychology
P6	F	43	UG	School of Business & management
P7	M	35	UG	School of Business & management
P8	M	40	UG and PG	Psychology

M = Male; F = Female; UG = Undergraduate; PG = Postgraduate

3.3 Interview Procedure

The researchers conducted the interview on an online video conferencing platform called Cisco Webex/ Google meet at a mutually convenient time for the researcher and the participant. Study employed funnelling approach for interviews. The questions asked were flexible and researcher constructs on-spot prompts to seek further deeper information (Smith & Osborn, 2008). Interviews held for about 15 to 20 minutes and researcher recorded each interview session. The order of the questions remained same to each participant; this help in maintaining the same context and add on to the validity and reliability of the data collected (Huffcutt & Arthur, 1994; Campion et al., 1997). The recordings of the interview is stored safely and is accessible only to the researchers. Researchers created verbatim transcripts manually for each of the interview sessions and analysed the transcripts using interpretative phenomenological analysis (IPA).

3.4 Interview Schedule

As study planned to use funnelling approach for the interview data collection, researcher framed the following items for the study. Framed interview items were face validated by the researchers and verified the same with senior teachers for content validity.

1. Were you aware of Hybrid teaching before?
2. What were your initial thoughts when university announced re-opening and planning to adopt hybrid teaching at the threshold of new normal?
3. How was your hybrid teaching experience?
4. What are the major challenges faced while conducting hybrid classes amid COVID-19?

3.5 Data Analysis (Interpretative Phenomenological Analysis)

Study used interview transcripts data to evolve with themes and subthemes as per the procedures of IPA (Smith & Shinebourne, 2012). The procedure followed for data analysis is as given below. Researchers read the verbatim transcripts thoroughly multiple times to get a fair account of the responses. Multiple reading gave ideas to emerge with major themes and unique points from each of the interview transcript. Further in-depth analysis of the details under each major theme gave ideas to come up with subthemes. Researchers emerged with suitable discrete titles to each of the major theme and subthemes. Researchers were able to note the detailed account of the hybrid teaching experiences of University teachers. Researchers further brain stormed on the titles arrived for themes and subthemes and found them valid. Finally, all authors have gone through the themes and sub themes and confirmed the same.

4 Results

Present study intended to understand University teachers' awareness on hybrid teaching, lived experiences of hybrid classes, and challenges faced by them during hybrid teaching amid COVID-19. Researchers have organised the results as per the research questions in order and mapped all the emerged themes and sub themes in the same order. Table 2 below presents the emerged themes and subthemes out of interpretative phenomenological analysis. Researchers have explained each themes and subthemes in detail following the Table 2.

Table 2. Titles of emerged Themes and subthemes.

1. Hybrid teaching understanding and awareness
(a) Lack of awareness
(b) Lack of experience
2. Teachers lived experiences of hybrid teaching amid COVID-19
(c) Transition from Online to Hybrid teaching
(d) Workplace Uncertainty
(e) Student engagement
(f) Self and family safety
(g) Faculty development programme
3. Challenges faced while Hybrid teaching
(h) Common challenges

4.1 Major Theme: Hybrid Teaching Understanding and Awareness

This major theme described the knowledge about hybrid teaching by the University teachers. The Conceptual understanding of hybrid teaching and its process. Awareness on the theoretical framework guiding the hybrid teaching. Participants initial understanding of hybrid teaching and their awareness to begin hybrid teaching. Study found that teachers are not clear on how to go about hybrid teaching and had their own assumed ideas.

Sub theme: Lack of awareness. Most of the university teachers are not aware of the hybrid teaching nor ever tried out hybrid teaching in their teaching career. They described a very trivial understanding of hybrid teaching

P1: [...] Hybrid teaching has been forced because of the situation, it is a combination of online and offline which is facilitated with a lot of technology and a lot of tools and applications.

P2: [...] for me hybrid actually means... during the entire semester...few face to face sessions and few online sessions... a few students joining online and a few here offline...is not hybrid.

P6: [...] I never heard earlier...I mean...practiced this....so I was aware of online teaching but not this...Hybrid mode of teaching...new experience for me.

Sub theme: Lack of experience. Majority of the University teachers' mentioned that they were always using traditional lecturing mode of teaching with technology integration sometimes. Pandemic has pushed them to teach online and therefore got used to it in the past 10 months and hybrid teaching is something new for us and never experienced.

P3 - [...] I was not using technology so much earlier. I used to prefer a chalk and board of teaching style but for online now, I have to be much more prepared. For every session, I have to plan on what I am going to do for the next one and half-hours or so. So a day before I have to spend at least two to three hours for each session to prepare to plan and think...

P2 - [...] when we shifted to hybrid this was again a new thing a lot of changes and modifications had to be done in the strategies and the way we handle things so little bit of anxiety and confusion as to how this will go about...

4.2 Major Theme: Teachers Lived Experiences of Hybrid Teaching Amid Covid-19

Majority of the teachers reported that they had varied experiences each day while doing hybrid teaching. Transition from online to hybrid was not easy they had to learn newer skills and unlearn certain practices. They found lot of workplace uncertainty in terms of schedules, makeshift arrangements, time tabling and so on. Student engagement has been a challenge. Individual and family safety has been a concern. There is need for more and more faculty development programmes pertaining suitable hyflex model for the given context.

Sub theme: Transition from online to hybrid teaching. Many believed that hybrid teaching means more of technology integration within their classes. They were anxious about which technology integration works well for hybrid teaching, as they have to handle students both online and offline at the same time.

P1 - [...] in the hybrid teaching I have started to use the jam board and the sketch board. I am not a tech pro, but whenever there is a need, I use it. So after exploring all these google classrooms, sharing of word documents, and these other types of tools, I feel that it has saved a lot of time, and I'm now more inclined to learn and explore new technology.

P6 - I feel sometimes teaching online they follow little less. Maybe because the technical issue... may not be at our end...but maybe at students end...they tend to lose the connection, sometimes the voice feeble...

P5 - [...] I do have issues but most of it are technological issues. Design of home Computer, are not robust enough to work. See I have back-to-back classes. When we use Microsoft teams for hours, a lot of lagging and fluctuation happens. Hence, our teaching and delivery is entirely on the mercy of our computers performance...

P4 - [...] I use an additional camera to capture the entire board, this camera captures the entire board compared to an inbuilt camera in a laptop...

Sub theme: Workplace Uncertainty. A significant number of University teachers reported that they are not sure how the hybrid classes will go on as it has numerous requirements. For online classes, they used their personal internet connections. Either home Wi-Fi or mobile hotspots had issues like fluctuating bandwidth, power outage, low battery backups, UPS issues and so on. Now in hybrid teaching University campus Wi-Fi has to be strong to conduct online-cum-offline classes. Cleanliness and social distancing at the campus possible? How to teach by wearing a mask, will all offline and online students hear it? What if students have symptoms of COVID-19 and we are not aware of it? How do staff travel to reach University campus as using public transport can be a risk. How do I set up the camera to be visible online while teaching offline? What about the power-fluctuations at campus? Will the laptop microphone help online students? What happens if I could not reach the morning first hour offline class due to some reason who would inform online students? These are some of the thoughts quite common to all the university teachers and were uncertain of the hybrid teaching possibilities at the beginning.

P5 - [...] hybrid teaching can only go on for a very short period, there is only a makeshift arrangement as supposed to regular things. Nothing can replace regular classes...

P4 - [...] I feel it should be just offline or online, having it together makes it difficult for teachers to address students present in the class and the students attending it online...

P3 - [...] if they are coming from some other cities. They come with negative RT PCR reports. If they are localities, it is relatively safe for us. We have given an orientation to follow the government norms with respect to social distancing, usage of masks, and hand sanitization...

P7 - [...] I feel that Online teaching would continue to some extent because of various Coursera and various other courses also....Online teaching has a scope...offline teaching also has a scope but I don't know whether Hybrid teaching has a scope in the future...

Sub theme: Student engagement. Most of the participants reported that engaging students in a hybrid class is extremely difficult. Teachers' are not sure; they must sit or stand while teaching. If they want their online students to see their face then they need to sit down. At the same time, they are not used to teaching by sitting in an offline classroom. It has put them in an uncomfortable situation. Sometimes online students are interacting randomly and are not clearly hearing the offline class discussions. Interestingly, few teachers are responding more to online students than offline during their hybrid classes as they feel they are away and students here can always ask their doubts face to face. When technology integrated online platforms are used for class activities online are doing better than offline students. Offline students somewhere feel that it is for online students upon imposing only they are doing. Few faculties reported that, over a period of time online students are not actively responding and are giving reasons like not audible, lack connectivity, platform links not working, and some are just unresponsive.

P1 - A few students... take efforts to explore the technology. However, few students... answer... my phone or my laptop does not support this application. Therefore, I am not able to submit the assignments, which you had asked for. So further, we cannot ask them.

P1 - [...] sometime in the google meet or Webex platform, I show them how to use google classroom or how the moodle platform can be used. Apart from that, I do not take any other efforts.

P6 - [...] their cameras are also off...without looking at their reactions...we do not know whether they have understood the concept or not...only...here only few students who are active in the online mode...others are passive listeners.

P7 - [...] when it comes to Hybrid and Online, content preparation is the major time taker and if it is Hybrid, okay a major problem is, there are very few listeners in the classroom and whether to focus on the offline students or the online students is a big challenge...

Sub theme: Self and family safety. Some of the teachers opted to teach online mode even when University announced to conduct hybrid teaching. It is due to their pre-existing health conditions, their age, location, family problems, and personal problems. They reported that taking hybrid classes could be a risk as infection can spread from students or faculty colleagues due to their weak immunities. It could bring infection to their family members living with them. Traveling to university at this stage can be a risk as the second wave of COVID-19 is fast spreading.

P1: [] we all are quite skeptical about the spread of the virus. The same happens with the students...So how will they feel; will they have the same enthusiasm? They still prefer online teaching to offline teaching.

P8: [] one student sent a mail to us stating that in his or her PG someone got infection. So as a precautionary measure she was moving from offline to online mode staying away from the family, I feel that for the people who are staying away from the family, the family are more anxious... it creates anxiety with them. That is something that will affect our wellbeing personally, because family was asking us, and... things like that. We are pacifying them. However, in other ways, it creates a trauma within ourselves. Therefore, at a professional level, it affects our mental wellbeing.

P1 and P8 have expressed their anxiety about their safety and student safety. They found students having contact with affected patients and it has affected their mental well-being and of their family. From the above transcripts, we infer that even at the wake of COVID-19 and the imposition of hybrid teaching from Universities have left the teachers in an uncertain future. Most of the university teachers are not aware of hybrid teaching processes and have anxiety on the way to handle hybrid teaching.

Sub theme: Faculty development programme (FDP). Faculty members had mixed responses towards FDP. Some reported that there is a need for more FDP to understand the way to conduct hybrid teaching. Few said the FDP conducted is merely talking about technology integration in the classroom than strategies to implement hybrid teaching. Few have reported that FDP conducted were useful and more such FDP are to be organised for developing the hybrid teaching competence among university teachers. Some even reported that only FDP's will not suffice, even students are to be trained on how to cope up with the hybrid teaching- learning processes.

P5 - [...] another thing that I would point out is, in our QIP, FDP that we are learning, we should have at least one subject expert utilising these tools. We are exposed to these tools...but how are we going to use a particular tool for a particular subject... because each subject need is different....then the teachers can relate better. Otherwise they will be very alienated....it will be like a third person's perspective and understanding instead of putting themselves into the shoot.

P2 - [...] when the University started planning for such kind of hybrid teaching then the orientation and the training was given only then I came to know that there is something called hybrid teaching...

4.3 Major Theme Challenges Faced While Hybrid Teaching

Majority of the teachers reported that they are not satisfied with the hybrid teaching. They reported that only a few students opted to come online and it is difficult to follow COVID-19 precautions all the time like teaching by wearing a mask. Sanitising frequently.

Handling offline and online students simultaneously is a tedious job and needs a lot of preparation time. Unlike working from home for online teaching, setting hybrid teaching in physical classrooms is time consuming and that leads to indiscipline issues sometimes. In spite of the use of a variety of online platforms to keep the student engaged it was found that many times online students don't respond properly to the class instructions this has resulted in lack of interest for offline students. Some of the students even asked why offline students should do asynchronous work. As conceptually, asynchronous work is a part of online mode of learning. Taking attendance in hybrid class is also difficult, time consuming as some students are offline, and some are online. Sometimes offline students also attend online classes when they are not well then it is very difficult to track the students and their work. If we use classroom external speakers then offline students are to mute their audio in the laptop otherwise it creates echo. So monitoring audio systems with teachers, offline students, and online students is again a difficult task that leads to wastage of time for technical set-up etc. Overall teachers reported that hybrid teaching is not efficient when compared to only online or traditional teaching.

> *P1 - [...] I have started using many sketchpads, digital sketchpads. Therefore, these digital sketchpads are just giving the feeling, just like how we are sitting the class and it gives opportunities to the students who are taking online classes to enjoy. It is just like a blackboard. Apart from that, I have also...*

> *P1 - [...] When we are taking the online classes we can keep a book ... we can look into the important points but when we stand. However, when it comes to hybrid classes and prepare the presentations 2–3 days before... I have to maintain... eye contact with the offline students...my voice modulation with the online student...have to clear the doubts ofboth...encourage...*

> *P8 - [...] when the students did not respond, I said like online students we are not marking attendance ...I observed them ...saying my mic is not working*

> *P5 - [...] online students are not able to understand why the discussion is going on because the student who is asking the question in the offline classroom, their question is not audible to the online students...*

> *P7 - [...] frankly speaking, Asynchronous is not supporting much....we are just giving the work....cover the syllabus....*

P1 reported that he made use of several technology tools, prepared for classes in advance, and works on eye contact, voice modulation, and encourages students to learn. P8 reported that, when students are not responding and giving excuses like mic not working and were made to respond by telling them they would be marked absent.

Sub theme: Common challenges. Some of the common challenges in engaging hybrid classes as reported by the teachers are as below. Teachers' need more preparation time for hybrid classes. It is difficult to attend online, and offline students doubts simultaneously in class hour. Monitoring students' performance in the class activities is difficult. Engaging them in class discussions, formative assessment of both the groups, providing a variety of asynchronous work and its monitoring, ensuring COVID-19 safety protocol, integrating appropriate technology tools, technical glitches such as low bandwidth, audio connection, network issues, power outage, power fluctuation, capturing attention of both online and offline students, Work-life balance, and future uncertainty.

P1 - Ma'am to be honest with you, from the teacher's point of view, I would tell that not all teachers are very well versed including me there are a lot of instances where I have explored technology and that has not been working properly. So sometimes students accept that we all are exploring technology and it is not easy to explore but sometimes they have this attitude that at least you do not know, then how will you teach us...

P1 - To be honest a lot of times Technology has not been supporting, because either the computer or the internet connection or the Wi-Fi connection is disrupted. Secondly I am not able to raise my voice very loud because I get a throat pain because there's a lot of voice modulation third we go to the class we sit in the chair Which is very uncomfortable fourth sometimes back to back classes it's very difficult. Fifth one we are afraid if anyone has any covid symptoms. Pandemic symptom so we have an eye on that also which is quite affecting the ability of the teachers...

P6 - [...] we have to use tools to engage them...but that preparation requires a lot of time...

P5 - [...] Because of hybrid learning, even if I forget to switch off the mike in my phone, I will have feedback. It immediately becomes very complex. If I plug my laptop to the HDMI....or even when online students are talking, what they talk can only be heard by me and not by the rest of the class. So if I connect my laptop to the class audio system, the whole setting up will take away some 10–15 minutes in the beginning...

P4 - [...] I think one of the primary challenges would be not knowing how many students are genuinely paying attention in the online class...you are not sure whether the students are solving the problems that you're giving them during the classes and it's also hard to assess the students. It also affects the student-teacher interaction...

P7 - [...] students are sitting in any corner now. I have to notice the screen, I have to look at the students, and I have to look at the chat box also. So, I don't know what the offline students are doing on the corners....they may be using mobile phones or they may be looking at something else also....that control I don't have....and I am not even able to check whether they are writing down in their books or not...

5 Discussion

The study explored the hybrid teaching experiences of University teachers in the context of rising from COVID-19 to reach new normal. The study focused on University teachers' awareness, factual experiences, and challenges faced by them during their hybrid classes. A detailed account of their experiences on hybrid teaching brought out from in-depth interviews would help the higher education teaching community. The majority of participants reported that they are not aware about hybrid teaching processes and the way it is going amid pandemic. There is no conceptual clarity among the University teachers about hybrid teaching. How is it different from blended learning? None of the participants had any hybrid teaching experiences until Universities forced it onto them. Though they knew about online-teaching, e-learning, blended learning but had no experience of handling an hybrid class and they understood about it when the University conducted a professional development programme (PDP) and the similar situation found in Chinese universities (Zhang et al., 2020). Though the PDP had several theoretical components, technology tools, and examples, yet teachers lacked practical experience. Therefore, teachers had their own beliefs about their real time hybrid classes. Participants repeatedly mentioned that there is uncertainty about many things be it internet connectivity, safe travelling to campus, local government decisions, offline students safety, and online students demands. Many of the participants are also worried about their personal safety and the safety of their family members while they resort to hybrid teaching (Suhail et al. 2020).

The core findings of the study brought out the need for faculty development programmes on hybrid teaching to enrich faculty members' understanding and awareness about hybrid learning. Many participants reported that there are plenty of technology tools available but learning them quickly and adopting them to hybrid classes is not easy and technical glitches they face while implementing several such tools are unforeseen (Fong et al. 2008). This brings down class morale and teacher satisfaction and hence resilience has to be built into the system (Tam and El-Azar, 2020). Most of the participants reported that, majority of the online students have gradually become unresponsive to classes and their presence is terribly fluctuating. Teachers are helpless as attending the queries of students in hybrid classrooms demands swift responses and good technological skill. Eventually teachers reported the challenges faced by them during hybrid classes, few of the major challenges are handling both online and offline student groups, technology integration, auditory issues, internet issues, engaging both the groups simultaneously, COVID-19 anxiety, class preparation time, and so on.

6 Conclusion

Higher education institutions have confined themselves to their campuses due to unprecedented pandemic situations. It has been a year since they resorted to online teaching and learning. Many countries across the globe are facing the second wave of COVID-19 and are in the process to rise back to new normal. However, third world countries and other lower and middle-income countries are suffering from numerous challenges in the wake of the pandemic. Safety issues, health issues, decrease in admissions, international collaboration has come to a standstill, dilapidated technological infrastructure, and regular uncertainty of work place. Universities are running out of finance (Thomas, 2020). In order to rise back from the situation few universities have started hybrid teaching where in few of the interested students attend classes offline and remaining attend via online mode. University teachers have to handle both the groups in their classes with a hybrid mode of teaching. The present study aimed at understanding university teachers' awareness about hybrid teaching, university teachers' lived experiences of hybrid classes, and the challenges faced by them while handling hyflex model of hybrid classes. Study came up with the finding out of qualitative research conducted. IPA answered the research questions as 3 main themes and 8 sub themes. Study found that, University teachers clearly lack awareness on hybrid teaching at conceptual and process level. Teachers' hybrid teaching experience included; anxiety over transition from online to hybrid teaching, work place uncertainty, uncertain student engagement experiences, self and family safety, and need for FDPs. Study revealed major challenges of hybrid teaching especially in hyflex model are; transition from online to hybrid mode, handling online and offline students group simultaneously, work place uncertainty, technology integration to classes, technological barriers, COVID-19 anxiety at personal and family level, student engagement, appropriate assessments, planning synchronous and asynchronous activities, class management, class attendance, and investment in digital infrastructure.

6.1 Suggestion for Future Research

In terms of future research, researchers can focus on developing techniques for smooth conduct of hybrid classes based on the micro-level observations and issues via qualitative and quantitative research. Researchers can also explore the options available for Universities to have technology enabled teaching and learning to address the crisis like this in future. Strategies and readiness for the education sector in the new normal. The future research may focus on resolving the classroom challenges of hybrid teaching and provide technological solutions to the smooth conduct of hybrid teaching and learning processes. Future researchers can also suggest best practices for practicing hybrid teaching.

Acknowledgements. There is no funding received for this research work. We solicit our gratitude to all the University teachers who participated in the interview.

Author Contribution. There is no conflict of interest among the authors. Corresponding author has edited the entire article and carried out data analysis. Other authors have worked on the review of literature, conducted the interviews to collect the data, and worked on the references.

References

Babatunde, A.O., Emrah, S.: Covid-19 pandemic and online learning: the challenges and opportunities. Interact. Learn. Environ., 1–13 (2020)

Bower, M., Dalgarno, B., Kennedy, G.E., Lee, M.J., Kenney, J.: Design and implementation factors in blended synchronous learning environments: outcomes from a cross-case analysis. Comput. Educ. **86**, 1–17 (2015)

Butz, N.T., Stupnisky, R.H.: Improving student relatedness through an online discussion intervention: the application of self-determination theory in synchronous hybrid programs. Comput. Educ. **114**, 117–138 (2017)

Camic, D.L.L., Panter, A.T., Rindskopf, D., Sher, K.J. (eds.): APA Handbook of Research Methods in Psychology. American Psychological Association. https://doi.org/10.1037/13620-005

Campion, M.A., Campion, J.E., Palmer, D.K.: A review of structure in the selection interview. Pers. Psychol. **50**, 655–702 (1997)

Guest, G., Bunce, A., Johnson, L.: How many interviews are enough? An experiment with data saturation and variability. Field Methods. **18**(1), 24 (2006). https://doi.org/10.1177/1525822X05279903

Huffcutt, A.I., Arthur, W.: Hunter and hunter (1984) revisited: interview validity for entry-level jobs. J. Appl. Psychol. **79**, 184–190 (1994)

Daniel, S.J.: Education and the COVID-19 pandemic. Prospects. **49**, 91–96 (2020). https://doi.org/10.1007/s11125-020-09464-3

Dowling, C., Godfrey, J.M., Gyles, N.: Do hybrid flexible delivery teaching methods improve accounting students' learning outcomes? Acc. Educ. **12**(4), 373–391 (2003)

Gunderson, L.H., Holling, C.S.: Panarchy – Understanding Transformations in Human and Natural Systems. Island Press, Washington, DC (2002)

Hege, B.A.: The online theology classroom: strategies for engaging a community of distance learners in a hybrid model of online education. Teach. Theol. Relig. **14**(1), 13–20 (2011)

Huang, B., Hwang, G.J., Hew, K.F., Warning, P.: Effects of gamification on students' online interactive patterns and peer-feedback. Distance Educ. **40**(3), 350–379 (2019)

Huang, R., Tlili, A., Chang, T.-W., Zhang, X., Nascimbeni, F., Burgos, D.: Disrupted classes, undisrupted learning during COVID-19 outbreak in China: application of open educational practices and resources. Smart Learn. Environ. **7**(1), 19 (2020)

Ilgu, A.K.: Faculty Perspectives on Benefits and Challenges of Hybrid Learning, vol. 26, p. 1. Sage (2015)

Jason, M.: Hybrid teacher leaders and the new professional development ecology. Prof. Dev. Educ. **38**(2), 291–315 (2012)

Johnson, N., Veletsianos, G., Seaman, J.: US Faculty and administrators' experiences and approaches in the early weeks of the COVID-19 pandemic. Online Learn. **24**(2), 6–21 (2020)

Fong, J., Kwan, R., Wang, F.L. (eds.): Hybrid Learning: A New Frontier, pp. 161–168 (2008)

ICHL: City University of Hong Kong, Hong Kong (2008)

Khan, A.I., Shaik, M.S., Ali, A.M., Bebi, C.V.: Study of blended learning process in education context. Int. J. Mod. Educ. Comput. Sci. **4**(9), 23–29 (2012)

McFarlin, B.K.: Hybrid lecture-online format increases student grades in an undergraduate exercise physiology course at a large urban university. Adv. Physiol. Educ. **32**(1), 86–91 (2008)

Marshall, M.: Covid virus pushing education to online learning/teaching creates big challenges: ethical, practical & financial issues for teachers/professors and administration leadership. PM World J. **10**(1), 1–4 (2020)

Murphy, M.P.A.: COVID-19 and emergency eLearning: consequences of the securitization of higher education for post-pandemic pedagogy. Contemp. Secur. Policy. **41**(3), 492–505 (2020)

Quezada, R.L., Talbot, C., Quezada-Parker, K.B.: From bricks and mortar to remote teaching: a teacher education program's response to COVID-19. J. Educ. Teach. (2020). https://doi.org/10.1080/02607476.2020.1801330

Raes, A., Vanneste, P., Pieters, M., Windey, I., Van Den Noortgate, W., Depaepe, F.: Learning and instruction in the hybrid virtual classroom: an investigation of students' engagement and the effect of quizzes. Comput. Educ. **143**, 103682 (2020)

Rasheed, R.A., Kamsin, A., Abdullah, N.A.: Challenges in the online component of blended learning: a systematic review. Comput. Educ. **144**, 103701 (2020)

Reasons, S.G., Valadares, K., Slavkin, M.: Questioning the hybrid model: student outcomes in different course formats. JALN. **9**(1), 83–94 (2005)

Skulmowski, A., Rey, G.D.: COVID-19 as an accelerator for digitalization at a German university: establishing hybrid campuses in times of crisis. Hum. Behav. Emerg. Technol. **2**(3), 212–216 (2020)

Smith, J.A., Osborn, M.: Interpretative Phenomenological Analysis. Qualitative Psychology: a Practical Guide to Research Methods, pp. 53–88. SAGE (2008)

Smith, J.A., Shinebourne, P.: Interpretative Phenomenological Analysis. H. Cooper (2012)

Suhail, A., Iqbal, N., Smith, J.: Lived experiences of Indian youth amid COVID-19 crisis: an interpretative phenomenological analysis. Int. J. Soc. Psychiatry. (2020). https://doi.org/10.1177/0020764020966021

Tam, G., El-Azar, D.: Three ways the coronavirus pandemic could reshape education march 13. In: World Economic Forum: Global Agenda (2020) File:///D:/COVID/3%20ways%20the%20coronavirus%20pandemic%20could%20reshape%20education%20World%20Economic%20Forum.Html

Thomas, C.J.: Coronavirus and Challenging Times for Educa-Tion in Developing Countries April 13, Brookings (2020) file:///D:/COVID/Coronavirus%20and%20challenging%20times%20for% 20education%20in%20developing%20countries.html

Zhang, W., Wang, Y., Yang, L., Wang, C.: Suspending classes without stopping learning: China's education emergency management policy in the COVID-19 outbreak. J. Risk Financ. Manag. **13**(55), 2–6 (2020). https://doi.org/10.3390/jrfm13030055

Prakasha, G.S., Muttungal, P., Lapina, M., Taran, V.: Flipped classroom strategy in online teaching: challenges faced by higher education teachers. In: Lapina, M., Prakasha, G., S., Grigoriev, S. (eds.) SLET 2022: International Conference on Innovative Approaches to the Application of Digital Technologies in Education and Research. SLET 2022. Lecture Notes in Networks and Systems, pp. 241–254. Springer, Cham (2024a)

Prakasha, G.S., Thamaraikannan, S., Lapina, M., Sanskriti, R.: Pre-service and in-service teachers' perceptions of using virtual reality tools in teaching. In: Lapina, M., Prakasha, G., S., Grigoriev, S. (eds.) SLET 2022: International Conference on Innovative Approaches to the Application of Digital Technologies in Education and Research. SLET 2022. Lecture Notes in Networks and Systems, pp. 340–356. Springer, Cham (2024b)

Lapina, M., Shiriaev, E., Babenko, M., Istamov, I.: High-speed convolution core architecture for privacy-preserving neural networks. Program. Comput. Softw. **50**, 417–424 (2024). https://doi.org/10.1134/S0361768824700282

Rusanov, M., Babenko, M., Lapina, M., Sajid, M.: Identification of exploited unreliable account passwords in the information infrastructure using machine learning methods. Big Data Cognit. Comput. **8**, 159 (2024). https://doi.org/10.3390/bdcc8110159

Active Digital Footprint in Assessing the Level of Development of Professional Competencies of Master's Students

Galina Shevchenko(✉), Alexander Shevchenko, and Alexandra Shevchenko

North Caucasus Federal University, Stavropol 355017, Russia
gishevchenko@ncfu.ru

Abstract. Currently, in the context of digital transformation of education, special attention is paid to the collection and analysis of the digital footprint of the student, the study of which helps to model the physiological, psychological and cognitive characteristics of students, allows you to effectively build personal development trajectories, digital profiles of students, control the learning process, increasing its effectiveness. Based on theoretical analysis, the article reveals the essence of the concept of «digital footprint» in the concept of education, the type of collected digital footprint depending on the information system used. The purpose of the work is to form professional competencies of students based on the analysis of the active digital footprint. The authors describe the components of the digital footprint and the use of various methods and tools for data processing, such as statistical analysis, a software platform for data research using an extensive set of analytical tools KNIME Analytics Platform. The results indicate that the analysis of the active digital footprint of master's students studying in the profile «Digital Technologies in Science and Education» contributes to the development of their professional competencies. The conclusion provides examples of practical applications of the digital footprint and draws conclusions about its main possibilities for teachers and students.

Keywords: Digital footprint · Digital footprint analysis · Visual workflow · Up-to-date set of competencies

1 Introduction

The formation of the concept of «digital footprint» took place in the context of the creation, testing and application of educational and methodological materials based on the use of artificial intelligence methods, which are one example of digital transformation in education [1–3].

Shamsutdinova T. in order to understand the essential characteristics of the digital footprint, suggests structuring its components into: technical and technological; personality-psychological; behavioral; activity-based; competence-based; communicative; reflexive. The presented components «are closely intertwined and interact with each

M. Lapina et al. (Eds.): DET 2024, LNNS 1616, pp. 406–415, 2025.
https://doi.org/10.1007/978-3-032-04365-8_39

other, forming the possibility of considering one subject area from different points of view with different research goals» [4].

The notion of a «digital footprint» in relation to education is interpreted in different ways. According to some researchers, «it includes the content of the log logs of the digital system, others identify the digital footprint and the digital portfolio, implying that the footprint is formed by data on the student's achievements and the results of his educational activities, and often, when studying the digital footprint in education, they are limited to the content contained in the user's profile on social networks» [5].

According to Baranova, E., Vereshchagina, N., Shvetsov G. «a student's digital footprint begins to form from the moment of his admission to the university and contains personal data, information about admission, direction of study, educational program, academic performance during the entire period of study, as well as educational analytics data collected automatically when students work in the systems distance learning and open education platforms when students take online courses» [6].

Skripnik V. defines a digital footprint as the basis for forming a student's digital profile, which «will allow us to judge the dominant qualities of a person and determine promising trajectories of education and self-education» [7].

A student's digital footprint can be presented electronically in the form of coursework; reports on internships, laboratory work and practical classes; project descriptions; final qualifying works; tests, etc. Its accumulation and assessment contribute to:

1. Identification of categories of master's students included in the «risk» group.
2. Monitoring the state and dynamics of the contingent movement by year.
3. Development of educational programs, academic disciplines and components of the digital educational environment.
4. Assessment of the level of formation of competencies in the learning process [8].

2 Research Goals and Objectives

Currently, the terms «active» and «passive» are used in relation to digital traces.

An active digital trace is a trace containing data obtained with the direct and conscious participation of the user.

A passive digital trace, on the contrary, is collected by an interested party without the user's knowledge. It includes information unintentionally left by the user.

When studying a digital trace, metadata is also recorded for subsequent analysis. This should include a student ID, the time the digital trace was obtained, an educational event, etc.

Various methods are used to analyze, process and visualize traces. This work focuses on statistical methods and tools that allow not only to analyze data and predict conclusions, but also to see the results of work in numbers [9, 10].

The purpose of the study is to develop professional competence in master's students based on the analysis of an active digital trace. Research objectives:

1. To analyze the capabilities of predictive analytics of the KNIME Analytical Platform for analyzing the active digital footprint of students, represented by the results obtained in mastering the academic discipline, during the midterm assessment.

2. To propose criteria for assessing the levels of development of professional competencies of master's students based on the results of the analysis of the active digital footprint.
3. To identify topics in the studied discipline, the essence of which does not contribute to the formation of the declared professional competence of master's students at the proper level.

3 Literature Review

To analyze and interpret the active digital footprint of students, a data mining software platform was used, using the extensive suite of analytical tools, KNIME Analytics Platform [11], which allows visual workflows to be created using an intuitive drag-and-drop style graphical interface without the need for coding (see Fig. 1).

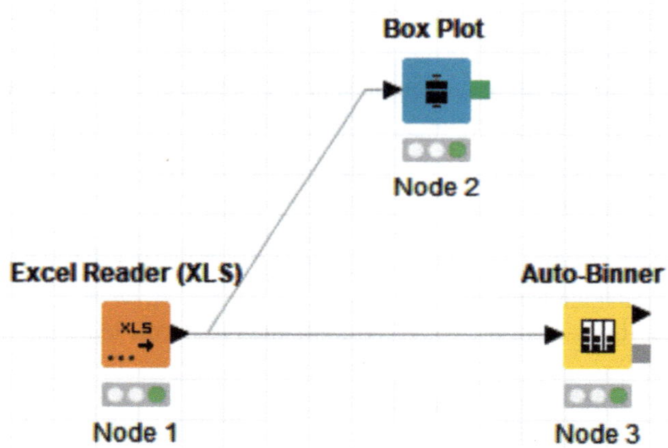

Fig. 1. Visual workflow.

Visual workflow (Workflow) consists of nodes (Node) that perform some function - reading data from a database or spreadsheets, transformation, data visualization. Arrows connecting nodes show the direction of data movement. The nodes are categorized. KNIME Analytics Platform supports the following types of nodes:

1. IO data input/output.
2. Manipulation transforms data, including filtering rows, columns, and sorting.
3. Views visualized data using Histogram, Pie Chart, Scatter Plot, etc.
4. The database provides the ability to connect to a database, read/write.
5. Workflow Control is used to create loops, iterate through groups at runtime, and more.

After the workflow was created, it can be launched for execution, during which, starting from the very first, the Workflow nodes are processed one by one. A traffic light reflects the current state of each node: red error; yellow ready for execution; green completed.

If an error occurred in the node during execution, the execution of the entire branch following it stops. Workflow can be restarted from any node. Almost every node has a configuration dialog where you can configure properties [11].

4 Methodology

There are various methods and tools for analyzing the active digital footprint. In this study, its results were analyzed in the KNIME analytical platform.

The active digital footprint was extracted from the number of correctly completed tests. Each test included 25 test tasks on the academic discipline «Educational data-engineering». The workflow consisted of three nodes. In this case, Node 1 was initially configured for the number of correct answers that the student showed when completing six tests.

Eleven master's students took part in the testing. It was conducted to assess the level of development of their professional competencies. The results of the study are presented in Table 1.

Table 1. Results of the number of correct answers for each student for all tests.

Student	Test 1	Test 2	Test 3	Test 4	Test 5	Test 6
Student 1	23	20	20	19	19	17
Student 2	13	15	14	12	11	14
Student 3	19	22	19	19	18	14
Student 4	20	21	21	19	22	20
Student 5	23	24	22	20	19	25
Student 6	25	21	24	18	25	23
Student 7	21	21	16	17	21	20
Student 8	24	23	23	25	23	24
Student 9	20	19	24	20	18	19
Student 10	21	22	22	23	22	22
Student 11	17	16	18	17	19	18

The Auto Binner node was used in the visual workflow to display the minimum and maximum value of the number of correct answers for each test. The results of these measurements are shown in the Table 2 in the Lower Bound and Upper Bound columns, respectively.

Table 3 shows the levels corresponding to the lower and upper limits of correct answer points for each test.

The work presents data corresponding only to the results of the lower and upper limits of the number of correct answers for each test. However, in the study, when determining

Table 2. Points for correct answers.

Test	Lower Bound	Upper Bound
Test 1	13	25
Test 2	15	24
Test 3	14	24
Test 4	12	25
Test 5	11	25
Test 6	14	25

Table 3. Scoring results for each test.

Test	Lower Bound	Upper Bound
Test 1	0.52	1.00
Test 2	0.60	0.96
Test 3	0.56	0.96
Test 4	0.48	1.00
Test 5	0.44	1.00
Test 6	0.56	1.00

the assessment of the level of formation of professional competencies, the answers to all test tasks of each test were taken into account.

The work presents data corresponding only to the results of the lower and upper limits of the number of correct answers for each test. However, in the study, when determining the assessment of the level of formation of professional competencies, the answers to all test tasks of each test were taken into account.

For this purpose, indicators were developed that clarify and reveal the competencies being formed. We presented them in the form of planned learning outcomes for the type of professional activity. The criteria for assessing the development of professional competencies of master's students, based on the current analysis of their digital footprint, included four levels:

1. very low level if the score is less than 0.53 of the maximum score for the completed task;
2. low level if the percentage of points scored is 0.53–0.71 of the maximum score for the completed task;
3. basic level if the score is 0.72–0.87 of the maximum score for the completed task;
4. high level if the score is 0.88–1.00 of the maximum score for the completed task.

The maximum score that a student who has completed all the tasks can get is 1.

The presented levels allow us to assess the degree of development of professional competence of each master's student.

It should be noted that the KNIME Analytics Platform, a software platform for analytical tools, allows you to select the required data from the source, perform various filtering and sorting, and visualize the result. The development and assessment of students' professional competencies in the con-text of continuous digital transformation of the educational environment [12–14] is complex task, the solution of which requires the combined efforts of both teachers of master's students.

5 Results

The professional competencies of master's students studying in the field of «Pedagogical Education» are formed within the framework of studying the disciplines of the curriculum. During the course of study, the master's student must master universal, general professional, professional and professional-profile competencies. The latter are established by the university taking into account the profile of the educational program. As for the discipline «Educational data-engineering», during its study, the Master's student must develop the ability to:

1. Use modern methods and technologies for organizing educational activities.
2. Diagnose and evaluate the quality of the learning process in various educational programs based on digital technologies and artificial intelligence methods.

An active digital trace, represented by the results of electronic tests, was used to assess the acquired abilities. The KNIME Analytics Platform software platform was used to systematize and analyze it. A visual workflow was created on the platform, configured for the digital trace data of master's students in the academic discipline «Educational data-engineering».

As a result, reliable statistical parameters were obtained: minimum, lower quartile, median, upper quartile and maximum. They are called reliable because they are insensitive to extreme outliers and show what level of development of competencies in the form of professional activity the obtained result should be attributed to. The obtained reliable statistical parameters are presented in Table 4.

Table 4. Reliable statistical parameters obtained from the study.

Test	Minimum	lower quartile	Median	Upper quartile	Maximum
Test 1	13	19	21	23	25
Test 2	15	19	21	22	24
Test 3	14	18	21	23	24
Test 4	12	17	19	20	25
Test 5	11	18	19	22	25
Test 6	14	17	20	23	25

A graphical interpretation of the obtained reliable statistical indicators, implemented using a chart with grouping in the Microsoft Excel spreadsheet processor, is presented in the figure (see Fig. 2).

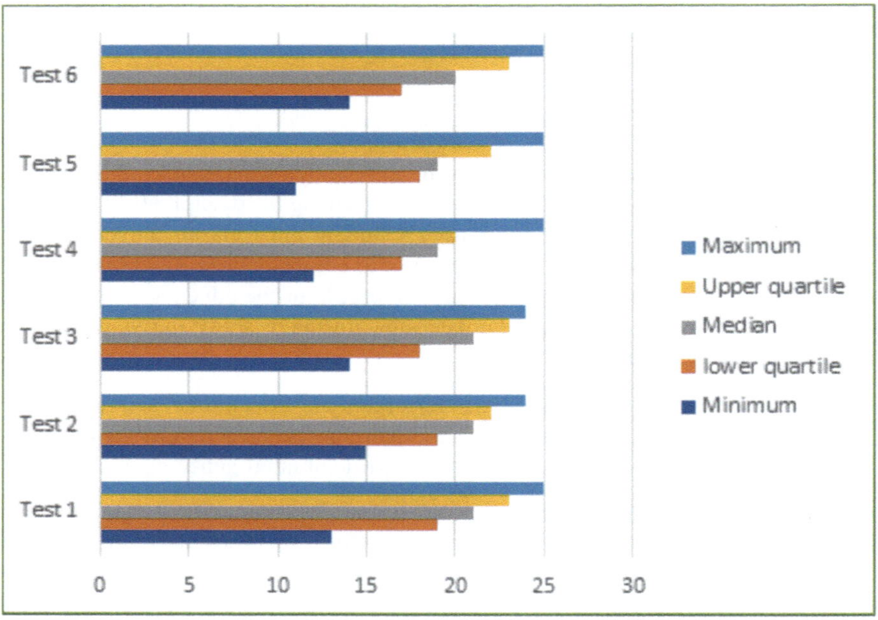

Fig. 2. Reliable statistical parameters obtained from the study.

The figure and table «Reliable statistical parameters obtained from the study» show that the analysis of the active digital footprint of master's students allows us to assess not only the level of development of their professional competencies, but also to identify topics that require adjustment. Since the content of these topics does not contribute to the development of the competence being tested at the proper level. In our case, these are the second and third topics.

6 Discussion

The presented experience, approaches and obtained results indicate the feasibility of using the selected methods and tools for analyzing the educational process. To visualize reliable statistical parameters obtained from the active digital footprint, we used the graphical tools of the Microsoft Excel spreadsheet. Because it makes it easy and convenient to professionally design diagrams.

Indicators of the level of development of professional competencies of master's students and the methods of their extraction from the results of digital footprint analysis are given in many works [15, 16]. The authors consider the methods of correlating digital footprint indicators with formalized requirements of employers, which are reflected in the texts of vacancies in the field of information technology.

The study, based on statistical analysis methods and modern practices of Russian higher education, shows one of the ways to improve the quality of education and identify the level of development of professional competencies based on the active digital footprint of students.

The study was limited to solving problems that provide participants in the educational process with proven means of evaluating effectiveness in digital reality, as one of the necessary results of students mastering the content of an academic discipline.

The use of theoretical and empirical methods allowed us to establish not only the essence of the concepts of "digital footprint" in the context of education, the type of digital footprint collected, but also to identify topics whose content did not allow us to achieve the proper level of development of professional competence. The article presents data on the results of a study conducted with master's students

7 Conclusion

The use of an active digital footprint in the educational process of a university can be carried out when monitoring class attendance; determining the activity and involvement of master's students in the learning process; calculating average grades; determining students' reflection on themselves and the subject; identifying the risks of dropping out, etc.

Such data will be useful to the teacher, since educational analytics provides him with more complete information on the quality of educational content and the activities carried out by him in order to ensure continuous modernization and improvement of the educational process.

Understanding the potential of the active digital footprint of master's students will allow:

1. When developing the content and programs of academic disciplines, take into account the educational needs and individual characteristics of each master's student; when forming educational trajectories, take into account the relationship between learning style, academic performance, extracurricular activities, behavior and self-expression in social networks [8, 17];
2. Master's students will be able to realize their potential in the digital educational environment, build development trajectories, and form a relevant set of competencies taking into account the situation on the labor market [18–21].

It should be noted that without solving the problem that leads to the necessary results when using digital recording and interpretation methods of facts in education, it is difficult to effectively build personal development trajectories, control the effectiveness of the learning process, and digital profiles of students.

The proposed method and tools for analyzing the digital footprint of master's students can be used to make informed management decisions.

References

1. Stepanenko, A., Feshchenko, A.: Digital footprint of the student: search, analysis, interpretation. Open Distance Educ. **4**(68), 58–62 (2017). https://doi.org/10.24412/2225-8264-2023-4-12-20
2. Nesterov, S., Smolina, E.: Digital footprint concept and digital footprint analysis in education. Syst. Anal. Des. Manag. **3**(26), 309–314 (2023). https://doi.org/10.18720/SPBPU/2/id23-492
3. Balyakin, A., Mamonov, M., Nurbina, M., Taranenko, S.: Digital footprint in education: from science to society. Scientific and pedagogical review. Pedagog. Rev. **5**(45), 9–19 (2021). https://doi.org/10.23951/2307-6127-2022-5-9-19
4. Shamsutdinova, T.: Cognitive model of electronic learning trajectories based on digital footprint. Open Educ. **24**(2), 47–54 (2020). https://doi.org/10.21686/1818-4243-2020-2-47-54
5. Bogdanova, A., Lapchik, M., Fedorova, A.: Educational practices of collecting and analyzing data from the digital footprint of schoolchildren. Bull. Siber. Inst. Bus. Inf. Technol. **4**(12), 12–20 (2023). https://doi.org/10.24412/2225-8264-2023-4-5-11
6. Baranova, E., Vereshchagina, N., Shvetsov, G.: Digital tools for learning activities analysis. News Herzen State Pedagog. Univ. **198**, 56–65 (2020). https://doi.org/10.33910/1992-6464-2020-198-56-65
7. Skripnik, V.: Practice of application of provisions of the trace in education. News Baltic State Acad. Fishing Fleet Psychol. Pedagog. Sci. **1**(59), 109–110 (2022). https://doi.org/10.46845/2071-5331-2022-1-59-109-111
8. Drozdova, V., Shevchenko, G., Shagrova, G.: Shevchenko a.: modernization of the content of the academic discipline based on the analysis of the digital footprint. In: Lapina, M., Prakasha, G.S., Grigoriev, S. (eds.) Conference 2022, LNCS, vol. 1222, pp. 448–455. Springer, Cham (2024). https://doi.org/10.1007/978-3-031-78776-8_45
9. Baranova, E., Shvetsov, G., Noskova, T.: Educational data mining methods for the analysis of student's digital footprint. In: Conference 2021, LNCS, vol. 2920, pp. 44–58 (2021) CEUR-WS.org. CEUR-WS.org/Vol-2920/paper_4.pdf
10. Ardeev, A., Bagdasaryan, L., Poddubnaya, N., Kulikova, T.: Methods for collecting and analyzing the digital footprint of project activities in the educational process. In: Lapina, M., Prakasha, G.S., Grigoriev, S. (eds.) Conference 2022, LNCS, vol. 1222, pp. 284–294. Springer, Cham (2024). https://doi.org/10.1007/978-3-031-78776-8_28
11. Guide to Working with KNIME Files. https://docs.knime.com/latest/analytics_plat-form_file_handling_guide/index.html#introduction. Accessed 10 Feb 2025
12. Zakharova, I., Bagnyuk, Y., Vorobyeva, M., Pavlova, E.: Diagnostics of professional competence of IT students based on digital footprint data. Inform. Educ. **4**, 4–11 (2020). https://doi.org/10.32517/0234-0453-2020-35-4-4-11
13. Shevchenko, G., Gorovaya, V., Shevchenko, A.: Monitoring of professionally important qualities of students in the conditions of electronic information and educational environment. Standard. Monit. Educ. **7**(1), 3–6 (2019). https://doi.org/10.12737/article_5c5406352247e1.51124218
14. Shevchenko, G., Kalashnikova, A., Rybakova, A., Brynza, S.: Tasks, functions and basic activities of an online tutor. In: Rugelj, J., Lapina, M. (eds.) Conference 2020, LNCS, vol. 2861, pp. 285–291. CEUR-WS.org/Vol-2861/paper_33.pdf
15. Lapchik, D., Fedorova, G., Gaidamak, E.: Digital footprint in the educational environment as a regulator of student vocational guidance to the teaching profession. J. Siberian Fed. Univ. Human. Soc. Sci. **14**(9), 1388–1398 (2021). https://doi.org/10.17516/1997-1370-0827
16. Dyachenko, M., Leonov, A.: Digital footprint in education as a driver of professional growth in the digital age. E-Management. **5**(4), 23–30 (2022). https://doi.org/10.26425/2658-3445-2022-5-4-23-30

17. Mantulenko, V.: Prospects for digital footprint usage in the higher education. Teacher 21st Century 3, 32–42 (2020). https://doi.org/10.31862/2073-9613-2020-3-32-42.
18. Lapina, M., Tokmakova, M., Movzalevskaya, V., Karabasheva, M., Balakrishnan, D.: Analysis of modern methods and approaches using it in education and pedagogy. In: Conference 2025, LNCS, vol. 1222, pp. 40–52. Springer, Cham (2025). https://doi.org/10.1007/978-3-031-78776-8_5
19. Bogatyreva, Y., Privalov, A., Romanov, R., Konopko, E.: The organization of scientific research work of students in the conditions of the digital environment in the modern universities. In: Rugelj, J., Lapina, M. (eds.) Conference 2019, LNCS, vol. 2494., CEUR-WS.org, (2019)
20. Agibova, I., Kulikova, T., Poddubnaya, N., Fedina, O.: Development of digital competence of a future teacher in the context of informatization and digitalization of modern teacher education. In: Conference 2020, LNCS. ARPHA Proceedings, vol. 3, pp. 13–26 (2020). https://doi.org/10.3897/ap.2.e0013
21. Vayndorf-Sysoeva, M., Pchelyakova, V.: Prospects for using the digital footprint in educational and scientific processes. Vestnik Minin Univ. **9**(3) (2021). https://doi.org/10.26795/2307-1281-2021-9-3-1

AI-Driven Translation of Sign Language for Differently Abled

Samprit Tanuj Patel[1], Vijay Anant Athavale[1](✉), and Vitalii Lapin[2]

[1] Walchand Institute of Technology, Solapur, India
vijay.athavale@gmail.com
[2] North Caucasus Federal University, Stavropol, Russia

Abstract. The AI Tool/Mobile Application for hearing impaired that will help to bridge communication gaps between the hearing and hearing-impaired communities of India, a country that contains more than 63 million hearing-impaired citizens. Such a system utilizes complex technologies including speech recognition, NLP, and computer vision to provide real-time translation, thus enhancing interaction in such vital sectors as health, education, and public services. The AI-based approach therefore tackles the shortcomings of the traditional sign language interpreters: scalable and instant translation, which will make sure that the users, can access crucial services by themselves. The proposed application for Indian Sign Language (ISL) will significantly enhance the quality of life in the lives of the hearing-impaired by promoting social participation and positive interaction in various aspects of the community. It translates from the spoken languages, like English and Hindi, into ISL gestures and vice versa, translating ISL gestures into text or speech.

Keywords: Artificial Intelligence · Sign Language · Machine Learning · Natural Language Processing · Hearing Impaired

1 First Section

Effective communication represents one of the most critical aspects involved in interpersonal relations; however, despite this, numerous people around the world face significant barriers due to hearing impairment. In various communities, sign language is needed in everyday communication among deaf or hard of hearing individuals. The gap between sign language users and nonusers often means that misunderstandings may occur with exclusion from some social activities as well as reduced availability of important services.

India has about 63 million people with hearing impairments, which creates a big communication barrier because many people do not know Indian Sign Language (ISL) [1]. This gap makes it hard for hearing-impaired individuals to access important services like education, healthcare, and jobs, often leaving them excluded [5]. With India to be a developed nation, inclusivity is necessary, according to the National Education Policy 2020, which encourages inclusive education for all [3, 17]. We present an innovative

M. Lapina et al. (Eds.): DET 2024, LNNS 1616, pp. 416–426, 2025.
https://doi.org/10.1007/978-3-032-04365-8_40

solution for addressing this challenge by breaking the language barrier-it involves seamless communication through the conversion of sign language into text and audio, and vice versa. This instrument will facilitate reducing the communication divide and help promote more improved inclusion and accessibility for members in this community with respect to major life fields [7].

Artificial intelligence methods and machine learning algorithms are currently used both in manufacturing, to improve automation efficiency, and in medicine, medicine, and financial structures for big data analysis. With the growing popularity of language models and the launch of open source GPT, artificial intelligence methods have entered even more areas of human activity [25, 26].

In fact, it is a new technology that uses the powerful technologies of computer vision, natural language processing, and speech recognition for the real-time translation between sign language and spoken language [6]. The solution developed aims to empower hearing as well as hearing-impaired people by developing an accessible interface that can be installed on mobile devices and all the currently available platforms. It further develops involvement in routine contexts of education, health, and public services but also advances social inclusion and understanding [4]. A technical framework that outlines the implementation methodologies and prospecting effects of our application explains much about its significance to a more accessible and equitable society for all [5].

2 Background

Literature review on the translation technologies of Indian Sign Language reveals an increasing fascination with the application of artificial intelligence, computer vision, and natural language processing to reduce the communication barriers that are contributed by the hearing impaired [23, 24]. With the passage of several decades, the awareness of needing enhancements in newer avenues for inclusion of the hearing-impaired population has witnessed a good number of efforts that are envisaged at connecting the sign languages with the spoken one. It must be noted that human translators, though crucial throughout history, have been underrepresented in areas of the countryside and underdeveloped areas in order to propel the development of automated translations.

Preliminary work in ISL translation is done on static gesture recognition using traditional machine learning methods such as Support Vector Machines (SVM) and k-Nearest Neighbours (k-NN). It is not so helpful for the sign language as the sign language is highly dynamic and complex, and most of the time, a sequence of gestures along with facial expression and body movements makes up the sign language. That's how the most advanced models are gradually developed, starting with using deep learning techniques, such as convolutional neural networks, which analyze images and videos [19, 21].

Researchers have discussed the use of CNNs for the purpose of recognizing sign language gestures. One of the researcher have proposed a deep learning framework to recognize statically created ISL gestures in real-time by training a CNN model using a dataset of ISL hand signs. It demonstrated high accuracy for the recognition of isolated hand gestures; however it was unable to recognize the gesture in successive sequences, which eventually points out the necessity for the temporal analysis [9, 16].

To overcome this limitation, RNNs and LSTM networks had been devised that were capable of representing temporal dependencies. Those networks also aided in the detection of dynamic gestures wherein the sequence from one frame to the next frame is essential. In one of the research, an LSTM-based approach was used for the purpose of detecting uninterrupted ISL gestures using video input [10]. Translation accuracy has been significantly improved using the interaction of temporal properties, especially for longer and more complex sentences in ISL [20].

Latest NLP techniques are also provided to language translation. As ISL has a structure other than all the spoken languages like English or Hindi, most of these are focused on systems capable of restructuring sentences into an application appropriate format for ISL interpretation [22]. In the last few years, SpaCy and Transformer models such as BERT and GPT are more predominantly applied for the above purpose. The context of the original message is preserved through the use of large datasets of paired text and gesture sequences, enabling a translation of spoken language to simpler, ISL-friendly text, using the models.

Further, the works have been on those datasets that are rich in diversity with a huge number of gestures, facial expressions, and various signing speeds. Several ISL datasets have been formed, most of such datasets are limited in scale and richness that poses a challenge to training models that generalize well to realistic settings [1, 11]. There is the call to enrich datasets of events where there are moves that are more complex than just the simple gesture: continuous sign language as well as subtle elements of context.

Audio-to-ISL translation has considerably gained strength with advancements in speech recognition. Take, for example, Google Speech-to-Text and Mozilla DeepSpeech technologies, the correct transcription of the spoken word by means of which it creates constituent components transcribable into text, requiring further processing for ISL translation. Such problems surely exist that need to be overcome, for example, regional accents in speech recognition systems or how they can pick up on background noises and complex vocabularies within diverse Indian languages.

Moreover, various research work has been conducted on the suitability of offering real-time translation service provision within communities. Such research work on ISL-based translation systems for applying to public service departments such as healthcare centers and government offices demonstrated how AI-based solutions would be a prime need for enhancing access [12]. This study, therefore demonstrates that tools used in real-time translation not only improve the effectiveness of communication but empower those with hearing losses to feel themselves in control and assured to conduct an interaction with a public service.

Generally, the contributions of computer vision in this development of ISL translation systems have been immense. With this, OpenCV and MediaPipe thus become tools necessary in video input acquisition and manipulation for the translation of ISL gestures into text or speech. These frameworks have already become successful in recognizing hand movements and facial expressions; the difficulty lies in developing them concerning both speed and accuracy of real-time translation [18].

Existing literature, then, derives quite clearly from the emphasis on ISL translation technology that fully displays interdisciplinary character embracing the advance gained in artificial intelligence, computer vision, natural language processing, and speech recognition. Continuous scholarly effort across these areas has led to the development of tools that claim increased ability to provide real-time, contextualized ISL translation. One area of necessity for continued inquiry remains aspects such as dataset generation, real-time precision, and the blending of multiple languages within a singular system.

3 Solution

The AI tool/mobile application for Indian Sign Language (ISL) (Fig. 1) Generator is an innovative solution designed to bridge the communication gap between the hearing and hearing impaired communities.

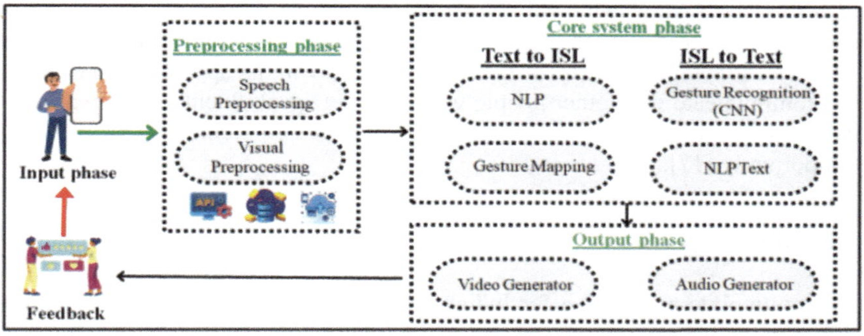

Fig. 1. Architecture.

There is a pressing necessity for the evolution of technology that could ease the conversation between millions of Indians suffering from hearing impairments and the rest of the world, who does not understand their Indian Sign Language, ISL. It shall translate Spoken Languages such as English and Hindi into ISL gestures and vice versa; convert ISL gestures to text or speech. This further helps application development to foster communication, inclusion, and independence for all sectors and spheres-spheres that include health, education, and other public services-where ISL-interpreting professionals are relatively scarce. There are sophisticated technologies for speech recognition and natural language processing and computer vision that can enable instant translation through the application, thereby facilitating smooth interaction and social inclusion.

The application is a step toward ending this miserable communication problems that characterize service within the hearing-impaired in public service areas such as hospitals, education institutions, and governmental offices. This absence of ISL interpreters creates big barriers and denies access to full availability of fundamental information and services in health care, education, and employment opportunities. Historically, human interpreters have been the inducing catalyst for resolving these challenges though their access has always been significantly restricted, so dependency on them for routine communications is neither scalable nor practicable. This context underlines the necessity for an artificial intelligence-based solution that could provide real-time translations independent of human interpreters. An application of such a nature would no doubt grant an easier life to the users who suffer from such hearing impairments by letting them participate more freely in discussions and making use of services that have not been accessible hitherto due to language barriers.

This would be a domain wherein the solution will discover unimaginable worth: healthcare, education, public services, and daily interaction. Especially for the health sector, it could facilitate coordination with patients who are deaf so that medical doctors can communicate crucial medical information without any amount of vagueness. The application would hence improve the learning experience of sign language-interpreting students with lessons to ISL. Therefore, teachers who don't understand ISL, by this application would come out much better teaching deaf students. Public services will also benefit since through this application, the deaf will be able to get access to all government services or knowledge of public transport without necessarily requiring human interpreters. Moreover, even in the most thriving natural communication environments, like grocery shopping, traveling, or socializing in general, the app will help the deaf be able to communicate with other people who may not speak fluent ISL. Therefore, the ability to enable society to be accessible, inclusive, and equitable will bridge this gap very importantly [17].

4 Methodology

The AI tool/mobile application for Indian Sign Language (ISL) Generator employs a multi-step process that integrates several advanced technologies to deliver accurate real-time translations between spoken language and ISL. The solution begins with capturing audio and video inputs, which are then processed using various techniques to convert them into ISL gestures or spoken language. Here is a breakdown of the key steps involved in the solution (Fig. 2):

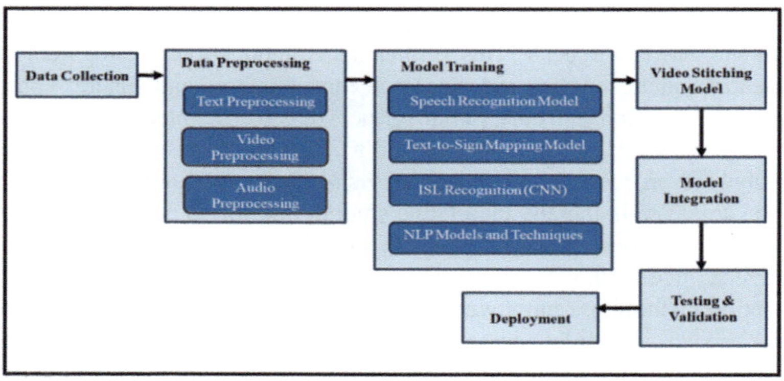

Fig. 2. Methodology.

1. **Audio to ISL Translation Process**

Step 1 - Capture Audio: To get the process started, capturing speech from a live source, such as a microphone, is necessary. The app is developed using different technologies which utilize platform-specific APIs for the recording of sound into WAV or MP3 formats that are usually clean and usable. Noise reduction techniques and voice activity detection (VAD) are the strategies used to remove background noise while maintaining high quality.

Step 2 - Convert Audio to Text: Following the audio capturing, the automatic speech recognition (ASR) technology comes in and transcribes speech to text. The widely known ASR apps like Google Speech-to-Text API, Mozilla DeepSpeech and Kaldi processing the audio and generating the text output.

Step 3 - Text Analysis: For the reason that Indian Sign Language (ISL) differs from English or Hindi in its grammar word order, the transcribed text should be modified accordingly to match ISL grammar. The achievement of this goal would be possible with the help of such a branch of artificial intelligence as natural language processing (NLP) libraries like spaCy or transformer models such as BERT or GPT. These natural language processing (NLP) libraries such as spaCy, on the other hand, or transformer models like BERT or GPT, are usually utilized for analysis. These libraries perform maneuvers such as removing useless parts like articles, rephrasing words and, at the same time, conveying the essence of the sentence, parsing sentences by using the tools.

Step 4 - Text to Video Mapping: Next, the numerical elaboration of the processed text is done through sequence-to-sequence (Seq2Seq) models. By the means of such systems like LSTM-based or Transformer architectures, speech recognition models evaluate how close a given sequence of words is to ISL gesture sequences [20]. The model systematically fed on a train dataset of paired text and ISL gestures for several instances, ensuring that the sentence endures no loss of context between input and output.

Step 5 - Retrieval of ISL Gesture Videos: A database containing a number of pre-recorded ISL gesture videos is organized such that they are counter posed and supplied with apposite context. Methods such as K-Nearest Neighbours (KNN) or similarity-based searches among others are employed to retrieve a particular ISL video for each corresponding word or phrase. This undoubtedly guarantees that the chosen ISL gestures are inline with the real meaning of the text.

Step 6 - Create a Single Video: As soon as the sign videos are retrieved, they are put together in a single, flowing sequence by means of video processing software such as FFmpeg or OpenCV. Being in perfect sync is a key binding for transitions between the gestures that should be natural. This is critical to establish that the expression flows well.

Stage 7 - Show Video on Screen: The finally produced ISL video is shown on the user's display through the internet or the mobile device's video display engine. It serves as a visual output for the hearing-impaired person which is easy to follow.

2. **ISL to Audio Translation**

Step 1 - Capture Video: To translate ISL gestures into text or speech, the system captures video of a person performing the gestures using libraries like OpenCV or MediaPipe. The video is then broken down into individual frames, which are preprocessed for further analysis.

Step 2 - Frame-Level Analysis: Each video frame is analyzed using convolutional neural networks (CNNs) to detect static ISL signs. Key points of the hands and face are extracted using pose estimation techniques, enabling the system to recognize specific hand shapes and gestures.

Step 3 - Temporal Analysis: ISL involves dynamic movements over time, so recurrent neural networks (RNNs) or LSTMs are used to analyze the sequence of frames and detect transitions between signs. This helps the system capture the flow of gestures and their contextual nuances.

Step 4 - Video to Text Mapping: Using Seq2Seq models, the system maps the recognized gestures to corresponding words or phrases, converting the gesture sequence into text. Attention mechanisms can be applied to improve translation quality by focusing on the most relevant parts of the gesture sequence.

Step 5 - Arrange Set of Words: The generated text is arranged into grammatically correct sentences using NLP tools. Since ISL often follows a different word order compared to spoken languages, models like BERT or GPT are used to restructure the sentence according to the grammar of English or Hindi [22].

Step 6 - Convert Text to Audio: The text output is converted into speech using text-to-speech (TTS) models such as Google TTS or Amazon Polly. The TTS system produces human-like speech, which can be customized in terms of voice tone, pitch, and accent.

Step 7 - Display Text & Play Audio: Finally, the system displays the text on the screen while playing the audio output, ensuring both visual and auditory outputs are synchronized for the user.

5 Benefits

The AI tool/mobile application for Indian Sign Language has been designed to be very efficient across all domains and offers numerous benefits (Fig. 3). It is the perfect medium that is not only Unicode, but also gives sight to the disabled people and non-signers a chance to communicate in a normal and inclusive way so allowing abortive interactions to become smoother and more inclusive. Through the application, the hearing impaired persons can talk to others and access services freely since it provides real-time translations. It is more economical and it allows human receptacles to do necessary things instead. Moreover, the application is available for free for the use of the school students who are of the hearing-impaired through the translation of lectures in ISL so that they can be taught.

This tool communicates between the hearing-impaired and those who know little about the International Sign Language in the most non-disruptive manner. As more people use this application, awareness of hearing impairments and ISL will become stronger, which in turn will create a society that is more sympathetic and compassionate. Therefore, the application helps deaf people who can't hear others in a conversation to be independent and access the services independently by providing them with real-time translations. This independence not only enhances their self-confidence but also makes them as full members as they actively participate in social activities and school events that can eventually improve their total quality of life. The application is a life-essential attribute arming the deaf people to be able to follow the lectures by translating them

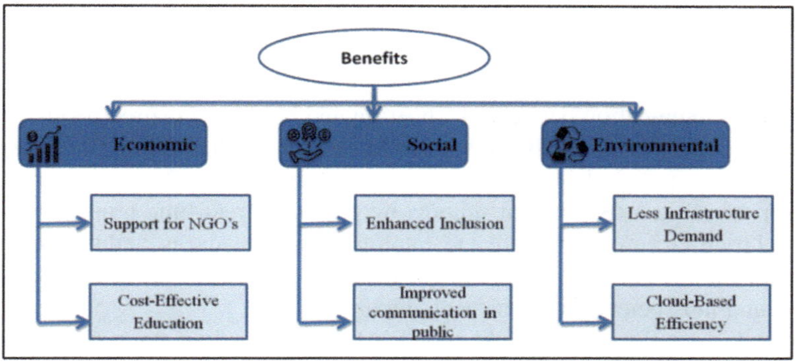

Fig. 3. Benefits of the application.

into ISL. This means that they are on an equal level with their non-deaf schoolmates. In addition, educational opportunities for more students with disabilities occur, and the inclusive education system as a whole becomes more feasible. The deaf driven process if they are in comfortability and have enough trust then they are more likely to participate in activities and functions of the community. This wider engagement can lead to more secured relationships among people and a more diverse and richer community fabric.

The application reduces interpreting service-related expenses by replacing human interpreters. This is especially useful for public services, educational institutions, and healthcare providers, as it enables them to utilize their resources more productively. People suffering from hearing loss receive a bigger job market with an employers' widening of communication capability. Enterprises can fill positions with these people who otherwise cannot communicate soundly. This way, they can promote diversity and inclusion in the workplace. When the individuals are confident in the use of such services as healthcare and education, they will become more active consumption and participation figures in the local economy. The application's capacity to expand services means that governments can reach more people without an equal increase in costs by employing human interpreters. It is clear that scalability is necessary for public services in order to serve different groups, especially with the rise of cities.

The advent of easier communication through the use of mobile applications can go a long way in the observation of reduced physical meetings, requiring only interpreters. In the long run, savings in transportation are quite significant. This, in turn, will be assured by cutting on fuel and increasing the adoption of new transport technologies. Since the app encourages the use of digital tools for communication, it fosters therefore a transfer to more sustainable, technology-driven solutions. This digital transformation would enable the avoidance of paper-based materials that are usually used for traditional, regular educational and communication settings and can conserve other resources. The application can speed up processes in public services and educational institutions, thus helping in reducing the utilization of various resources (such as time and materials) which are normally linked with manual interpreting and communication methods. In turn, this improved efficiency leads to more sustainable operations.

6 Conclusion

The AI tool for ISL translation is a truly promising innovation in the direction of bridging the gap in communication between the hearing and hearing-impaired people. The app, which was created through real-time spoken English or Hindi to ISL gestures translation, is designed with the goal of being inclusive and accessible. The plans for the future of the app entail its expansion in schools by providing real-time translation services, vocational training courses, assistive technology in public spaces, and the improvement of communication in healthcare settings, which are all meant to bring about a more inclusive and fair society for the hearing impaired.

In conclusion, the proposed AI tool for Indian Sign Language (ISL) translation represents a landmark in the process of connecting the deaf community with the rest of the world. This program works by employing modern technologies such as computer vision, natural language processing, and speech recognition to fix the problems faced by millions of hearing-impaired individuals in India.

As has been seen from the very start of this paper above, the tool does not only make real-time communication more easily possible but it also allows in turn the users themselves to get involved in important areas including education, health, and public services. Through the application support of accessibility and decrease in using of live human interpreters, the device allows easier access to crucial services, on the other hand, it is the social equity and involvement which are promoted.

The employment of this method is also in agreement with national strategies such as the National Education Policy (NEP) 2020, which stresses the significance of inclusive education. The requirement for various and equally accessible answers, in order to set ourselves in a more developed and fair society, appears to be now at the top of the agenda [17].

The successful implementation of this AI-driven tool will eventually be a life-changing experience, ensuring that deaf people, who suffer from hearing impairments, can travel through their life more confident and independent on their own. Through promoting conversation and understanding we create the opportunity for a society that is more inclusive where all members can partake even if they are not hearing. The future for the hearing-impaired in communication is full of potential, and the application is the core of that road.

References

1. World Health Organization (WHO): World Report on Hearing (2021). https://www.who.int/publications/i/item/9789240063932
2. Kumar, R., Bhardwaj, S.: Understanding sign language in the context of Indian deaf community: a review. J. Indian Sign Lang. Res. **6**(1), 10–25 (2019)
3. National Education Policy (NEP): Ministry of Education, Government of India (2020). https://www.education.gov.in/sites/upload_files/npi/results/NEP-2020-English-2.pdf
4. Hindman, L.H., Elmore, M.L.: The impact of communication barriers on deaf individuals in educational settings. Int. J. Incl. Educ. **24**(12), 1299–1313 (2020)
5. Ghosh, S.: The role of technology in enhancing communication access for deaf individuals in India. J. Access. Des. All. **8**(1), 51–68 (2018)

6. Vashishtha, M., Prakash, S.: Emerging technologies in sign language recognition: a review. J. Ambient. Intell. Humaniz. Comput. **12**(4), 4081–4092 (2021)
7. Cohen, J.: Bridging the gap: enhancing communication between hearing and hearing-impaired individuals through technology. Commun. Technol. Soc. Change. **34**(3), 229–245 (2017)
8. Zhou, H., Wang, S.: A survey of deep learning techniques for sign language recognition. J. Comput. Sci. Technol. **35**(5), 913–934 (2020)
9. Kumar, A., Singh, S.: Deep learning framework for real-time recognition of Indian sign language gestures. Int. J. Comput. Appl. **182**(14), 1–6 (2018). https://doi.org/10.5120/ijca2018916823
10. Gupta, R., Sharma, P.: Recognition of continuous Indian sign language gestures using LSTM networks. J. Ambient. Intell. Humaniz. Comput. **10**(10), 3985–3995 (2019). https://doi.org/10.1007/s12652-019-01453-x
11. Paul, S., Saha, D.: Indian sign language dataset: a comprehensive database for research on Indian sign language recognition. In: Proceedings of the 2017 International Conference on Computer Vision and Image Processing, pp. 163–172 (2017). https://doi.org/10.1007/978-3-319-74860-3_16
12. Bhaskar, A., Tiwari, P.: AI-based Indian sign language translation for public services: enhancing accessibility in hospitals and government offices. J. Disabil. Technol. **12**(2), 115–127 (2020). https://doi.org/10.1080/20421338.2020.1789793
13. Venkatesh, R., Kumar, M.: Sign language translation using NLP and deep learning: an overview. J Artif Intell Res. **70**, 479–505 (2021). https://doi.org/10.1613/jair.1.12516
14. Mohammed, A.A., Farooq, U.: Speech recognition technologies for Indian languages: a comprehensive review. Int. J. Speech Technol. **23**(2), 267–280 (2020). https://doi.org/10.1007/s10772-020-09645-8
15. Yadav, A., Singh, R.: Real-time gesture recognition for Indian sign language: challenges and future directions. J. Comput. Sci. Technol. **36**(3), 496–515 (2021). https://doi.org/10.1007/s11390-021-0141-3
16. Bansal, A., Athavale, V.A., Singh, A., Aditi, M.A.S., Singh, K.D.: 3D-CNN empowered assistive machine learning model for the hearing impaired. In: International Conference on Sustainable Emerging Innovations in Engineering and Technology (ICSEIET), Ghaziabad, India, vol. 2023, pp. 487–490 (2023). https://doi.org/10.1109/ICSEIET58677.2023.10303045
17. Shashidharan, M., Bansal, R., Hothi, B.S., Athavale, V.A., Mahajan, Y., Anwar, S.: A review on national education policy 2020 and its influence on academics. J. Legal Ethic. Regul. Issues. **24**, 1 (2021)
18. Gupta, S.C., Kumar, D., Athavale, V.: A review on human action recognition approaches. In: 2021 10th IEEE International Conference on Communication Systems and Network Technologies (CSNT), pp. 338–344. IEEE (2021, June)
19. Athavale, V.A., Gupta, S.C., Kumar, D., Savita, S.: Human action recognition using CNN-SVM model. Adv. Sci. Tech. **105**, 282–290 (2021)
20. Sood, V., Kumar, D., Athavale, V.A., Gupta, S.C.: Epilepsy seizure detection by using Bayesian optimize bi-LSTM Model. In: 2020 Sixth International Conference on Parallel, Distributed and Grid Computing (PDGC), Waknaghat, India, pp. 291–296 (2020). https://doi.org/10.1109/PDGC50313.2020.9315779
21. Khairul Rezki, M., Mazdadi, M.I., Indriani, F., Muliadi, M., Saragih, T.H., Athavale, V.A.: Application of SMOTE to address class imbalance in diabetes disease classification utilizing C5.0, random forest, and SVM. J. Electron. Electromed. Eng. Med. Inform. **6**(4), 343–354 (2024) https://www.jeeemi.org/index.php/jeeemi/article/view/434
22. Manchanda, B., Athavale, V.A.: Various statistical techniques used in NLP. Int. J. Comput. Appl. Inform. Technol. **9**, 172. https://www.ijcait.com/IJCAIT/91/913.pdf. Accessed 18 Oct 2024

23. Grigoryan, K., Olefirenko, E., Basan, E., Lapina, M., Mecella, M.: Analysis of security problems in groups of intelligent sensors. In: Lecture Notes in Computer Science (Including Subseries Lecture Notes in Artificial Intelligence and Lecture Notes in Bioinformatics), 12689 LNCS, pp. 26–37 (2021)
24. Basan, A.S., Basan, E.S., Lapina, M.A., Lapin, V.G.: Behavior-based assessment of trust in a cyber-physical system. In: Communications in Computer and Information Science (including subseries Lecture Notes in Artificial Intelligence and Lecture Notes in Bioinformatics), 1395 CCIS, pp. 190–201 (2021)
25. Lapina, M., Shiriaev, E., Babenko, M., Istamov, I.: High-speed convolution core architecture for privacy-preserving neural networks. Program. Comput. Softw. **50**, 417–424 (2024). https://doi.org/10.1134/S0361768824700282
26. Rusanov, M., Babenko, M., Lapina, M., Sajid, M.: Identification of exploited unreliable account passwords in the information infrastructure using machine learning methods. Big Data Cognit. Comput. **8**, 159 (2024). https://doi.org/10.3390/bdcc8110159

Research of Approaches to the Application of Programming Elements in Solving Problems of Probability Theory and Mathematical Statistics in Higher Education Institutions

Svetlana Solomatina[1], Nazar Podruchnji[1](✉), Marina Tokmakova[1], Walaa H. Elashmawi[2], and Elena Dombrovskaya[3]

[1] North Caucasus Federal University, 355017 Stavropol, Russia
podrucnyjnazar@gmail.com
[2] Suez Canal University, Ismailia, Egypt
[3] Financial University Under the Government of the Russian Federation, Moscow, Russia

Abstract. Modern education in probability theory and mathematical statistics (PTMS) increasingly includes programming elements that allow you to automate routine calculations, visualize distributions and random processes, conduct numerical experiments, and solve complex analytically unsolvable problems.

The paper examines approaches to integrating programming into the educational process, analyzes popular tools (Python, R, MATLAB), and offers methodological recommendations. The main methods of teaching mathematics, estimating the complexity and time of solving problems in probability theory and mathematical statistics are highlighted. An experiment was conducted to solve problems manually and using a program.

Keywords: Programming · Probability Theory · Mathematical Statistics · Higher Education · Programming Integration in Curriculum

1 Introduction

Statistical methods for assessing students' knowledge allow us to analyze data obtained as a result of testing, exams, or other forms of control, and draw conclusions about the level of material acquisition, learning efficiency, and other aspects of the educational process. These methods help to objectively interpret the results and make informed decisions to introduce new technologies into the learning process. In the process of teaching mathematics in higher educational institutions, you can use [1]:

1. Descriptive statistics, which includes the main statistical functions, such as the arithmetic mean, median, mode, variance, and standard deviation.
2. Analysis of the distribution of estimates in the form of histograms, visualizations of the distribution of estimates, checking the distribution of estimates relative to the normal (Gaussian) distribution.

M. Lapina et al. (Eds.): DET 2024, LNNS 1616, pp. 427–435, 2025.
https://doi.org/10.1007/978-3-032-04365-8_41

3. Correlation analysis, for example, Pearson correlation coefficient, rank correlations (Spearman or Kendall).
4. Comparison of groups by Student's t-test, analysis of variance.
5. Factor analysis is used to identify hidden factors that affect academic performance (for example, motivation, level of training, quality of teaching).
6. Regression analysis, linear regression and logistic regression.
7. A method of classifying students into groups, for example, by the level of knowledge or time sufficient to master the material, based on previous tests.

As a result, after the conducted research, it is necessary to analyze the dynamics of learning, compare the results before and after training, evaluate the effectiveness of the course or program for further implementation or study of this methodology. Another important factor in learning ability is the evaluation and analysis of time series, depending on the training methods used and the time of training [5].

Artificial intelligence methods and machine learning algorithms are currently used both in manufacturing, to improve automation efficiency, and in medicine, medicine, and financial structures for big data analysis. With the growing popularity of language models and the launch of open source GPT, artificial intelligence methods have entered even more areas of human activity [26, 27].

An analysis of the literature [6–8] has shown that the introduction of programming in solving mathematical problems of varying complexity has a positive effect on learning ability.

2 Basic Methods of Teaching Mathematics

Teaching mathematics requires a systematic approach that combines theory, practice, and the development of logical thinking. When studying mathematics in higher educational institutions, most of the time is devoted to independent study, the ability to use additional and reference literature [14, 15].

The main focus of training is on theoretical training with subsequent application to practice. Currently, online technologies, courses, video lessons, mathematical programs (Mathematica, MATLAB, Python) and mobile applications are increasingly used, which allows you to consolidate your theoretical and practical skills independently [3, 16, 18, 24, 25].

To solve problems in probability theory and mathematical statistics (PTMS), it is important to pay attention to solving applied problems in chemistry, physics, economics, computer science, etc.

3 Evaluation of the Complexity of Solving PTMS Tasks

When evaluating complexity criteria, tasks are divided by complexity type. At the beginning, it is suggested to solve simple problems, with a further increase in the level of difficulty. Simple PTMS tasks include finding classical probability, conditional probability, and using simple formulas. Statistical problems with a small amount of data, using specific formulas and theorems [9–11]. Distribution problems: Binomial, Poisson, normal, exponential, and other distributions. Tasks related to knowledge of distribution properties and ability to work with tables or software are considered to be of an average level of complexity. High-level tasks include statistics, evaluating the parameters of experiments performed, testing hypotheses, and working with big data. Such calculations are often difficult to perform manually, and knowledge and skills of working with mathematical packages and programs are required. Combined tasks that combine several types of tasks (for example, calculating probability using a chain of events and statistical verification) are classified as very high levels of complexity. Such tasks allow you to identify strong students for an in-depth study of mathematics [4, 17].

Examples of Difficulty Assessment

Example 1: A Simple Task

Condition: What is the probability of exactly two heads falling out with three coin flips?

Type: Classical probability.

Calculation volume: Small (the formula for binomial distribution is used).

Difficulty: Low.

Example 2: A Medium-Level Task

Condition: The store receives an average of 5 customers per hour. What is the probability that no more than 3 clients will arrive in an hour?

Type: Poisson distribution.

Calculation volume: Average (requires calculating the sum of probabilities).

Difficulty: Medium.

Example 3: Complex Task

Condition: Test the hypothesis that the mean values of the two samples are equal using the t-test. Data: two samples of 30 values each.

Type: Statistical hypothesis testing.

Calculation volume: Large (you need to calculate averages, variances, and t-statistics).

Difficulty: High.

Estimation of Time for Solving Tasks by PTMS

The estimation of time spent on solving mathematical problems depends on the complexity of the problem, the level of preparation of the student, and the tools used. Let's look at the time estimation methods for a manual solution and using software in Table 1.

When manually solving PTMS problems, it is important to be able to divide tasks into stages: determining the type of problem (classical probability, conditional probability, distributions, etc.), choosing a suitable formula or method for solving it, performing calculations, and checking the result. This saves you time to get the desired result [12, 13].

Let's look at how long it will take to solve each of the presented tasks in Examples 1–3. Estimates of the time required to manually solve problems of varying complexity are presented in Table 2.

Table 1. Factors affecting time.

Problem complexity	Simple tasks are solved faster than tasks with complex distributions or event chains	Example 1–3
Experience of the decider	An experienced person recognizes the type of task faster and applies the necessary formulas	It is necessary to consolidate the acquired knowledge and skills not only in the classroom, but also independently using IT technologies
Amount of calculations performed	Tasks with a large amount of data or complex calculations require more time	For example, statistical problems in economics, physics, chemistry, and biology with a large number of experiments and samples

Table 2. Time estimation for solving problems of varying complexity using the manual method.

Difficulty level	Task characteristics	Time spent (min)	Total time (min)
Low	Classical probability, small data	• Understanding: 1 • Formula: 1 • Calculations: 2–3 • Check: 1	5–6
Average	Distribution properties, tables	• Understanding: 2–3 • Formula: 1 • Calculations: 3–4 • Verification: 1–2	7–10
High	Statistics, hypotheses, big data	• Understanding: 2 • Hypotheses: 1 • Formula: 1 • Calculations: 10–15 • Check: 2	16–21

It can be concluded that the time to complete tasks can decrease or increase depending on the amount of information, the experience of the solver, and the method of solution. The accuracy of a manual solution depends on care, knowledge of the theoretical foundations, and computational skills.

We reviewed the time spent on solving problems of various levels of complexity manually. Let's now consider estimating the solution time using software. Important factors that affect time may include:

- The type of program. Simple calculators (for example, for calculating factorials) work faster than specialized programs (for example, MATLAB, Python with libraries).

- The user's training level. Knowing the program's syntax and features speeds up the process.
- Difficulty of the task. Programs can handle large-scale calculations faster, but setting up a model or algorithm can take time.

Using the built-in functions, for example, in Python [2, 18–20] with the scipy.stats library, you can quickly calculate probabilities for standard distributions (binomial, normal, etc.), taking from a few seconds to 1–2 min (depending on the complexity of the task). When using specialized programs such as MATLAB or R, you can quickly plot distributions or perform complex calculations.

Non-standard tasks require writing code. Time depends on the complexity of the algorithm and the programmer's experience, and the time can vary from 5–10 min to an hour. However, the task itself takes several seconds to complete. The estimation of problem-solving time in various ways is presented in Table 3.

Table 3. Estimate of the time spent manually solving tasks and using programs.

Criteria	Manual solution	Using programs
Time for simple tasks	5–10 min	1–2 min using special programs and built-in functions; Up to 30 min for independent code writing and debugging
Time for complex tasks	15–30 min (or more)	5–10 min using special programs and built-in functions or for experienced programmers; More than 60 min when writing code independently by an inexperienced specialist
Accuracy	Depends on mindfulness	High (with the correct code)
Flexibility	Limited knowledge of the decider	High (you can solve any task)
Training	Knowledge of probability theory is required	Knowledge of theory and programming is required
Solving many similar problems	Time increases when solving a large number of similar tasks	When solving similar tasks, you can use a ready-made program to save time. In other words, the temporary efficiency increases, taking into account long-term use

It can be concluded that the introduction of programming and IT technologies can increase the speed of solving many PTMS application tasks. It allows you to quickly solve many problems with different conditions, compare the obtained values, and check the results. But for simple tasks and teaching students, it is better to use the manual method, which allows you to better consolidate your theoretical knowledge.

4 Experiment

To obtain statistical data, an experiment was conducted in a group of IT students majoring in the discipline "Probability Theory and Mathematical Statistics" on the topic "Basic numerical characteristics of discrete random variables".

Ten problems were selected for calculating the mathematical expectation E (X), the variance D(X), and the mean square deviation σ(X).

At the beginning, students were asked to solve these problems manually. Of the 150 people who wrote the paper, approximately 20% completed the work in less than 50 min, 60% of students took from 60 to 80 min, and a little more than 20% passed the paper after 100 min. Thus, it took an average of 70 min to complete all the work, and at least 7 min to complete a single task. At the same time, 53 students were rated "excellent", 81 – "good" and 16 people – "satisfactory". Most of the mistakes were made due to carelessness and due to large calculations. Students performed the first tasks efficiently, then made more and more computational errors.

Then the students were asked to solve the same problems by writing program code in Python, which accelerated the solution of similar problems, but many students had to learn additional software constructs. Most of the students managed to write the code in 40 min, and all the proposed tasks took no more than 60 min to complete. Tasks were processed electronically, which also reduced the time required to complete the work. When the number of tasks increases, the calculation time practically does not increase. It is important to note that students, in addition to mathematical knowledge of probability theory, received knowledge related to programming, which plays an important role in teaching IT students. However, it is worth noting that not everyone was able to develop high-quality code, and weak students asked for additional help, which was not required when solving problems manually.

It is important to note that students were interested in this approach to learning, and entire projects were created (Fig. 1) to develop mathematical calculators in JavaScript [22]. In this case, the technologies presented in Table 4 were used.

Many students noticed that the numpy library can be used to calculate the mathematical expectation, variance, and mean square deviation in Python, which accelerated the process of writing the program [2].

Below is a diagram (Fig. 2) the number of problems to be solved (n), proposed in the experiment, from the time (t) of solving manually and when writing the program.

As we can see, the manual solution time increases exponentially, while the number of errors increases. When solving tasks programmatically, the time increases slightly. After writing the program, you just need to work with the compiler and complete tasks in electronic form. At the same time, the program does not get tired and does not make mistakes.

Undoubtedly, the proposed approach to training in the discipline "Probability Theory and Mathematical Statistics" will have a positive result in the learning ability of students of technical, engineering and IT specialties, but it is not suitable for students whose mathematics and programming is not a specialized subject.

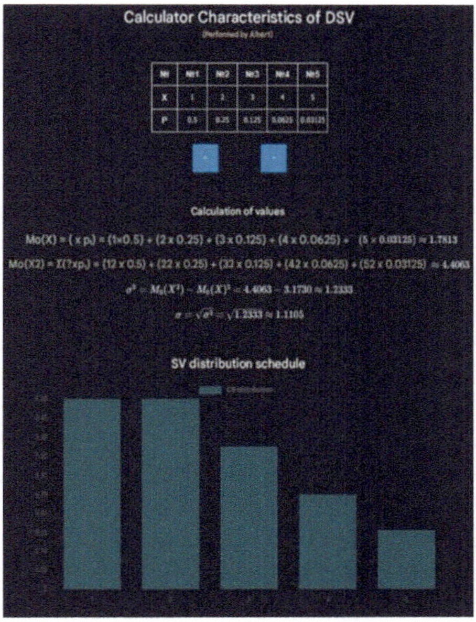

Fig. 1. Developing a mathematical calculator.

Table 4. Technologies used in mathematical calculator projects.

Technology	Appointment	Usage example
React	Creating a user interface	Developing interactive components
Redux	Managing app state	Storing calculation data
KaTeX	Displaying mathematical formulas	Rendering probability formulas
Math.js	Performing mathematical calculations	Calculations of statistical indicators
Chart.js	Plotting charts	Visualization of distributions

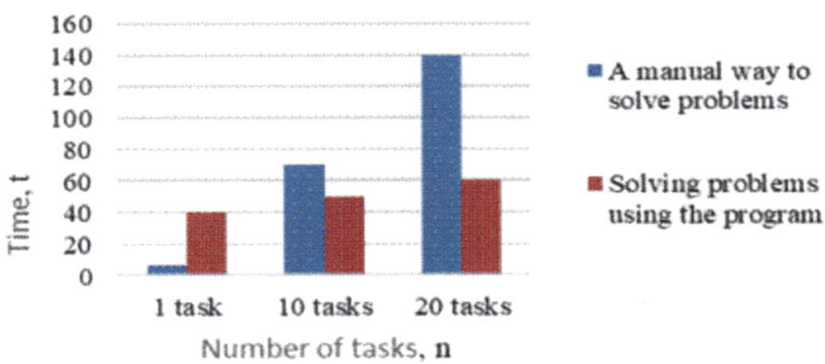

Fig. 2. Time dependence of the number of tasks to be solved.

5 Conclusion

Programming greatly expands the possibilities of learning PTMS, allowing students to gain a deeper understanding of probabilistic models through practice. They allow you to expand the scope of your acquired skills. The best approach is to combine the theoretical framework with modern computing tools, such as Python.

Integrated learning with the integration of programming and IT technologies into various math topics works better, as students understand the subject better if they see its connection to practice. Real-life and industrial examples make learning more interesting and motivating. Interest is the main engine of learning: when the topic is useful and understandable, students learn more actively. When teaching complex topics, it is very important to set students up to learn, so as not to lose motivation to learn. To do this, it is necessary to show how knowledge is applied in life, to anticipate difficulties and explain how to overcome them, to help students come to a decision on their own. If the problem is clearly formulated and analyzed, students will be able to solve it using the acquired knowledge. When teaching a course in probability theory and mathematical statistics, teachers should change their previous teaching methods, boldly use teaching reform methods to improve the quality of teaching, so that students can understand the basic idea of probability theory and mathematical statistics, master the skills of probability and statistical inference, and improve their ability to analyze and solve problems using not only mathematical skills, but also but also a programmatic approach.

As part of continuing research on this topic, it is proposed to consider the analysis of the effectiveness of different programming languages in education, the development of specialized courses (for example, "Statistics in Python"), the introduction of programming not only in probability theory and mathematical statistics, but also in other branches of mathematics in order to improve the quality of knowledge among students [23].

References

1. Wasserman, L.: All of Statistics: A Concise Course in Statistical Inference. Springer Science & Business Media, Cham (2013)
2. VanderPlas, J.: Python Data Science Handbook. O'Reilly Media, Inc (2016)
3. Ross, S.M.: Introduction to Probability and Statistics for Engineers and Scientists, 6th edn. Academic (2020)
4. Blei, D.M., et al.: Variational inference: a review for statisticians. J. Am. Stat. Assoc. **112**(518), 859–877 (2017)
5. Nolan, D., Temple Lang, D.: Computing in the statistics curriculum. Am. Statist. **64**(2), 97–107 (2010)
6. Devore, J.L., et al.: Modern Mathematical Statistics with Applications, 6th edn. Springer Nature, Berlin (2021)
7. Manoukian, E.B.: Mathematical Nonparametric Statistics. Taylor & Francis (2022)
8. Kothandapani, H.P.: A benchmarking and comparative analysis of python libraries for data cleaning: evaluating accuracy, processing efficiency, and usability across diverse datasets. Eigen. Rev. Sci. Technol. **5**(1), 16–33 (2021)
9. Kothinti, R.R.: Data analytics for sustainable development: harnessing big data to track and achieve global sustainability goals. IRE J. **7**(11), 750–760 (2024)

10. Vajda, S.: Mathematical Programming. Courier Corporation (2009)
11. Biehler, R., Frischemeier, D., Reading, C., Shaughnessy, J.M.: Reasoning about data. In: International Handbook of Research in Statistics Education, pp. 139–192 (2018)
12. Balmes, I.L.: Correlation of mathematical ability and programming ability of the computer science students. Asia Pac. J. Educ. Arts Sci. **4**(3), 85–88 (2017)
13. Caluza, L.J., Trecene, J.K.: Predicting academic performance of information technology students using C4. 5 classification algorithm: a model development. Int. J. Inf. Sci. Appl. **10**(1), 7–21 (2018)
14. Borkar, S., Rajeswari, K.: Predicting students academic performance using education data mining. Int. J. Comput. Sci. Mob. Comput. **2**(7), 273–282 (2013)
15. Elgamal, A.F.: An educational data mining model for predicting student performance in programming course. Int. J. Comput. Appl. **70**(17), 22–30 (2013)
16. Rae, A.W.M.: Descriptive and inferential statistics. In: Colman, M.A., et al. (eds.) Companion Encyclopedia of Psychology, vol. 2, (2019)
17. Barlow-Jones, G.: Problem solving as a predictor of programming performance. In: Annual Conference of the Southern African Computer Lecturers' Association, pp. 209–216. Springer, Berlin (2017)
18. Sentance, S., Csizmadia, A.: Computing in the curriculum: challenges and strategies from a teacher's perspective. Educ. Inf. Technol. **22**, 469–495 (2016)
19. Wing, J.M.: Computational thinking's influence on research and education for all. Ital. J. Educ. Technol. **25**, 7–14 (2017)
20. Tian, B.P., Wang, Y.: Exploration and thoughts for teaching of probability and statistics in the institute of technology. Colg. Math. **21**, 17–21 (2005)
21. Lapina, M., Gorbunova, N., Tokmakova, M., Movzalevskaya, V., Elashmawi, W.H.: Features of the project approach in information security training. In: Lapina, M., Prakasha, G.S., Grigoriev, S. (eds.) SLET 2022: International Conference on Innovative Approaches to the Application of Digital Technologies in Education and Research. SLET 2022. Lecture Notes in Networks and Systems, pp. 313–323. Springer, Cham (2024). https://doi.org/10.1007/978-3-031-78776-8
22. Lapina, M., Prakasha, G.S.: ICT integrated project-based learning practicum among undergraduate students of Russia and India. Информатика и образование 2022
23. Rugelj, J., Lapina, M.: Game design based learning of programming. In: CEUR Workshop Proceedings SLET- 2019 – Proceedings of the International Scientific Conference Innovative Approaches to the Application of Digital Technologies in Education and Research, vol. 2494, pp. 29–42 (2019) https://ceur-ws.org/Vol-2494/invited_paper_4.pdf
24. Prakasha, G.S., Muttungal, P., Lapina, M., Taran, V.: Flipped classroom strategy in online teaching: challenges faced by higher education teachers. In: Lapina, M., Prakasha, G.S., Grigoriev, S. (eds.) SLET 2022: International Conference on Innovative Approaches to the Application of Digital Technologies in Education and Research. SLET 2022. Lecture Notes in Networks and Systems, pp. 241–254. Springer, Cham (2024)
25. Rugelj, J., Lapina, M.: Game Design Based Learning of Programming. In: CEUR Workshop Proceedings SLET- 2019 – Proceedings of the International Scientific Conference Innovative Approaches to the Application of Digital Technologies in Education and Research, vol. 2494, pp. 29–42 (2019)
26. Lapina, M., Shiriaev, E., Babenko, M., Istamov, I.: High-speed convolution core architecture for privacy-preserving neural networks. Program. Comput. Softw. **50**, 417–424 (2024). https://doi.org/10.1134/S0361768824700282
27. Rusanov, M., Babenko, M., Lapina, M., Sajid, M.: Identification of exploited unreliable account passwords in the information infrastructure using machine learning methods. Big Data Cognit. Comput. **8**, 159 (2024). https://doi.org/10.3390/bdcc8110159

The Problem of Estimating the Teacher's Time Spent on Preparing Digital Educational Content

Yuriy Sosnovskiy[1](✉), Victor Milyukov[1], Roman Schostak[1], Veronica Ilyina[2], and Vitalii Lapin[3]

[1] Institute of Physics and Technical Sciences, Crimean Federal V. Vernadsky University, Crimea 265007, Russia
mailsender256@mail.ru

[2] Tauride Academy, Crimean Federal V. Vernadsky University, Crimea 265007, Russia

[3] North Caucasus Federal University, Pushkin Street, 1, Stavropol 355017, Russia

Abstract. The situation with the COVID-19 pandemic and the massive transfer of elements of the educational process to electronic and remote formats have spurred the development of digital technologies and their widespread implementation in higher education. Obviously, independent work and part of the lectures will remain in digital format. The situation looks more complicated with seminars, practical, and even more so laboratory classes, but the percentage of digital technology use in them will steadily increase. The learning technology itself is becoming hybrid, multimodal, combining elements of a classical approach – classroom classes with a teacher, as well as a large percentage of distance, electronic, and online education. All this brings to the fore the problem of the formation and support of digital educational content of higher education institutions. The main problem of many electronic educational systems, including moodle, is a sharp increase in the labor costs of teachers in the preparation, placement and maintenance of digital educational content. This problem is of an international nature, the main focus of research is on the introduction of increasingly new technological solutions, but they do not always help reduce the burden on the teacher. This problem came to the fore during the COVID-19 pandemic, especially in the teaching of high-tech disciplines (IT, natural sciences, medicine, etc.). As a result, updating the methods of calculating teachers' labor costs is required. A possible solution to this problem may be a conditional division into areas with a standard CSC, an advanced and advanced CSC level. This will allow you to control the process of increasing time (and financial) costs and introduce the most advanced technical IT solutions into the educational process. A formula is also proposed for calculating the teacher's time spent on the preparation, development and placement of the CSC, taking into account the three levels of complexity of the CSC. The provisions of the methodological package and the information support package for teachers have been developed in order to reduce the time spent working with the COC, which will improve the quality of digital educational content.

Keywords: Teacher Workload · Digital Content Creation · Educational Technology · Time Estimation · Instructional Design · Curriculum Development

M. Lapina et al. (Eds.): DET 2024, LNNS 1616, pp. 436–447, 2025.
https://doi.org/10.1007/978-3-032-04365-8_42

1 Analysis of Materials on the Topic

The authors note the peculiarities of the socio-psychological aspects of digital learning, the risks of the system of value-semantic orientations of the individual [1], which is relevant not only for the student, but also for the teacher. There are also a number of problems with IT equipment [2, 3], a decrease in the level of motivation of participants in the educational process [4, 5].

The cancellation of classes and the introduction of additional vacations, the transfer of part of the educational material to independent study, as well as the rapid transfer to an online format could lead to a decrease in objective indicators of the quality of education. For example, [6] provides convincing data that the proportion of students who received an unsatisfactory grade during the 2020–2021 academic year increased significantly, with a drop in grades ranging from 8.7% to 15.3%. The increase in the number of students who did not receive certification in certain subjects at different levels ranged from 82% to 475%, and the increase in the number of students with incomplete Grades was huge, from 1142% to 2470%. One of the problems was the reduction of the time of contact work with the teacher [6].

Also, the widespread introduction of distance learning highlights the disadvantages of the digital tools used, and there are reports of insufficient opportunities for collaborative learning [2]. An interesting remark concerns the increased time required to master the educational material [7].

The authors of a large-scale study of the organization of the educational process within the limits of online restrictions [8] propose a number of measures, starting with increasing the leadership role of a teacher in secondary schools, and up to providing additional educational resources and materials to senior schoolchildren and students. There is also a need for technical support in the use of IT solutions, the choice of e-learning platforms, etc.

The learning technology itself is becoming hybrid [9], adding remote interaction with students and invited specialists to classroom classes with a teacher. Synchromodal and blended learning is made possible by the development of electronic video conferencing and augmented reality systems and is used in real-world interaction with students [10, 11].

The analysis of the electronic information and educational systems used by universities of the Russian Federation allows us to identify two main directions for the implementation of EIE.

More often than others, moodle is used as a standard electronic educational system, which is confirmed by the authors' own experience, as well as a number of sources [12, 13], etc.

The advantages of LMS moodle are well-known, such as free distribution and open-source codes, the availability of a large number of modules for various tasks and an exhaustive set of various documentation. At the same time, the large scale of the system places high demands on support and maintenance services, making it difficult to update the system. As a result, there are many cases when users complain about the insufficient level of usability of the moodle interface [14–16], which leads to an additional increase in the time spent by the teaching staff.

On the one hand, their own EIOS, created by educational organizations independently or based on deep processing of opensource solutions, satisfy additional individual requirements of a particular university exactly as the management sees it. Such systems are characterized by one or another specialization, partially closed solutions are used. Due to the lack of a significant number of installations, the system turns out to be unique, and a number of problems arise when trying to implement it in another NGO:

- lack of comprehensive documentation;
- difficulties in supporting someone else's software product;
- adaptation of the LMS to the requirements of another educational organization;
- lack of guarantees for further development and updating of the LMS.

There are proposals to provide access to the LMS installed on the servers of the NGO of the author of the system, however, for a large educational organization this is associated with significant risks (the problem of fault tolerance of the LMS, the data transmission channel, compliance with the Federal Law of the Russian Federation on Personal Data, as well as issues of dependence and competition).

A common problem for moodle-based EIOS and proprietary software solutions is a sharp increase in the labor costs of teaching staff in the preparation, placement and maintenance of digital educational content (DCS), which was especially pronounced during the COVID-19 pandemic, revealed a number of features of the use of LMS by teachers. As shown in a large-scale survey of participants in the educational process [17], 100% of teachers noted the high time required to prepare and upload materials to the LMS.

2 International Experience and Features of the Process of Widespread Implementation of LMS and CSC in the Educational Process

Due to the spread of the coronavirus (COVID-19), students are increasingly relying on online learning. The problems that arise in this case affect many areas. According to the author, there are several key considerations that teachers should consider when creating a CSC and supporting online student learning. Martin (2020, Australia) offers a five-point guide for educators to optimize online learning in the time of COVID-19. The five key considerations include instruction (explicit, orderly, and well-organized); content (high-quality and appropriate to students' level); motivation (self-regulation, parents' involvement, and tasks that separate students from online environment); relationships (interpersonal relationships through various communication channels and sufficient face-to-face online instructions); and mental health (reaching out to students who may need help and informing them about who to contact when they need mental health support) [18].

Today, university teachers are gaining a wide range of new pedagogical experience through the introduction of online learning approaches and related software and hardware. (WBTLA, web-based teaching and learning approaches). With the use of WBTLA, there have been issues related to usability in the technical, pedagogical, and contextual aspects of teaching and learning. The results of a survey of participants in the educational process show (Malaysia) that although teachers agreed on the technical and pedagogical usability and the degree of associated difficulties they could overcome, their perceptions and beliefs about contextual suitability and the degree of associated difficulties are less clear. This leaves open the issue of improving teaching technologies using e-learning, WBTLA, and requires one way or another its solution from each relevant educational institution, which should invest both in technology and in teacher training [19].

The importance and, at the same time, the complexity of the widespread introduction of online technologies into education is emphasized, certain technical principles of the organization of CCS and information systems have been developed, which, according to the authors of the Department of Education (Department for Education UK, Great Britain), should be acceptable in many educational institutions and will solve the tasks of reducing the burden on participants in the educational process.: Providing access to remote provision, replicating the classroom remotely, interactive platforms, inking platforms to applications, continuing the planned curriculum, finding solutions in remote provision etc. [20].

A lot of work has been done to analyze the impact of the distance learning format on the academic workload of teachers in the USA. The study showed that online learning required at least 14% more time than traditional learning, most of which was spent on developing a learning center. The weekly impact on academic workload also varied significantly depending on the type of teaching activity, increasing strongly during consultation and assessment periods. As a particular method of solving the problem of overloading teaching staff, it is proposed to reduce the number of study groups, for which the calculation of the optimal composition of study groups for online courses is given [21].

It is noted the importance of feedback from students during the course, understanding their views on the course, will prevent mistakes and take into account their wishes when developing new electronic cases. For example, some students were satisfied with their online courses but still struggled with balancing online courses and work responsibilities. Other students found that enrolling in an online program related to their jobs was very beneficial [22, 23].

3 The Main Components of the Digital Educational Content Preparation Process

Today, in the Russian Federation, there are standards for the time spent on teacher training for various types of classes [24, 25]. All these standards are designed to use the teacher's conditional "afternoon" and to prepare materials for classical types of classes.

However, the preparation of digital educational content, especially in high-tech disciplines (IT, natural sciences, medicine, etc.) requires updating methods for calculating the labor costs of teaching staff. The main components of the process of preparing and posting digital educational content, divided by required competencies and, in particular, time costs, are shown in Fig. 1.

Color (digital) coding and explanations to Fig. 1: the color (1) corresponds to the classical elements of teacher preparation for classes. Blocks (2) correspond to the additional burden due to the specifics of digital content, however, they fit within the framework

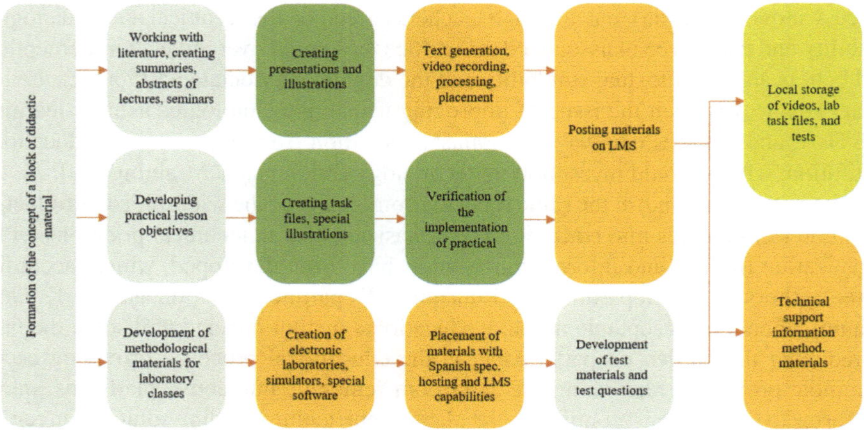

Fig. 1. The main components of the process of preparation and placement of digital educational content.

of the requirements for teaching staff in terms of digital technology proficiency and do not require special competencies from teaching staff to prepare CCS. Blocks (3), in addition to additional labor intensity, imply special competencies of teaching staff in the framework of information technology, special software, programming basics, system administration, etc. Block (4) becomes relevant and significant when creating video content (video lectures, seminars, etc.), because network and cloud storage are not reliable and local storage of video content is required, which takes up a large amount of disk space and requires additional financial expenses.

Examples of recording the COC of group (3), implemented with the participation of the authors of the work, are shown in Figs. 2, 3 and 4.

Fig. 2. Video recording using a teleprompter, embedding accompanying images.

The development of feedback cases in an electronic or distance learning format is necessary, it allows you to fill in the missing channel of communication between the student audience and the teacher. Such work requires technical means – special settings of the testing system and methodological ones – the compilation of feedback tests based

Fig. 3. Video recording using a special LED board for mathematical and physical formulas.

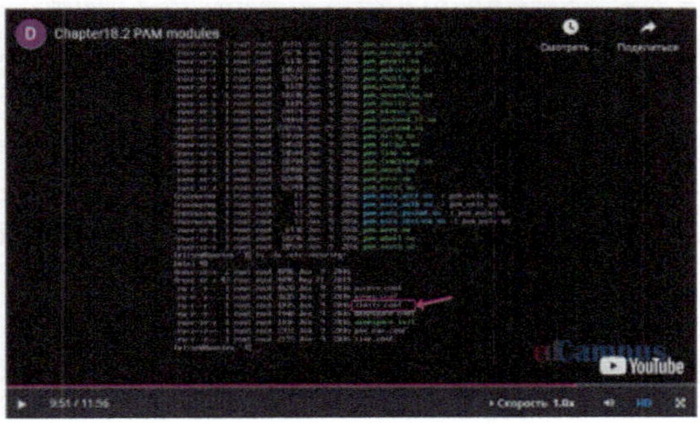

Fig. 4. Screen video recording – screencast.

on competencies in sociology to verify the results. A fragment of the results of this form is shown in Fig. 5.

The authors conducted an experiment to involve teachers in the development of a feedback questionnaire embedded in the training course independently (5 participants), as well as with the help of sociologists (4 participants). The results of the time spent are shown in Table 1.

Thus, the involvement of specialists in the field of sociology to develop standard feedback case templates, with subsequent scaling, will reduce costs in the production of this type of data center by up to 3 times.

The problem of differences in the structural divisions of a large educational organization. Many non-specialized educational organizations with more than a thousand students face the fact that they have different areas of training. For some, teaching a standard approach to digitizing educational materials and providing illustrations is sufficient to ensure the quality of training. For other disciplines, such as engineering and IT, this approach is insufficient and requires additional efforts to implement video effects, develop video illustrations for individual tasks, etc. A number of disciplines allow and require the active introduction of gamification elements and interactive tasks, which significantly increases the level of discipline development, but leads to a significant increase in the time spent by the teacher.

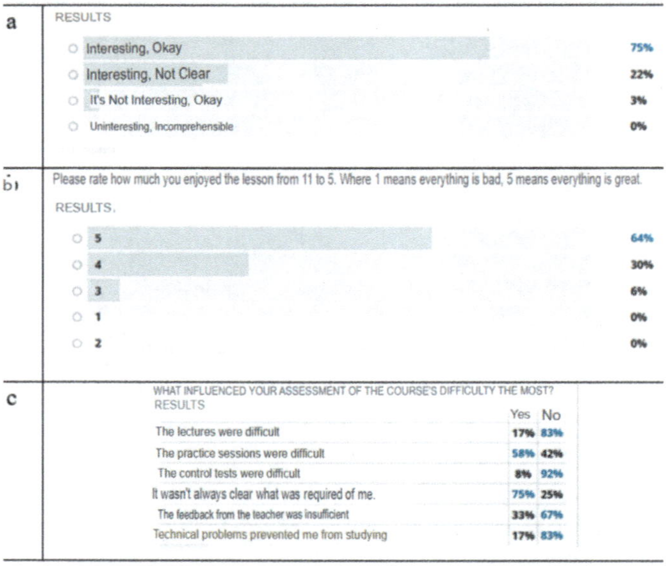

Fig. 5. The design and results of the feedback block on the sections of the course (data from the authors).

Table 1. Time spent on developing a feedback block for teaching staff and sociologists.

Feedback case development, contractor	Time, min	
	Development	Placement
Teaching staff	15 ± 0,7	5 ± 1,0
Profile specialist	5 ± 0,7	–

The solution to the problem may be a conditional division into areas with standard CSC, advanced and advanced CSC levels, which will allow, on the one hand, to control the process of increasing time (and financial) costs, and, on the other hand, to introduce the most advanced technical IT solutions into the educational process where required.

4 Estimating the Teacher's Time Spent Developing Digital Educational Content

The following standard set of educational materials was selected for the study, which differ in the level of implementation of IT solutions in the CSC:

- lecture materials;
- materials on practical work (interactive tasks, support forum);
- tasks for independent work;

– feedback from students (creating a questionnaire, processing incoming individual messages).

Studies have been conducted to determine the time spent in the production of a video fragment – comparing the number of video effects on the duration of rendering, as well as the total production time of the video fragment.

Some materials are shown in Fig. 6. Applying a large number of effects does not affect the duration of rendering a video fragment. However, the total production time of a video fragment shows a positive correlation with the number of video effects.

Fig. 6. Special effects processing, code acceleration, intro, concealment effects.

Table 2 shows the coefficients of the technical time spent on creating a data center for three different disciplines with different levels of implementation of IT solutions in the data center.

Table 2. The coefficient of increasing the time for the formation of the COC, M ± m.

Discipline / level	SOC standard level (C)		TSOK advanced level (Ad)		Additional basement, advanced level (i)			Rounded cost increase factor t for 1 h
	Preparation/recording	Processing, placement	Preparation/recording	Processing, placement	Text content	Interactive	Gamification Elements	H(t), min
History, Sociology (C)	0,97 ±0,14	2,02 ±0,30			0,98 ±0,11			4
Quantum mechanics, theory. Mechanics (Ad)			1,48 ±0,16	3,08 ±0,22				4,5
OS security, system programming, microprocessor systems (i)	1,00 ±0,11	3,06 ±0,27	1,50 ±0,12	2,02 ±0,11		3,02 ±019	3,94 ±0,32	14,5

The data shown in Table 2 were obtained based on the work on the development of CSCs of various levels of complexity by the university's Center for e–learning and distance learning technologies. Teachers in the relevant disciplines indicated in Table 2 were involved in the study. The total number of study participants was 15 people - 5 in each category.

Based on the analysis, the formula for calculating the teacher's time spent on the preparation, development and placement of digital educational content is presented in (1).

$$T(\text{min.}) = \frac{C + Ad + i + p}{\sum n_k} \cdot 60 \quad (1)$$

where: n_k - the number of specialized specialists and teaching staff involved; *c, Ad, i, p* – coefficients from Table 1.

The above formula (1) and the data in Table 2 are the basis for estimating the time it takes to prepare a data center for disciplines that are conditionally divided into three levels in terms of the volume of use of modern IT solutions - standard (C), advanced (Ad) and advanced (i). Such a division of disciplines will formalize the created groups of templates, involve specialized specialists in the creation of case templates (see the case discussed above with a feedback questionnaire) and, ultimately, to ensure the high quality of the COC, to support the motivation of teachers and to control the increase in financial costs for this process.

5 Quality Assurance Methodology of the CSC

As it was shown above, the process of developing, hosting and supporting a data center requires a lot of time on the part of the teacher, which can also be compared with the financial equivalent, as well as the availability of additional IT competencies, which often turn out to be insufficient.

The authors see a particular solution to this problem as the creation of a set of methodological, organizational and technical measures that will simplify work with the COC, thereby increasing the average quality level of the COC, providing support to advanced teachers who are willing to spend their personal time, but they lack competencies in the field of IT, software, etc.:

1. The division of disciplines in the curricula of an educational organization into standard, advanced and in-depth levels in terms of the volume and complexity of the digital technologies used in the development of CCS
2. Using a basic course template for a "quick start".
3. Interview the teaching staff and identify the required competencies to create templates for course elements
4. Creation of course development teams, course templates, and individual blocks (evaluation, feedback, gamification elements, etc.)
5. Creation of teams of testers/tutors of the course from among students/graduate students/young scientists undergoing internship in the field of study, in order to track errors and increase the usability of educational content.

The complex of information support for teaching staff:

1. Ensuring the work of the EIAS support team, which answers teachers' questions about the COC no later than 12 h.
2. Creation of an online FAQ resource, a help desk, where standard questions about the EIOS used and the answers to them are quickly moved.

3. Weekly webinars on practical work with the BIOS and the software registry for the preparation of data centers, saving video webinars and creating navigation.
4. Creation of an interactive chatbot for quick navigation of reference information for non-specialists in the field of IT.

Thus, this list of actions is aimed at increasing the motivation of the teaching staff, simplifying the development of new digital technologies and their implementation in the process of preparing CCS, reducing the cost of producing CCS through the use of template courses and ready-made modules.

6 First Section

The analysis of a large number of messages and sources shows the importance of the problem of increasing the burden on the teacher during the transition to electronic learning technologies. There are proposals to control this excessive workload in one way or another by reducing the number of students in groups, introducing the most modern LMS, etc. However, this does not solve the problem of excessive time spent developing digital educational content.

The authors propose a division of training courses into three categories depending on the depth of use of IT solutions, a practical assessment of the time spent on the production of COC is carried out, a table and a formula for calculating the teacher's time spent on the preparation of COC are proposed.

In addition, a set of methodological, organizational and technical measures is proposed that will simplify work with the COC and increase the average quality level of the COC, provide support to advanced teachers who are willing to spend their personal time, but they lack IT competencies.

References

1. Yarullina, L.R.: Digital learning in higher school: psychological risks and effects. World Sci. Pedagog. Psychol. **6**(8) (2020) https://mir-nauki.com/PDF/42PSMN620.pdf
2. Barbour, M.K.: The landscape of k-12 online learning: examining what is known. In: Handbook of Distance Education, vol. 3, pp. 574–593 (2013)
3. Garrison, D.R., Cleveland-Innes, M., Fung, T.S.: Exploring causal relationships among teaching, cognitive and social presence: student perceptions of the community of inquiry framework. Internet High. Educ. **13**(1–2), 31–36 (2010). https://doi.org/10.1016/j.iheduc.2009.10.002
4. Basuony, M.A.K., EmadEldeen, R., Farghaly, M., El-Bassiouny, N., Mohamed, E.K.A.: The factors affecting student satisfaction with online education during the COVID-19 pandemic: an empirical study of an emerging Muslim country. J. Islam. Market. (2020). https://doi.org/10.1108/JIMA-09-2020-0301
5. Pardo, A., Jovanovic, J., Dawson, S., Gašević, D., Mirriahi, N.: Using learning analytics to scale the provision of personalised feedback. Br. J. Educ. Technol. **50**(1), 128–138 (2019). https://doi.org/10.1111/bjet.12592
6. Kwakye, I., Kibort-Crocker, E.: Facing learning disruption: examining the effects of the COVID-19 pandemic on K-12 students. Educ. Insights, 20 (2021) https://wsac.wa.gov/sites/default/files/2021-03-30-COVID-Learning-Disruption-Report.pdf

7. Pardo, A., Jovanovic, J., Dawson, S., Gašević, D., Mirriahi, N.: Using learning analytics to scale the provision of personalised feedback. Br. J. Educ. Technol. **50**(1), 128–138 (2019). https://doi.org/10.1111/bjet.12592
8. Yan, L., Whitelock-Wainwright, A., Guan, Q., Wen, G., Gašević, D., Chen, G.: Students' experience of online learning during the COVID-19 pandemic: a province-wide survey study [published online ahead of print, 2021 May 4]. Br. J. Educ. Technol. (2021). https://doi.org/10.1111/bjet.13102
9. Barab, S., Squire, K.: Design-based research: putting a stake in the ground. J. Learn. Sci. **13**(1), 1–14 (2004). https://doi.org/10.1207/s15327809jls1301_1
10. Bell, J., Sawaya, S., Cain, W.: Synchromodal classes: designing for shared learning experiences between face-to-face and online students. Int. J. Des. Learn. **5**(1) (2014). https://doi.org/10.14434/ijdl.v5i1.12657
11. Heilporn, G., Lakhal, S., Bélisle, M.: An examination of teachers' strategies to foster student engagement in blended learning in higher education. Int. J. Educ. Technol. High. Educ. **18**, 25 (2021). https://doi.org/10.1186/s41239-021-00260-3
12. Al-Ajlan, A., Zedan, H.: Why Moodle. Future Trends of Distributed Computing Systems, IEEE International Workshop., 58–64 (2008). https://doi.org/10.1109/FTDCS.2008.22
13. Lopes, A.: Teaching with Moodle in Higher Education. In: Proceedings of INTED2011 – International Technology, Education and Development Conference. Cd ISBN: 978-84-614-7423-3, and in Abstracts Cd ISBN: 978-84-614-7422-6 (2011)
14. Kiget, N., Wanyembi, G., Peter, A.I.: Evaluating usability of E-learning systems in Universities (IJACSA). Int. J. Adv. Comput. Sci. Appl. **5**(8) (2014)
15. Putnik, Z., Ivanovic, M., Komlenov Mudrinski, Ž., Welzer, T., Hölbl, M., Beranič, T.: Usability and privacy aspects of Moodle - students' and teachers' perspective. Informatica. **37**, 2013 (2013)
16. Hasan, L.: Usability Problems on Desktop and Mobile Interfaces of the Moodle Learning Management System (LMS), pp. 69–73 (2018). https://doi.org/10.1145/3194188.3194192
17. Evgenievich, E., Petrovna, M., Evgenievna, T., Aleksandrovna, O., Yevgenyevna, S.: Moodle LMS: positive and negative aspects of using distance education in higher education institutions. Propósitos y Representaciones. **9**(SPE2), e1104 (2021). https://doi.org/10.20511/pyr2021.v9nSPE2.1104
18. Martin, Andrew. (2020). How to Optimize Online Learning in the Age of Coronavirus (COVID-19): a 5-Point Guide for Educators.
19. Nordin, A.B., Alias, N.: Web-based teaching and learning approach (WBTLA) usability in institutions of higher learning in Malaysia. Malay. Online J. Educ. Technol. **1**(2), 44–55
20. Department for Education UK: Guidance. Remote education good practice. Access: https://www.gov.uk/government/publications/remote-education-good-practice/remote-education-good-practice
21. Tomei, L.: The impact of online teaching on faculty load: computing the ideal class size for online courses. J. Technol. Teacher Educ. **14**(3), 531–541 (2006) Waynesville, NC USA: Society for Information Technology & Teacher Education. https://www.learntechlib.org/primary/p/5626/. Accessed 7 Mar 2022
22. Blackmon, S.J.: Claire major student experiences in online courses a qualitative research synthesis the quarterly review of. Distance Educ. **13**(2), 77–85 (2012)
23. Михалкина, Е.В., Скачкова, Л.С., Гапоненко, Н.В.: Анализ норм труда профессорско-преподавательского состава вуза. Экономика образования. **2** (2013) https://cyberleninka.ru/article/n/analiz-norm-truda-professorsko-prepodavatelskogo-sostava-vuza

24. Goleva, E.V.: Methodology for calculating the annual academic load of the teaching staff of an educational institution of higher education. **5** (2014) https://cyberleninka.ru/article/n/metodika-rascheta-godovoy-uchebnoy-nagruzki-professorsko-prepodavatelskogo-sostava-obrazovatelnoy-organizatsii-vysshego
25. Valentinovich, R.E.: Rationing the workload of teachers: problems and search for solutions. Univ. Manag. Pract. Anal. **4**(104) (2016) https://cyberleninka.ru/article/n/normirovanie-nagruzki-prepodavateley-problemy-i-poisk-resheniy

Employing the Decomposition Method Within the Context of Pedagogical Approaches to Mathematics Education in Higher Education Institutions

Albina A. Temerbekova(✉)

Gorno-Altaisk State University, Gorno-Altaisk, Russia
tealbina@yandex.ru

Abstract. The article presents the decomposition method and examines its application in higher education institutions to develop students' professional competencies. The foundation for implementing this method is the project-based learning approach, widely recognized within engineering and technological education in the field of mathematics teaching methodology. The theoretical and methodological aspects of applying the decomposition method in interactive learning environments for university students are explored. Examples illustrating the decomposition method's application in studying contemporary educational platforms and various didactic tools for mathematics instruction within methodological coursework are provided. Research studies employing the decomposition method in constructing educational projects used across diverse educational systems and frameworks are analyzed. The article also highlights prospective directions for utilizing the decomposition method in developing professional competencies among university students.

Keywords: Education · Competencies · Methodology · Decomposition Method

1 Introduction

The Strategy for the Development of the Information Society in the Russian Federation establishes a new format for teaching within the higher education system, emphasizing the necessity of addressing complex project-based tasks through the proficient mastery of professional competencies at various levels [1].

The formation and development of students' professional competencies during their university education are most effectively facilitated through engagement with specific subject matter. Extensive educational practice indicates that studying mathematical disciplines notably enhances and reveals students' potential abilities, including analytical thinking, critical reasoning, modeling capabilities, and the ability to build personal practical experience in the discipline. Currently, higher education systems encounter numerous challenges, particularly educational and methodological deficits tied to the imperative of sustaining national scientific-technological and industrial sovereignty.

M. Lapina et al. (Eds.): DET 2024, LNNS 1616, pp. 448–457, 2025.
https://doi.org/10.1007/978-3-032-04365-8_43

The decomposition method proposed herein is intended to cultivate relevant strategies and effective techniques for processing and comprehending academic content. Mastering these techniques will significantly enhance students' learning processes and support the development of professional competencies essential for their future careers.

2 Literature Review

To establish a clear framework, it is essential first to define the concept of "decomposition." In this context, decomposition is understood as the division of a primary methodological task into subtasks (or intermediate stages), employed within project-based activities and grounded in the deductive method. These subtasks constitute the elementary and fundamental components of an academic course. In terms of goal-setting, decomposition entails dividing the overarching objective of a projected system into specific, individual subtasks [2, p. 220].

The decomposition method is philosophically grounded, as it essentially represents a transition from general concepts to specific instances, or from complex to simpler structures. This method can be effectively applied at various stages of the educational process. Within this context, distinct concepts, definitions, and mathematical statements (axioms, postulates, and theorems) can be considered as smaller subsystems within a unified educational framework, each addressing specific educational sub-goals or subtasks at the university level.

When methodological courses are studied through the application of the decomposition method, students—as prospective mathematics educators—develop a systemic perspective on the processes involved in mathematics education. They gain insight into the contemporary state of mathematical knowledge, including the methods and instructional forms currently practiced in schools. Throughout their methodological training, future teachers acquire and enhance skills in obtaining, analyzing, and synthesizing mathematical information, adapting this knowledge to their personal teaching experience, and applying it to novel, subject-specific educational scenarios aimed at addressing professional challenges.

The discipline of mathematics teaching methodology itself is structured into instructional blocks or units defined by the educational program, which are subsequently subdivided into more specific subsections. This hierarchical arrangement of instructional content within mathematics education methodology creates a unified, goal-oriented structure, realized through coherent integration of teaching content, forms, and methods.

The decomposition method is actively utilized at Gorno-Altaisk State University within the pedagogical process, specifically in the educational programs 01.03.01 «Mathematics» and 02.03.01 «Mathematics and Computer Science» at the Federal State Budgetary Educational Institution of Higher Education «Gorno-Altaisk State University». This method is employed in the delivery of courses such as «Mathematics», «Mathematics and Computer Science», and within methodological disciplines including «Mathematics and Computer Science», «Mathematics Teaching Methodology», «Mathematics Education at Various Educational Levels», «Mathematics and Informatics Teaching Methods», among others.

The methodological support currently existing within educational modules undoubtedly comprises an adequate array of resources for facilitating the educational process.

Such resources are effectively integrated into instructional practice and encompass a sufficient variety of educational materials. The utilization of these resources is intrinsically linked to project-based activities, conducted both during classroom sessions and independent student work. Contemporary pedagogical research extensively discusses various interpretations and applications of the project- based learning approach, as reflected in scholarly works [3–5].

The use of the project method constitutes one of the most significant instruments in the methodological preparation of future mathematics teachers. It represents an exceptionally effective tool within the educational process in mathematics teaching methodology.

The application of the decomposition method is fundamentally based on the principles inherent in the project-based approach. As one potential direction for its development, the unification of functional components within educational projects employed both in instructional activities and in children's technical creativity has been proposed [6].

In carrying out methodological educational projects, the decomposition method plays a crucial role, as it underpins the division of project activities into logically structured, interrelated, and content-aligned stages:

1. Preparatory stage, characterized by defining the objectives of studying the methodological course, establishing fundamental concepts, and goal-setting.
2. Analytical stage, aimed at identifying methods of presentation and establishing criteria for evaluating project outcomes.
3. Practical stage, involving the identification of problematic tasks, discussing alternative solutions, critically analyzing possible solutions, selecting optimal solutions, and discussing the presentations.
4. Presentation stage, characterized by presenting project outcomes, including the preparation of presentation materials and demonstrating the results of project activities.
5. Evaluation stage, encompassing assessment, self-assessment, analysis of obtained results, and characterization of the achieved outcomes [7].

Project-based activities, grounded in the decomposition method, «make the educational process closely aligned with practical professional activities, addressing challenges related to individualization of the educational process and enhancing motivation, while also creating conditions conducive to the development of professional competencies» [8, p. 458]. For example, the visualization of project activities as a component of project work can also be viewed as «a means of developing students' professional competencies» [9, 10]. Project-based learning is considered a pedagogical technology divided into discrete instructional segments, the final implementation of which culminates in the creation of a tangible project outcome (scientific project, research resource, didactic materials, functional model, etc.).

3 Research Methods

Several practical examples illustrating the application of the decomposition method within the mathematics teaching methodology course are presented below.

1. Application of the decomposition method in conducting a logical-didactic analysis of topics from mathematics curricula («Algebra, » «Algebra and Introductory Analysis», «Geometry (Plane Geometry)», «Geometry (Solid Geometry)». A logical-didactic analysis of topics in school mathematics constitutes a sequential chain involving specific instructional components, including goal-setting, logical-mathematical analysis, and methodological (didactic) analysis. Each component is detailed through its corresponding content and illustrative examples. Particular emphasis should be placed on the logical-mathematical component, which encapsulates the mathematical foundation of the logical-didactic analysis. It encompasses identifying principal approaches for the logical structuring of educational processes at the university level, establishing the conceptual and terminological framework for the study, characterizing problem-based materials, and formulating algorithms and rules for working with mathematical objects. The primary objective of methodological analysis is to design the didactic cycle, support and adjust the complexity of theoretical content, select effective methods and tools for visualizing educational materials, substantiate teaching techniques and diagnostic methods, and determine the appropriate forms and tools for organizing the educational process at higher education institutions.
2. Application of the decomposition method in solving methodological problems within methodological courses. Any methodological task within mathematics teaching methodology can be represented through a decomposition structure with clearly defined content. Modeling a classroom lesson, which students encounter during their teaching practice, also embodies the use of the decomposition method. At the beginning of a school lesson, the educational content (theoretical and practical) undergoes decomposition, whereas at the end of the lesson, these elements are integrated into a cohesive unit, thereby reflecting an overall compositional process. Structuring a school lesson as part of a comprehensive educational process in their future profession allows students, as prospective teachers, to understand that each subgoal associated with individual lesson stages should be achieved through specific learning actions and with the application of certain educational tools, including interactive technologies.
3. Specific topics from the school mathematics curriculum are employed when teaching mathematics methodology to students. For instance, applying the vector-coordinate method in solving geometric problems represents a decomposition process. Initially, the problem is divided into subtasks, which are subsequently integrated to achieve a comprehensive solution. Such algorithmization simplifies the search for solutions to mathematical problems. With an appropriate approach, nearly all solid geometry problems in the school mathematics curriculum can be effectively solved using the vector-coordinate method.

Within the school mathematics curriculum, these types of problems are exemplified by the solid geometry tasks included in the Unified State Examination, which contemporary students are expected to solve proficiently. The value of mastering skills required to solve tasks at this level is determined by the following educational objectives related to the study and practical application of this method:

- acquiring knowledge and understanding of the role of mathematics, specifically analytic geometry, and its application across various fields of study;

- effective use and practical application of the conceptual and terminological framework of the vector-coordinate method in solving mathematical problems and their real-world applications;
- proficiency in analytic methods for solving geometric problems through the vector-coordinate method;
- identification and exploration of interdisciplinary connections relevant to tasks from other areas of mathematical knowledge, such as algebra, physics, probability theory, mathematical analysis, mathematical logic, and others.

4. When working with the conceptual and terminological framework within mathematics methodology topics, it is essential to analyze various concepts and definitions that prospective teachers will subsequently utilize with school students. In this context, the decomposition method can be employed to structure the conceptual basis of specific topics within the school mathematics curriculum. For example, when developing the terminological framework related to quadrilaterals, it is advisable to consider them according to the scope and content of mathematical concepts, explicitly identifying their immediate genus and specific differences. Consequently, the entire course would comprise concepts grouped into particular clusters, collectively forming the conceptual basis for individual educational topics and ultimately integrating into a unified conceptual and terminological framework for the school mathematics curriculum.
5. The decomposition method is applied in the completion of bachelor's and master's students' graduation theses. The graduation thesis represents the culmination of the student's scientific research conducted throughout their university studies. Structuring research and developing the methodological framework of the study involve the integrated application of several decompositional elements, particularly noticeable in the formulation of research goals and objectives. Each chapter or section is dedicated to resolving specific research tasks systematically reflected within the methodological framework, starting from identifying research contradictions to formulating conclusions summarizing the comprehensive research activities conducted during the student's professional training at the university.

The integration of previously decomposed elements into a unified whole constitutes the student's graduation thesis. Master's and doctoral students often make structural mistakes when determining the structure of their dissertations, typically due to insufficient comprehension of the roles and purposes of methodological components described in the introductory part of their research.

The structure of a graduation thesis can be schematically represented. The proposed schematic serves as a recommendation, and certain elements within its content may be adjusted or combined as necessary (see Fig. 1).

Please try to avoid rasterized images for line-art diagrams and schemas. Whenever possible, use vector graphics instead A figure caption is always placed below the illustration. Short captions are centered, while long ones are justified. The macro button chooses the correct format automatically.

6. In studying mathematics teaching methodology, the accumulated pedagogical experience holds significant importance for future mathematics educators. This experience

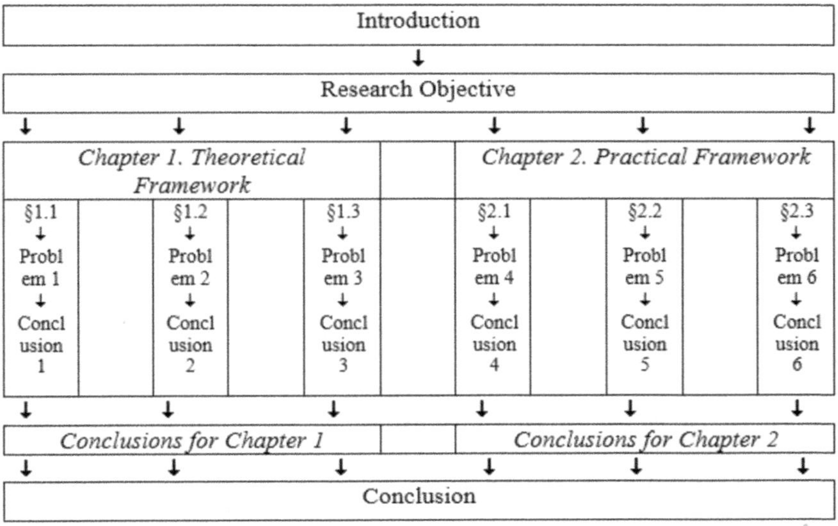

Fig. 1. Schematic Representation of the Research Structure.

is notably enhanced by the student portfolio compiled during the study of methodological disciplines, containing methodological materials of varying complexity and purpose. These may include lesson plans, didactic materials, assessment criteria and diagnostic tools, instructional videos, various cards, models, educational constructs, films, and other resources. Organization of project outputs within educational modules, logical-didactic materials for mathematics topics from the school curriculum, and extracurricular activity plans for future professional practice may also be included in the student's methodological portfolio.

The primary purpose of a digital portfolio is the "systematization of accumulated research experience and scholarly knowledge aimed at personal development and orientation towards future professional activities" [12, p. 124]. In the context of digitalizing the educational process, methodological complexes have been developed to prepare future teachers for creating a subject- specific informational and educational environment through the use of digital student portfolios [13, 14].

7. An effective way to actively incorporate the decomposition method into methodological courses at universities is by integrating digital technologies into the educational process. For instance, when solving tasks using the widely adopted software 1C: Mathematical Constructor, numerous psychological and pedagogical benefits of the instructional process are realized. Utilizing this mathematical constructor enhances spatial visualization skills, develops graphical memory through interaction with educational objects, and strengthens the ability to critically analyze one's learning activities. Furthermore, it fosters creativity in constructing mathematical objects. Collectively, these factors significantly contribute to the effective development of methodological competencies in future mathematics teachers, thereby enhancing the overall

quality of university education. The principal advantages of this software over others include its modeling capabilities, extensive set of tools, and dynamic visualization of geometric objects.

For example, when studying plane geometry, suppose it is necessary to construct an incircle within a triangle. A prospective teacher may solve this problem using the software "1C: Mathematical Constructor." To effectively utilize the decomposition method, this task is divided into the following subtasks, each implemented through separate fundamental subtasks:

1. Construct bisectors of two angles in the triangle and determine their intersection point D. (Fundamental subtask: determining the intersection point of angle bisectors in a triangle.)
2. Construct a perpendicular from point D to one side of the triangle and identify the intersection point F on that side. (Fundamental subtask: finding the foot of a perpendicular dropped from a point to a given line.)
3. Construct a circle with its center at point D (the intersection of angle bisectors) and radius equal to the distance DF. (Fundamental subtask: constructing a circle on a plane given its center and radius.)

The result of constructing an incircle within a triangle using GeoGebra software is presented (see Fig. 2).

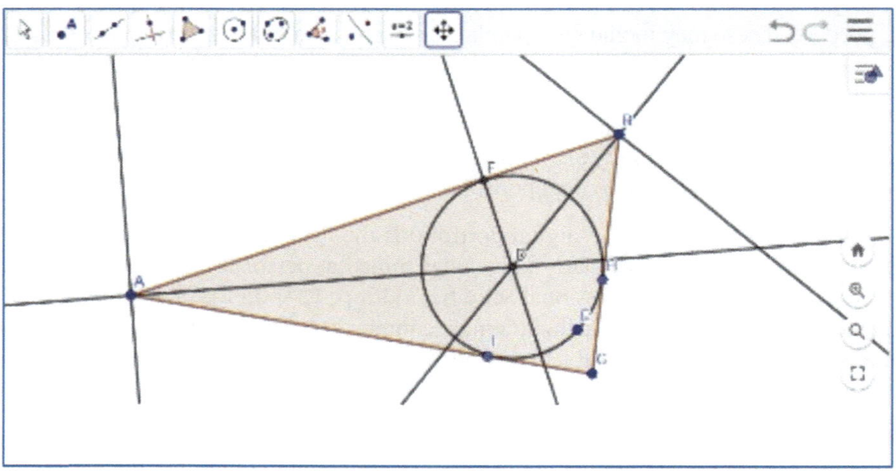

Fig. 2. Result of constructing the in circle of a triangle using GeoGebra software.

Consider the task of constructing a circumscribed circle around a triangle. To achieve this using GeoGebra software:

1. Construct perpendicular bisectors to two sides of the triangle and determine their intersection point. (Fundamental subtask: identifying the intersection point of the perpendicular bisectors of a triangle.)

2. Find the radius of the circle by measuring the distance from the intersection point of the perpendicular bisectors to one of the triangle's vertices. (Fundamental subtask: determining the distance between two points.)
3. Construct a circle with its center at the intersection point of the perpendicular bisectors and the determined radius. (Fundamental subtask: constructing a circle on a plane given its center and radius.)

The result of constructing a circumscribed circle around a triangle using GeoGebra software is presented (see Fig. 3).

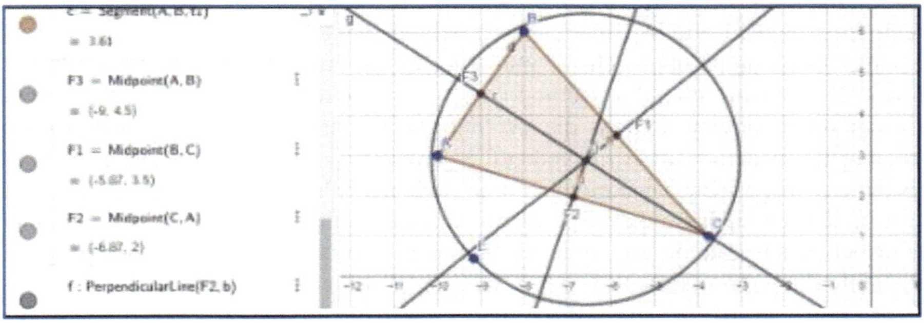

Fig. 3. Construction of a circumscribed circle around a triangle using GeoGebra software.

The construction tasks described above, executed with GeoGebra software, inherently exhibit a decompositional nature. They are divided into fundamental subtasks on specific topics and subsequently integrated into a cohesive final solution.

In the process of implementing such decomposition activities, attention was focused on how students comprehend and practically apply the knowledge acquired using this method. According to a conducted survey, 74% of respondents indicated the effectiveness of this approach due to an improved understanding of educational content and the resulting increased interest in studying the discipline. Approximately one-fifth (21%) of students highlighted the necessity of preliminary training in using GeoGebra software before practical application. As a result of the conducted research, students' academic performance in analytic geometry increased by 15%.

Additionally, students emphasized the efficiency of geometric constructions, accuracy of the obtained results, aesthetic appeal of geometric images and combinations, as well as the enhancement of their algorithmic skills necessary for achieving effective solutions to geometric tasks.

It should be noted that the previously discussed applications of the decomposition method within mathematics teaching methodology courses reveal new potential for successfully preparing future mathematics teachers for professional practice. The theoretical significance of the conducted analysis regarding the use of the decomposition method in the educational process at universities allows one to conclude that numerous possibilities exist for implementing this method within instructional settings.

Introducing mathematical literacy into various educational domains within Russian education is currently recognized as «a strength and a matter of competitiveness»

[15, p. 46]. While maintaining the fundamental nature of contemporary mathematics education, such an approach can modernize the educational process towards greater humanization, optimizing it through the integration of digital learning technologies and artificial intelligence. Additionally, this method holds considerable potential [16, 17] for establishing interdisciplinary connections among different courses during students' study of specific areas of mathematical knowledge.

4 Conclusion

Summarizing the above considerations, it can be concluded that the decomposition method, as applied within mathematics teaching methodology courses, represents a versatile instructional approach for preparing future mathematics educators. By systematically dividing complex methodological tasks into simpler, fundamental steps, this method effectively supports the resolution of comprehensive educational projects. Utilizing the decomposition method offers several psychological advantages: it fosters increased confidence and motivation in students; enhances focus at each individual stage of solving methodological problems; facilitates improved educational communication between instructors and students, particularly in collaborative tasks; and promotes critical thinking through step-by-step correction of inaccuracies and errors.

The practical significance of this research lies in identifying the levels of motivation and engagement among prospective mathematics teachers when applying the decomposition method to various stages of the instructional process, as well as evaluating the overall effectiveness of student learning within mathematics teaching methodology courses.

References

1. On Development Strategy in Российской the Russian Federation for 2017–2030. http://www.garant.ru/products/ipo/prime/doc/71570570.garant.ru/products/ipo/prime/doc/. Accessed 30 Jan 2025
2. Temerbekova, A.: Methodological features of using the decomposition method when performing logical and didactic analysis of a topic. World Sci. Cult. Educ. **6**(91), 219–221 (2021)
3. Andersen, A.: The European Project Semester: a useful teaching method in Engineering Education. In: Campos, L.C., Dirani, E.A.T., Manrique, A.L., Hattum-Janssen, N. (eds.) Project Approaches to Learning in Engineering Education. Sense Publishers, Rotterdam. https://www.sensepublishers.com/media/803-project-approaches-to-learningin-engineering-education.pdf
4. Mullen, J., Byun, C., Gadepally, V., Samsi, S., Reuser, A., Kepner, J.: Learning by doing, high performance computing education in the MOOC era. J Parallel Distrib Comput. **105**, 105–115 (2017). https://doi.org/10.1016/j.jpdc.2017.01.015
5. Powell, P.: Assessment of team-based projects in project-led education. Eur. J. Eng. Educ. **29**(2), 221–230 (2004). https://doi.org/10.1080/03043790310001633205
6. Kudryavtsev, N., Temerbekova, A.: Special features of the project interfaces method as a mechanism of developing a project approach to the educational process and children's technical creativity. Novosibirsk State Pedagog. Univ. Bull. **8**(6), 167–182 (2018). https://doi.org/10.15293/2226-3365.1806.11

7. Temerbekova, A.A., Kudryavtsev, N.G., Leushina, I.: Project-based approach in implementing relevant educational models in the Republic of Altai (Exemplified by the Regional Technology Park Quantorium-04). In: Solovev, D.B., Savaley, V.V., Bekker, A.T., Petukhov, V.I. (eds.) Proceeding of the International Science and Technology Conference «FarEastCon 2020». Smart Innovation, Systems and Technologies, vol. 227, pp. 801–2013. Springer, Singapore (2021). https://doi.org/10.1007/978-981-16-0953-4_78
8. Kondaurova, I.G., Filipovich, L.A.: Proektnaya deyatel'nost' kak vid uchebno-issledovatel'noy deyatel'nosti studentov vuza [Project activity as a type of educational and research activity of university Xstudents]. In: Scientific and Practical Conference from the International Conference Center. Moscow, Kemerovo, June 22, 2023, pp. 454–459. Kuzbass. GSHA, Kemerovo (2023)
9. Kudryavtsev, N., Temerbekova, A., Malova, V.: Visualization of project activity as one of the components of the methodology of working with the project team. In: Nanotechnologies: Education, Science, Innovation September 23, 2020, pp. 188–190 (2020)
10. Temerbekova, A.A.: Osnovnye tendentsii razvitiya vysshego obrazovaniya v vek tsifrovykh tekhnologii [Main trends in the development of higher education in the digital age]. Informatsiya i obrazovanie: granitsy kommunikatsii INFO'21: sbornik nauchnykh trudov Gorno-Altaisk: BIC GAGU. **13**(21), 11–13 (2021)
11. Nagorny, D.O., Shcherbakov, S.M.: Project activity in higher education: features, problems, management technologies. Inform. Digit. Econ. **2**(4), 167–180 (2021). https://doi.org/10.18334/ide.2.4.113393
12. Temerbekova, A.A., Dzhanabilova, S.A.: Various approaches to defining the concept of "Digital Student Portfolio". Sci. Pedagog. Rev. **3**(21), 122–128 (2018). https://doi.org/10.23951/2307-6127-2018-3-122-128
13. Temerbekova, A.A.: Preparation of future teachers for the formation of a subject information and educational environment through a student's digital portfolio. In: Grynshkun, V.V. (ed.) Informatization of Continuing Education-2018: Proceedings of the International Scientific Conference: in 2 volumes, Moscow, October 14–17, 2018, vol. 2, pp. 723–726. Peoples' Friendship University of Russia (RUDN), Moscow (2018)
14. Meshcheryakova, K.S.: Formirovanie graficheskoi kul'tury obuchayushchikhsya na urokakh matematiki sredstvom tsifrovykh tekhnologii [Formation of graphic culture of students in math lessons through digital technologies], vol. 15, pp. 417–419 (2023). https://doi.org/10.59131/2411-9814_2023_15(23)_417
15. Pozdnyakova, E.V.: Mathematical activity as the basis of modelling key universal educational actions of pupils in basic education. Continuum. Math. Comput. Sci. Educ. **2**(26), 42–56 (2022). https://doi.org/10.24888/2500-1957-2022-2-42-56
16. Dobrina, E.A.: Vozmozhnosti ispol'zovaniya mezhdubetnykh svyazov matematicheskogo analiza i analiticheskoi geometrii na zanyatiyakh po matematike (dlya studentov matematicheskikh spetsialnosti srednego professional'nogo obrazovaniya) [Possibilities of using inter-subject relations of mathematical analysis and analytical geometry in mathematics classes (for students of mathematical specialties of secondary professional education)], vol. 3, pp. 89–97 (2022). https://doi.org/10.24888/2073-8439-2022-59-3-89-97
17. Temerbekova, A.A., Yandykova, P.A., Solov'eva, L.A.: Ispol'zovanie informatsionnykh tekhnologii v, uchebnom protsesse pri izuchenii geometricheskikh obektov [The use of information technologies in the educational process when studying geometric objects], vol. 15, pp. 422–425 (2023). https://doi.org/10.59131/2411-9814_2023_15(23)_422

OLAP Model for Monitoring Student Success in Distance Education

Irina Tumanova[1] and Vladimir Tumanov[2(✉)]

[1] Private Educational Institution Orthodox Classical Gymnasium Named After Konstantin Bogorodsky, Noginsk, Moscow Region, Russian Federation
[2] Moscow Regional Medical College No. 3, Noginsk, Moscow Region, Russian Federation
tve9000@gmail.com

Abstract. The article considers an OLAP model for analyzing the student success in distance learning in natural science courses and adjusting the learning process based on individual indicators. The proposed model allows you to get a number of reports on the success of mastering the material by students on topics in a group: a report on mastering the topic of the course allows you to visualize the picture of the progress of the group as a whole, the distribution of tasks completed by the test. Reports on the individual success of the student include a visualization of the performance of a specific test of the student and a report on the assimilation of the subject for a certain period of study. Data analysis based on this model allows for pedagogical monitoring of learning success, building individual student trajectories and correcting the learning process.

The proposed OLAP model allows you to get a visual picture of the development of the subject as a group of students, as well as a picture of the individual achievements of the student.

Keywords: OLAP model · fact tables · measurement tables · student success · distance learning

1 Introduction

Methods of data mining are used in the digital educational space of distance learning to analyze individual student performance and the group as a whole in order to increase the effectiveness of training and improve the quality of education [1].

OLAP models are used in distance education as a tool for analyzing educational technologies and their effectiveness. Building individual educational trajectories of students based on the developed OLAP models can be used to monitor, analyze and predict future student performance. Also, the given models can be successfully used in blended learning, as discussed by the authors in [2–7]. The given works offer OLAP models that are more focused on predicting the success of training both individually and in groups, but do not have a high degree of detail on the subject matter of the discipline, but can be rationally used in the educational process. Further development of OLAP models for pedagogical monitoring of the success of students in distance and mixed education remains an actual pedagogical task.

M. Lapina et al. (Eds.): DET 2024, LNNS 1616, pp. 458–465, 2025.
https://doi.org/10.1007/978-3-032-04365-8_44

2 OLAP Student Success Monitoring Model

As a software environment for the implementation of the OLAP model, you can choose a data mart [8]. This model is intended for a teacher and helps to build an individual learning path. The most convenient in this case is a multidimensional data model or OLAP model.

A data mart [8] can be selected as a software environment for implementing an OLAP model. As an example, let's consider an OLAP model implemented for analyzing the academic performance of students in distance learning, which is shown in Fig. 1.

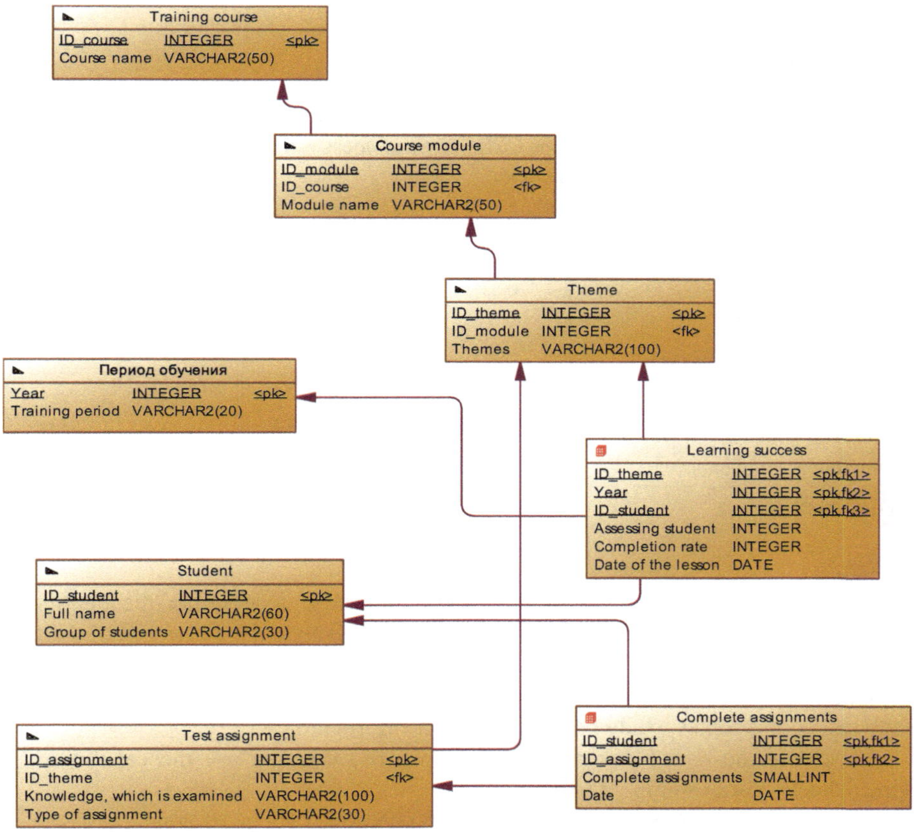

Fig. 1. OLAP model for analyzing the student success in distance learning.

The presented educational model includes two fact tables (OLAP cubes) and six dimension tables. Thus, the model shown in the figure describes the criteria for a student's success in studying a discipline. Using the specified data, a teacher can successfully build an individual trajectory for students to master this subject.

The fact table "Learning success" contains data on the results of the performance of the control and measurement materials on the subject of the course during the period of study of the course section. The general data include the title of the topic, the period of study, the type of control and measurement materials, the list of tasks (couple of values: number of task - what is checked).

For each trainee data include its identifier in the system, result of execution of each task (couple of values: task is completed - task not done), number of correctly and incorrectly executed tasks, completed tasks, evaluation according to the selected criterion (Table 1). The criteria for the evaluation of the test can be as follows: 10 tasks - «excellent rating», 8–9 tasks - «good rating», 6–7 tasks - «satisfactory», 5 or less - «partially satisfactory».

Table 1. Student success in mastering the topic of the course.

Knowledge, which is examined	Topic: Dynamics of a material point Period of study Type CMM (Control measuring materials): self-test operation									
Date	1	2	3	4	5	6	7	8	9	10
# Task	K1	K2	K3	K4	K5	K6	K7	K8	K9	K10
Students Student 1 (credit)	1	0	1	1	1	0	0	1	1	0
Assessment					2					
Complete assessments					5/5					
Student 2 (credit)	1	1	1	1	1	1	1	1	1	0
Complete assessments					8/2					
Assessment					4					

The dimension table «Test assignment» contains the topic identifier, the identifier of the type of control and measurement materials and the code decipherment of «what is being checked in this task» (Table 2).

Other tables in the data mart contain student lists, course descriptions, course sections, course topics, study periods and other information.

To access the data kiosk and generate reports, you can use the capabilities of spreadsheets. For example, MS Excel.

The Fig. 2 shows the structure of a physics course in college (fragment).

The Table 3 shows a fragment of the educational trajectory of the student (fragment). The table shows the results of the learning process for students of the subject studied, evaluation of the skills acquired in the process of learning.

From the received report you can see how the student is mastering theoretical and practical skills in the possession of a classical course of physics. This report allows the adjustment of the student's assimilation of basic skills and practical skills in both distance and face-to-face training.

Table 2. Knowledge, skills, competencies, which is examined.

Topic	Test assignment	Knowledge, which is examined (Kn)
Dynamics	1	Newton's first law, net force
	2	Measurement units for physical quantities (force, mass)
	3	The law of gravity
	4	Newton's first law, reference frame, inertial reference frames
	5	Newton's Second Law, direction of the force acting
	6	Elastic force, Hooke's law, stiffness coefficient
	7	Motion graphics
	8	Laws of Gravitation and curvilinear motion
	9	Kinematics and dynamics
	10	Frictional force, coefficient of friction
	11	Bodyweight, weightlessness

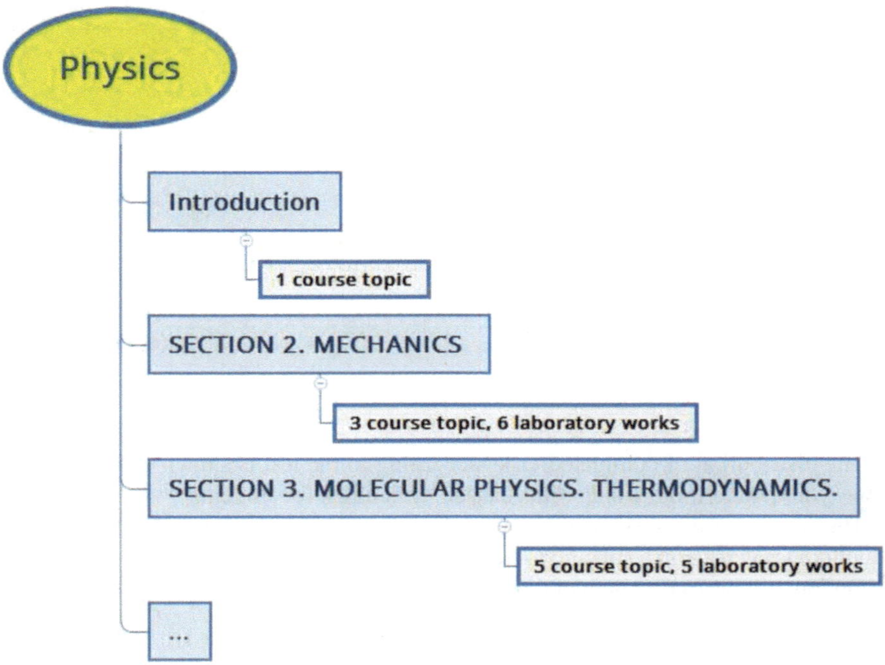

Fig. 2. Structure of the physics course in college.

The course topic mastery report allows you to visualize a picture of the performance of a group of students as a whole. Example in Fig. 3 shows that the topic has not been mastered.

Table 3. Educational trajectory of the student.

Student	Name	Result
Sect. 1	Introduction	
Topic 1.1.		
Theory		
Test 1.1.	Physical laws	7/3
Assessment		4
No competence received		1.1.K1, 1.1.K6, 1.1.K9
Total by section		4
Sect. 2.	Механика	
Topic 2.1.		
Theory		
Test 2.1.	Mechanical motion	8/2
Assessment		4
No competence received		2.1. K3, 2.6. K6
Links to previous topics		1.1.K6
….		
Practice		
Laboratory work 2.1	Study of body movement under the action of constant force.	5/2
Assessment		3
No competence received		2.1K3, 2.6. K6
…		
Total by section		3
…		

The distribution of completed tasks according to the test is shown in Fig. 4, which allows you to determine which tasks caused difficulties for students.

The report on the results of the test allows you to get the point of the student's individual trajectory on a specific topic on a specific date (Fig. 5).

On Fig. 6 shows a fragment of the student's individual trajectory for the study period.

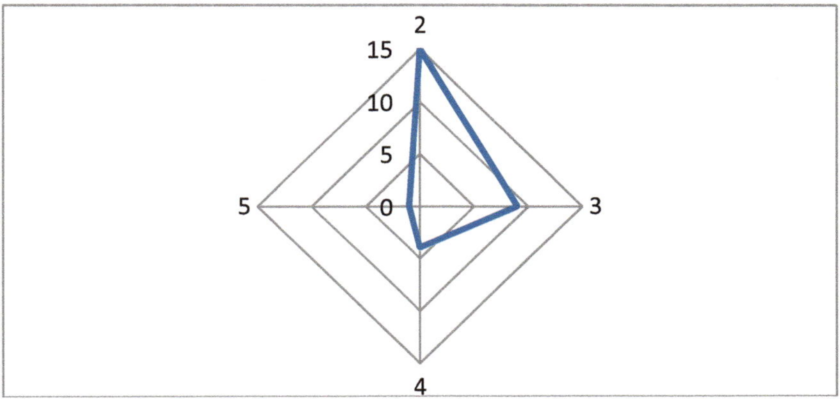

Fig. 3. Visualization of the performance of a group of students as a whole.

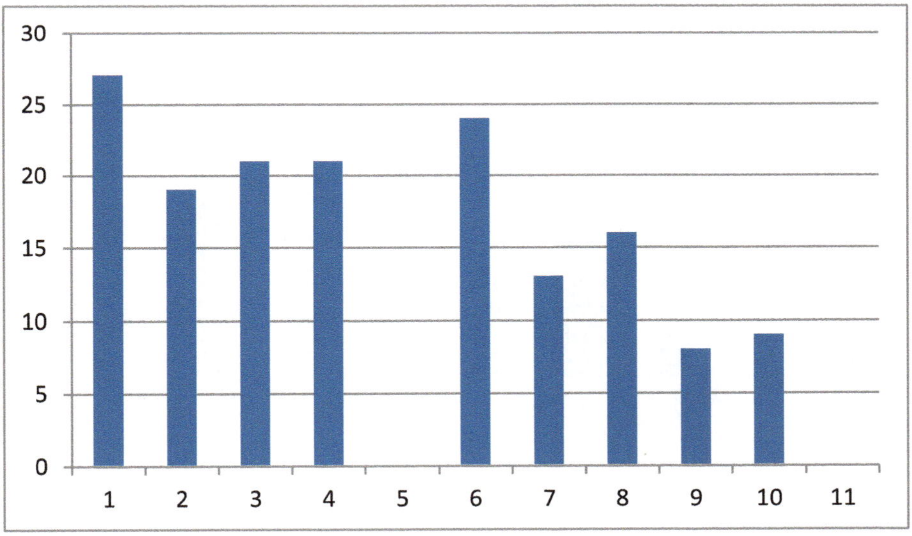

Fig. 4. The distribution of completed tasks according to the test.

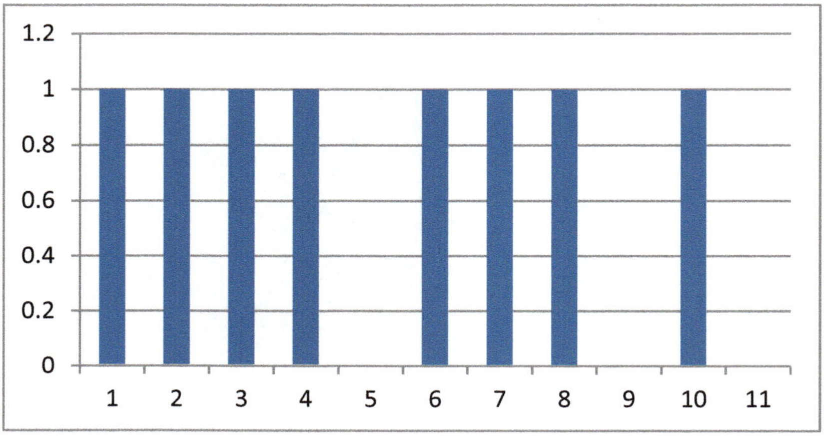

Fig. 5. Point of the student's individual trajectory on the topic.

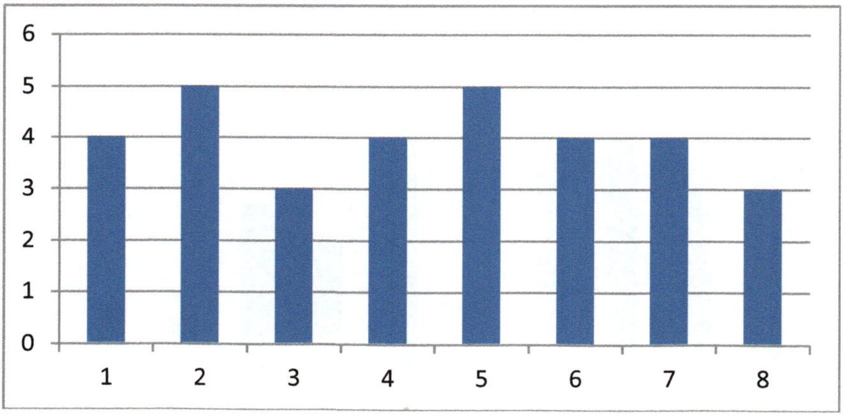

Fig. 6. A fragment of the student's individual trajectory for the study period.

3 Conclusions

The proposed OLAP model allows you to get a visual picture of the development of the subject as a group of students, as well as a picture of the individual achievements of the student.

Data analysis based on this model allows for pedagogical monitoring of learning success, building individual student trajectories and correcting the learning process.

The proposed model allows to monitor the process of students' development of necessary competence achievements of students in the development of the subject, and also adjust.

The teacher motivates students to learn, self-educate, and gain deeper knowledge if he or she uses modern pedagogical educational technologies in the classroom [9–12], and the students become highly professional specialists.

References

1. Martin, F., Abdou, N.: Using learning analytics to assess student learning in online courses. J. Univ. Teach. Learn. Pract. **13**, 1–20 (2016)
2. Ola, A.G., Bai, X., Omojokun, E.: Modeling of students' profile and learning chronicle with data cubes. Res. High. Educ. J. **24**, 1–6 (2014)
3. Ovsyanitskaya, L.: The OLAP-cubes technology application for Analysising Pedagogocal monitoring data. Izvestia: Herzen Univ. J. Human. Sci. **171**, 271–279 (2014) (in Russion)
4. Korchemkina, Y.V., Belousova, N.A., Maltsev, V.P.: Application of OLAP-technologies for construction and implementation an individual trajectory of student. Mod. High Technol. **10**, 196–200 (2018) (in Russion)
5. Ozerova, G.P., Pavlenko, G.F.: Prediction of student performance in blended learning by utilizing learning analytics data. Sci. Educ. Today. **9**, 73–87 (2019)
6. Charikov, P.N., Fokin, S.Y.: Study of the student progress monitoring system of a university department for a distance learning system based on a web portal and for OLAP samples using to automate analysis. J. Phys. Conf. Ser. **1515**, 1–10 (2020)
7. Najm, I.A., Dahr, J.M., Hamoud, A.K., Alasady, A.S., Awadh, W.A., Kamel, M.B.M., Humadi, A.M.: OLAP mining with educational data Mart to predict students' performance. Informatica. **46**, 11–19 (2022)
8. Tumanov, V.E.: Data Warehouses: Tutorial. IPR Media, Moscow (2022) (in Russion)
9. Lapina, M., Tokmakova, M., Movzalevskaya, V., et al.: Analysis of modern methods and approaches using it in education and pedagogy. In: International Conference on Innovative Approaches to the Application of Digital Technologies in Education and Research, 2022, vol. 1222, pp. 40–52. Springer, Cham (2025). https://doi.org/10.1007/978-3-031-78776-8_5
10. Zolotarev, V., Arkhipova, A., Lapina, M., et al.: Assessing player interaction for a social networking cooperative educational game. In: International Conference on Innovative Approaches to the Application of Digital Technologies in Education and Research, 2022, vol. 1222, pp. 114–121. Springer, Cham (2025). https://doi.org/10.1007/978-3-031-78776-8_12
11. Taran, V., Savelova, L., Abdullaev, J.: Increasing learning motivation through the use of educational and technological tools. In: Lapina, M., Prakasha, G.S., Grigoriev, S. (eds.) International Conference on Innovative Approaches to the Application of Digital Technologies in Education and Research. SLET 2022. Lecture Notes in Networks and Systems, vol. 1222. Springer, Cham (2025). https://doi.org/10.1007/978-3-031-78776-8_7
12. Lutsenko, L., Testova, I., Chirkov, M., Taran, V., Mezentseva, O.: Methodical specifics of the subject distance camp organization "young cryptograph". In: Lapina, M., Prakasha, G.S., Grigoriev, S. (eds.) International Conference on Innovative Approaches to the Application of Digital Technologies in Education and Research. SLET 2022. Lecture Notes in Networks and Systems, vol. 1222. Springer, Cham (2025). https://doi.org/10.1007/978-3-031-78776-8_21

Training Communication and Pragmatic Skills with AI-Powered Tools in Linguistics and Media Studies

Liana Vartanova[1], Tatiana Marchenko[1(✉)], Olga Lepilkina[1], and Munis Xamidov[2]

[1] North-Caucasus Federal University, 1 Pushkin Street, Stavropol 355017, Russia
tatiana-marchenko-25@yandex.ru
[2] Samarkand State University Named After Sharof Rashidov, Samarkand 140104, Uzbekistan

Abstract. The article discusses methods of training communication skills in linguistics and media studies with the use of AI-powered technologies. The study works out a model to train pragmatic interaction, pursuing and counteracting strategies and tactics, analyze its strong and weak points. The model is based on preliminary analysis of communication principle of balanced communication. The use of AI-assisted options in linguistics and media studies is given with account of their application and best practices of use. The selection of a prospective tool to be used in the developing the model was reasoned by its options of maintaining dialog interaction. The model rests on the accepted communication principles and target skills to be obtained within linguistics and media studies. On this basis, a comprehensive model has been developed. The model incorporates six stages: preliminary, introductory, preparatory, training, feedback and reflection ones. Each stage covers activities starting with communicative and technical aspects, proceeding with the functional and the reflection one. Prospective challenges in introduction and testing of the model in the academic context are analyzed, the ways to cope with them are proposed. The results of the study suggest that the introduction of AI-technologies in the process of training communication pragmatic skills requires comprehensive AI-powered tools so that various relevant skills could be trained.

Keywords: communication skills · chatbot · development of communication skills

1 Introduction

Blurring the borders of communication and increase in social media interaction both individually (social networks) and at an institutional level (official media sources) actualize the relevance of comprehensive studying the transformation of communication processes, strategies and techniques pursued by interlocutor in various formal and informal contexts. The changes concern all levels of interpersonal interaction and cross-cultural communication in its diversity. A number of specific features are claimed to shape the communication process and determine it balanced or misbalanced flow. These features

M. Lapina et al. (Eds.): DET 2024, LNNS 1616, pp. 466–473, 2025.
https://doi.org/10.1007/978-3-032-04365-8_45

include the type (online / offline, personal / official, intracultural / intercultural, etc.) and nature of interaction (monolog / dialog / polylog), the semiotic coding systems involved (verbal / visual / audio / converged types). They also include the form and genre of communication (speech / article / review, etc.), the number of interlocutors, the target audience (if any is supposed), the extent of involvement into communication processes and the degree of dependence on them. The list of characteristics features is not limited and can be expanded with the reference to unique conditions of communication. In each act, they modify both the process itself and its final result – reaching the goals of either balanced or misbalanced cooperation.

Training for communication with account of the above-mentioned multi-code conditions is becoming increasingly relevant since their variety is getting more and more complex. The background of any communicative act serving the base for tuning communication in all its domains and achieving mutual understanding are the speech quality principles. They incorporate clarity of expression that is closely associated with correctness and logic, relevance of the message, the use of expressive and emphatic means that is balanced by the requirement of simplicity. A generally accepted principle on specifying communicative and pragmatic attitudes of verbal interaction is the one of cooperation. According to P. Grice, the principle is complex and is constructed by fours maxims: quantity, quality, relation / relevance and manner of expression. A related concept explaining the success or failure in communication is the principle of politeness laid by G. Leech. It fall into a set of maxims (tact, generosity, approbation, modesty, agreement and sympathy), which predetermine further interpretation of communication behavior.

The study aims to specify the linguopragmatic aspects of coordinated communication based on the integrity principle, identify speech tactics that contribute to tuning the communicative act and train students' communication skills with an AI-powered tool (a neural language model chatbot service).

2 Purpose and Objectives of the Research

The study presents a step-by-step model of training communication skills (turn-taking exchange) with an AI-assisted tool capable of generating human-like text responses and participating in contextual conversation. The target group are the students studying at courses with a significant focus on communication skills, namely linguistics and media studies.

The goal of the study has determined the range of tasks to be completed. They include:

- a review of operational potential of a neural language model chatbot service for communication skills training;
- the assessment of chatbot along with the academic purposes of linguistics and media studies;
- selection of a set of pragmatic interaction principles relevant in intra- and intercultural communication;
- developing the model to customize a chatbot character in line with available functional features of the selected tool and the interaction principle to be trained;

– critical analysis of strong and weak points in training communication skills an AI-assisted tool.

3 Literature Review and Methodology

The studies of IT- and AI- assisted training in this field of linguistics and media courses fall into several areas:

1. Analysis of practices related to integration of interactive tools into language and media education and the efficiency of the results obtained. This area of studies offers an overview of the variety and the quality of content developed by specialists in the relevant field, psychological conditions due to changes in the mode of training, challenges in incorporating the technological component and greater individualization [1]. The analytical nature of these works gives insight into feedback and prospects for enhancement [2].
2. Case studies on development of a single or multiple language skills. These, for instance, include writing [3], speaking [4], vocabulary [5], listening [6], reading [7] and translation [8]. The case studies are mainly based on academic contexts with the use of one technological solution. These works also include reviews of evaluative feedback on the part of students [9–11].
3. Analytical reviews of IT- and AI-tools with account of their functionality, strong and weak points, and conditions of their use.

The concept of communication integrity a deliberate desire for consonance and agreement. It is an intended communicative behavior actualized in pursuing certain communication strategies and tactics aimed at reaching mutual understanding and concordance.

Communicative strategies and tactics are perceived and studied in theory as an intended plan of communicative actions incorporating the goal, major (strategies) and minor (tactics) steps that are made to achieve it. The use of a specific strategy repertoire modifies the choice of language means, visual and audio sources that provide the achievement of the prospective result. Simultaneously, this choice is governed by the key communication principle listed above. This interrelation makes a communicant search for balance to meet the expectations interlocutors and achieve an individual goal set for a particular communication act. It preconditions the use of the consistency principle in training communication skills as it presents the interaction as an ever-flexible process incorporating cognitive, linguistic and cultural empirics.

Another relevant principle is contrastivity. In communication, contrast works as a background against which a set of prospective steps taken in interaction is opposed to previous experience, norms of appropriate behavior, acceptability or non-acceptability under current conditions, etc. A complex of these terms shapes the framework within which all integral elements are related. The case when one of them fails to meet the expectations features the break in well-tuned communication. The simplest examples being the use of a colloquial or slang word in academic discourse, reference to inappropriate genre peculiarities, etc. Contrastivity reveals the sensitive point for it to be addressed and corrected. It makes this principle helpful for didactic purposes as it can demonstrate and make obvious an integral element of communication, which can be tuned through training.

Besides, all types of observations (direct, indirect and immediate) added to the efficiency of contextualized assessment of communication.

4 Results

The study is based on contextual interpretation of communication behavior actualized in offline / online communication in the form of major– strategies – and minor plans - tactics. A communicative strategy is claimed to be a set of speech actions pursued within a relevant context of one or more sentences. It is governed by the intention to manage, either positively or negatively, a communicative situation. Whereas a tactic is a minor communication step under a wide strategy. It is actualized at a lower level of a sentence, less frequently a number of sentences. The act of communication can feature several strategies and even more tactics. Pragmatics behind the communicants intentions and the analysis of the conditions accompanying the communication reduce and specify the choice of language tools to pursue one's own strategies and tactics and counteract / yield to the interlocutor's ones. It is the moment when not only the analysis but also the detection plays a great role.

A set of strategies and tactics actualized and detected by a communicant depends on their personal experience, views, cultural norms, rules, and, notably, the communicative competence that can be trained. There are two main types of communication strategies – the action and the re-action ones – that are actualized through a number of subdominant ones. The terms arise from the interpretation of communication process as an inter-action, namely the action between several participants (real or imagery). The action type involves the implementation of assertive behavior aimed at establishing and maintain contact. This type of strategy is typical of the initial stage of communication. Originally the action strategy realizes the phatic function of the language and does not require sophisticated skills, yet at higher and more specific levels it manifests proficient communicative competence of the linguistic personality.

The strategy is most obvious when juxtaposed to speech aggression or conflict interaction. The latter are instances that result from misbalanced communication and could be immediate reactions or planned communication actions. The reaction to ether positive or negative communication feed forward is probable in two ways: instant reaction or pursuing the principle of interaction integrity through a number of strategies and tactics. Following the principle helps avoid communicative failures and gain balance. The options and practical moves to achieve it are supposed to be trained in real life and AI-assisted learning.

The action strategy falls into a number of tactics. Information delivery appears to be basic since in many cases communication misbalance and miscomprehension arise due to insufficient information concerning the act of communication, cultural values and background, social awareness, etc. The strategy is actualized through the tactics of establishing the unity, giving illustration, assertive analysis, providing an analogy and referring to best practices. Each of the listed tactics is ruled by the principle of cooperation. Biased communicative behavior results in the emergence of counter-effective communicative behavior.

Another action strategy is presentation, which presupposes the manifestation of one's personal commitment to the principles of politeness and cooperation. It is actualized by the tactics of approval, implicit expression, explicit expression and reference to an authorized source. The latter could be quotations, precedent statements, sayings, proverbs, references to culturally significant texts and people.

The preventive strategy is aimed at tackling the impetus for confrontation and communication misbalance. References to negative experience, alternatives and condemning negative practices represent tactical moves within the strategy. The preventive nature of these tactics relies on the awareness of already existing problems in communicative interaction, with the focus on modifying the communication context and overcoming the challenges. The implementation of strategy and tactics is only possible if a communicant possesses proficient analytical skills.

Ignoring and disapproval tactics form the re-action strategy. They presuppose the demonstration of indifference, intentional distraction and switching attention to a different aspect, and mitigation. Indifference as communication behavior does not necessarily indicate the unwillingness of the participant in communication to tackle the violation of the politeness and cooperation principles. It can also arise from the inability (insufficient skills) to detect their violation. In the case of a competent communicant, this may indicate an unwillingness to focus attention on the deviant aspect of communication, thereby not increasing its status and relevance. Tactics of direct and indirect censure represent the strategy of disapproval. The identified tactics involve implicit or explicit condemnation of provocative communicative behavior.

Communication skills are best trained with AI tools that imitate real-life communication, namely chatbots. The tool that was chosen to work out a model to train different pragmatic strategies is Character.ai, a web application based on a large language model [12]. Its specific features and customization opportunities drove the choice. The application imitates human speech and is capable of participating in communication on an array of topics since it is trained on various data with an emphasis on dialogue interaction of users on the Internet. User-friendly and intuitive interface requires no preliminary training, which is beneficial in academic context of limited hours. Another benefit is the prompt-free start as the interlocutor indulges into interaction with already configured character and does not need to produce the prompt.

From the didactic point of view, Character.ai meets the criteria of communication diversity. The key features include:

- An option to interact with a model of a real one (the suggested ones). It works either way as one can communicate with a celebrity character or play the role of some famous person. The ready-made characters come with a detailed description of their background that adds to the personality profile and suggests prospective communication strategies. The option provides extra motivation on the part of the student, and offers almost unlimited range of character varying in communication features and behavior.
- An option to create a fictional character. Character profile includes a name, a tagline with a brief and clear-cut description of the character, a detailed description of the character outlining specific features, a greeting used by the character at the beginning of each chat, a definition –with account of specific instructions on how the character

should behave and respond to the user's messages, and visibility, which makes the created character available for other users.
- One-to-one mode of interaction with the application makes available keeping an individual pace in training with account of the proficiency level, strategies and tactics pursued.
- Customization of phonetic (voice and its characteristics) and prosodic peculiarities of the character.
- Customization of communication strategies and tactics through intended correction of character's feedback, which can be applied both by a teacher creating a character for training target communication aspect and a student tuning the process.

With the reference to functional features of the application, a model for the development of pragmatic tools has been worked out. It falls into several stages of preliminary tuning and the training process itself:

- Preliminary stage. It includes setting the goal and the outcomes of the course, choosing the action and reaction pragmatic strategies to be trained, selecting or creating a character-interlocutor with distinctive communication settings capable of pursuing the interaction in the expected way. The training material is compiled. The set is expected to contain a brief account of definitions, a complex of communication contexts with the description of topics to be discussed. Each character is selected to match the goals and pursue certain communication strategies.
- Introductory stage. Students are introduced to the communicative behavior concept, the notion of strategies and tactics, action and reaction strategies and tactics. This stage is covered within contact hours. Students get familiar with strategic interpretation of communication based on sample texts from fiction and media sources. The conception is training pragmatic strategies through dialog interaction is offered.
- Preparatory stage. Students get the briefing on the application and the mode of training, the training material is handed out. At this stage, students are set the timing and the feedback form. They get familiar with the application and are assigned the tasks that can be both unified (supposed to be trained by every member of the group) and individual (customized in line with specific communication requirements).
- Training stage. The approbation of the developed model is carried out individually. Students can be conventionally divided into groups in accordance with the strategies trained. The content of the training course is almost identical, with the exception of some aspects preconditioned by the average level of language proficiency varying from pre-intermediate up to the advanced one.
- Feedback stage. This step is focused on consolidation. The students provide their feedback on tackling the communicative strategies, challenges in overcoming counteraction strategies and general communication assessment. The development of the skills can be evaluated according to a number of gradable and non-gradable criteria, e.g. communication efficiency, the flow of interaction, pursuing and counteracting communication strategies.
- Reflection stage. Overall analysis of the efficiency of the model as a whole and its constituent parts. The stage results in the feed-forward to update the model.

The choice of texts for practical analysis prior to AI-assisted training was largely driven by fact that when training the verbal component of communication, the focus of attention is on personal comprehension and interpretation of extralinguistic conditions and communicative activity aimed at achieving the tasks of communication. All these aspects are comprehensively combined in media texts as they reflect the complex nature of semiotically diverse interaction, cover up-to-date topics and socially relevant issues This fact allows tracing in detail the speech behavior of the participant in the communicative process under consideration.

5 Discussion

Flexibility and customization offered by AI-powered application makes it an efficient communication training tool. Testing of the model can reveal a number of drawbacks. They can be prospectively analyzed in order to provide appropriate feed forward and further tune the model and the course.

The fluency in training according to the model is individual since some student can require extra time to get familiar with the application and the mode of work, create a character meeting individual training requirements or specify the existing one.

Training requires a certain period to be assigned for instructions. If it expands, it can result in the decrease in the amount of time spent on mastering communication aspects and strategies. Overall timing could also pose a challenge since as any communication process it is flexible and does not have any strict limit. The case when work with the application is time-consuming, extra hours for independent work could be provided in order to develop operational skills. Along with that individual consulting on challenging issues could be offered. There is an opportunity to choose an alternative program with analogous functionality and user-friendly interface.

Reaching the proficiency of communication skills (pragmatic aspect in particular) calls for preliminary work with special literature on pragmatics of communication, enhancement of general linguistic skills.

Overall, the strong point is expected to be a higher degree of motivation and commitment to improve communication skills in the medium and with the reference to the tools gaining more and more spread in both personal and academic contexts.

6 Conclusion

The development of a communication-training model integrating AI-powered tools appears to be up-to-date, since the they offer an unlimited range of flexible options tuned in line with tasks and requirements. Work with AI-powered tools increases the motivation of students for training activities. Availability of large language model solutions allows students to gain more experience in operating AI-powered tools, develop independence and initiative, train a wide scope of subject-related and generic skills, enhance leadership qualities, elaborate the integration of IT tools with professional skills.

The model offered in the study is open to tuning in line with various inputs and requires update due to ongoing advance in AI-based applications. The content of the course is open to modifications, enhancement of specific strategies aimed at tackling counter-productive and conflict-provoking aspects of communication, e.g. aggression, misunderstanding, imposing one's point of view, etc.

References

1. Prakasha, G.S., Muttungal, P., Lapina, M., Taran, V.: Flipped classroom strategy in online teaching: challenges faced by higher education teachers. In: Lapina, M., Prakasha, G.S., Grigoriev, S. (eds.) International Conference on Innovative Approaches to the Application of Digital Technologies in Education and Research. SLET 2022. Lecture Notes in Networks and Systems, vol. 1222, pp. 241–253. Springer, Cham (2025)
2. Çelik, Ö., Yavuz, F.: The effect of using mobile applications on literal and contextual vocabulary instruction. Int. J. Learn. Teach. **10**(2), 126–136 (2018)
3. Marchenko, T., Bredikhin, S., Lapin, V.: Application of text analysis instruments in foreign language training. In: Lapina, M., Prakasha, G.S., Grigoriev, S. (eds.) International Conference on Innovative Approaches to the Application of Digital Technologies in Education and Research. SLET 2022. Lecture Notes in Networks and Systems, vol. 1222, pp. 323–332. Springer, Cham (2025)
4. Gamlo, N.: The impact of mobile game-based language learning apps on EFL learners' motivation. Engl. Lang. Teach. **12**(4), 49–56 (2019)
5. Lin, C.C., Yu, Y.C.: Effects of presentation modes on mobile-assisted vocabulary learning and cognitive load. Interact. Learn. Environ. **4**(25), 528–542 (2017)
6. Hwang, G.J., Hsu, T.C., Hsieh, Y.H.: Impacts of different smartphone caption / subtitle mechanisms on English listening performance and perceptions of students with different learning styles. Int. J. Hum. Comput. Interact. **35**(4–5), 333–344 (2019)
7. Naderi, N., Akrami, A.: EFL learners' Reading comprehension development through MALL: telegram groups in focus. Int. J. Instr. **11**(2), 339–350 (2018)
8. Marchenko, T., Kolobkova, A.: Design and implementation of a CAT training course. In: SLET-2020: International Scientific Conference on Innovative Approaches to the Application of Digital Technologies in Education, pp. 324–331. CEUR Work (2020)
9. Hao, Y., Lee, K.S., Chen, S.T., Sim, S.C.: An evaluative study of a mobile application for middle school students struggling with English vocabulary learning. Comput. Hum. Behav. **95**, 208–216 (2019)
10. Dembsey, J.M.: Closing the grammarly gaps: a study of claims and feedback from an online grammar program. Writing Center J. **1**(36), 63–100 (2017)
11. Park, J.: An AI-based English grammar checker vs. human raters in evaluating EFL learners' writing. Multimedia-assisted. Lang. Learn. **1**(22), 112–131 (2019)
12. Character.ai: https://character.ai/. Accessed 20 Jan 2025

Pairing a Neural Network with a Science Fiction Novel: New Opportunities for Psychological Diagnostics of Students and Schoolchildren

Vladimir N. Yevstifeyev[1,2](✉), Ibragim E. Suleimenov[1], Dinara K. Matrassulova[1], and Dina B. Shaltykova[1]

[1] National Engineering Academy of the Republic of Kazakhstan, 050010 Almaty, Kazakhstan
yvladimirn@gmail.com

[2] Institute "KazNIPIEnergoprom" JSC, 050004 Almaty, Kazakhstan

Abstract. A new approach to the psychological testing of students is proposed, based on pairing the text of a science fiction novel with a neural network. The text of the novel, specially written for this work, is full of hyperlinks, the transitions on which provide a connection with the neural network. More precisely, each link is a neuron of the first layer of the network. When following this link, the state of the neuron changes from zero to one. As a result, a well-defined digital image of the reader interacting with the novel is created, fixed in binary variables and analyzed by the neural network.

Keywords: Neural Network · Science Fiction · Learning Trajectory · Education Quality

1 Introduction

Kazakh universities, as well as universities of many other contrives are forced to focus on students with an average level of ability [1]. In such conditions, the least advantageous position is occupied by gifted and talented students who really have increased abilities. Universities do not provide for their requests, as a result of which this category of students often loses interest in studying, goes into destructive entrepreneurship, and so on [1].

Consequently, the task of identifying such students and providing them with the opportunity to independently form the learning path is an acute task. There are all the necessary prerequisites for this, in particular, there are numerous Internet resources, when a talented student can independently receive education, using education at the university, only as a kind of secondary resource. It should be noted that at least in Kazakhstan, many gifted students really study independently (self-education), formally studying at second-class universities only for the sake of obtaining a formal status [2]. An additional argument in favor of receiving non-formal education in countries such as Kazakhstan is a sharp drop in the quality of education, associated, among other things, with the degradation of teaching staff, which has been taking place for a very long time [3, 4].

M. Lapina et al. (Eds.): DET 2024, LNNS 1616, pp. 474–480, 2025.
https://doi.org/10.1007/978-3-032-04365-8_46

The first need here is the clear demonstration of the existence of appropriate opportunities. It is also necessary to provide for talented student concrete tools that will allow him to objectively assess his own abilities, i.e., to answer the question - is it worth looking for your own non-trivial path or should you stay within the framework of formal education. A person entering independent non-formal learning must make sure that he is talented enough, and also that he is able to choose his own learning path.

Standard psychological tests used, for example, for the purpose of choosing a profession, are not suitable for solving such a problem. Some of them are focused on identifying a propensity for a particular type of activity [5], others on the ability to find non-standard solutions to various problems [6, 7], but the task under consideration requires the implementation of an integrated approach. Such an approach can be implemented using a large set of tests for various purposes, but this requires a very serious work of a professional psychologist of a high qualification, who becomes an academic coach for a particular young person. It is not obvious that many young people are ready to resort to services of this kind. At a minimum, in order to use the tools of academic coaching, young people must be motivated.

Therefore, it is necessary to develop new non-trivial tools aimed at finding an individual response to the "challenge of mass character", which, we emphasize once again, is in dire need of talented students. It is important that the use of such tools is not only useful, but also interesting.

One such tool is proposed in this paper.

The purpose of this article is creation of the science fiction novel coupled with a neural network, designed to diagnose students in relation to their creative abilities, creativity and ability to make independent decisions. We emphasize that the text of this novel was specially written to solve the problem under consideration.

2 Main Part

Science fiction in the period of its formation (the era of the Second Industrial Revolution) largely performed the function of popularizing the achievements of science. At present, of course, it is for the most part "sword and magic" literature, but this does not mean that it cannot return to its above-mentioned function.

Popularization of the achievements of science, especially modern achievements of information technology, remains a very important task. Therefore, there is a wide field for activity here. Moreover, there is an opportunity to combine the tools of science fiction with information technology. It is possible to ensure the pairing of a text of any nature with a neural network [8] and further with an artificial intelligence system.

This method is as follows. The text, in this case the text of the novel, is saturated with hyperlinks associated with the highlighted words/terms, the transitions through which reflect the reader's interest in a particular issue. For example, in the novel that was written for this work, there is an episode where the main characters approach the house of Tycho Brahe, located on Novy Svet Street in Prague.

A set of links is attached to this episode, by clicking on which the reader receives information about the life and work of Tycho Brahe himself, a link to a map where this house is in the vicinity of Prague Castle, etc.

Each of the links contained in the text is a neuron of the first layer of the network used. If the user has made a transition, then the state of the neuron changes by one. If there was no transition, it remains equal to zero. Thus, as a result of the interaction of the reader with the text, its well-defined digital image is formed, which can then be analyzed using typical neural network methods.

To train the neural network, this work used standard psychological tests, including those that allow students to identify the creative abilities of students, and they were also asked to solve several non-standard logical problems from open resources.

Figure 1 shows a page from a novel intended for distribution via the Internet. Red circles mark hyperlinks, which are designed to reveal the inclinations of the reader.

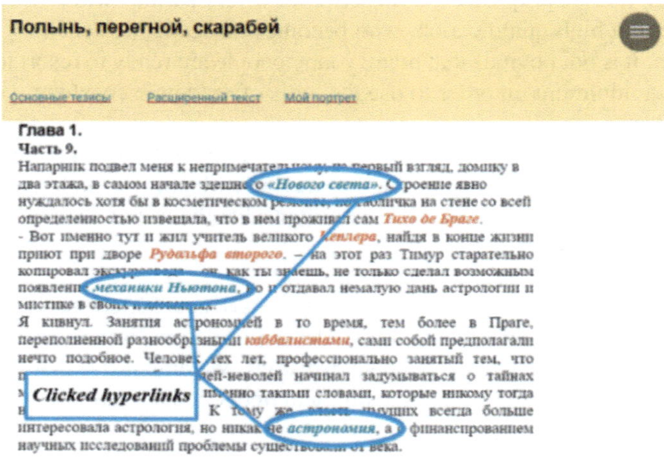

Fig. 1. An example of a novel page containing hyperlinks (links that were clicked by a specific user are highlighted).

An Internet resource built based on a science fiction novel, coupled with a neural network, provides for the formation of a personal account, in which the user has the ability to track their clicks on links, as well as see the test results. This resource is currently being developed in test mode.

The plot of the novel is built on an idea that has a consistent scientific justification. Namely, as shown in works [9, 10], the intellect, mind and consciousness of a person have a dual nature: they simultaneously contain both individual and collective components.

The formation of the collective component can also be explained at a qualitative level.

Let us consider several people entering a conversation. It is customary to say that in this case there is an exchange of information between individuals. In fact, this is an approximation, and a rather rough one at that. There is an exchange of signals between neurons that form a common neural network: in any case, interpersonal communication is seen as an exchange of signals between neurons, even if they belong to different individuals.

The fact that the exchange of information between neurons located in different brains is sporadic does not affect this conclusion. A neural network can also function when communication channels between neurons appear only now of signal transmission. Therefore, it is permissible to assert that humanity forms a global communication neural network, which can be identified with the noosphere, understood according to V.I. Vernadsky.

The behavior of the collective component of human consciousness has not been sufficiently studied, therefore, there is a wide scope for various kinds of hypotheses, as well as the formation of plot lines in science fiction novels.

Specifically, in the science fiction novel written for this work, the collective component of human consciousness provides the main characters with the opportunity to contact with figures of science and technology from bygone eras, with Tycho de Brahe mentioned above. They directly communicate with the "shadows of the past", with prominent figures of past eras, who, in accordance with the basic hypothesis underlying the plot, do not die to the end, but move into a collective transpersonal information environment (noosphere). This ensures the popularization of modern achievements of science and allows you to identify a tendency to perceive non-trivial ideas of various kinds, in accordance with the above method.

More precisely, the system of hyperlinks is chosen so that it is possible to detect not only the nature of the reader's interest in certain provisions from various branches of science, but also to identify the ability to think outside the box.

This is ensured by the fact that the plot of the novel itself explains very non-trivial ideas related to the dual nature of human consciousness. It becomes possible to detect how often the reader follows those links that reflect aspects of the formation of dual consciousness and the global communication environment (noosphere).

In other words, the selected system of references allows us to analyze whether a particular reader has understood the relevant provisions stated in a popular form, whether he is ready to accept non-trivial ideas, whether he is ready to believe that modern information technologies really allow transforming the world in the most significant way.

Specifically, in the text of the novel (more precisely, in the dialogues of the main characters) links are inserted, including to real scientific works, as well as publications of a philosophical nature. Links are provided to publications that prove that the dual component of human consciousness is becoming increasingly important for the further development of science and technology [11–14].

We emphasize that the issue related to the dual nature of human consciousness and intellect, in the future, also acquires a pronounced applied interest. Namely, as shown in [15], the idea of the dual nature of consciousness allows us to translate the concept of digital immortality into a practical plane.

In the cited work, it was shown that the personality structure is very complex, and its individual components, which are most closely adjacent to the collective consciousness, can be transferred to a digital medium in the foreseeable future. In essence, the solution of this problem is reduced only to the decoding of the algorithm for the functioning of the neural network formed by the brain. Of course, there is no need to talk about the direct implementation of such an approach soon, however, such an idea can certainly inspire people capable of independent creativity.

Accordingly, the indication of interest in relevant concepts can be used to identify those readers who are capable of perceiving non-trivial ideas. We emphasize that this approach is quite justified, since the perception of non-trivial ideas and the ability to generate non-trivial ideas, yourself are closely related.

To work out the neural network associated with the text of a science fiction novel, 74 volunteers were recruited from among high school students and first-year university students.

Intermediate test results are shown in Fig. 2, which displays the statistics of clicks on links for a particular user. This statistic was further used by the neural network, which was organized as follows.

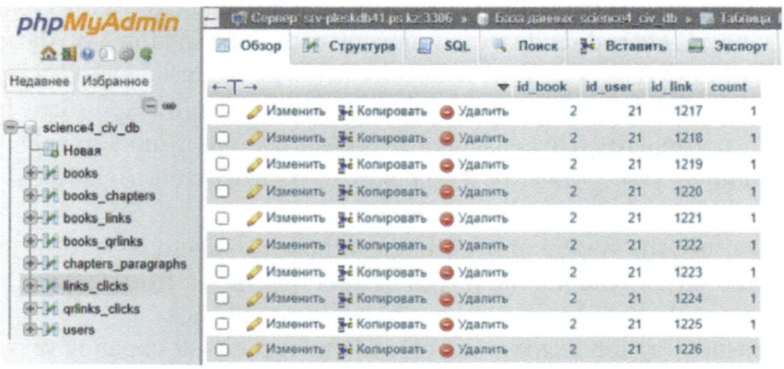

Fig. 2. Statistics of transitions in the database.

For the classification problem, we used a multilayer perceptron model with two hidden layers of 128 neurons each. The ReLu function was used as the activation function on the hidden layers, and the cross entropy was used as the error function. We also added a Dropout layer to prevent model overfitting.

In order to turn the output values of the neural network into the probability that the input object belongs to a particular class, the soft-max function was used. At the input we will submit the data obtained during the experiment, at the output there are 6 output neurons. As an optimizer, we used not the usual stochastic gradient descent, but its modification ADAM - Adaptive Moment Estimation. The accuracy is approximately 0.8857.

The results of testing carried out using this neural network are displayed on diagrams, a specific example is shown in Fig. 3.

These diagrams are like those proposed in [8]. The most significant, from the point of view of assessing the possibility of choosing non-formal education, is a variable called passionarity.

This term is not generally accepted, but it is gaining more and more supporters, including in the field of education, because it reflects the focus on success, the ability to overcome difficulties.

From our point of view, for students studying at Kazakhstani universities, passionarity is extremely important, since any talented student will have to overcome the significant resistance of the conservative environment oriented towards the average level, and this

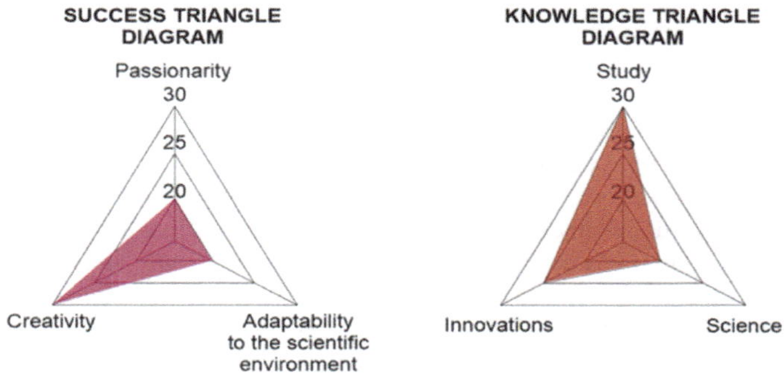

Fig. 3. Example of test results.

concerns overcoming the resistance of the environment formed not only by students, but also by teachers.

Accordingly, a diagram that reflects the potential success of a student focused on non-formal education cannot but contain parameters that reflect these factors. The other two variables on the left side of Fig. 3 are self-explanatory, one being the capacity for creativity and the other being adaptability to the existing science and technology community. We emphasize that adaptability and passionarity, in essence, are not diametrically opposed personality characteristics. A passionate personality, aimed at a very definite success, very often can really show increased adaptability to the prevailing conditions in order to achieve a specific goal.

The second part of Fig. 3 reflects the user's ability to navigate the scientific and technical environment formed at the university. There are three axes here, one of them is related to one's own learning, the other is related to doing science, and the third is related to participation in innovative activities.

3 Conclusion

This approach fully complies with the concept of the knowledge triangle, according to which education at universities, in modern conditions, should be based on the modernized Humboldt principle. Recall that, in accordance with the Humboldt principle, education at the university should be inextricably linked with the pursuit of science. This was the basis of the classical universities during their heyday. In modern conditions, however, it is necessary to raise the question of modernizing this principle and supplementing it with a component related to the participation of students in the generation of innovations. The ability of a particular user to each of these varieties is reflected on the right side of Fig. 3.

Acknowledgments. This research is funded by the Science Committee of the Ministry of Science and Higher Education of the Republic of Kazakhstan (Grant No. AP23490107).

References

1. Mun, G., Masalimova, A., Suleymenova, K., Tasbulatova, Z., Vituleva, Y., Suleimenov, I.: Von Humboldt principles and the realities of Post-Soviet Universities. Bull. KazNU Philos. Cult. Polit. Sci. Ser. **69**(3), 21–30 (2019)
2. Suleimenov, I.E., Vitulyova, Y.S., Bakirov, A.B., Kabdushev, S.B., Yegemberdieva, Z., Mun, G.A.: The use of artificial intelligence systems in higher education: in search of an answer to the "challenge of mass character". Proc. Sci. Tech. Soc. "KAHAK". **4**(67), 39 (2019)
3. Suleimenov, I.E., Guichard, J.P., Baikenov, A.S., Obukhova, P.V., Suleimenova, K.: Degradation of higher education in Kazakhstan as an example of post-transitional crisis. Int. Lett. Soc. Humanist. Sci. **54**, 26–33 (2015)
4. Suleimenova, K.I., Obukhova, P.V., Shaltykova, D.B., Suleimenov, I.E.: Post-transition period and quality of higher education: ways to overcome the crisis phenomena. Int. Lett. Soc. Humanist. Sci. **8**(8), 49–56 (2013)
5. Career Test Truity: https://www.truity.com/test/holland-code-career-test. Accessed 15 Jul 2024
6. Creativity Challenges: https://creativityboost.net/creative-challenges. Accessed 15 Jul 2024
7. Test on Creative Thinking: https://ustaliy.ru/test-na-kreativnoe-myshlenie-proverte-svoyu-logiku-i-erudicziyu. Accessed 15 Jul 2024
8. Kabdushev, S., Shaltykova, D., Vitulyova, Y., Bakirov, A., Suleimenov, I.: Neural network technology for choosing a learning path. In: Proc. 2023 15th Int. Conf. Comput. Autom. Eng. (ICCAE), pp. 13–16 (2023)
9. Suleimenov, I.E., Matrassulova, D.K., Moldakhan, I., Vitulyova, Y.S., Kabdushev, S.B., Bakirov, A.S.: Distributed memory of neural networks and the problem of the intelligence's essence. Bull. Electr. Eng. Inform. **11**(1), 510–520 (2022)
10. Bakirov, A.S., Vitulyova, Y.S., Zotkin, A.A., Suleimenov, I.E.: Internet users' behavior from the standpoint of the neural network theory of society: prerequisites for the meta-education concept formation. Int. Arch. Photogramm. Remote Sens. Spatial Inf. Sci., 83–90 (2021). https://doi.org/10.5194/isprs-archives-XLVI-4-W5-2021-83-2021
11. Seth, A.K., Bayne, T.: Theories of consciousness. Nat. Rev. Neurosci. **23**, 439–452 (2022)
12. Melloni, L., Mudrik, L., Pitts, M., Koch, C.: Making the hard problem of consciousness easier. Science. **372**(6545), 911–912 (2021). https://doi.org/10.1126/science.abj3259
13. Gabrielyan, O.A., Vitulyova, Y., S., Suleimenov, I.: E.: multi-valued logics as an advanced basis for artificial intelligence. Wisdom. **1**(21), 170–181 (2022)
14. Suleimenov, I.E., Gabrielyan, O.A., Bakirov, A.S.: Neural network approach to the interpretation of ancient Chinese geomancy Feng Shui practices. Eur. J. Sci. Theol. **19**(2), 39–51 (2023)
15. Bakirov, A., Suleimenov, I., Vitulyova, Y.: To the question of the practical implementation of "digital immortality" technologies: new approaches to the creation of AI. Proc. Future Technol. Conf. Lect. Notes Netw. Syst. **559**, 368–377 (2023). https://doi.org/10.1007/978-3-031-18461-1_25

Formation of a Digital Educational Environment Based on the Content of Social Networks

Elena Zolotova[1(✉)], Alexandra Zolotova[1], Julia Dyadkova[1], Marina Pukhova[2], and B. Deepanraj[3]

[1] North Caucasus Federal University, 355017 Stavropol, Russia
zolotowa@mail.ru
[2] Financial University Under the Government of the Russian Federation, Moscow, Russia
[3] Prince Mohammad Bin Fahd University, Dhahran, Saudi Arabia

Abstract. The article examines the role of social networks such as the Telegram channel, VKontakte and other platforms as a tool for creating a digital educational environment. The channel's visual content, including illustrations, photographs, videos, stickers and emojis, is analyzed in order to identify the features of its use and impact on communication with the audience and educational organizations. The study, using the Skolkovo LIVE channel as an example, shows that social networks effectively use visual elements to structure information, attract attention and create a positive image in the digital environment. Particular attention is paid to the dominant role of illustrations, the use of corporate identity and sticker packs, as well as the use of emojis to visually highlight key points.

Keywords: Telegram Channel · Visual Content · Digital Educational Environment · Illustrations · Emojis · Stickers · Communication · Branding · Innovations

1 Introduction

The introduction of new digital technologies into the media space has expanded the concept of mass media to include not only traditional websites and social networks, but also instant messaging platforms known as messengers [1]. A clear definition of this term is necessary for a clear understanding of the subject area of the study. A messenger is a software application designed for the prompt exchange of text messages, audiovisual materials and other types of multimedia content. The application can be installed on various devices, including personal computers, smartphones and tablet computers, provided that there is access to the Internet. The modern digital era has radically changed approaches to education and communication, making social networks and messenger's key tools for shaping the educational environment [2]. The active development of platforms such as Telegram, VKontakte, Instagram and YouTube has opened up new opportunities for disseminating knowledge, engaging the audience and creating interactive educational content [3]. Of particular interest is the analysis of visual formats (illustrations, videos, stickers, emojis), which not only simplify the perception of

M. Lapina et al. (Eds.): DET 2024, LNNS 1616, pp. 481–488, 2025.
https://doi.org/10.1007/978-3-032-04365-8_47

information, but also contribute to the formation of stable associations with the brand of educational organizations Table 1. The Skolkovo LIVE Telegram channel, studied in this paper, serves as a vivid example of the effective use of digital tools to popularize science and innovation. Since its creation in 2017, the channel has attracted more than 24,800 subscribers, and its publications combine popular science content with elements of corporate style. This emphasizes the importance of adapting educational strategies to the specifics of digital platforms and their audience [4, 5].

Table 1. Comparison of formats in different social networks.

Format	Telegram	VK	Instagram	YouTube
Illustrations	59% of content, popular science	Infographics, announcements	Main format (stories, posts)	Video thumbnails
Video	Short videos (1–7 min)	Long videos (7+ min)	Short reels (up to 60 s)	Lectures (15+ min)
Interactive	Polls with emoji, stickers	Voting, comments	Questions in stories	Discussions under videos
Texts	Formatting (bold, spoiler)	Long posts with hashtags	Photo captions	Video descriptions

The purpose of the study is to identify patterns in the use of visual content in social networks to form a digital educational environment, as well as to assess its impact on communication and branding. The work used methods of quantitative analysis (share of illustrations, videos and text) and qualitative comparison (cross-platform publications).

Messengers are transforming beyond their original function as simple means of communication. They are evolving into effective channels for transmitting information, allowing for the rapid dissemination of relevant information. Platforms with advanced functionality, such as Telegram, provide users with the ability to create and manage their own thematic channels for broadcasting news and various content [6].

The functionality for creating Telegram channels was implemented in September 2015, but their popularity has grown significantly since 2019. In addition to the overall increase in Telegram's user base, driven by the need for convenient and confidential communication, the demand for channels as a means of quickly disseminating information has increased. Telegram channels are a chat format in which updates are regularly published, providing instant information to subscribers. The Skolkovo Innovation Center, focused on attracting young people, who make up a significant part of the Telegram audience, naturally used this platform for communication. It is noted that in the last two years, the structure of Telegram's audience has undergone changes, becoming more balanced in terms of age groups.

The Skolkovo LIVE Telegram channel was created on September 17, 2017, and at the time of the study, it had more than 24,880 subscribers. The average number of views of each publication is 2,400, which is 10% taking into account rounding.

2 Study

The study analyzed 522 publications in the Telegram channel. The analysis showed that 304 publications (58% of the total) contain visual content. In addition to original content, the Telegram channel actively uses reposts from third-party sources, which may indicate a desire to expand audience reach and broadcast a variety of information.

When analyzing the visual characteristics of the Telegram channel, it is necessary to take into account the possibility of individual customization of the application interface by users, which excludes the possibility of analyzing the general background of the channel design Table 2. The standard Telegram design involves the use of light colors, which corresponds to the corporate style of Skolkovo. The visual design of the channel is limited to the main illustration of the profile, where the organization's logo is placed.

Table 2. Analysis of visual content of the Telegram channel "Skolkovo LIVE".

Content format	Illustrations	Photos	Videos	Stickers/Emojis	Text formats
Number of publications	178	65	61	44	218
Share of total (%)	59%	21%	20%	–	42%

The presence of 7 publications duplicating the content of the Skolkovo community on the VKontakte social network was established. The analysis showed that, with the exception of one publication, the content is identical on both platforms. The exception is the publication dedicated to Mikhail Mishustin's visit to Skolkovo, which is a repost from the Russian Government Telegram channel. Despite the common news hook, the text and visual design of the publications in Telegram and VKontakte differ significantly. The remaining 6 duplicate publications are event announcements, in particular, Biotechmed 2023, made in the format of illustrations-badges. Similar announcements were considered during the previous analysis of the Skolkovo community content on VKontakte, which allows for a comparative analysis of visual communication strategies.

As part of the study of the visual content of the Skolkovo LIVE Telegram channel, the following main formats were identified: illustrations, photographs, and videos. The analysis showed that illustrations are the dominant format, accounting for 59% (178 out of 304 units) of the total number of publications containing visual content.

It is noted that illustrations are often used in groups that are a sequence of images that reveal a specific topic. For example, the publication "How Cookies Got Their Name" demonstrates the use of illustrations to create popular science content, which is a characteristic feature of the Telegram channel and is not observed in the Skolkovo community on VKontakte or on the official website (see Fig. 1). When analyzing this publication, the use of the corporate green color of Skolkovo as a frame and background was revealed, which allows for the creation of a recognizable visual style.

Fig. 1. Example of popular science content.

3 Results

The analysis showed that the illustrations in the Telegram channel are stylistically consistent with the illustrations used in the VKontakte community, indicating a unified approach to the visual design of the brand. Technologies are also a frequent object of visualization, in particular, robotics, cybernetics and other developments of the Skolkovo Innovation Center.

Photographs make up 21%–65% of the total number of publications with visual content. The main object of photographs are people - 59 units, captured in the process of working at events, developing or implementing innovative technologies. Photos demonstrating products, technology exhibitions and mechanisms are presented in significantly smaller quantities - 6 units. Unlike illustrations, photographs, as a rule, are not grouped in large quantities in one publication, limited to a maximum of three images, more often one. Videos occupy 20% of the total amount of visual content in the materials studied. We also divided the videos into those that are less than 1 min and more than 1 min. They occupy 12 and 49 units, respectively. In this case, 4 video materials were in the form of "circle" messages that can be sent via Telegram.

An analysis of the use of video content revealed that in one case, when reposting from the Telegram channel "Government of Russia", the topic of the publication is duplicated with a post in the Skolkovo community on VKontakte. However, the video material dedicated to this topic has different durations on different platforms: 7 min 11 s on VKontakte and 1 min on Telegram.

There are publications in which video content is used in combination with static images or other video materials. As a rule, these are short videos in which people comment on events, talk about their inventions or express their opinions on technologies.

Despite the fact that visual content is not present in all publications, the authors of the Telegram channel pay significant attention to the text design. The Telegram platform provides text formatting options, including highlighting fragments in bold, italics, strikethrough, and hiding text under a spoiler. The authors of Skolkovo LIVE regularly use these tools to highlight key semantic fragments of the text, while the headings are always formatted in bold.

The Telegram platform offers a Premium subscription function that provides users with advanced features, including the use of animated emoji in publications and the addition of their own sticker packs for channel design. The Skolkovo LIVE Telegram channel uses a branded sticker pack called Skolkovo, which includes 44 emoji visually adapted to the style and theme of the innovation center's publications (see Fig. 2). The use of this sticker pack is an important element of visual communication and contributes to the formation of a recognizable style for the channel [4].

Fig. 2. Individual sticker pack "Skolkovo".

It should be noted that the use of stickers and emoji is a common practice in new media. In modern online journalism, stickers and emoji are considered a separate format of visual content, although inferior in importance to traditional forms such as photographs, illustrations, videos and infographics [7, 8]. This visualization format also includes memes, GIF animations and typography used as a visual effect. Working with text, as noted earlier, is a characteristic feature of the Skolkovo LIVE Telegram channel.

The Premium subscription functionality allows you to customize the color of the channel name and quotes, which in the Skolkovo LIVE Telegram channel also corresponds to the corporate style of the organization using a characteristic light green shade (see Fig. 3).

Along with the branded Skolkovo sticker pack, the channel authors use the animated News Emoji sticker pack, available to users with a Telegram Premium subscription. This sticker pack contains emojis that imitate dialogue boxes, which are used to visually design quotes and interview fragments (see Fig. 4). The use of these emojis allows you to visually highlight direct speech and give it a dynamic character.

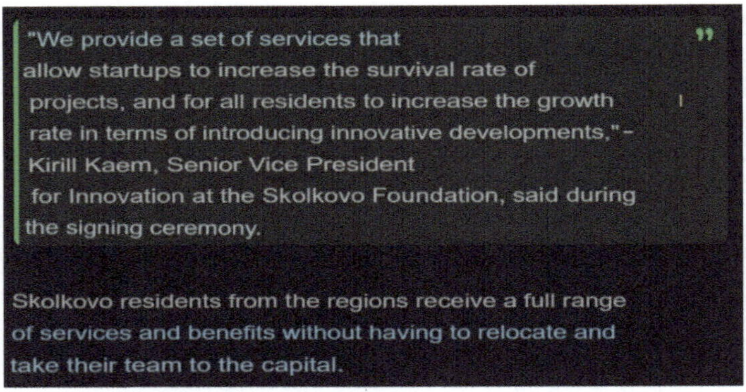

Fig. 3. An example of visual highlighting of a quote.

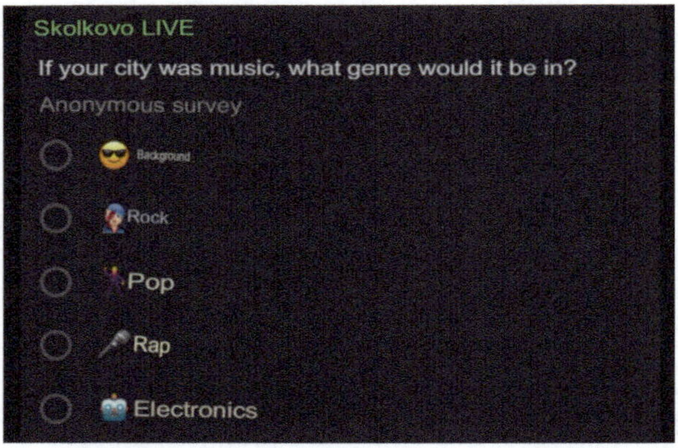

Fig. 4. An example of visual highlighting of a quote.

It is worth noting the use of an interactive format - a short entertainment survey asking subscribers to determine the musical genre that best suits their city of residence. Despite the lack of visual content in the title and main text, each answer option is illustrated with an emoji. This facilitates the perception of information and makes the survey more attractive to the audience compared to a text without visual accompaniment (see Fig. 4).

An important feature of the design of publications in the Telegram channel "Skolkovo LIVE" is the conclusion of each post with the brand slogan "Skolkovo LIVE - innovations are here", accompanied by the logo of the innovation center, taken from the previously mentioned sticker pack (see Fig. 5). This element is a constant and helps to strengthen the brand identity of the channel.

Fig. 5. Slogan “Skolkovo LIVE”.

4 Conclusion

Social networks have become an integral part of the digital educational environment, offering flexible tools for communication, learning and branding. Successful cases, such as Skolkovo LIVE, demonstrate the importance of visual content and adaptation to the specifics of the platforms. The integration of Telegram, VKontakte, Instagram and YouTube allows us to create a comprehensive education support system that meets the needs of a modern audience. Unlike the traditional approach, where visual content accompanies text content, not every publication in the Skolkovo LIVE Telegram channel contains photos or videos. However, this does not negatively affect the dynamics of the narrative. The authors strive for brevity of text messages, highlighting key semantic fragments and using emoji to visually structure the text. Emoji, acting as micro-illustrations, allow you to visually separate paragraphs, format lists and focus on important details.

An analysis of visual content formats revealed the dominant role of illustrations, making up 59% of the total number of visual elements. Photo materials take second place (21%), and video content is third (20%).

Thus, the Skolkovo LIVE Telegram channel is an effective visual communication tool that allows the Skolkovo Innovation Center to maintain contact with the audience, broadcast key messages and form a positive image in the digital environment.

References

1. Zhigalova, O.P.: Formation of the educational environment in the context of the digital transformation of society. Sci. Notes Zabaikalsky State Univ. **14**(2), 69–74 (2019)
2. Mironenko, E.S.: Digital educational environment: concept and structure. Soc. Space. **4**, 6–6 (2019)
3. Kartseva, E.A.: Telegram and its influence: new trends in educational art practices. BBK. **N76**, 200 (2024)
4. Denisevich, T.I.: Telegram channel of an educational institution as a tool for managing educational interaction. BBK 74.204 P84, 132
5. Prakasha, G.S., Lapina, M., Balakrishnan, D., Sajid, M. (eds.): Digital Technologies in Modeling and Management: Insights in Education and Industry. IGI Global (2024). https://doi.org/10.4018/978-1-6684-9576-6
6. Terentyeva, E.V., Argunov, G.O.: Functions and services of the telegram channel as a tool for the development of the Russian information and educational environment. World Sci. Cult. Educ. **2**(111), 87–89 (2025)
7. Prakasha, G.S., Lapina, M., Balakrishnan, D., Sajid, M. (eds.): Educational Perspectives on Digital Technologies in Modeling and Management. IGI Global (2024). https://doi.org/10.4018/979-8-3693-2314-4

8. Avadaeva, I.V., Anisimova-Tkalich, S.K., Vezetiu, E.V., Vovk, E.V., Goldenova, V.S., Grebennikova, V.M., et al.: Methodological Foundations for the Formation of a Modern Digital Educational Environment (2018)
9. Kochetkova, O.V., Kochetkov, A.B.: The role of modern information technologies in the formation of an innovative educational environment of the university. Fundam. Res. **1–1**, 85–89 (2013)
10. Lelchitsky, I.D., Silchenko, A.P., Shcherbakova, S.Y., Baranova, N.A.: Digital educational workshop as a functional and content model of modern education. Bull. Tver State Univ. Ser. Pedagog. Psychol. **3**, 92–100 (2021)
11. Tripathi, V.V., Prakasha, G.S., Lapina, M.: Cognitive engagement scale (CES) in an online environment: construction and validation. In: Lapina, M., Prakasha, G.S., Grigoriev, S. (eds.) SLET 2022: International Conference on Innovative Approaches to the Application of Digital Technologies in Education and Research. SLET 2022. Lecture Notes in Networks and Systems, pp. 73–87. Springer, Cham (2024). https://doi.org/10.1007/978-3-031-78776-8
12. Skulkin, A.A.: Formation of a digital educational space: adaptation of digital pedagogy. World Sci. Cult. Educ. **1**(86), 277–280 (2021)
13. Akopyan, A.A., Medvedeva, Y.Y., Semenova, S.V.: Sources of enterprise competitiveness in the B2B market. Actual problems of science and technology. Materials, 388 (2022)
14. Krasilnikov, A.A., Lubyshev, E.A.: Formation of a digital educational environment of an educational organization in the context of the introduction of information and computer technologies. In: V V Collection: Topical Issues of Practice and Science. Abstracts of XXVI International Scientific and Practical Conference, pp. 443–452 (2021)
15. Berestneva, A.V.: Aspects of the application of the moodle system in education as an element of the formation of an effective educational environment. BBK. **C23**, 129 (2017)
16. Yaremchuk, I.G.: Information technology as the foundation for the formation of a digital educational environment of a medical college. In: Education of the Future: Proceedings of the IV International Scientific and Practical Conference with International Participation. November 17–18, 2023, p. 241. GGNTU, Grozny (2023)
17. Isaykina, M.A., Maksimova, E.A., Grigorieva, T.Y.: Chapter 2. Formation of a digital educational environment. In: Digitalization as a Priority Direction for the Modernization of Russian Education, pp. 44–66 (2019)
18. Krasinsky, I.V.: Experience of using digital educational resources in teaching the history of Belarus in preparing applicants for the CT/CE courses at the Belarusian state Pedagogical University. Stud. Human. **1**, 10 (2025)

Author Index

M. Lapina et al. (Eds.): DET 2024, LNNS 1616, pp. 489–490, 2025.
https://doi.org/10.1007/978-3-032-04365-8

GPSR Compliance

The European Union's (EU) General Product Safety Regulation (GPSR) is a set of rules that requires consumer products to be safe and our obligations to ensure this.

If you have any concerns about our products, you can contact us on ProductSafety@springernature.com

In case Publisher is established outside the EU, the EU authorized representative is:

Springer Nature Customer Service Center GmbH
Europaplatz 3
69115 Heidelberg, Germany

Batch number: 09019469

Printed by Printforce, the Netherlands